Lecture Notes in Computer Science 12050

Oded Goldreich et al.

Computational Complexity and Property Testing

On the Interplay Between Randomness and Computation

With Contributions by
Itai Benjamini, Scott Decatur, Maya Leshkowitz, Or Meir, Dana Ron, Guy Rothblum, Avishay Tal, Liav Teichner, Roei Tell, and Avi Wigderson

Springer

Volume Editor
Oded Goldreich
Weizmann Institute of Science
Rehovot, Israel

ISSN 0302-9743 ISSN 1611-3349 (electronic)
Lecture Notes in Computer Science
ISBN 978-3-030-43661-2 ISBN 978-3-030-43662-9 (eBook)
https://doi.org/10.1007/978-3-030-43662-9

LNCS Sublibrary: SL1 – Theoretical Computer Science and General Issues

Cover illustration: Artwork by Harel Luz, Tel Aviv, Israel

This Springer imprint is published by the registered company Springer Nature Switzerland AG
The registered company address is: Gewerbestrasse 11, 6330 Cham, Switzerland

Preface

This volume contains a collection of studies in the areas of complexity theory and property testing. These studies were conducted at different times, mostly during the last decade. Although most of these works have been cited in the literature, none of them were formally published before.

Indeed, this volume is quite unusual, but not without precedence. In fact, in 2011, I published a similar volume, titled *Studies in Complexity Theory and Cryptography* (LNCS, Vol. 6650), and my impression is that it was well received. Still, these volumes raise two opposite questions regarding the publication of the foregoing studies: (1) why were these studies not published (formally) before? and (2) why are they being published now?

I believe that the second question is answered *a posteriori* by the popularity of the first volume. Although many of the works included in it were known before they appeared in that volume, my impression is that their dissemination benefited from this publication. Furthermore, I feel that it is somewhat more appropriate to refer to publication in a volume of the current type rather than to a posting on forums such as ECCC.

The latter assertion is related to the first question; that is, why were these works not published (formally) before, and why not publish them (especially, the more recent ones) in an ordinary venue now? While there are specific reasons in some of the specific cases, I believe that the answer is more general. In a nutshell, I think that the standard mechanism of conferences and journals has become dysfunctional. The source of trouble is over-preoccupation with competition, and neglect of the original goal of providing accessibility and dissemination.

Specifically, the relevant scientific community seems to act as a reviewing panel rather than as an (active) audience; it seems too preoccupied with the question of whether the submission is "competitive" (with respect to the publication venue)[1] and tends to neglect the actual contents of the submission (beyond, of course, whatever is necessary to determine competitiveness). In other words, the energy and resources of the community are devoted to determining competitiveness, and whatever does not serve that goal gets too little attention. Furthermore, under the current mind-frame, having a submission accepted to such a venue merely means "taking the slot" from other submissions.

The point is that I want the works included in this volume to be read, because I think they are interesting. I want them to be read out of interest in their contents, not as means towards ranking them. I do not want to feel that the main effect of submitting a

---

[1] Originally, "competitive" in the context of these publication venues was understood as fitting some absolute (or relative) standards, but with time "competitive" has evolved to mean worthy of the "award" of being accepted by the venue (as discussed below). I object to the dominance of the question of "competitiveness" even under the former interpretation, let alone under the latter one.

work to a venue is competing against the works of others. I would not feel good about it, regardless of whether I win or lose. I want to contribute to the community, not to compete with its members.

A short detour: On excellence and competitions. Let me stress that I do acknowledge that any realistic struggle for excellence gives rise to a competition, at least implicitly. But this does not mean that struggle and competition are identical; ditto regarding achievement and success. Of course, my issue is not with the semantics of (the colloquial meaning of) these words, but rather with fundamentally different situations that can be identified by referring to these words.

Loosely speaking, by *struggle* I mean both the huge investment of intellectual energy towards *achieving* some goals and the inherent conflict that arises between individuals (or groups) who attempt to achieve the same goals and positions relative to a given setting. That is, the achievements are the goals of the struggle, and the focus of this situation is on the achievements. In contrast, by *competitions* I mean artificial constructs that are defined on top of the basic setting, while not being inherent to it, and *success* typically refers to winning these competitions. That is, success is merely the outcome of the competition, and the focus of this situation is on the competition.

Of course, once these competitions are introduced, the setting changes; that is, a new setting emerges in which these competitions are an inherent part. Still, in some cases—most notably in scientific fields—one can articulate in what sense the original (or basic) setting is better than the modified setting (i.e., the setting modified by competitions). These issues as well as related ones are the topic of my essay *On Struggle and Competition in Scientific Fields.*[2]

Needless to say, in reality we never encounter pure struggles for excellence, devoid of competitional aspects, nor are we likely to encounter—at least in academia—pure competitions devoid of any contents. Reality is always mixed, although it is often useful to analyze it using pure notions. But, currently, the standard publication venues seem extremely biased towards the competition side. Under these circumstances, one may seek alternative vehicles for communicating one's work.

About the contents of this volume. The works included in this collection address a variety of topics in the areas of complexity theory and property testing. Within complexity theory the topics include constant-depth Boolean circuits, explicit construction of expander graphs, interactive proof systems, monotone formulae for majority, probabilistically checkable proofs (PCPs), pseudorandomness, worst-case to average-case reductions, and zero-knowledge proofs. Within property testing the topics include distribution testing, linearity testing, lower bounds on the query complexity (of property testing), testing graph properties, and tolerant testing. A common theme in this collection is *the interplay between randomness and computation*.

About the nature of these works. In the previous volume (LNCS, Vol. 6650), I partitioned the works to three categories labeled 'research contributions', 'surveys', and 'programmatic' papers. The current collection contains a few works for which these

---

[2] See the web-page http://www.wisdom.weizmann.ac.il/~oded/on-struggle.html as well as *SIGACT News*, Vol. 43, Nr. 1, March 2012.

categories feel too rigid. So I decided to avoid such categories and listed all works in chronological order (of original completion time).

About the revisions. All papers were revised by me in the last few months. In some cases the revision is extremely significant, and in other cases it is very minimal. One benefit of editing this collection is that it provided me with motivation to look back at past works and to reflect on their contents and form in retrospect.

Two outliers. A look at the table of contents reveals two outliers. The first is the work "A Probabilistic Error-Correcting Scheme that Provides Partial Secrecy" (co-authored by Scott Decatur and Dana Ron), which was posted in 1997. This work would have fit better in the previous volume (LNCS, Vol. 6650), and the only reason that it was not included in it is an accidental omission. The second outlier is a (solo) work by Roei Tell, titled "Note on Tolerant Testing with One-Sided Error", which deviates from my original plan of including only works co-authored by me. Still, given that this study grew out of a discussion between us, and that I have supervised its writing (as Roei's PhD adviser), I felt that it was okay to bend the rules a bit.

Acknowledgements. The research underlying the papers included in this volume was partially supported by various Israel Science Foundation (ISF) grants (i.e., Nr. 460/05, 1041/08, 671/13, and 1146/18). The papers were revised and edited while I was enjoying the hospitality of the computer science department of Columbia University.

February 2020 Oded Goldreich

Contents

A Probabilistic Error-Correcting Scheme that Provides Partial Secrecy

Scott Decatur, Oded Goldreich, and Dana Ron

Abstract. In the course of research in computational learning theory, we found ourselves in need of an error-correcting encoding scheme for which relatively few bits in the codeword yield no information about the plain message. Being unaware of a previous solution, we came-up with the scheme presented here.

Clearly, a scheme as postulated above cannot be deterministic. Thus, we introduce a probabilistic coding scheme that, in addition to the standard coding theoretic requirements, has the feature that any constant fraction of the bits in the (randomized) codeword yields no information about the message being encoded. This coding scheme is also used to obtain efficient constructions for the *Wire-Tap Channel* Problem.

Appeared (under the title "A Probabilistic Error-Correcting Scheme") as record 1997/005 of the *IACR Cryptology ePrint Archive*, 1997. In the current revision, the introduction was intentionally left almost intact, but the exposition of the main result (esp., its proof) was elaborated and made more reader-friendly.

1 Original Introduction (Dated April 1997)

We believe that the following problem may be relevant to research in cryptography:

> Provide an error-correcting encoding scheme for which relatively few bits in the codeword yield no information about the plain message.

Certainly, no deterministic encoding may satisfy this requirement, and so we are bound to seek probabilistic error-correcting encoding schemes. Specifically, in addition to the standard coding theoretic requirements (i.e., of correcting upto a certain threshold number of errors), we require that obtaining less than a threshold number of bits in the (randomized) codeword yield no information about the message being encoded.

Below we present such a probabilistic encoding scheme. In particular, the scheme can (always) correct a certain constant fraction of errors, and has the property that fewer than a certain constant fraction of the bits (in the codeword) yield no information about the encoded message. Thus, using this encoding scheme over an insecure channel tampered by an adversary who can read and modify (only) a constant fraction of the transmitted bits, we establish correct and private communication between the legitimate end-points.

O. Goldreich (Ed.): Computational Complexity and Property Testing, LNCS 12050, pp. 1–8, 2020.
https://doi.org/10.1007/978-3-030-43662-9_1

The new coding scheme is also used to obtain *efficient constructions* for the *Wire-Tap Channel* Problem (cf., [10]). Related work has been pointed out to us recently by Claude Crépeau. These include [1,3,4,8]. In particular, the seemingly stronger version of the problem, considered in this work, was introduced by Csiszár and Körner [4]. Maurer has shown that this version of the problem can be reduced to the original one by using bi-directional communication [8]. Crépeau (private comm., April 1997) has informed us that, using the techniques in [1,3], one may obtain an alternative efficient solution to the Wire-Tap Channel Problem again by using bi-directional communication.[1]

Our own motivation to study the problem had to do with computational learning theory. Indeed, the solution was introduced and used in our work on *computational sample complexity* [5].

2 Main Result

We focus on good error correcting codes (and encoding schemes), which are codes of constant rate and constant relative distance. Recall that a standard (binary) (error-correcting) code of rate $\rho > 0$ and relative distance $\delta > 0$ is a mapping $C : \{0,1\}^* \rightarrow \{0,1\}^*$ that satisfies $|C(x)| = |x|/\rho$ and $\min_{x \neq y : |x|=|y|}\{\mathrm{wt}(\mathrm{C}(x) \oplus \mathrm{C}(y))\} \geq \delta \cdot |\mathrm{C}(x)|$, where $\mathrm{wt}(z) \stackrel{\mathrm{def}}{=} |\{i \in [|z|] : z_i = 1\}|$ is the Hamming weight of z and $\alpha \oplus \beta$ denotes the bit-by-bit exclusive-or of the strings α and β.

We are interested in good codes that have efficient encoding and decoding algorithms, where the latter are applicable to error rates below $\delta/2$. That is, for some constant $\eta \in (0, \delta/2]$, we may hope to have a decoder such that for every x and $e \in \{0,1\}^{|C(x)|}$ of Hamming weight at most $\eta \cdot |C(x)|$, given a corrupted codeword $G(x) \oplus e$, recovers the original message x.

The non-standard (for coding theory) aspect that we consider here is *partial secrecy*. Specifically, for some constant $\varepsilon > 0$, any ε fraction of the bits of the codeword should yield no information about the original message. Obviously, this is not possible with an actual code, and so we settle for probabilistic encoding schemes as implicitly defined next.

Theorem 1 (a probabilistic error correction scheme with partial privacy): *There exist constants $\rho, \eta, \varepsilon > 0$ and a pair of probabilistic polynomial-time algorithms, denoted (E, D), such that the following holds.*

1. Constant Rate: $|E(x)| = |x|/\rho$, *for all* $x \in \{0,1\}^*$.
2. Error Correction: *for every* $x \in \{0,1\}^*$ *and every* $e \in \{0,1\}^{|E(x)|}$ *such that* $\mathrm{wt(e)} \leq \eta \cdot |\mathrm{E(x)}|$, *it holds that*

$$\mathrm{Prob}[D(E(x) \oplus e) = x] \;=\; 1.$$

Furthermore, Algorithm D is deterministic.

[1] Added in revision: We stress that, in contrast, our solution uses uni-directional communication. On the other hand, our solution holds only for a limited range of parameters; see discussion at the end of Sect. 3.

3. Partial Secrecy: *A substring containing* $\varepsilon \cdot |E(x)|$ *bits of* $E(x)$ *yields no information on* x. *That is, for* $I \subseteq [|\alpha|] = \{1, ..., |\alpha|\}$, *let* α_I *denote the substring of* α *corresponding to the bits at locations in* I (i.e., for $I = \{i_1, i_2, ..., i_t\}$ such that $i_j < i_{j+1}$, it holds that $\alpha_I = \alpha_{i_1}\alpha_{i_2}\cdots\alpha_{i_t}$). *Then, for every* $n \in \mathcal{N}$, $x, y \in \{0,1\}^n$, *and* $\varepsilon \cdot (n/\rho)$*-subset* $I \subseteq [n/\rho]$, *it holds that* $E(x)_I$ *is distributed identically to* $E(y)_I$; *that is, for every* $\alpha \in \{0,1\}^{|I|}$,

$$\text{Prob}[E(x)_I = \alpha] = \text{Prob}[E(y)_I = \alpha].$$

Furthermore, $E(x)_I$ *is uniformly distributed over* $\{0,1\}^{|I|}$.

In addition, on input x, *algorithm* E *uses* $O(|x|)$ *coin tosses.*

Items 1 and 2 are standard requirements of coding theory, first met by Justesen [7]. What is non-standard in Theorem 1 is Item 3. Indeed, Item 3 is impossible if one insists that the encoding algorithm (i.e., E) is deterministic.

Proof: The key idea is to encode the information by first augmenting it with a sufficiently long random padding, and then encoding the result using a good error correcting code (i.e., one of constant rate and constant relative distance).

To demonstrate this idea, consider an $2n$-by-m matrix M defining a good (linear) error-correction code. That is, the string $z \in \{0,1\}^{2n}$ is encoded by $z \cdot M$. Further suppose that the submatrix defined by the last n rows of M and any $\varepsilon \cdot m$ of its columns is of full-rank (i.e., rank $\varepsilon \cdot m$). Then, we define the following probabilistic encoding, E, of strings of length n. To encode $x \in \{0,1\}^n$, we first select $y \in \{0,1\}^n$ uniformly at random, let $z = xy$ and output $E(x) = z \cdot M$.

Clearly, the error-correction features of M are inherited by E. To see that the secrecy requirement holds consider any sequence of $\varepsilon \cdot m$ bits in $E(x)$. The contents of these bit locations is the product of z by the corresponding columns in M; that is, $z \cdot M' = x \cdot A + y \cdot B$, where M' denotes the submatrix corresponding to these columns in M, and A (resp., B) is the matrix resulting by taking the first (resp., last) n rows of M'. By hypothesis B is full rank, which implies that $y \cdot B$ is uniformly distributed. Hence, $z \cdot M'$ is uniformly distributed (regardless of x).

We stress that the foregoing argument relies on the hypothesis that *the submatrix defined by the last* n *rows of* M *and any* $\varepsilon \cdot m$ *of its columns is of full-rank.* Let us call such a matrix nice. So what we need is a *construction of a good linear code that is generated by a nice matrix and has an efficient decoding algorithm.* Such a construction can be obtained by mimicking Justesen's construction [7]. Basically, we construct outer and inner encoding schemes, which correspond to the outer and inner codes used in [7], and apply composition (and analyze it) analogously. Specifically, after reviewing Justesen's code, we show that the corresponding outer and inner encoding schemes satisfy the error correction and secrecy requirements of the theorem, and then we show that composing these schemes yields a composed scheme that satisfies these requirements too. This establishes the theorem.

Justesen's Code. Recall that Justesen's Code is obtained by composing two codes: An *outer* linear code over a large alphabet is composed with an *inner* binary linear code that is used to encode single symbols of the large alphabet. The outer code is the Reed-Solomon Code; that is, the n-bit long message is encoded by viewing it as the coefficients of a polynomial of degree $t-1$ over a field with $\approx 3t$ elements, where $n \approx t\log_2(3t)$, and letting the codeword consists of the values of this polynomial at all field elements. Using the Berlekamp-Welch Algorithm [2], one can efficiently retrieve the information from a codeword provided that at most t of the symbols (i.e., the values of the polynomial at t field elements) were corrupted.

Our Outer Encoding. We obtain a variation of this outer-code as follows: Given $x \in \{0,1\}^n$, we pick a minimal $t \in \mathcal{N}$ such that $2n < t\log_2(3t)$, and view x as a sequence of $\frac{t}{2}$ elements in $\mathrm{GF}(3t)$.[2] We uniformly select $y \in \{0,1\}^n$ and view it as another sequence of $\frac{t}{2}$ elements in $\mathrm{GF}(3t)$. We consider the degree $t-1$ polynomial defined by these t elements, where x corresponds to the high-order coefficients and y to the low-order ones. Clearly, we preserve the error-correcting features of the original outer code. Furthermore, any $t/2$ symbols of the codeword yield no information about x. To see this, note that the values of these $t/2$ locations are obtained by multiplying a t-by-$t/2$ Vandermonde with the coefficients of the polynomial. We can rewrite the product as the sum of two products the first being the product of a $t/2$-by-$t/2$ Vandermonde with the low order coefficients. Thus, a uniform distribution on these coefficients (represented by y) yields a uniformly distributed result (regardless of x). (In other words, the generating matrix of the corresponding linear code is nice.) Hence, we have obtained a randomized outer-encoding that satisfies both the error-correction and secrecy requirements of the theorem, but this is encoding is over the alphabet $\mathrm{GF}(3t)$.

Our Inner Encoding. Next, we obtain an analogue of the inner code used in Justesen's construction. Here, the aim is to encode information of length $\ell \stackrel{\text{def}}{=} \log_2(3t)$ (i.e., the representation of an element in $\mathrm{GF}(3t)$) using codewords of length $O(\ell)$. Hence, we do not need an efficient decoding algorithm, since Maximum Likelihood Decoding via exhaustive search is affordable (because $2^\ell = O(t) = O(n)$). Furthermore, any code that can be specified by $O(\log n)$ many bits will do (since we can try and check all possibilities in $\mathrm{poly}(n)$-time), which means that we can use a randomized argument provided that it utilizes only $O(\log n)$ random bits. For example, we may use a linear code specified by a (random) 2ℓ-by-4ℓ Toeplitz matrix.[3] Using a probabilistic argument one can show that, with positive probability, a random 2ℓ-by-4ℓ Toeplitz matrix is as required in the motivating discussion (i.e., it generates a good code and is nice (i.e., its rows generate a code of distance $\Omega(\ell)$ and the submatrix induced by any $\Omega(\ell)$ columns and the last ℓ

[2] Here we assume that $3t$ is a prime power. Actually, we use the first power of 2 that is greater than $3t$. Clearly, this inaccuracy has a negligible effect on the construction.

[3] A Toeplitz matrix, $T = (t_{i,j})$, satisfies $t_{i,j} = t_{i+1,j+1}$, for every i, j.

rows is of full rank)).[4] In the rest of the discussion, we fix such a nice Toeplitz matrix. We shall use it to randomly encode ℓ-bit strings (i.e., elements of $\mathrm{GF}(3t)$) by applying the matrix to a random 2ℓ-bit long padding of the ℓ-bit long input. Hence, we obtain a randomized inner-code that satisfies both the error-correction and secrecy requirements of the theorem.

The Composition. We now get to the final step in mimicking Justesen's construction: the composition of the two codes. That is, we have outer and inner encoding schemes that satisfy both the error-correction and secrecy requirements of the theorem, and we need to show that their composition satisfies these features too. Let us first spell out what this composition is.

Recall that we want to encode $x \in \{0,1\}^n$, which is viewed as $x \in \mathrm{GF}(3t)^{t/2}$, where $n \approx (t/2)\log_2(3t)$. Applying the outer encoding scheme, with randomization $y \in \{0,1\}^n$, we obtain a $3t$-long sequence over $\mathrm{GF}(3t)$, denoted $x_1, ..., x_{3t}$. (Specifically, the Reed-Solomon code is applied to the $2n$-bit long string xy, viewed as a t-long sequence over $\mathrm{GF}(3t)$, resulting in the sequence $(x_1, ..., x_{3t}) \in \mathrm{GF}(3t)^{3t}$.) Next, applying the inner encoding scheme to each of the x_i's, viewed as an ℓ-bit long string, we obtain a $3t$-long sequence of 4ℓ-bit inner codewords. That is, using the inner code (i.e., the Toeplitz matrix) and additional $3t$ random ℓ-bit strings, denoted $y_1, ..., y_{3t}$, we encode each of the above x_i's by a 4ℓ-bit long string that is the result the multiplying the Toeplitz matrix with the vector $x_i y_i$. Hence, letting M denote the fixed 2ℓ-by-4ℓ Toeplitz matrix and $C : \mathrm{GF}(3t)^t \to \mathrm{GF}(3t)^{3t}$ denote the Reed-Solomon code, we have $E(x) = (x_1 y_1 \cdot M, ..., x_{3t} y_{3t} \cdot M)$, where $(x_1, ..., x_{3t}) \leftarrow C(xy)$ and $y, y_1, ..., y_{3t}$ are uniformly and independently distributed in the relevant domains (i.e., $y \in \{0,1\}^n$ and $y_1, ..., y_{3t} \in \{0,1\}^\ell$).

Clearly, E preserves the error-correcting features of Justesen's construction [7], and the rate also remains constant (although cut by a factor of 4). The secrecy condition is proved analogously to the way in which the error correction feature is established in [7], where the analogy is between revealed codeword bits and corrupted codeword bits. Specifically, the symbol x_i remains secret if few bits in its encoding are revealed, whereas the relatively few x_i's that cannot be guaranteed to remain secret do not harm the secrecy of x. Details follow.

Establishing the Secrecy of E. We consider the partition of the codeword into consecutive 4ℓ-bit long subsequences corresponding to the codewords of the inner code. Given a set I of locations (as in the secrecy requirement), we consider the relative locations in each subsequence, denoting the induced locations in the ith subsequence by I_i. We classify the subsequences into two categories depending

[4] The proof uses the fact that any (non-zero) linear combination of rows (or columns) in a random Toeplitz matrix is uniformly distributed. The first condition is proved by observing that the probability that a non-zero combination of the rows of the 2ℓ-by-4ℓ matrix has Hamming weight smaller than ℓ' is upper-bounded by $(2^{2\ell}-1) \cdot \sum_{i=0}^{\ell'-1} \binom{4\ell}{i} \cdot 2^{-4\ell}$, which is $o(1)$ for some $\ell' = \Omega(\ell)$. The second condition is proved by observing that the probability that there exist ℓ'' columns that yield a submatrix (of the last ℓ rows) that is not full rank is upper-bounded by $\binom{4\ell}{\ell''} \cdot (2^{\ell''}-1) \cdot 2^{-\ell}$, which is $o(1)$ for some $\ell'' = \Omega(\ell)$.

on whether or not the size of the induced I_i is above the secrecy threshold for the inner-encoding. By a counting argument, only a small fraction of the subsequences have I_i's with size above the threshold.

For the typical (i.e., relatively small) I_i's, we use the secrecy feature of the inner-encoding, and infer that no information is revealed about the corresponding x_i's. Hence, the only information about x that may be present in $E(x)_I$ is present in the non-typical subsequences (i.e., those associated with large I_i's). Using the secrecy feature of the outer-encoding, we conclude that these few subsequences (or even the corresponding x_i's themselves) yield no information about x. The secrecy condition of the composed encoding follows. ■

3 An Efficient Wire-Tap Channel Encoding Scheme

The *Wire-Tap Channel* Problem, introduced by Wyner [10], generalized the standard setting of a Binary Symmetric Channel. Recall that a Binary Symmetric Channel with crossover probability p, denoted $\mathtt{BSC}_p$, is a randomized process which represents transmission over a noisy channel in which each bit is flipped with probability p (independently of the rest). Thus, for a string $\alpha \in \{0,1\}^n$, the random variable $\mathtt{BSC}_p(\alpha)$ equals $\beta \in \{0,1\}^n$ with probability $p^d \cdot (1-p)^{n-d}$, where d is the Hamming distance between α and β (i.e., the number of bits on which they differ). In the *Wire-Tap Channel Problem* there are two (independent) noisy channels from the sender: one representing the transmission to the legitimate receiver, and the other representing information obtained by an adversary tapping the legitimate transmission line and incurring some noise as well. In Wyner's work [10] the wire-tap channel introduces additional noise on top of the legitimate channel (and so may be thought of as taking place at the receiver's side). Here we consider a seemingly more difficult setting (introduced in [4]) in which the wire-tap channel is applied to the original packet being transmitted (and so may be thought of as taking place at the sender's side).

Wyner studied the information theoretic facet of the problem [10], analogously to Shannon's pioneering work on communication [9]. Below we consider the computational aspect of the problem for the special case of very noisy tapping-channel.

Theorem 2 (efficient wire-tap channel encoding): *Let $\mathtt{BSC}_p(\alpha)$ be a random process that represents the transmission of a string α over a Binary Symmetric Channel with crossover probability p.*[5] *Then, for an encoding scheme (E, D) as in Theorem 1, where η and ε are the constants associated with the error-correction and secrecy guarantees of E, the following holds.*

1. Error Correction*: Decoding succeeds with overwhelmingly high probability. That is, for every $x \in \{0,1\}^*$,*

$$\text{Prob}[D(\mathtt{BSC}_{\frac{\eta}{2}}(E(x))) = x] \;=\; 1 - \exp(-\Omega(|x|)).$$

[5] Recall that the *crossover probability* is the probability that a bit is complemented in the transmission process.

2. Secrecy: *The wire-tapper gains no significant information. That is, for every* $x \in \{0,1\}^*$

$$\sum_{\alpha \in \{0,1\}^{|E(x)|}} \left| \text{Prob}[\texttt{BSC}_{\frac{1}{2}-\frac{\varepsilon}{4}}(E(x)) = \alpha] - 2^{-|E(x)|} \right|$$

is exponentially vanishing in $|x|$.

Proof: Item 1 follows by observing that, with overwhelmingly high probability, the channel $\texttt{BSC}_{\frac{\eta}{2}}$ complements less than a η fraction of the bits of the codeword. Item 2 follows by representing $\texttt{BSC}_{(1-\gamma)/2}(\alpha)$ as a two-stage process: In the first stage each bit of α is *set* (to its current value) with probability γ, independently of the other bits. In the second stage each bit that was not set in the first stage, is assigned a uniformly chosen value in $\{0,1\}$. Next, letting $\gamma = \varepsilon/2$, we observe that, with overwhelmingly high probability, at most $2 \cdot \gamma \cdot |E(x)| = \varepsilon \cdot |E(x)|$ bits were set in the first stage. Suppose we are in this case. Then, applying Item 3 of Theorem 1, the bits of $E(x)$ that are set in Stage 1 do not reveal any information regarding x, whereas the values of the bits set in Stage 2 are independent of x. ■

Discussion: As mentioned above, the setting considered in Theorem 2 is actually due to Csiszár and Körner [4]. Clearly, a solution cannot exist unless the channel of Item 1 is more reliable than the one of Item 2. A special case of the results in [4] asserts that a solution always exists when the channel of Item 1 is more reliable than the one of Item 2. However, the latter result is non-constructive. In contrast, the result of Theorem 2 is constructive and efficient, but it requires a significant gap between the reliability of the two channels. In particular, the crossover probability of the channel in Item 1 (denoted $\frac{\eta}{2}$) is typically very small (i.e., of the order of 0.01); whereas the crossover probability of the channel in Item 2 (denoted $\frac{1}{2} - \frac{\varepsilon}{4}$) is typically very close to 1/2 (i.e., of the order of 0.49).

Crépeau (private comm., April 1997) has informed us that alternative solutions, which utilize bi-directional communication, may be obtained by using the techniques in [1,3,8]. We stress that when using bi-directional communication one can cope with an arbitrary pair of channels (and specifically the channel in the secrecy condition may be more reliable than the channel in the error-correcting condition) – see [8].

Postscript

When preparing this revision, we found out that probabilistic error-correcting schemes of the type we suggested were later suggested by Guruswami and Smith [6], and termed *stochastic codes*. Their motivation was obtaining encoding schemes with optimal rate for computationally restricted channels.

Acknowledgments. We are grateful to Moni Naor and Ronny Roth for helpful discussions. We also wish to thank Claude Crépeau for pointing out and explaining to us some related work (i.e., [1,3,4,8]).

References

1. Bennett, C.H., Brassard, G., Crépeau, C., Maurer, U.: Generalized privacy amplification. IEEE Trans. Inf. Theory **41**(6), 1915–1923 (1995)
2. Berlekamp, E., Welch, L.: Error correction of algebraic block codes. US Patent 4,633,470 (1986)
3. Cachin, C., Maurer, U.M.: Linking information reconciliation and privacy amplification. J. Cryptol. **10**(2), 97–110 (1997). https://doi.org/10.1007/s001459900023
4. Csiszár, I., Körner, J.: Broadcast channels with confidential messages. IEEE Trans. Inf. Theory **24**, 339–348 (1978)
5. Decatur, S., Goldreich, O., Ron, D.: Computational sample complexity. In: 10th COLT, pp. 130–142 (1997). (Later appeared in SIAM J. Comput. **29**(3), 854–879 (1999))
6. Guruswami, V., Smith, A.: Optimal rate code constructions for computationally simple channels. J. ACM **63**(4), 35:1–35:37 (2016)
7. Justesen, J.: A class of constructive asymptotically good algebraic codes. IEEE Trans. Inf. Theory **18**, 652–656 (1972)
8. Maurer, U.M.: Perfect cryptographic security from partially independent channels. In: 23rd STOC, pp. 561–571 (1991)
9. Shannon, C.E.: A mathematical theory of communication. Bell Syst. Tech. J. **27**, 379–423, 623–656 (1948)
10. Wyner, A.D.: The wire-tap channel. Bell Syst. Tech. J. **54**(8), 1355–1387 (1975)

Bridging a Small Gap in the Gap Amplification of Assignment Testers

Oded Goldreich and Or Meir

Abstract. Irit Dinur's proof of the PCP theorem via gap amplification (*J. ACM*, Vol. 54 (3) and *ECCC* TR05-046) has an important extension to Assignment Testers (a.k.a PCPPs). This extension is based on a corresponding extension of the gap amplification theorem from PCPs to Assignment Testers (a.k.a PCPPs). Specifically, the latter extension states that the rejection probability of an Assignment Tester can be amplified by a constant factor, at the expense of increasing the output size of the Assignment Tester by a constant factor (while retaining the alphabet). We point out a gap in the proof of this extension, and show that this gap can be bridged.

We stress that the gap refers to the amplification of Assignment Testers, and the underlying issue does not arise in the case of standard PCPs. Furthermore, it seems that the issue also does not arise with respect to the applications in Dinur's paper, but it may arise in other applications.

This paper appeared as Comment Nr 3 on TR05-046 of *ECCC*. (The comment was posted in Oct. 2007.) With the exception of Sect. 4, the current revision is intentionally minimal. Still it does include some minor corrections and clarifications of the original text as well as some stylistic improvements. Section 4 is a digest that was added in the current revision.

1 Background

(We make references to specific items as numbered both in the journal version of Dinur's work [3] and in the version posted on ECCC [2], since both versions are cited in the literature.)

1.1 Assignment Testers and Gap Amplification

We begin by recalling the definition of Assignment Testers, as stated in [2,3] (following [4]), while commenting that this notion is closely related to the notion of PCPs of Proximity (as defined in [1]). The specific formulation of Assignment Testers that we use refers to binary constraints over a large (constant-sized) alphabet. Specifically, we refer to the notion of a constraint graph, which is actually a graph $G = (V, E)$ augmented with binary constraints $\mathcal{C} = (c_e)_{e \in E}$ that are associated with its edges. These constraints refer to an assignment α of values in Σ_0 to the vertices (i.e., $\alpha : V \to \Sigma_0$), and $\mathrm{UNSAT}_\alpha(G)$ denotes the fraction of

O. Goldreich (Ed.): Computational Complexity and Property Testing, LNCS 12050, pp. 9–16, 2020.
https://doi.org/10.1007/978-3-030-43662-9_2

edge-constraints that are violated by this assignment (i.e., $\mathrm{UNSAT}_\alpha(G)$ equals $|\{e = \{u,v\} \in E : c_e(\alpha(u),\alpha(v))=0\}|/|E|$).

Definition 1 (Assignment Testers, See [2, Definition 3.1] and [3, Definition 2.8]): *An assignment-tester with alphabet Σ_0 and rejection probability $\varepsilon > 0$ is a polynomial-time algorithm that, on input a circuit Φ over Boolean variables X, outputs a constraint graph $G = \langle (V,E), \Sigma_0, \mathcal{C}\rangle$ such that $X \subseteq V$ and the following hold with respect to any assignment $a : X \to \{0,1\}$.*

- (Completeness): *If $a \in SAT(\Phi)$, then there exists $b : (V \backslash X) \to \Sigma_0$ such that $UNSAT_{a \cup b}(G) = 0$.*
- (Soundness): *If $a \notin SAT(\Phi)$, then for all $b : (V \backslash X) \to \Sigma_0$,*

$$UNSAT_{a \cup b}(G) \geq \varepsilon \cdot \mathtt{dist}(a, SAT(\Phi)).$$

Indeed, $SAT(\Phi)$ denotes the set of assignments that satisfy Φ, and $\mathtt{dist}(a,S)$ denotes the relative Hamming distance of the assignment a from the set S.

The main technical result of [2,3] is a gap amplification theorem for PCPs. The following important extension of this theorem to Assignment Testers is also provided in [2,3]:

Theorem 2 (Gap Amplification for Assignment Testers, See [2, Theorem 8.1] and [3, Theorem 9.1]): *There exists $t \in \mathbb{N}$ such that for every assignment-tester T with constant-size alphabet Σ and rejection probability ε, there exists an assignment-tester T' with the same alphabet Σ such that*

- *the rejection probability of T' is at least* $\min(2\varepsilon, 1/t)$*; and*
- *the output size of T'* (i.e., the number of edges in the constraint graph that T' outputs) *is at most a constant factor larger than the output size of T.*

Furthermore, T' consists of composing T with a fixed polynomial-time algorithm; that is, T' first invokes T, and then applies a fixed polynomial-time transformation on the constraint graph that T outputs.

1.2 Overview of the Proof of Theorem 2

The assignment tester of Theorem 2 is constructed in two steps: First, for a fixed constant $d \in \mathbb{N}$ and an arbitrary constant $t \in \mathbb{N}$ (to be determined later), an intermediate assignment tester with alphabet $\Sigma^{d^{t/2}}$ and rejection probability $p = \Omega(\min(\sqrt{t} \cdot \varepsilon, 1/t))$ is constructed. Then, a composition theorem of Dinur and Reingold [4] is applied to the intermediate assignment tester in order to reduce its alphabet's size, resulting in an assignment tester with alphabet Σ and rejection probability $\Omega(p) = \Omega(\min(\sqrt{t} \cdot \varepsilon, 1/t))$. The parameter t is fixed to some sufficiently large natural number that yields the desired rejection probability.

The subject of this paper is a gap in the first step of the foregoing construction; namely, the construction of the intermediate assignment tester. Specifically, we show that under certain circumstances, the intermediate assignment tester has output size that is quadratic in the output size of the input assignment tester, failing to establish Theorem 2. Such an increase in the output size can not be afforded by the applications of Theorem 2 presented in [2] and [3]. We comment that those circumstances do not seem to occur in the applications of Theorem 2 presented in of [2]. In this paper we show that the proof of Theorem 2 can be corrected so the theorem holds *under any circumstances.*

Outline of the construction of the intermediate assignment tester. Let Φ be a circuit over Boolean variables X.

1. First, the intermediate assignment tester runs the given assignment tester on input Φ, yielding a constraint graph $G = \langle (V, E), \Sigma, \mathcal{C} \rangle$.
   For any vertex $v \in V$, let $\deg_G(v)$ denote the degree of v in G.
2. Next, the intermediate assignment tester constructs the constraint graph $H = (\text{prep}(G))^t$, where $\text{prep}(G)$ is the graph in which every vertex v of G is replaced by a $\deg_G(v)$-vertex expander graph, denoted $[v]$, whose vertices represent "copies" of v and whose edges correspond to equality constraints. We denote the set of vertices of H by V_H. (In addition, a larger expander, with trivial constraints, is superimposed on V_H; but we ignore it here.)[1]
   We stress that the $X \cap V_H = \emptyset$, since each $x \in X \subseteq V$ was replaced by $[x]$.
3. Finally, the intermediate assignment tester constructs and outputs a constraint graph H', whose set of vertices is $V_H \cup X$ and whose edges are the edges of H as well as some additional "consistency edges" that check consistency between X and V_H (i.e., between the value assigned to each $x \in X$ and the values assigned to the corresponding vertices in $[x]$). The edges are re-weighted such that the latter consistency edges form half of the edges of H'.
   For every $v \in V_H \cup X$, let $\deg_{H'}(v)$ denote the degree of v in H'.

We stress that the definition of the latter consistency edges, which are added in the last step, was not fully specified above (i.e., we did not say that each $x \in X$ is connected too all vertices in $[x]$). The question of how to actually define these edges is the issue that we address in this paper.

2 The Gap

The gap in the proof refers to the way that the consistency edges between X and V_H are defined. Specifically, we show that if the graph G is highly non-regular, then connecting each $x \in X$ to all vertices in $[x]$ yields a graph H' that may contain too many consistency edges. For simplicity, let us assume that $t = 1$, but note that the argument holds for any value of t. For $t = 1$, it holds that $H = \text{prep}(G)$ and $V_H = \bigcup_{v \in V} [v]$, where $[v]$ is the set of vertices that represent "copies" of the vertex v of G.

[1] We also ignored the constraints placed on the edges of H, which are a key issue in [2,3]. We do so since our focus is on the added constraints of Step 3.

The consistency edges as defined in [2,3]. The natural way to define the consistency edges, which is the way taken in [2,3], is based on the natural randomized testing procedure (to be described next). This procedure is given oracle access to an assignment $A : V_H \cup X \to \Sigma$ to H', and is allowed to make two queries to A, which it selects uniformly at random (see below). The procedure then decides whether to accept or reject A. Assuming that $t = 1$, the aforementioned procedure is as follows:

1. Select $x \in X$ uniformly at random.
2. Select $z \in [x]$ uniformly at random (recall that $[x]$ is the set of vertices in H that are copies of x).
3. Accept if and only if $A(x) = A(z)$.

The consistency edges are defined using the procedure as follows: For every possible outcome of the coin tosses ω, let v_1^ω and v_2^ω denote the vertices that the procedure queries on coin tosses ω. Then, a consistency edge is placed between v_1^ω and v_2^ω, and this edge accepts an assignment $A : V_H \cup X \to \Sigma$ if and only if the procedure accepts on coin tosses ω when given oracle access to A (i.e., $A(v_1^\omega) = A(v_2^\omega)$). Note that for every $x \in X$, it holds that $\deg_{H'}(x)$ equals to the number of consistency edges connected to x using the foregoing procedure. The problem is now as follows:

- Since the procedure chooses $x \in X$ uniformly at random (at Step 1), every variable $x \in X$ must have the same degree in H'. That is, for every two variables $x, y \in X$, it holds that $\deg_{H'}(x) = \deg_{H'}(y)$.
- Since the procedure chooses $z \in [x]$ uniformly at random (at Step 2), every variable $x \in X$ must satisfy $\deg_{H'}(x) \geq |[x]| = \deg_G(x)$.
- Combining the previous two items, it follows that the degree of every variable $x \in X$ is at least $\max_{x \in X} \{\deg_G(x)\}$, and therefore the number of consistency edges added by the foregoing procedure is at least $|X| \cdot \max_{x \in X} \{\deg_G(x)\}$.

Now, suppose that $|X| = \Omega(\text{size}(G))$ and that there exists $x_0 \in X$ for which $\deg_G(x_0) = \Omega(\text{size}(G))$; this can be the case if G is highly non-regular. In such a case, the number of consistency edges that will be added in the construction of H' will be at least $|X| \cdot \deg_G(x_0) = \Omega(\text{size}(G)^2)$, and therefore we will have $\text{size}(H') = \Omega(\text{size}(G)^2)$, contradicting the claim of Theorem 2. Note that this problem does not occur if G is a regular graph (i.e., $\deg_G(x) = \deg_G(y)$ for every $x, y \in X$), since in such case we have that

$$N \stackrel{\text{def}}{=} |X| \cdot \max_{x \in X} \{\deg_G(x)\} = \sum_{x \in X} \deg_G(x) \leq \text{size}(G)$$

and therefore we will have $\text{size}(H') = 2 \cdot \max(\text{size}(H), N) = O(\text{size}(G))$, as required. Ditto if G is almost regular (i.e., $\deg_G(x) = \Theta(\deg_G(y))$ for every $x, y \in X$).

It seems that the assignment testers to which this construction is applied in [2, 3] are regular, and in such a case the gap we discuss does not occur. However, envisioning possible applications in which the regularity condition does not hold or is hard to verify, we wish to establish the result also for such (general) cases.

3 Bridging the Gap

We turn to describe how the gap can be bridged. In order to bridge the gap, we modify the foregoing randomized procedure as follows. For every $x \in X$, let $[x]'$ to be an arbitrary subset of $[x]$ that has size $\min(|[x]|, \text{size}(H)/|X|)$. The modified procedure is the same as the original procedure, except for that in Step 2, it chooses z uniformly at random from *the set* $[x]'$ *instead of in* $[x]$. Observe that this modification indeed solves the problem, since now the degree of every variable $x \in X$ in H' is at most $\text{size}(H)/|X|$, and therefore the total number of consistency edges is at most $\text{size}(H) = O(\text{size}(G))$.

The reason that the modified procedure works is roughly as follows: Consider some assignment to $X \cup V_H$. Ideally, we would like that if a variable $x \in X$ is assigned a value that is inconsistent with most of $[x]$, then this variable violates $\Omega(1/|X|)$-fraction of the edges of H'. Suppose now that some variable $x \in X$ is assigned a value that is inconsistent with most of the vertices in $[x]$. Then, either that x is inconsistent with most vertices in $[x]'$ or most of the vertices in $[x]'$ are inconsistent with most of the vertices in $[x]$. In the first case, at least $\Omega(1/|X|)$-fraction of the edges are violated, since the modified procedure chooses x with probability $1/|X|$ and then chooses with probability at least $1/2$ a vertex $z \in [x]'$ that is inconsistent with x. In the second case, the inconsistency of the two majorities yields sufficiently many violated edges by virtue of the mixing property of the expander $[x]$. Details follow.

Indeed, the case where x is consistent with most of $[x]'$ is more problematic, since the procedure is likely to choose $z \in [x]'$ that is consistent with x, whereas this value is inconsistent with the majority in $[x]$. Indeed, this is possible only when $[x]' \neq [x]$, which in particular implies that $|[x]| > s \stackrel{\text{def}}{=} \text{size}(H)/|X| = |[x]'|$. Hence, there is an $(s/2)$-subset of $[x]$ (i.e., the majority in $[x]'$) that is inconsistent with most of $[x]$, and therefore (by the mixing property of the expander $[x]$) at least $\Omega(s/2)$ inner edges of $[x]$ are violated. It follows that the fraction of violated edges that are incident at x is at least

$$\frac{\Omega(s)}{\text{size}(H')} = \frac{\Omega(\text{size}(H)/|X|)}{O(\text{size}(H))} = \Omega(1/|X|)$$

as required. Below we give a rigorous proof of this argument.

We first describe the modified procedure for an arbitrary value of t (rather than just $t = 1$). Recall that in this case the assignment we deal with, denoted $A : V_H \cup X \to \Sigma^{d^{t/2}} \cup \{0,1\}$, assigns each $v \in V_H$ a sequence of $d^{t/2}$ values, one per each vertex u at distance at most $t/2$ from v. We denote by $A(v)_u$ the value that v attributes to u.

The modified consistency-testing procedure for arbitrary t.

1. Select $x \in X$ uniformly at random.
2. Select $z \in [x]'$ uniformly at random (recall that $[x]'$ is an arbitrary s-subset of $[x]$, where $s = \min(|[x]|, \text{size}(H)/|X|)$).

3. Take a $\lfloor t/2 \rfloor$-step random walk in prep(G) starting from z, and let w be the endpoint of the walk. Accept if and only if $A(w)_z = A(x)$.

We now use the procedure to define the consistency edges as before, and then re-weight the edges of H' such that the consistency edges form half of the edges of H'. Observe that this modification *solves the problem*: Indeed, this construction requires placing at most size$(H)/|X|$ consistency edges on H' for each variable in X, but this sums-up to only $O(\text{size}(H)) = O(\text{size}(G))$ consistency edges.

It remains to show that the intermediate assignment tester that uses the modified consistency-testing procedure has rejection probability $\Omega(\min(\sqrt{t} \cdot \varepsilon, 1/t)$, where ε is the rejection probability of the original assignment tester. In order to do it, we prove a result analogous to [2, Lemma 8.2] and [3, Lemma 9.2]. The reason that we prove this result is that Dinur [2,3] proves the analogous result for her construction of H', while we prove it for the modified version of this construction. The following lemma also differs from [2, Lemma 8.2] and [3, Lemma 9.2] in some (hidden) constant factors.

Lemma 3 (analogous to [2, Lemma 8.2] and [3, Lemma 9.2]): *Let $\varepsilon < 1/t$ be the rejection probability of the original assignment tester, and H' be the constraint graph output by the new assignment tester. Then, for any assignment $a : X \to \{0,1\}$, the following holds.*

- *If $a \in SAT(\Phi)$, then there exists $b : V_H \to \Sigma^{d^{t/2}}$ such that $UNSAT_{a \cup b}(H') = 0$.*
- *If $\delta = \mathtt{dist}(a, SAT(\Phi)) > 0$, then for every $b : V_H \to \Sigma^{d^{t/2}}$ it holds that $UNSAT_{a \cup b}(H') = \Omega(\sqrt{t} \cdot \varepsilon) \cdot \delta$.*

Proof. The first item of the lemma can be proved using the same proof as in [2,3]. Turning to the second item, assume that $\delta = \mathtt{dist}(a, \text{SAT}(\Phi)) > 0$ and fix an assignment $b : V_H \to \Sigma^{d^{t/2}}$ to H. We shall prove that $\text{UNSAT}_{a \cup b}(H') = \Omega(\sqrt{t} \cdot \varepsilon) \cdot \delta$. As in [2,3], let b_1 be the assignment to prep(G) decoded from b using a plurality vote, and let b_0 the assignment to G decoded from b_1 using plurality vote.

The case where $\mathtt{dist}(b_0|_X, a) \le \delta/2$ can be proved using the same proof as in [2,3], since in this case $\mathtt{dist}(b_0|_X, \text{SAT}(\Phi)) \ge \delta/2$ (by the triangle inequality), and the argument focuses on this fact only (while ignoring a). Specifically, in this case, by the definition of G, it holds that $\text{UNSAT}_{b_0}(G) \ge \varepsilon \cdot \delta/2$. Now, the reasoning of [2,3] (which rely on the properties of preprocessing and graph powering that take place in H), implies that $\text{UNSAT}_b(H) = \Omega(\sqrt{t} \cdot \varepsilon) \cdot \delta$. Since the edges of H form half of the edges of H', it follows that $\text{UNSAT}_{a \cup b}(H') = \Omega(\sqrt{t} \cdot \varepsilon) \cdot \delta$, as required.

We turn to the case where $\mathtt{dist}(b_0|_X, a) > \delta/2$, where we must rely on a and refer to the current modified construction. We shall prove that $\text{UNSAT}_{a \cup b}(H') = \Omega(\delta)$, which implies the required result. Recall that $b_0(v)$ is defined by plurality vote of b_1 in $[v]$. In contrast, we define b_0' to be an assignment for G such that for every $v \in V$ the value $b_0'(v)$ is the plurality vote among the values assigned by b_1 to the vertices in $[v]'$ (i.e., $b_0'(v)$ is the value that maximizes $\Pr_{u \in [v]'}[b_1(u) =$

$b'_0(v)]$). Indeed, the key question is whether these two assignments are close or not, and we consider the two possible cases: (1) $\texttt{dist}(b_0|_X, b'_0|_X) \le \delta/4$ and (2) $\texttt{dist}(b_0|_X, b'_0|_X) > \delta/4$.

1. Suppose that $\texttt{dist}(b_0|_X, b'_0|_X) \le \delta/4$. We show that in such case $a \cup b$ violates at least $\delta/16$ of the consistency edges of H', by considering the action of the modified consistency-testing procedure defined above. Using the triangle inequality, it holds that $\texttt{dist}(b'_0|_X, a) > \delta/4$; hence, in this case, with probability at least $\delta/4$, the procedure chooses in Step 1 a vertex $x \in X$ such that $b'_0(x) \ne a(x)$. Recall that the value $b'_0(x)$ is defined to be the most popular value assigned by b_1 to the vertices of $[x]'$, and therefore with probability at least $\frac{1}{2}$ the procedure chooses in Step 2 a vertex $z \in [x]'$ such that $b_1(z) \ne a(x)$. Similarly, conditioned on $b_1(z) \ne a(x)$, with probability at least $\frac{1}{2}$, the procedure chooses in Step 3 a vertex w such that $b(w)_z \ne a(x)$. Putting all of these together, it follows that in this case the consistency-testing procedure rejects $a \cup b$ with probability at least
$$\frac{\delta}{4} \cdot \frac{1}{2} \cdot \frac{1}{2} = \frac{\delta}{16}$$
and therefore $\mathrm{UNSAT}_{a \cup b}(H') = \Omega(\delta)$, as required.
2. Suppose that $\texttt{dist}(b_0|_X, b'_0|_X) > \delta/4$. We show that $\mathrm{UNSAT}_b(H) = \Omega(\delta)$ holds also in this case, but this time the bound is due to the violation of the equality constraints of $\mathrm{prep}(G)$. Recall that $\mathrm{prep}(G)$ is constructed by replacing every vertex v of G with a set $[v]$ (of copies v) of size $\deg_G(v)$, placing the edges of an expander on $[v]$ and associating those edges with equality constraints. Observe that the inequality $b_0(x) \ne b'_0(x)$ can only hold for variables $x \in X$ for which $[x]' \ne [x]$, and in this case it follows that $|[x]'| = \mathrm{size}(H)/|X|$, since $|[x]'| = \min(|[x]|, \mathrm{size}(H)/|X|)$ and $[x]' \subseteq [x]$. Now, observe for every $x \in X$ that satisfies $b_0(x) \ne b'_0(x)$, it holds that $\Omega(|[x]'|)$ equality edges of $[x]$ (i.e., edges between the majority vertices of $[x]'$ and the majority vertices of $[x]$) are violated by b_1, due to the mixing property of the expander that was used for the construction of $\mathrm{prep}(G)$. It follows that in this case the number of edges of $\mathrm{prep}(G)$ that are violated by b_1 is at least
$$\begin{aligned}|\{x \in X : b_0(x) \ne b'_0(x)\}| \cdot \Omega(\frac{\mathrm{size}(H)}{|X|}) &= \texttt{dist}(b_0|_X, b'_0|_X) \cdot \Omega(\mathrm{size}(H)) \\ &= \Omega(\delta \cdot \mathrm{size}(H)).\end{aligned}$$
Hence, $\mathrm{UNSAT}_b(H) = \Omega(\delta)$, and $\mathrm{UNSAT}_{a \cup b}(H') = \Omega(\delta)$ follows, as required.

This completes the proof of the lemma. ■

4 Digest

The issue at hand is the adaptation of the gap amplification procedure from the PCP setting to the setting of assignment-testers (resp., PCPPs). This adaptation boils down to augmenting the construction of the intermediate constraint graph, which is the pivot of the gap amplification procedure. The natural augmentation calls for connecting each input variable (i.e., a variable representing a bit in the input assignment (resp., a location in the input oracle)) to all auxiliary copies of this variable (i.e., auxiliary variables (resp., locations in the proof oracle)) that were produced in the preprocessing step.

The problem is that this natural augmentation may yield too large of an overhead in the case that the original constraint graph is not (almost) regular. The solution is to connect each input variable only to a number of copies that does not exceed the average degree of input variables in the original graph. The reason that this works is that expanders were placed (in the preprocessing step) among the copies of each variable, and so a mismatch between the majority of the "connected copies" and the majority of all copies will violate a large number of edges (i.e., a number that is linearly related to the number of connected copies). Hence, the number of violated edges is linearly related to the minimum between the degree of the input variable in the original constraint graph and the average degree of all input variables in that graph. The latter term suffices since the distance between assignments treats all input variables equally.

The last assertion also explains why it is unlikely that the given assignment tester will contain input variables of significantly different degrees. Hence, we expect that in most applications, all input variables will have (almost) the same degree, and in that case the original analysis of [2,3] suffices. Still, it feels better not to augment the definition of assignment testers so to require that all input variables have (almost) the same degree.

References

1. Ben-Sasson, E., Goldreich, O., Harsham, P., Sudan, M., Vadhan, S.: Robust PCPs of proximity, shorter PCPs and applications to coding. SIAM J. Comput. **36**(4), 889–974 (2006). Preliminary version in STOC 2004, pp. 120–134
2. Dinur, I.: The PCP theorem by gap amplification. In: ECCC TR05-046
3. Dinur, I.: The PCP theorem by gap amplification. J. ACM **54**(3), 12-es (2007). Preliminary version in STOC 2006, pp. 241–250
4. Dinur, I., Reingold, O.: Assignment testers: towards combinatorial proofs of the PCP theorem. SIAM J. Comput. **36**(4), 975–1024 (2006). Preliminary version in FOCS 2004, pp. 155–164

On (Valiant's) Polynomial-Size Monotone Formula for Majority

Oded Goldreich

Abstract. This exposition provides a proof of the existence of polynomial-size monotone formula for Majority. The exposition follows the main principles of Valiant's proof (*J. Algorithms*, 1984), but deviates from it in the actual implementation. Specifically, we show that, with high probability, a full ternary tree of depth $2.71 \log_2 n$ computes the majority of n values when each leaf of the tree is assigned at random one of the n values.

This is a drastic revision of a text that was posted on the author's web-site in May 2011.[1] The original text was somewhat laconic, and the current revision is aimed to be more reader friendly.

1 The Statement

It is easy to construct quasi-polynomial-size monotone formulae for majority by relying on divide-and-conquer approaches: For example, consider the recursion $\mathtt{TH}_t(x'x'') = \vee_{i=0}^{t}(\mathtt{TH}_i(x') \wedge \mathtt{TH}_{t-i}(x''))$, where $\mathtt{TH}_t$ is the threshold function that is satisfied if the string has at least t ones (i.e., $\mathtt{TH}_t(z) = 1$ if and only if $\mathtt{wt}(z) \geq t$ (see notation below)). Using $\mathtt{MAJ}(x) = \mathtt{TH}_{|x|/2}(x)$, this yields a size recursion of the form $S(n) = O(n) \cdot S(n/2)$, which solves to $S(n) = O(n)^{\log_2 n}$.

It is less obvious how to construct polynomial-size formulae (let alone monotone ones; cf. [6] and the references there-in). This exposition presents a variant of Valiant's classic proof of the existence of such formulae [6].

Theorem 1 (the classic theorem): *There exist polynomial-size monotone formulae for computing majority.*

The existence of polynomial-size (monotone) formulae is known to be equivalent to the existence of logarithmic-depth (monotone) circuits of bounded fan-in.[2] Hence, we shall focus on proving the existence of logarithmic-depth monotone formulae (of bounded fan-in) for majority.

We note that two radically different proofs are known for Theorem 1: The first proof uses the rather complicated construction of sorting networks of logarithmic

[1] See http://www.wisdom.weizmann.ac.il/~oded/PDF/mono-maj.pdf.
[2] One direction is almost trivial, for the other direction see [5].

O. Goldreich (Ed.): Computational Complexity and Property Testing, LNCS 12050, pp. 17–23, 2020.
https://doi.org/10.1007/978-3-030-43662-9_3

depth [1,4].[3] The second proof, due to Valiant [6] and revised below, uses the probabilistic method (cf. [2]). Specifically, it combines a random projection of the n input bits to $m = \text{poly}(n)$ locations with a simple formula that is applied to the resulting m-bit long string. Valiant's original proof uses a full binary tree with alternating AND and OR gates, whereas we shall use a full ternary tree with 3-WAY MAJORITY gates.

Notation. Suppose, for simplicity that n is odd, and consider the majority function $\texttt{MAJ} : \{0,1\}^n \to \{0,1\}$ defined as $\texttt{MAJ}(x) = 1$ if $\texttt{wt}(x) > n/2$ and $\texttt{MAJ}(x) = 0$ otherwise, where $\texttt{wt}(x) = |\{i \in [n] : x_i = 1\}|$ denotes the Hamming weight of $x = x_1 \cdots x_n$.

2 The Proof

We prove the existence of logarithmic-depth monotone formulae (of bounded fan-in) for majority in two steps.

The First Step: A Randomized Reduction. This step consists of reducing the worst-case problem (i.e., of computing MAJ on all inputs) to several average-case problems, one per each possible weight of the original input. That is, we shall use a (simple) randomized reduction of the computation of $\texttt{MAJ}(x)$ to the computation of $\texttt{MAJ}(R(x))$, where $R(x)$ denotes the output of the reduction on input x. Specifically, each position in $R(x) \in \{0,1\}^{\text{poly}(|x|)}$ holds one of the bits of x, which is selected uniformly at random (and independent of the other bits). The key observation is that if the error probability (of the average case solver on $R(x)$) is sufficiently low (i.e., lower than $2^{-|x|}$), then this randomized reduction yields a non-uniform reduction that is correct on all inputs. (Hence the existence of such a non-uniform reduction is proved by using the probabilistic method.)

The Second Step: A Simple Boolean Formula Solves All Average-Cases with Exponentially Vanishing Error Probability. In the second step, we show that a very simple (monotone) formula suffices for solving MAJ on the average (w.r.t each of the distributions $R(x)$). Specifically, we shall use formulae obtained by iterating the three-way majority function; that is, the resulting formula isessentially a ternary tree of logarithmic depth with (3-way) majority gates in internal nodes and distinct variables in the leaves. A Boolean formula is obtained by a straightforward implementation of the three-way majority gates by depth-three formulae (of fan-in two).

Composing the (monotone) reduction with the latter formulae, we obtain the desired (monotone) formulae. Since the randomized reduction is merely a

[3] Sorting networks may be viewed as Boolean circuits with bit-comparison gates (a.k.a comparators), where each comparator is a (2-bit) sorting device. Observe that a comparator can be implemented by a monotone circuit (i.e., $\texttt{comp}(x,y) = (\min(x,y), \max(x,y)) = ((x \wedge y), (x \vee y)))$, and that the middle bit of the sorted sequence equals the majority value of the original sequence.

randomized projection (i.e., each output bit is assigned at random one of the input bits), the complexity of the final formulae equals the complexity of the formulae constructed in the second step.

The formula used in the second step is more intuitive than the one used by Valiant, and consequently the randomized reduction used in the first step is more intuitive. For a detailed comparison, see Sect. 3. We mention that our construction was instrumental for the subsequent work of Cohen *et al.* [3], in which the use of small threshold functions is essential.

2.1 The Randomized Reduction

Given an n-bit long input $x = x_1 \cdots x_n$, we consider a sequence of $m = \text{poly}(n)$ independent identically distributed 0–1 random variables $R(x) = (y_1, ..., y_m)$ such that $\Pr[y_j = 1] = \texttt{wt}(x)/n$ for each $j \in [m]$. In other words, for each $j \in [m]$, an index $i_j \in [n]$ is selected uniformly at random (independently of all other choices) and y_j is set to x_{i_j}.

Both these equivalent descriptions are used in our analysis. Specifically, the first description is pivotal to the average-case analysis of the proposed Boolean formula (presented in Sect. 2.2), whereas the second analysis is instrumental to seeing that composing this reduction with a Boolean formula yields a Boolean formula of the same complexity.

A sanity check. Note that, for $m = \Omega(n^3)$, the following holds for every $x \in \{0,1\}^n$:

$$\Pr[\texttt{MAJ}(R(x)) = \texttt{MAJ}(x)] > \Pr\left[\left|\frac{\texttt{wt}(R(x))}{m} - \frac{\texttt{wt}(x)}{n}\right| < \frac{1}{2n}\right] > 1 - 2^{-n}, \quad (1)$$

(by Chernoff bound). This means that computing the majority of n-bit long strings in the worst-case is randomly reduced to computing the majority of m-bit long strings on the average-case. Note that we do not reduce the worst-case problem to an average-case problem with a single distribution but rather to $n+1$ different distributions, where the resulting distribution depends on the weight of the input. The point is that the average-case problem seems easier to analyze, as illustrated in Sect. 2.2, where we capitalize on the fact that $R(x)$ is a sequence of identically and independently distributed 0–1 random variables (each being 1 with probability $\texttt{wt}(x)/|x|$).

Towards the Actual Application. Note that any formula $F : \{0,1\}^m \rightarrow \{0,1\}$ that satisfies the analogous inequality (i.e., $\Pr[F(R(x)) = \texttt{MAJ}(x)] > 1 - 2^{-n}$, for every $x \in \{0,1\}^n$) yields a formula that correctly computes the majority of all n-bit strings (see Fact 2). This holds provided that average-case means being correct with probability greater than $1 - 2^{-n}$; we cannot afford straightforward error-reduction, because it involves taking majority (which is the task we wish to solve).

Fact 2 (trivial derandomization): *Let $R: \{0,1\}^n \to \{0,1\}^m$ be a randomized process and $F: \{0,1\}^m \to \{0,1\}$ be a function. Suppose that, for every $x \in \{0,1\}^n$, it holds that $\Pr[F(R(x)) = \mathtt{MAJ}(x)] > 1 - 2^{-n}$. Then, there exists a choice of coin tosses ω for the random process R such that for every $x \in \{0,1\}^n$ it holds that $F(R_\omega(x)) = \mathtt{MAJ}(x)$, where R_ω denotes the residual function obtained by fixing the coins of R to ω.*

Turning back to the specific process R defined before, note that, for every fixed ω, the residual function R_ω just projects its input bits to fixed locations in its output sequence; that is, letting $\omega = (i_1, ..., i_m) \in [n]^m$, it holds that $R_\omega(x_1 x_2 \cdots x_n) = x_{i_1} x_{i_2} \cdots x_{i_m}$. Hence, $F \circ R_\omega$ preserves the depth (resp., size) complexity and monotonicity of F.

Proof: Using $\Pr[F(R(x)) \neq \mathtt{MAJ}(x)] < 2^{-n}$ (for every x), and applying a union bound, we get

$$\Pr_\omega[\exists x \in \{0,1\}^n \ F(R_\omega(x)) \neq \mathtt{MAJ}(x)] < 1,$$

and it follows that there exists ω such that $F(R_\omega(x)) = \mathtt{MAJ}(x)$ holds for every $x \in \{0,1\}^n$. ■

Digest: The probabilistic method was used here to infer the existence of ω such that $F \circ R_\omega = \mathtt{MAJ}$ from the fact that $\Pr_\omega[(\forall x \in \{0,1\}^n) \ F(R_\omega(x)) = \mathtt{MAJ}(x)] > 0$.

2.2 Solving the Average-Case Problem

We now turn to the second step, which consists of presenting a monotone formula F of logarithmic depth that satisfies the hypothesis of Fact 2 (w.r.t the simple process R defined above). Generalizing the foregoing hypothesis, we wish F to satisfy the following condition: *If $Y_1, ..., Y_m$ are independent identically distributed 0–1 random variables such that for some $b \in \{0,1\}$ it holds that $\Pr[Y_1 = b] \geq 0.5 + 1/2n$, then $\Pr[F(Y_1, ..., Y_m) = b] > 1 - 2^{-n}$.*

The construction uses a full ternary tree of depth $\ell = \log_3 m$, where internal vertices compute the majority of their three children. (For simplicity, we assume that m is a power of three.) Specifically, let $\mathtt{MAJ}_3$ denote the three-variable majority function, and define $F_1(z_1, z_2, z_3) = \mathtt{MAJ}_3(z_1, z_2, z_3)$ and

$$F_{i+1}(z_1, ..., z_{3^{i+1}}) \stackrel{\text{def}}{=} \mathtt{MAJ}_3(F_i(z_1, ..., z_{3^i}), F_i(z_{3^i+1}, ..., z_{3^i+3^i}), F_i(z_{2 \cdot 3^i+1}, ..., z_{3^{i+1}})). \tag{2}$$

for every $i \geq 1$. Finally, we let $F(z_1, ..., z_m) = F_\ell(z_1, ..., z_m)$.

The intuition is that each level in F increases the bias of the corresponding random variables (which are functions of $Y_1, ..., Y_m$) towards the majority value; that is, the probability that an internal vertex in the corresponding ternary tree evaluates to to b (where $\Pr[Y_1 = b] > 0.5$ for every $j \in [m]$), increases when going up the tree (i.e., away from the leaves). This effect is due to the bias-increasing property of $\mathtt{MAJ}_3$, which is stated next.

Fact 3 (three-way majority amplifies bias): *Let Z_1, Z_2, Z_3 be three independent identically distributed 0–1 random variables, and let $p \stackrel{\text{def}}{=} \Pr[Z_1 = 1]$. Then:*

1. $p' \stackrel{\text{def}}{=} \Pr[\mathtt{MAJ}_3(Z_1, Z_2, Z_3) = 1] = 3 \cdot (1-p) \cdot p^2 + p^3$.
2. Letting $\delta \stackrel{\text{def}}{=} p - 0.5$, *it holds that* $p' = 0.5 + (1.5 - 2\delta^2) \cdot \delta$.
3. $p' < 3p^2$.

We stress that the three parts hold for every $p \in [0, 1]$, but we shall use Part 2 with $p > 0.5$ and use Part 3 with $p \ll 0.5$. Note that Part 2 implies that if $p \in (0.5, 1)$, then $p' > 0.5 + \delta = p$.

Proof: The three parts follow by straightforward calculations. Specifically, Part 1 merely gives the expression for $\Pr[Z_1 + Z_2 + Z_3 \in \{2, 3\}]$, and the other parts merely manipulate this expression (e.g., for Part 2 we use $p' = (3-2p) \cdot p^2 = (3 - 1 - 2\delta) \cdot (0.25 + \delta + \delta^2)$, which implies $p' = 0.5 + 1.5\delta - 2\delta^3$, and for Part 3 we use $p' < 3 \cdot (1-p) \cdot p^2 + 3 \cdot p^3 = 3p^2$). ■

Analyzing $F_\ell(Y_1, ..., Y_m)$ *using Fact* 3. Fact 3 asserts that $\mathtt{MAJ}_3$ increases the bias of (independent and identically distributed) 0–1 random variables towards the majority value. The question is how fast does the bias (i.e., $p - 0.5$) tend to the extreme (i.e., 0.5) when the foregoing process is iterated. Applying Part 2 to the majority value, we see that as long as p is bounded away from 1 the value $p - 0.5$ increases by a constant factor. This will be useful towards increasing p from a value slightly above 0.5 (i.e., $p = 0.5 + 1/2n$) to a constant value in $(0.5, 1)$. At this point Part 3 becomes handy, provided that we apply it to the minority value. Doing so we drastically reduce the probability of the minority value (essentially squaring it). Details follow.

Part 2 of Fact 3 implies that if $p = 0.5 + \delta > 0.5$ and $\delta \leq \delta_0 < 0.5$, then $p' \geq 0.5 + (1.5 - 2\delta_0^2) \cdot \delta$, which means that the bias (i.e., $p - 0.5$) increases by a multiplicative factor in each iteration (until it exceeds δ_0). (Note that we assumed $p \geq 0.5 + 1/2n$, but similar considerations hold for $p \leq 0.5 - 1/2n$.)[4]. This means that we can increase the bias (i.e., $p - 0.5$) from its initial level of at least $1/2n$ to any constant level of $\delta_0 < 1/2$, by using $\ell_1 = \lceil c_1 \cdot \log_2(2\delta_0 n) \rceil$ iterations of $\mathtt{MAJ}_3$, where $c_1 = 1/\log_2(1.5 - 2\delta_0^2)$.[5]

The best result is obtained by using an arbitrary small $\delta_0 > 0$. In this case, we may use $c_1 \approx 1/\log_2(1.5) \approx 1.70951129$. Using $\ell_2 = O(1)$ additional iterations (and Part 2), we may increase the bias from δ_0 to any larger constant that is smaller than 0.5. Specifically, we shall increase the bias to 0.4 (using $\ell_2 = \lceil \log_{1.18}(0.4/\delta_0) \rceil$).

At this point, we use Part 3 of Fact 3, while considering the probability for a wrong majority value. In each such iteration, this probability is reduced from a current value of $1-p$ to less than $3 \cdot (1-p)^2$. Thus, using $\ell_3 = \lceil \log_2 n \rceil$ additional iterations, the probability of a wrong value reduces from $1 - (0.5 + 0.4) < 1/6$ to $3^{2^{\ell_3}-1} \cdot (1/6)^{2^{\ell_3}} < 2^{-2^{\ell_3}} \leq 2^{-n}$.

[4] One way to see this is to define $p = \Pr[Z_1 = 0]$.
[5] Suppose that i iterations are sufficient for increasing the bias from $1/2n$ to δ_0. Then, $(1.5 - 2\delta_0^2)^i \cdot (1/2n) \geq \delta_0$ holds, which solves to $i \geq \log_{1.5-2\delta_0^2}(2\delta_0 n)$. Hence, we may use the minimal such $i \in \mathbb{N}$.

Conclusion. Letting $\ell = \ell_1 + \ell_2 + \ell_3 < 2.71 \log_2 n$ and $m = 3^\ell$, we obtain a formula $F = F_\ell$ on m variables that, given $R(x)$, computes $\mathtt{MAJ}(x)$ with overwhelmingly high probability. That is:

Theorem 4 (majority formulae via three-way majority): *For $x \in \{0,1\}^n$, let $\ell = 2.71 \log_2 n$ and $m = 3^\ell$, and consider the random process $R : \{0,1\}^n \to \{0,1\}^m$ such that each bit in $R(x)$ equals 1 with probability $\mathtt{wt}(x)/n$, independently of all other bits. Then, $\Pr[MAJ(x) = F_\ell(R(x))] > 1 - 2^{-n}$, where F_ℓ is as defined in Eq. (2).*[6]

Using Fact 2, Theorem 4 yields a formula (i.e., $F_\ell \circ R_\omega$) that computes $\mathtt{MAJ}(x)$ correctly on all inputs x, but this formula uses the non-standard $\mathtt{MAJ}_3$-gates. Yet, a $\mathtt{MAJ}_3$-gate can be implemented by a depth-three monotone formula (e.g., $\mathtt{MAJ}_3(z_1, z_2, z_3)$ equals $(z_1 \wedge z_2) \vee (z_2 \wedge z_3) \vee (z_3 \wedge z_1)$), and hence we obtain a standard monotone formula F' of depth $3\ell < 8.13 \log_2 n$. Recall that if $Y_1, ..., Y_m$ are independent identically distributed 0–1 random variables such that for some b it holds that $\Pr[Y_1 = b] \geq 0.5 + 1/2n$, then $\Pr[F'(Y_1, ..., Y_m) \neq b] < 2^{-n}$. Thus, for every $x \in \{0,1\}^n$ it holds that $\Pr_\omega[F'(R_\omega(x)) \neq \mathtt{MAJ}(x)] < 2^{-n}$ and $\Pr_\omega[(\forall x \in \{0,1\}^n)\ F'(R_\omega(x)) = \mathtt{MAJ}(x)] > 0$ follows. Hence, there exists a choice of ω such that $F' \circ R_\omega$ computes the majority of n-bit inputs.

3 Comparison to Valiant's Proof

Interestingly, Valiant [6] obtains a somewhat smaller formula by using an iterated construction that uses the function $V(z_1, z_2, z_3, z_4) = (z_1 \vee z_2) \wedge (z_3 \vee z_4)$ as the basic building block (rather than $\mathtt{MAJ}_3$). Since V is not a balanced predicate (i.e., $\Pr_{z \in \{0,1\}^4}[V(z) = 1] = 9/16$), the random process used in [6] maps the string $x \in \{0,1\}^n$ to a sequence of independent identically distributed 0–1 random variables, $(y_1, ..., y_m)$, such that for every $j \in [m]$ the bit y_j is set to zero with some constant probability β (and is set to x_i otherwise, where $i \in [n]$ is uniformly distributed). The value of β is chosen such that if Z_1, Z_2, Z_3, Z_4 are independent identically distributed 0–1 random variables satisfying $\Pr[Z_1 = 1] = p \stackrel{\text{def}}{=} (1 - \beta)/2$, then $\Pr[V(Z_1, Z_2, Z_3, Z_4) = 1] = p$.

It turns out that V amplifies the deviation from p slightly better than $\mathtt{MAJ}_3$ does (w.r.t 1/2).[7] More importantly, V can be implemented by a monotone formula of depth two (and fan-in two), whereas $\mathtt{MAJ}_3$ requires depth three. Thus, Valiant [6] performs $2.65 \log_2 n$ iterations (rather than $2.71 \log_2 n$ iterations), and obtains a formula of depth $5.3 \log_2 n$ (rather than $8.13 \log_2 n$).

Acknowledgments. We thank to Alina Arbitman for her comments and suggestions regarding the original write-up.

[6] Actually, we have $\Pr[MAJ(x) = F_\ell(R(x))] > 1 - 2^{-n-\Omega(n)}$, because we can use $\ell_3 = (2.71 - c_1) \cdot \log_2 n - O(1) = \Omega(\log n)$, where $c_1 \approx 1/\log_2(1.5) \approx 1.70951129$. (Alternatively, note that the analysis of the last ℓ_3 iterations actually yields an error probability of $3^{2^{\ell_3}-1} \cdot (0.1)^{2^{\ell_3}} < (0.3)^{2^{\ell_3}} < 2^{-1.7n}$. Furthermore, 0.1 can be replaced by any positive constant.)

[7] This is surprising only if we forget that V takes four inputs rather than three.

References

1. Ajtai, M., Komlos, J., Szemerédi, E.: An $O(n \log n)$ sorting network. In: 15th ACM Symposium on the Theory of Computing, pp. 1–9 (1983)
2. Alon, N., Spencer, J.H.: The Probabilistic Method, 4th edn. Wiley, Hoboken (2016)
3. Cohen, G., Damgård, I.B., Ishai, Y., Kölker, J., Miltersen, P.B., Raz, R., Rothblum, R.D.: Efficient multiparty protocols via log-depth threshold formulae. In: Canetti, R., Garay, J.A. (eds.) CRYPTO 2013. LNCS, vol. 8043, pp. 185–202. Springer, Heidelberg (2013). https://doi.org/10.1007/978-3-642-40084-1_11
4. Paterson, M.S.: Improved sorting networks with $O(\log N)$ depth. Algorithmica **5**(1), 75–92 (1990)
5. Spira, P.M.: On time hardware complexity trade-offs for Boolean functions. In: The 4th Hawaii International Symposium on System Sciences, pp. 525–527 (1971)
6. Valiant, L.G.: Short monotone formulae for the majority function. J. Algorithms **5**(3), 363–366 (1984)

Two Comments on Targeted Canonical Derandomizers

Oded Goldreich

Abstract. We revisit the notion of a *targeted canonical derandomizer*, introduced in our prior work (*ECCC*, TR10-135) as a uniform notion of a pseudorandom generator that suffices for yielding $\mathcal{BPP} = \mathcal{P}$. The original notion was derived (as a variant of the standard notion of a canonical derandomizer) by providing both the distinguisher and the generator with the same auxiliary-input. Here we take one step further and consider pseudorandom generators that fool a single circuit that is given to both (the distinguisher and the generator) as auxiliary input. Building on the aforementioned prior work, we show that such pseudorandom generators of constant seed length exist if and only if $\mathcal{BPP} = \mathcal{P}$, which means that they exist if and only if the previously defined targeted canonical derandomizers (of exponential stretch, as in the prior work) exist. We also relate such targeted canonical derandomizer to targeted hitters, which are the analogous canonical derandomizers for $\mathcal{RP}$.

An early version of this work appeared as TR11-047 of *ECCC*. The current revision is quite minimal in nature. Still, some typos were corrected and some phrasings were improved.

Caveat: Throughout the text, we abuse standard notation by letting $\mathcal{BPP}$, $\mathcal{P}$ etc. denote classes of promise problems. We are aware of the possibility that this choice may annoy some readers, but believe that promise problem actually provide the most adequate formulation of natural decisional problems. Actually, the common restriction of general studies of feasibility to decision problems is merely a useful methodological simplification.

1 Introduction

In prior work [4], we presented two results that relate the existence of certain pseudorandom generators to certain derandomizations of the class $\mathcal{BPP}$. The first result referred to the standard notion of a uniform canonical derandomizer (as introduced in [5]) and asserted that such pseudorandom generators of exponential stretch exist if and only if $\mathcal{BPP}$ is effectively in $\mathcal{P}$ (in the sense that it is infeasible to find an input on which the polynomial-time derandomized algorithm errs).[1]

[1] More accurately, for any $S \in \mathcal{BPP}$ and every polynomial p, there exists a deterministic polynomial-time A such that no probabilistic p-time algorithm F can find (with probability exceeding $1/p$) an input on which A errs; that is, the probability that $F(1^n)$ equals an n-bit string x such that $A(x) \neq \chi_S(x)$ is at most $1/p(n)$, where χ_S is the characteristic function of S.

O. Goldreich (Ed.): Computational Complexity and Property Testing, LNCS 12050, pp. 24–35, 2020.
https://doi.org/10.1007/978-3-030-43662-9_4

The second result referred to a new notion of a canonical derandomizer, which was introduced in [4] and called a *targeted canonical derandomizer.* This notion is the subject of the current work. We mention that it was shown in [4] that targeted canonical derandomizers (of exponential stretch) exist if and only if $\mathcal{BPP} = \mathcal{P}$.

The foregoing notion of a targeted canonical derandomizer was derived as a variant of the standard notion of a canonical derandomizer, which is required to produce sequences that look random to any (linear size) non-uniform circuit. Specifically, a targeted canonical derandomizer is only required to fool uniform (deterministic) linear-time algorithms that obtain any auxiliary input (of linear length), but the generator is given the same auxiliary input. (This auxiliary input represent the main input given to a generic probabilistic polynomial-time algorithm that we wish to derandomize.)

In this work we revisit the notion of a targeted canonical derandomizer. Specifically, we take this approach to its logical conclusion, and consider pseudorandom generators that fool a single circuit that is given to them as auxiliary input. (This circuit represents the combination of the probabilistic polynomial-time algorithm that we wish to derandomize coupled with the main input given to that algorithm. In terms of the notion of a targeted canonical derandomizer as defined in [4], we replace the free choice of a uniform (deterministic) linear-time algorithm by a fixed choice of an evaluation algorithm for circuits.)

We stress that constructing such generators is not trivial. In fact, building on the ideas of [4], we show that *such pseudorandom generators* (of logarithmic seed length (equiv., exponential stretch)) *exist if and only if* $\mathcal{BPP} = \mathcal{P}$. Furthermore, such pseudorandom generators may use a seed of constant length (i.e., a two-bit long random seed).

Applying the same approach to hitting set generators, we derive a notion of a *targeted hitter*, which is adequate for derandomizing $\mathcal{RP}$. Specifically, a targeted hitter is a deterministic polynomial-time algorithm that, on input a circuit that accepts most strings of a certain length, finds a string that satisfies this circuit. Clearly, such a targeted hitter implies that $\mathcal{RP} = \mathcal{P}$, which in turn implies $\mathcal{BPP} = \mathcal{P}$ (see, e.g., [3, §6.1.3.2]). Thus, *targeted hitters exist if and only if targeted canonical derandomizers exist.*

Organization. For sake of self-containment, we recall (in Sects. 2 and 3) the preliminaries and background that forms the basis for the current work. These parts are reproduced from [4], and contain the prior notion of targeted canonical derandomizers (see Definition 3.2). The new part of this work start in Sect. 3.3, which presents the new notion of targeted canonical derandomizers (see Definition 3.3). Sections 4 and 5 present the (modest) technical contributions of this work, and Sect. 6 de-constructs them.

2 Preliminaries

In Sect. 2.1, we review the notion of promise problems, while adapting it to the context of search problems, and in Sect. 2.2 we define "BPP search problem" (while warning that the definition is not straightforward). The entire section is reproduced from [4].

Standard Notation. For a natural number n, we let $[n] \stackrel{\text{def}}{=} \{1, 2, ..., n\}$ and denote by U_n a random variable that is uniformly distributed over $\{0,1\}^n$. When referring to the probability that a uniformly distributed n-bit long string hits a set S, we shall use notation such as $\Pr[U_n \in S]$ or $\Pr_{r\in\{0,1\}^n}[r \in S]$.

2.1 Promise Problems

We rely heavily on the formulation of promise problems (introduced in [2]). We believe that, in general, the formulation of promise problems is far more suitable for any discussion of feasibility results. The original formulation of [2] refers to decisional problems, but we shall also extend it to search problem.

In the setting of decisional problems, a promise problem, denoted $\langle P, Q\rangle$, consists of a promise (set), denoted P, and a question (set), denoted Q, such that the problem $\langle P, Q\rangle$ is defined as *given an instance* $x \in P$, *determine whether or not* $x \in Q$. That is, the solver is required to distinguish inputs in $P \cap Q$ from inputs in $P \setminus Q$, and nothing is required in case the input is outside P. Indeed, an equivalent formulation refers to two disjoint sets, denoted Π_{YES} and Π_{NO}, of YES- and NO-instances, respectively. We shall actually prefer to present promise problems in these terms; that is, as pairs $(\Pi_{\text{YES}}, \Pi_{\text{NO}})$ of disjoint sets. Indeed, standard decision problems appear as special cases in which $\Pi_{\text{YES}} \cup \Pi_{\text{NO}} = \{0,1\}^*$. In the general case, inputs outside of $\Pi_{\text{YES}} \cup \Pi_{\text{NO}}$ are said to violate the promise.

Unless explicitly stated otherwise, all decisional problems discussed in this work are actually promise problems, and $\mathcal{P}, \mathcal{BPP}$ etc. denote the corresponding classes of promise problems. For example, $(\Pi_{\text{YES}}, \Pi_{\text{NO}}) \in \mathcal{BPP}$ if *there exists a probabilistic polynomial-time algorithm* A *such that for every* $x \in \Pi_{\text{YES}}$ *it holds that* $\Pr[A(x)=1] \geq 2/3$, *and for every* $x \in \Pi_{\text{NO}}$ *it holds that* $\Pr[A(x)=0] \geq 2/3$.

2.2 BPP Search Problem

Typically, search problems are captured by binary relations that determine the set of valid instance-solution pairs. For a binary relation $R \subseteq \{0,1\}^* \times \{0,1\}^*$, we denote by $R(x) \stackrel{\text{def}}{=} \{y : (x, y) \in R\}$ the set of valid solutions for the instance x, and by $S_R \stackrel{\text{def}}{=} \{x : R(x) \neq \emptyset\}$ the set of instances having valid solutions. Solving a search problem R means that given any $x \in S_R$, we should find an element of $R(x)$; in addition, for $\notin S_R$, we should indicate that no solution exists.

The definition of "BPP search problems" is supposed to capture search problems that can be solved efficiently, when random steps are allowed. Intuitively, we do not expect randomization to make up for more than an exponential blow-up, and so the naive formulation that merely asserts that solutions can be found in probabilistic polynomial-time is not good enough. Consider, for example, the relation R such that $(x, y) \in R$ if $|y| = |x|$ and for every $i < |x|$ it holds that $M_i(x) \neq y$, where M_i is the ith deterministic machine (in some fixed enumeration of such machines). Then, the search problem R can be solved by a probabilistic polynomial-time algorithm (which, on input x, outputs a uniformly distributed $|x|$-bit long string), but cannot be solved by any deterministic algorithm (regardless of its running time).

What is missing in the naive formulation is any reference to the "complexity" of the solutions found by the solver, let alone to the complexity of the set of all valid solutions. To correct this lacuna, we just postulate that the set of all valid instance-solutions pairs is easily recognizable. Actually[2], we generalize the treatment to search problems with a promise, where the promise allows to possibly discard some instance-solution pairs. (At first reading, the reader may assume that $R_{\text{NO}} = \{0,1\}^* \setminus R_{\text{YES}}$.)

Definition 2.1 (BPP search problems): *Let R_{YES} and R_{NO} be two disjoint binary relations. We say that $(R_{\text{YES}}, R_{\text{NO}})$ is a* BPP-search problem *if the following two conditions hold.*

1. *The decisional problem represented by $(R_{\text{YES}}, R_{\text{NO}})$ is solvable in probabilistic polynomial-time; that is, there exists a probabilistic polynomial-time algorithm V such that for every $(x,y) \in R_{\text{YES}}$ it holds that $\Pr[V(x,y)=1] \geq 2/3$, and for every $(x,y) \in R_{\text{NO}}$ it holds that $\Pr[V(x,y)=1] \leq 1/3$.*
2. *There exists a probabilistic polynomial-time algorithm A such that, for every $x \in S_{R_{\text{YES}}}$, it holds that $\Pr[A(x) \in R_{\text{YES}}(x)] \geq 2/3$, where $R_{\text{YES}}(x) = \{y : (x,y) \in R_{\text{YES}}\}$ and $S_{R_{\text{YES}}} = \{x : R_{\text{YES}}(x) \neq \emptyset\}$.*

We may assume, without loss of generality, that, for every x such that $R_{\text{NO}}(x) = \{0,1\}^*$, it holds that $\Pr[A(x) = \perp] \geq 2/3$, because algorithm A can avoid outputting invalid solutions (i.e., elements of $R_{\text{NO}}(x)$) by checking them using algorithm V.

Note that the algorithms postulated in Definition 2.1 allow for finding valid solutions (i.e., elements of $R_{\text{YES}}(x)$) as well as distinguishing valid solutions (i.e., elements of $R_{\text{YES}}(x)$) from invalid ones (i.e., elements of $R_{\text{NO}}(x)$), but they do not offer a way of increasing the probability of finding valid solutions. This in the case because these algorithms do not provide a way of distinguishing elements of $R_{\text{YES}}(x)$ from elements of $\{0,1\}^* \setminus (R_{\text{YES}}(x) \cup R_{\text{NO}}(x))$. The latter fact will cause some technical difficulties.

3 Definitional Treatment

For sake of clarity and perspective, we start by reviewing the standard definition of (non-uniformly strong) canonical derandomizer (cf., e.g., [3, Sec. 8.3.1]). Next, we review the notion of a targeted canonical derandomizer that was introduced in [4, Sec. 4.4], and finally we present the new definition of a targeted canonical derandomizer. The first two subsections are reproduced from [4].

[2] See motivational discussion in [4, Sec. 3.1].

3.1 The Standard (Non-uniformly Strong) Definition

Recall that in order to "derandomize" a probabilistic polynomial-time algorithm A, we first obtain a functionally equivalent algorithm A_G that uses a pseudo-random generator G in order to reduce the randomness-complexity of A, and then take the majority vote over all possible executions of A_G (on the given input). That is, we scan all possible outcomes of the coin tosses of $A_G(x)$, which means that the deterministic algorithm will run in time that is exponential in the randomness complexity of A_G. Thus, it suffices to have a pseudorandom generator that can be evaluated in time that is exponential in its seed length (and polynomial in its output length).

In the standard setting, algorithm A_G has to maintain A's input-output behavior on all (but finitely many) inputs, and so the pseudorandomness property of G should hold with respect to distinguishers that receive non-uniform advice (which models a potentially exceptional input x on which $A(x)$ and $A_G(x)$ are sufficiently different). Without loss of generality, we may assume that A's running-time is linearly related to its randomness complexity, and so the relevant distinguishers may be confined to linear time. Similarly, for simplicity (and by possibly padding the input x), we may assume that both complexities are linear in the input length, $|x|$. (Actually, for simplicity we shall assume that both complexities just equal $|x|$, although some constant slackness seems essential.) Finally, since we are going to scan all possible random-pads of A_G and rule by majority (and since A's error probability is at most $1/3$), it suffices to require that for every x it holds that $|\Pr[A(x) = 1] - \Pr[A_G(x) = 1]| < 1/6$. This leads to the pseudorandomness requirement stated in the following definition.

Definition 3.1 (canonical derandomizers, standard version [3, Def, 8.14])[3]: *Let* $\ell:\mathbb{N}\rightarrow\mathbb{N}$ *be a function such that* $\ell(n) > n$ *for all* n*. A* canonical derandomizer of stretch ℓ *is a* deterministic *algorithm* G *that satisfies the following two conditions.*

(generation time): *On input a* k*-bit long seed,* G *makes at most* $\text{poly}(2^k \cdot \ell(k))$ *steps and outputs a string of length* $\ell(k)$*.*

(pseudorandomness): *For every* (deterministic) *linear-time algorithm* D*, all sufficiently large* k *and all* $x \in \{0,1\}^{\ell(k)}$*, it holds that*

$$|\Pr[D(x, G(U_k)) = 1] \; - \; \Pr[D(x, U_{\ell(k)}) = 1]| \; < \; \frac{1}{6} \tag{1}$$

The algorithm D represents a potential distinguisher, which is given two $\ell(k)$-bit long strings as input, where the first string (i.e., x) represents a (non-uniform) auxiliary input and the second string is sampled either from $G(U_k)$ or from $U_{\ell(k)}$. When seeking to derandomize a linear-time algorithm A, the first string (i.e., x) represents a potential main input for A, whereas the second string represents a possible sequence of coin tosses of A (when invoked on a generic (primary) input x of length $\ell(k)$).

[3] To streamline our exposition, we preferred to avoid the standard additional step of replacing $D(x, \cdot)$ by an arbitrary (non-uniform) Boolean circuit of quadratic size.

3.2 The Original Notion of Targeted Generators

Our main focus in [4] was on the standard notion of a uniform canonical derandomizer (as introduced in [5]), which was shown to exist (with exponential stretch) if and only if $\mathcal{BPP}$ is *effectively* in $\mathcal{P}$ (in the sense that it is *infeasible* to find an input on which the polynomial-time derandomized algorithm errs). Still, seeking a notion of a canonical derandomizer that can be shown to exist if and only if $\mathcal{BPP} = \mathcal{P}$ (*proper*), we suggested the following notion of a targeted canonical derandomizer, where both the generator and the distinguisher are presented with the same auxiliary input (or "target").

Definition 3.2 (targeted canonical derandomizers, [4, Def. 4.10]): *Let $\ell:\mathbb{N}\to\mathbb{N}$ be a function such that $\ell(n) > n$ for all n. A* targeted canonical derandomizer of stretch ℓ *is a deterministic algorithm G that satisfies the following two conditions.*

(generation time): *On input a k-bit long seed and an $\ell(k)$-bit long auxiliary input, G makes at most* $\mathrm{poly}(2^k \cdot \ell(k))$ *steps and outputs a string of length $\ell(k)$.*
(pseudorandomness (targeted)): *For every* (deterministic) *linear-time algorithm D, all sufficiently large k and all $x \in \{0,1\}^{\ell(k)}$, it holds that*

$$|\Pr[D(x, G(U_k, x)) = 1] - \Pr[D(x, U_{\ell(k)}) = 1]| < \frac{1}{6} \qquad (2)$$

Definition 3.1 is obtained from Definition 3.2 by mandating that G ignores x (i.e., $G(s,x) = G'(s)$). On the other hand, Definition 3.2 is a special case of related definitions that appeared in [8, Sec. 2.4]. Specifically, Vadhan [8] studied auxiliary-input pseudorandom generators (of the general-purpose type [1,9]), while offering a general treatment in which pseudorandomness needs to hold for a specific set of targets (i.e., $x \in I$ for some specific set $I \subseteq \{0,1\}^*$, whereas in Definition 3.2 we mandate $I = \{0,1\}^*$).[4]

The notion of a targeted canonical derandomizer is not as odd as it looks at first glance. Indeed, the generator is far from being general-purpose (i.e., it is tailored to a specific x), but this merely takes to (almost) the limit the insight of Nisan and Wigderson regarding relaxations that are still useful towards derandomization [6]. Indeed, even if we were to fix the distinguisher D, constructing a generator that just fools $D(x,\cdot)$ is not straightforward, because we need to find a suitable "fooling set" deterministically (in polynomial-time). The latter sentence (which is also reproduced from [4]), leads to the new definition.

3.3 The New Notion of Targeted Generators

Indeed, we suggest to consider canonical derandomizers that fool a single distinguisher, which is presented to them as input. The distinguisher is presented as a (deterministic) circuit, which determines the length of the sequence that the generator ought to produce. Thus, we no longer use a stretch function in our

[4] His treatment vastly extends the original notion of auxiliary-input one-way functions put forward in [7].

definitions. Instead, the seed length (denoted k) may be a (shrinking) function of the length of the output sequence (denoted ℓ). However, since it turns out that we may just use a fixed seed length (for all possible output lengths), we simplify our exposition by just using a fixed seed length.[5]

Definition 3.3 (targeted canonical derandomizers, revised): *A* targeted canonical derandomizer (with seed length k) *is a deterministic algorithm G that satisfies the following two conditions.*

(generation time): *On input a k-bit long seed and a circuit C with ℓ input bits, algorithm G makes at most* poly($|\langle C\rangle|$) *steps and outputs a string of length ℓ, where $\langle C\rangle$ denotes the description of the circuit C.*

(pseudorandomness (targeted)): *The* (ℓ-bit input) *circuit C cannot distinguish $G(U_k, \langle C\rangle))$ from U_ℓ; that is,*

$$|\Pr[C(G(U_k, \langle C\rangle)) = 1] - \Pr[C(U_\ell) = 1]| < \frac{1}{6} \tag{3}$$

A special case of Definition 3.3 (which mandates $\ell = |\langle C\rangle|$) can be obtained as a special case of a variant of Definition 3.2 in which the length of the generator's output (i.e., ℓ) equals the length of its auxiliary input (i.e., x) rather than being determined by the length of the seed; that is, $|G(x, s)| = |x|$ (rather than $|G(x, s)| = \ell(|s|)$). This is done by replacing the generic (linear-time) D with a specific (linear-time) algorithm—the circuit evaluation algorithm E (i.e., $E(\langle C\rangle, y) = C(y)$).[6] In this case Eq. (2) with $D(x, y)$ replaced by $E(\langle C\rangle, y)$ coincides with Eq. (3) (in which $C(y)$ appears instead).

Indeed, Definition 3.3 takes the approach of [6] to its logical conclusion: The derandomization of algorithm A with respect to input x just yields a single circuit $C_x(\cdot) = A(x, \cdot)$ that we need to fool, and Definition 3.3 (even more than Definition 3.2) is tailored to just do that. Indeed, the existence of a generator (as in Definition 3.3), even with a seed length that is logarithmic in the circuit size, implies $\mathcal{BPP} = \mathcal{P}$ (see proof of Theorem 4.1).

4 The Main Result

Building on the ideas of [4], we prove the following

Theorem 4.1 (main equivalence): *Targeted canonical derandomizers* (as per Definition *3.3*) *exist if and only if $\mathcal{BPP} = \mathcal{P}$. Furthermore, seed length two suffices.*

[5] Indeed, in general, one may allow k to be a function of the size of the circuit, provided that $k < \ell$, where ℓ denotes the length of the output of the generator (equiv., the length of the input of the circuit).

[6] This requires using a slightly redundant description of circuits so that evaluating them can be done in linear-time.

It follows that targeted canonical derandomizers as per Definition 3.3 exist if and only if generators as in Definition 3.2 (with exponential stretch) exist, because the latter also exist if and only if $\mathcal{BPP} = \mathcal{P}$ (see [4, Thm. 4.11]).

Proof: Using any targeted canonical derandomizer, we show that $\mathcal{BPP} = \mathcal{P}$ by following the standard derandomization path, while using the targeted canonical derandomizer and feeding it with the circuit that results from combining the randomized algorithm with the relevant input. That is, let A be a probabilistic polynomial-time algorithm for deciding a promise problem $\Pi = (\Pi_{\text{YES}}, \Pi_{\text{NO}}) \in \mathcal{BPP}$, and let G be a targeted canonical derandomizer with seed length $k = O(1)$. We first consider the probabilistic polynomial-time A' that, on input x, constructs the circuit $C_x(\cdot) = A(x, \cdot)$, and outputs $A(x, G(U_k, \langle C_x \rangle))$. Clearly, if $x \in \Pi_{\text{YES}}$, then

$$\begin{aligned}\Pr[A'(x,U_k)=1] &= \Pr[C_x(G(U_k,\langle C_x\rangle))=1] \\ &> \Pr[C_x(U_\ell)=1] - \frac{1}{6} \\ &\geq \frac{1}{2}\end{aligned}$$

Likewise, if $x \in \Pi_{\text{NO}}$, then $\Pr[A'(x, U_k) = 1] < \frac{1}{3} + \frac{1}{6} = \frac{1}{2}$. Next, scanning all possible k-bit long random inputs to $A'(x)$ and ruling by majority, we obtain the desired deterministic algorithm, and $\Pi \in \mathcal{P}$ follows. (Note that this argument holds also if k is logarithmic in $|\langle C_x \rangle|$.)

Turning to the opposite direction, using the hypothesis $\mathcal{BPP} = \mathcal{P}$, we construct a targeted canonical derandomizer (with constant seed length). We do so by following the approach of [4]; that is, we first show that constructing a targeted canonical derandomizer is a BPP-search problem, which is reducible to a decisional BPP problem (by [4, Thm. 3.5]), which yields a deterministic construction (since $\mathcal{BPP} = \mathcal{P}$ by the hypothesis). Details follow.

The BPP-search problem. We first detail a BPP-search problem, denoted $(R^{\text{PRG}}_{\text{YES}}, R^{\text{PRG}}_{\text{NO}})$, that captures the desired construction (for seed length $k = 2$). This promise problem refers to pairs of the form $(\langle C \rangle, \overline{s})$ such that, for some ℓ, the string $\langle C \rangle$ describes a circuit with ℓ input bits and $\overline{s} = (s_1, ..., s_4)$ is a quadruple of ℓ-bit long strings. We place a pair $(\langle C \rangle, \overline{s})$ in $R^{\text{PRG}}_{\text{YES}}$ if the difference between $\Pr_{i\in[4]}[C(s_i)=1]$ and $\Pr[C(U_\ell)=1]|$ is smaller than $1/7$ and place it in $R^{\text{PRG}}_{\text{NO}}$ if the difference is at least $1/6 > 1/7$. That is:

- $(\langle C \rangle, \overline{s}) \in R^{\text{PRG}}_{\text{YES}}$ if and only if $|\Pr_{i\in[4]}[C(s_i)=1] - \Pr[C(U_\ell)=1]| < 1/7$.
- $(\langle C \rangle, \overline{s}) \in R^{\text{PRG}}_{\text{NO}}$ if and only if $|\Pr_{i\in[4]}[C(s_i)=1] - \Pr[C(U_\ell)=1]| \geq 1/6$.

Indeed, we intentionally left a gap between the pairs in the two cases. This gaps allows for a probabilistic polynomial-time that distinguishes elements of $R^{\text{PRG}}_{\text{YES}}$ from elements of $R^{\text{PRG}}_{\text{NO}}$. Hence, showing that $(R^{\text{PRG}}_{\text{YES}}, R^{\text{PRG}}_{\text{NO}})$ is a BPP-search problem amounts to detailing a suitable probabilistic polynomial-time algorithm for finding valid solutions. Such an algorithm is given a circuit C with ℓ input bits, and needs to find $\overline{s}$ such that $(\langle C\rangle,\overline{s}) \in R_{\text{YES}}^{\text{PRG}}$. This can be done by first approximating $p_C \stackrel{\text{def}}{=} \Pr[C(U_\ell)=1]$, and then finding $(s_1,...,s_4)$ such that $\frac{1}{4} \cdot \sum_{i=1}^{4} C(s_i) = p_C \pm 0.14$, while relying on the fact that if $p_C > 0.01$ (resp., $p_C < 0.99$), then we can find a string in $C^{-1}(1)$ (resp., $C^{-1}(0)$) by repeated sampling. Details follow.

1. Using a constant number of samples, approximate $p_C \stackrel{\text{def}}{=} \Pr[C(U_\ell)=1]$ such that the approximation $\tilde{p}_C$ satisfies $\Pr[|\tilde{p}_C - p_C| > 0.01] < 0.1$.
2. Determine $i_C \in \{0,1,2,3,4\}$ such that $\frac{i_C}{4}$ is in the interval $[\tilde{p}_C - \frac{1}{8}, \tilde{p}_C + \frac{1}{8}]$. Specifically, let $i_C = \lfloor 4 \cdot \tilde{p}_C \rceil \in \{0,1,2,3,4\}$, where $\lfloor \alpha \rceil$ denotes the integer closest to $\alpha \in \mathbb{R}$.
3. Next, find $\overline{s} = (s_1,s_2,s_3,s_4)$ such that C evaluates to 1 on exactly i_C of them. This is done as follows:
   (a) If $i_C \in \{1,2,3\}$, then (using a constant number of samples) find strings $x_0, x_1 \in \{0,1\}^\ell$ such that $C(x_\sigma) = \sigma$ for every $\sigma \in \{0,1\}$. If $i_C = 0$ (resp., $i_C = 4$), then just find a string x_0 (resp., x_1) as above.
   (b) For $i = 1,...,4$, let $s_i = x_1$ if $i \leq i_C$ and let $s_i = x_0$ otherwise (i.e., if $i > i_C$).
   (Indeed, $\overline{s}$ contains i_C occurrences of x_1 and $4 - i_C$ occurrences of x_0.)

   Output $\overline{s} = (s_1,s_2,s_3,s_4)$.

Note that the probability that either Step 1 or Step 3a fails can be upper-bounded by 1/3. Otherwise, we have $|\tilde{p}_C - p_C| > 0.01$ and $\sum_{i=1}^{4} C(s_i) = i_C$, Recalling that $|\frac{i_C}{4} - \tilde{p}_C| \leq \frac{1}{8}$, it follows that (*in this case*) it holds that $|\Pr_{i\in[4]}[C(s_i) = 1] - p_C| \leq 0.125 + 0.01 < 0.14 < 1/7$. Hence, with probability at least 2/3, we output a sequence $\overline{s}$ such that $(\langle C\rangle,\overline{s}) \in R_{\text{YES}}^{\text{PRG}}$.

Next we reduce the foregoing BPP-search problem to a BPP decisional problem, by just invoking the following result of [4].

Reducing search to decision (appears as [4, Thm. 3.5]): *For every BPP-search problem* $(R_{\text{YES}}, R_{\text{NO}})$, *there exists a binary relation* R *such that* $R_{\text{YES}} \subseteq R \subseteq (\{0,1\}^* \times \{0,1\}^*) \setminus R_{\text{NO}}$ *and solving the search problem of* R *is deterministically reducible to some decisional problem in* $\mathcal{BPP}$, *denoted* Π.

We stress that the reduction solves R rather than R_{YES}, where R is sandwiched between R_{YES} and $(\{0,1\}^* \times \{0,1\}^*) \setminus R_{\text{NO}}$. This means that invoking the reduction on an input x in $S_{R_{\text{YES}}}$, we may obtain a solution $y \in R(x)$. Although it may be the case that $(x,y) \notin R_{\text{YES}}$, we are guaranteed that $(x,y) \notin R_{\text{NO}}$, which is typically good enough (as is the case here).[7]

---

[7] Indeed, when seeking to use this reduction with respect to a search problem R_{YES}, one should define R_{NO} such that the following two conditions hold:
1. On the one hand, for every x, obtaining a solution outside $R_{\text{NO}}(x)$ is almost as good as obtaining a solution in $R_{\text{YES}}(x)$.
2. On the other hand, one can solve the decision problem $(R_{\text{YES}}, R_{\text{NO}})$ in probabilistic polynomial-time.
This is exactly what we have done when defining $R_{\text{NO}}^{\text{PRG}}$.

Applying [4, Thm. 3.5] to the BPP-search problem $(R_{\rm YES}^{\rm PRG}, R_{\rm NO}^{\rm PRG})$, we obtain a deterministic reduction of the construction of the desired pseudorandom generator to some promise problem in $\mathcal{BPP}$; indeed, the key observation is that whenever C has a solution (i.e., there exists $\overline{s}$ such that $(\langle C\rangle, \overline{s}) \in R_{\rm YES}$) the reduction yields a sequence $\overline{s} = (s_1, ..., s_4)$ such that $(\langle C\rangle, \overline{s}) \notin R_{\rm NO}^{\rm PRG}$, which implies that $|\Pr_{i\in[4]}[C(s_i)=1] - p_C| < 1/6$.

Next, using the hypothesis $\mathcal{BPP} = \mathcal{P}$, we obtain a deterministic polynomial-time algorithm that, on input $\langle C\rangle$, finds a sequence $\overline{s} = (s_1, ..., s_4)$ such that $|\Pr_{i\in[4]}[C(s_i)=1] - p_C| < 1/6$. The generator is defined by letting $G(i) = s_i$ (for every $i \in [4] \equiv \{0,1\}^2$), and the theorem follows. ■

Remark 4.2 (on the distribution produced by the foregoing targeted canonical generator): *We observe that the targeted canonical generator constructed in the proof of Theorem 4.1 produces a distribution with at most two elements in its support, which can be shown to be the very minimum support size for any targeted canonical generator.*[8] *Furthermore, this generator uses a seed of length* $k = 2$, *which is an artifact of* $k = \lceil \log_2(3) \rceil$, *where* $1/3$ *is twice the desired distinguishing gap. In general, when dealing with a distinguishing gap of* $\delta < 1/2$, *it suffices to use a seed of length* $k = \lceil \log_2(1/2\delta) \rceil$, *and such a generator suffices for derandomizing algorithms of error probability* $0.5 - \delta$. *Thus, we may use* $\delta = 0.26$, *and obtain a generator that uses a single bit seed* (and suffices for derandomizing algorithms of error probability 0.24).

5 Targeted Hitters

In this section we merely detail the last paragraph of the introduction. We first adapt Definition 3.3 to the notion of hitting set generators, while observing that when targeting a single circuit there is no need to output a set of possible strings (since we may test these strings and just output one string that satisfies the circuit).

Definition 5.1 (targeted hitters): *A* targeted hitter *is a deterministic algorithm* H *that satisfies the following two conditions.*

(generation time): *On input a circuit* C *with* ℓ *input bits, algorithm* H *makes at most* $\text{poly}(|\langle C\rangle|)$ *steps and outputs a string of length* ℓ.

(hitting (targeted)): *If* $\Pr[C(U_\ell) = 1] > 1/2$, *then* $x \leftarrow H(\langle C\rangle)$ *satisfies* C (i.e., $C(x) = 1$).

Indeed, any targeted canonical derandomizer (as per Definition 3.3) yields a targeted hitter. While the converse is less clear, it can be shown to hold by combining Theorem 4.1 with the fact that any targeted hitter implies $\mathcal{RP} = \mathcal{P}$, which in turn implies $\mathcal{BPP} = \mathcal{P}$ (e.g., since $\mathcal{BPP} = \mathcal{RP}^{\mathcal{RP}}$, see, e.g., [3, §6.1.3.2]). Thus we get:

[8] Note that in case $\Pr[C(U_\ell) = 1] = 1/2$ it must hold that $|\Pr[C(G(U_k, \langle C\rangle)) = 1] \in (1/3, 2/3)$ and so $G(U_k, \langle C\rangle)$ must have support size at least two. This holds for any non-trivial distinguishing gap (i.e., any constant $\delta < 1/2$).

Theorem 5.2 (summary): *The following four conditions are equivalent:*

1. *There exist targeted canonical derandomizers* (as per Definition *3.3*).
2. *There exist targeted hitters.*
3. $\mathcal{BPP} = \mathcal{P}$.
4. $\mathcal{RP} = \mathcal{P}$.

Indeed, the proof outlined above takes the route (2) ⇒ (4) ⇒ (3) ⇒ (1) ⇒ (2). We comment that we do not see a direct proof of (4) ⇒ (2) (i.e., a proof that does not pass via (4) ⇒ (3)).

6 Reflections (or De-construction)

The new definition of a targeted canonical derandomizer (i.e., Definition 3.3) implicitly combines two tasks that need to be performed in deterministic polynomial-time:

1. Approximating the acceptance probability of circuits; that is, given a circuit C with ℓ input bits, the task is to (deterministically) approximate $\Pr[C(U_\ell)=1]$ up to $\pm 1/6$.
2. Finding an input that evaluates to the majority value; that is, given a circuit C with ℓ input bits, the task is to (deterministically) find an ℓ-bit string x such that $C(x) = \sigma$, where $\Pr[C(U_\ell)=\sigma] > 1/2$.

Indeed, the second task coincides with the task underlying the definition of a targeted hitter, whereas the first task is directly implied (only) by a targeted canonical derandomizer. Furthermore, deterministic polynomial-time algorithms for performing both tasks yield a targeted canonical derandomizer.[9] Interestingly, our construction of a targeted canonical derandomizer (based on $\mathcal{BPP} = \mathcal{P}$) implicitly uses a deterministic reduction of the second task to the first task, which is in turn reduced to $\mathcal{BPP}$.

The foregoing reductions are actually implicit in the proof of [4, Thm. 3.5]. Specifically, using a sufficiently good approximation of the acceptance probability of a circuit, we may find an input that satisfies the circuit by extending a prefix of such an input bit-by-bit (while making sure that the fraction of satisfying continuations is sufficiently large). Indeed, note that each of the aforementioned tasks can be used to solve a generalized version of this task that refers to an arbitrary threshold $\epsilon > 0$, provided that the running-time is allowed to depend (polynomially) on $1/\epsilon$ (see Footnote 9). (In these generalizations, the constants 1/6 and 1/2 are replaced by the parameter $\epsilon > 0$; Footnote 9 refers to the second task and the setting $\epsilon = 1/6$.)

[9] Actually, we need a variant of the second task that calls for (deterministically) finding an ℓ-bit string x such that $C(x) = \sigma$, provided that $\Pr[C(U_\ell) = \sigma] \geq 1/6$. This task can be reduced to the original one. specifically, if $\Pr[C(U_\ell) = 1] \geq 1/6$ (resp., $\Pr[C(U_\ell)=0] \geq 1/6$), then we find an input that satisfies the circuit C' such that $C'(x_1, x_2, x_3, x_4) = \bigvee_{i\in[4]} C(x_i)$ (resp., $C'(x_1, x_2, x_3, x_4) = \bigwedge_{i\in[4]} C(x_i)$), here we use the fact that $(5/6)^4 < 1/2$.

Acknowledgments. The current work was triggered by questions posed to me during my presentation of [4] at the Institut Henri Poincare (Paris).

References

1. Blum, M., Micali, S.: How to generate cryptographically strong sequences of pseudo-random bits. SICOMP **13**, 850–864 (1984). Preliminary version in 23rd FOCS, pp. 80–91 (1982)
2. Even, S., Selman, A.L., Yacobi, Y.: The complexity of promise problems with applications to public-key cryptography. Inf. Control **61**, 159–173 (1984)
3. Goldreich, O.: Computational Complexity: A Conceptual Perspective. Cambridge University Press, Cambridge (2008)
4. Goldreich, O.: In a world of P=BPP. In: Goldreich, O. (ed.) Studies in Complexity and Cryptography. Miscellanea on the Interplay Between Randomness and Computation. LNCS, vol. 6650, pp. 191–232. Springer, Heidelberg (2011). https://doi.org/10.1007/978-3-642-22670-0_20
5. Impagliazzo, R., Wigderson, A.: Randomness vs. time: de-randomization under a uniform assumption. JCSS **63**(4), 672–688 (2001). Preliminary version in 39th FOCS (1998)
6. Nisan, N., Wigderson, A.: Hardness vs randomness. JCSS **49**(2), 149–167 (1994). Preliminary version in 29th FOCS (1988)
7. Ostrovsky, R., Wigderson, A.: One-way functions are essential for non-trivial zero-knowledge. In: 2nd Israel Symposium on Theory of Computing and Systems, pp. 3–17. IEEE Computer Society Press (1993)
8. Vadhan, S.: An unconditional study of computational zero knowledge. SICOMP **36**(4), 1160–1214 (2006). Preliminary version in 45th FOCS (2004)
9. Yao, A.C.: Theory and application of trapdoor functions. In: 23rd FOCS, pp. 80–91 (1982)

On the Effect of the Proximity Parameter on Property Testers

Oded Goldreich

Abstract. This work refers to the effect of the proximity parameter on the operation of (standard) property testers. Its bottom-line is that, except in pathological cases, the effect of the proximity parameter is restricted to determining the query complexity of the tester. The point is that, in non-pathological cases, the mapping of the proximity parameter to the query complexity can be reversed in an adequate sense.

A preliminary version of this work was posted in 2012 on *ECCC*, as TR12-012. The current revision is minimal. On top of slightly improving the presentation, two additions were made: Footnote 2 was augmented (in light of [3]), and Remark 3 was added (while including a reference to [2]).

1 Introduction

Property Testing is the study of super-fast (randomized) algorithms for approximate decision making. These algorithms are given direct access to items of a huge data set, and determine whether this data set has some predetermined (global) property or is far from having this property. Remarkably, this approximate decision is made by accessing a small portion of the data set. Thus, property testing is a relaxation of decision problems and it focuses on algorithms, called *testers*, that can only read parts of the input.

A basic consequence of the foregoing description is that the testers should be modeled as oracle machines and the input should be modeled as a function to which the tester has oracle access. This modeling convention is explicit in almost all studies of property testing, but what is sometimes not explicit is that the tester also gets ordinary inputs (i.e., inputs that are given as strings and are read for free by the tester). These inputs include (1) the proximity parameter, denoted ϵ, and (2) parameters that describe the domain of the function (where the very least the size of the domain is given as input).[1] Note that the description

[1] For example, if the domain is a finite field, then one may need to provide its representation (and not merely its size), especially when no standard representation can be assumed (e.g., as in the case that the field has 2^n elements). Another example refers to the bounded-degree graph model (cf. [5]), where one should also provide the degree bound rather than just its product with the number of vertices; actually, one typically provides the degree bound and the number of vertices (and does not provide their multiple).

O. Goldreich (Ed.): Computational Complexity and Property Testing, LNCS 12050, pp. 36–40, 2020.
https://doi.org/10.1007/978-3-030-43662-9_5

of the domain must be provided so to allow the tester to make adequate queries.[2] The proximity parameter must also be provided, for reasons detailed next.

Recall that the standard definition of a tester (see Sect. 2)[3] requires that it accepts (with probability at least 2/3) any function that has some predetermined property but rejects (with probability at least 2/3) any function that is ϵ-far from the set of functions having the property, where the distance between functions are defined as the fraction of the domain on which the functions disagree. Note that, except in degenerated cases, one may avoid querying the function on its entire domain D only if $\epsilon > 1/|D|$. Thus, the tester must know that this is the case (i.e., that $\epsilon > 1/|D|$), if it is to make less than $|D|$ queries. In general, the query complexity of the tester typically depends on ϵ, and so the tester must obtain ϵ in order to determine the number of queries it is allowed to make. The question addressed in this work is whether or not ϵ is needed for any other purpose.

The foregoing natural question has also a concrete motivation. Various studies of property testing seem to assume that the tester only uses the proximity parameter to determine its query complexity (see, e.g., [7,8]). We show that this assumption is essentially justified, where the phrase "essentially" allows to discard unnatural cases in which the query complexity is not monotonically non-increasing in ϵ. See Theorem 2 for a precise statement, which refers to Definition 1, and note the technicality discussed right after this definition.

2 Technical Treatment

An asymptotic analysis is enabled by considering an infinite (indexed) sequence of domains, functions, and properties. That is, for any $s \in \mathbb{N}$, we consider functions from D_s to R_s. Indeed, s may be thought of as a description of the domain D_s, and typically it is related to $|D_s|$ (e.g., $s = |D_s|$).

Definition 1 (property tester): *Let $\Pi = \bigcup_{s\in\mathbb{N}} \Pi_s$, where Π_s contains functions defined over the domain D_s and range on R_s. A* tester *for the property Π is a probabilistic oracle machine T that satisfies the following two conditions:*

1. *The tester accepts each $f \in \Pi$ with probability at least 2/3; that is, for every $s \in \mathbb{N}$ and $f \in \Pi_s$* (and every $\epsilon > 0$)*, it holds that $\Pr[T^f(s,\epsilon) = 1] \geq 2/3$.*
2. *Given $\epsilon > 0$ and oracle access to any f that is ϵ-far from Π, the tester rejects with probability at least 2/3; that is, for every $\epsilon > 0$ and $s \in \mathbb{N}$, if $f : D_s \to R_s$ is ϵ-far from Π_s, then $\Pr[T^f(s,\epsilon) = 0] \geq 2/3$, where f is ϵ-*far from *Π_s if, for every $g \in \Pi_s$, it holds that $|\{e \in D_s : f(e) \neq g(e)\}| \geq \epsilon \cdot |D_s|$.*

[2] This crucial fact was overlooked in [7], as pointed out in [1,8]. Actually, the assertion itself is inaccurate; one may consider alternative models (cf. [3]) in which the tester is given sampling access to the domain (instead of its description).

[3] We refer to the standard definition (as in, e.g., [4,9]), and not to the definition of a proximity-oblivious tester (cf. [6]).

If the tester accepts every function in Π with probability 1, then we say that it has one-sided error*; that is, T has one-sided error if for every $f \in \Pi_s$ and every $\epsilon > 0$, it holds that* $\Pr[T^f(s,\epsilon) = 1] = 1$. *A tester is called* non-adaptive *if it determines all its queries based solely on its internal coin tosses* (and the parameters s *and* ϵ)*; otherwise it is called* adaptive.

Our choice to define ϵ-far as being at (relative) distance *at least* ϵ rather than being at (relative) distance *greater than* ϵ simplifies the formulation of Theorem 2. Unfortunately, this choice is inconsistent with our own preference (see, e.g., [2, Def. 1.6]). This issue is addressed in Remark 3.

The query complexity of T, viewed as a function of s and ϵ, is an upper bound (which holds for all $f : D_s \to R_s$) on the number of queries that T makes on (explicit) input (s, ϵ).

For the sake of simplicity, we assume that $s = |D_s|$, which means that the function's domain is fully specified by its size. The following result holds also when s is an arbitrary specification of the function's domain, which also determines the domain's size (or just allows to determine $q(|D_s|, \epsilon)$ when given ϵ).

Theorem 2 (on the restricted use of the proximity parameter): *Let* $q : \mathbb{N} \times (0,1] \to \mathbb{N}$ *be a computable function that is monotonically non-increasing in its second variable; that is,* $q(s, \epsilon_2) \leq q(s, \epsilon_1)$ *for every* $s \in \mathbb{N}$ *and* $\epsilon_2 > \epsilon_1 > 0$. *Suppose that the property Π has a tester T of query complexity q. Then, Π has a tester $\widehat{T}$ of query complexity q that only uses the proximity parameter for determining its query complexity; that is, on input parameters s and ϵ, the tester $\widehat{T}$ first computes and records $q(s, \epsilon)$, then deletes everything except s and $q(s, \epsilon)$ from its records, and only then proceeds with its actual operation. Furthermore, if T has one-sided error and/or is non-adaptive, then so is $\widehat{T}$.*

A typical case of a function q for which the hypothesis holds is $q(s, \epsilon) \stackrel{\text{def}}{=} \lceil f_s(\epsilon) \rceil$ for some collection of continuous and monotonically decreasing functions $f_s : (0,1] \to \mathbb{R}$ (e.g., $f_s(\epsilon) = 100/\epsilon^2$ or $f_s(\epsilon) = 2^{5/\epsilon} \cdot \log_2 s$).

Proof: The claimed algorithm $\widehat{T}$ consists of consecutively invoking the following three modules, denoted $\widehat{T}_1, \widehat{T}_2$ and $\widehat{T}_3$, which are defined as follows.

1. $\widehat{T}_1(s, \epsilon) \stackrel{\text{def}}{=} q(s, \lceil s\epsilon \rceil / s)$. That is, on input parameters s and ϵ, algorithm $\widehat{T}_1$ computes $\rho \leftarrow \lceil s\epsilon \rceil / s$ and returns $B \leftarrow q(s, \rho)$.
   (Note that $\rho \geq \epsilon$ is the smallest multiple of $1/s$ that is at least as large as ϵ; this implies that *being ϵ-far from Π* is equivalent to *being ρ-far from Π*.)
2. On input s and B, the module $\widehat{T}_2$ determines and returns the minimal $k \in \mathbb{N}$ such that $q(s, k/s) = B$.
   (Indeed, $k/s \leq \rho$.)
3. On input (s, k) and access to the oracle $f : D_s \to R_s$, the module $\widehat{T}_3$ emulates the execution of $T^f(s, k/s)$; that is, $\widehat{T}_3$ invokes T on input parameters s and k/s, and provides it with oracle access to f.

On input parameters (s, ϵ) and oracle access to f, algorithm $\widehat{T}$ first obtains $B \leftarrow \widehat{T}_1(s, \epsilon)$ and deletes ϵ, then it obtains $k \leftarrow \widehat{T}_2(s, B)$, and finally it obtains and returns the verdict of $\widehat{T}_3^f(s, k)$.

Note that $\widehat{T}$ only uses ϵ to determine its query bound B (when invoking $\widehat{T}_1$ at the very beginning), and that its actual activity (performed by $\widehat{T}_3$) depend only on s and the query bound B (which in turn determine k). Furthermore, $\widehat{T}$ maintains many features of T (e.g., non-adaptivity and one-sided error probability). The monotonicity of q is used to infer that the query complexity does not increased when the proximity parameter is rounded-up to the next multiple of $1/s$. In typical cases, the overhead in the time complexity (which arises mainly from $\widehat{T}_2$) is insignificant, since k can be determined by a binary search.

In analyzing $\widehat{T}$, we first consider any $s \in \mathbb{N}$ and $f \in \Pi_s$. In this case, for every ϵ', it holds that $\Pr[T^f(s, \epsilon') = 1] \geq c$, where $c = 1$ if T has one-sided error and $c = 2/3$ otherwise. Clearly, this holds also for $\epsilon' = k/s$, where k is as determined by $\widehat{T}_1$ and $\widehat{T}_2$, and therefore $\Pr[\widehat{T}^f(s, \epsilon) = 1] = \Pr[T^f(s, k/s) = 1] \geq c$.

Suppose, on the other hand, that $f : D_s \to R_s$ is ϵ-far from Π. Recalling that $\rho = \lceil s\epsilon \rceil / s$ (i.e., ρ is obtained by rounding-up ϵ to the next multiple of $1/s$) and using the fact that *the distance between functions over* D_s *is a multiple of* $1/s$, it follows that f is ρ-far from Π. Using $k/s \leq \rho$, we conclude that f is k/s-far from Π. Hence, $\Pr[\widehat{T}^f(s, \epsilon) = 0] = \Pr[T^f(s, k/s) = 0] \geq 2/3$, and the theorem follows. ■

Comment. An alternative presentation may suggest to invoke T on proximity parameter ϵ' that is chosen as the minimum for which $q(s, \epsilon') = B$ holds. This, seemingly more elegant approach, requires assuming that for every $s, B \in \mathbb{N}$ the set $\{\epsilon \in (0, 1] : q(s, \epsilon) = B\}$ is either empty or has a minimum element. More annoyingly, this minimum may have an infinite binary expansion, and so an actual algorithm will need to use a truncation of it anyhow. Indeed, one may always assume that the value of the proximity parameter is a multiple of $1/s$.

Remark 3 (a cumbersome version that fits [2, Def. 1.6]): *As noted above, our definition of* ϵ-far *as being at* (relative) *distance* at least ϵ *simplifies the formulation of Theorem 2. However, many sources* (see, e.g., [2, Def. 1.6]) *define* ϵ-far *as being at* (relative) *distance* greater than ϵ. *Theorem 2 can be adapted to this variant, with the query complexity of* $\widehat{T}$ *possibly increasing to*

$$\widehat{q}(s, \epsilon) = q\left(s, \tfrac{\lfloor s\epsilon \rfloor + u}{s}\right) \leq q\left(s, \tfrac{s\epsilon - (1-u)}{s}\right) = q\left(s, \epsilon - \tfrac{1-u}{s}\right)$$

where $u \in (0, 1)$ *may be selected as arbitrarily close to 1* (and it may be a function of s).[4] *In the revised proof, we let* $\rho \leftarrow (\lfloor s\epsilon \rfloor + u)/s$ *and later pick the minimal* $k \in \mathbb{N}$ *such that* $q(s, (k + u)/s) = q(s, \rho)$.

[4] Indeed, we use $\lfloor s\epsilon \rfloor + u > (s\epsilon - 1) + u = s\epsilon - (1 - u)$ and the hypothesis that $q(s, \epsilon') \leq q(s, \epsilon'')$ for every $\epsilon' \geq \epsilon'' > 0$. Actually, typically $\widehat{q}(s, \epsilon) = q(s, \epsilon)$ holds, since $s\epsilon - (1 - u) \approx s\epsilon$, let alone that $\lfloor s\epsilon \rfloor + u > s\epsilon$ may hold. (This is indeed the case when ϵ is a multiple of $1/s$ (as advocated in the previous comment).).

References

1. Alon, N., Shapira, A.: A characterization of the (natural) graph properties testable with one-sided. SIAM J. Comput. **37**(6), 1703–1727 (2008)
2. Goldreich, O.: Introduction to Property Testing. Cambridge University Press, Cambridge (2017)
3. Goldreich, O.: Flexible models for testing graph properties. In: ECCC, TR18-104 (2018). See revision in this volume
4. Goldreich, O., Goldwasser, S., Ron, D.: Property testing and its connection to learning and approximation. J. ACM **45**, 653–750 (1998)
5. Goldreich, O., Ron, D.: Property testing in bounded degree graphs. In: Algorithmica, pp. 302–343 (2002)
6. Goldreich, O., Ron, D.: On proximity oblivious testing. SIAM J. Comput. **40**(2), 534–566 (2011)
7. Goldreich, O., Trevisan, L.: Three theorems regarding testing graph properties. Random Struct. Algorithms **23**(1), 23–57 (2003)
8. Goldreich, O., Trevisan, L.: Errata to [7]. Manuscript, August 2005. http://www.wisdom.weizmann.ac.il/~oded/p_ttt.html
9. Rubinfeld, R., Sudan, M.: Robust characterization of polynomials with applications to program testing. SIAM J. Comput. **25**(2), 252–271 (1996)

On the Size of Depth-Three Boolean Circuits for Computing Multilinear Functions

Oded Goldreich and Avi Wigderson

Abstract. This paper introduces and initiates a study of a new model of arithmetic circuits coupled with new complexity measures. The new model consists of multilinear circuits *with arbitrary multilinear gates*, rather than the standard multilinear circuits that use only addition and multiplication gates. In light of this generalization, the *arity of gates* becomes of crucial importance and is indeed one of our complexity measures. Our second complexity measure is the *number of gates* in the circuit, which (in our context) is significantly different from the number of wires in the circuit (which is typically used as a measure of size). Our main *complexity measure*, denoted $\mathtt{AN}(\cdot)$, is the maximum of these two measures (i.e., the maximum between the arity of the gates and the number of gates in the circuit). We also consider the depth of such circuits, focusing on depth-two and unbounded depth.

Our initial motivation for the study of this arithmetic model is the fact that its two main variants (i.e., depth-two and unbounded depth) yield natural classes of *depth-three Boolean circuits for computing multilinear functions*. The resulting circuits have size that is exponential in the new complexity measure. Hence, lower bounds on the new complexity measure yield size lower bounds on a restricted class of depth-three Boolean circuits (for computing multilinear functions). Such lower bounds are a sanity check for our conjecture that multilinear functions of relatively *low degree* over GF(2) are good candidates for obtaining exponential lower bounds on the size of constant-depth Boolean circuits (computing explicit functions). Specifically, we propose to move gradually from linear functions to multilinear ones, and conjecture that, for any $t \geq 2$, some explicit t-linear functions $F : (\{0,1\}^n)^t \to \{0,1\}$ require depth-three circuits of size $\exp(\Omega(tn^{t/(t+1)}))$.

Letting $\mathtt{AN}_2(\cdot)$ denote the complexity measure $\mathtt{AN}(\cdot)$, when minimized over all depth-two circuits of the above type, our main results are as follows.

- For every t-linear function F, it holds that $\mathtt{AN}(F) \leq \mathtt{AN}_2(F) = O((tn)^{t/(t+1)})$.
- For almost all t-linear function F, it holds that $\mathtt{AN}_2(F) \geq \mathtt{AN}(F) = \Omega((tn)^{t/(t+1)})$.
- There exists a bilinear function F such that $\mathtt{AN}(F) = O(\sqrt{n})$ but $\mathtt{AN}_2(F) = \Omega(n^{2/3})$.

The main open problem posed in this paper is proving that $\mathtt{AN}_2(F) \geq \mathtt{AN}(F) = \Omega((tn)^{t/(t+1)})$ holds for an explicit t-linear function F, with $t \geq 2$. For starters, we seek lower bound of $\Omega((tn)^{0.51})$ for an explicit t-linear function F, preferably for constant t. We outline an approach that reduces this challenge (for $t = 3$) to a question regarding matrix rigidity.

O. Goldreich (Ed.): Computational Complexity and Property Testing, LNCS 12050, pp. 41–86, 2020.
https://doi.org/10.1007/978-3-030-43662-9_6

An early version of this work appeared as TR13-043 of *ECCC*. The current revision is quite substantial (cf. [11]). In particular, the original abstract was replaced, the appendices were omitted, notations were changed, some arguments were elaborated, and updates on the state of the open problems were added (see, most notably, the progress made in [9]).

1 Introduction

The introduction contains an extensive motivation for the model of arithmetic circuits that is studied in the paper. Readers who are only interested in this model may skip the introduction with little harm, except for the definition of three specific functions that appear towards the end of Sect. 1.2 (see Eqs. (2), (3), and (4)).

1.1 The General Context

Strong lower bounds on the size of constant-depth Boolean circuits computing parity and other explicit functions (cf., e.g., [12,34] and [26,29]) are among the most celebrated results of complexity theory. These quite tight bounds are all of the form $\exp(n^{1/(d-1)})$, where n denote the input length and d the circuit depth. In contrast, exponential lower bounds (i.e., of the form $\exp(\Omega(n))$) on the size of constant-depth circuits computing any explicit function are not known, even when using a weak notion of explicitness such as only requiring the Boolean function to be in $\mathcal{E} = \bigcup_{c\in\mathbb{N}} \mathrm{Dtime}(f_c)$, where $f_c(n) = 2^{cn}$.

Providing exponential lower bounds on the size of constant-depth Boolean circuits computing explicit functions is a central problem of circuit complexity, even when restricting attention to depth-three circuits (cf., e.g., [16, Chap. 11]). It seems that such lower bounds cannot be obtained by the standard interpretation of either the random restriction method [6,12,34] or the approximation by polynomials method [26,29]. Many experts have tried other approaches (cf., e.g., [14,17])[1], and some obtained encouraging indications (i.e., results that refer to restricted models, cf., e.g., [23]); but the problem remains wide open.

There are many motivations for seeking exponential lower-bounds for constant-depth circuits. Two notable examples are separating $\mathcal{NL}$ from $\mathcal{P}$ (see, e.g., [11, Apdx A]) and presenting an explicit function that does not have linear-size circuits of logarithmic depth (see Valiant [32]). Another motivation is the derandomization of various computations that are related to $\mathcal{AC}_0$ circuits (e.g., approximating the number of satisfying assignments to such circuits). Such derandomizations can be obtained via "canonical derandomizers" (cf. [7, Sec. 8.3]), which in turn can be constructed based on strong average-case versions of circuit lower bounds; cf. [21,22].

It seems that the first step should be beating the $\exp(\sqrt{n})$ size lower bound for depth-three Boolean circuits computing explicit functions (on n bits). A next

[1] The relevance of the Karchmer and Wigderson approach [17] to constant-depth circuits is stated explicitly in [18, Sec. 10.5].

step may be to obtain a truly exponential lower bound for depth-three Boolean circuits, and yet another one may be to move to any constant depth.

This paper focuses on the first two steps; that is, it focuses on depth-three circuits. Furthermore, within that confined context, we focus on a restricted class of functions (i.e., multilinear functions of small degree), and on a restricted type of circuits that emerges rather naturally when considering the computation of such functions.

1.2 The Candidate Functions

We suggest to study specific *multilinear functions of relatively low degree* over the binary field, GF(2), and in the sequel all arithmetic operations are over this field. For $t, n \in \mathbb{N}$, we consider *t-linear* functions of the form $F : (\{0,1\}^n)^t \to \{0,1\}$, where F is linear in each of the t blocks of variables (which contain n variables each). Such a function F is associated with a t-dimensional array, called a tensor, $T \subseteq [n]^t$, such that

$$F(x^{(1)}, x^{(2)}, ..., x^{(t)}) = \sum_{(i_1,i_2,...,i_t)\in T} x^{(1)}_{i_1} x^{(2)}_{i_2} \cdots x^{(t)}_{i_t} \quad (1)$$

where here and throughout this paper $x^{(j)} = (x^{(j)}_1, ..., x^{(j)}_n) \in \{0,1\}^n$ for every $j \in [t]$. Indeed, we refer to a fixed partition of the Boolean variables to t blocks, each containing n variables, and to functions that are linear in the variables of each block. Such functions were called set-multilinear in [23]. Note that the input length for these functions is $t \cdot n$; hence, *exponential lower bounds mean bounds of the form* $\exp(\Omega(tn))$.

We will start with a focus on constant t, and at times we will also consider t to be a function of n, but n will always remain the main length parameter. Actually, it turns out that $t = t(n) = \Omega(\log n)$ is essential for obtaining exponential lower bounds (i.e., size lower bounds of the form $\exp(\Omega(tn))$ for depth-d circuits, when $d > 2$).

A good question to ask is whether there exists any multilinear function that requires constant-depth Boolean circuit of exponential size (i.e., size $\exp(\Omega(tn))$). We conjecture that the answer is positive.

Conjecture 1.1 (a sanity check for the entire approach): *For every $d > 2$, there exist t-linear functions $F : (\{0,1\}^n)^t \to \{0,1\}$ that cannot be computed by Boolean circuits of depth d and size* $\exp(o(tn))$, *where* $t = t(n) \leq \mathrm{poly}(n)$.

We believe that the conjecture holds even for $t = t(n) = O(\log n)$, and note that, for any fixed t, there exist explicit t-linear functions that cannot be computed by *depth-two* Boolean circuits of size $2^{tn/4}$ (see [11, Apdx C.3]).

Merely proving Conjecture 1.1 may not necessarily yield a major breakthrough in the state-of-art regarding circuit lower bounds, although it *seems* that a proof will need to do something more interesting than mere counting. However, disproving Conjecture 1.1 will cast a shadow on our suggestions, which may nevertheless maintain their potential for surpassing the $\exp((tn)^{1/(d-1)})$ barrier. (Showing an upper bound of the form $\exp((tn)^{1/(d-1)})$ on the size circuits of depth d that compute any t-linear function seems unlikely (cf. [23], which proves an exponential in t lower bound on the size of depth-three arithmetic circuits (when $n=4$)).)

Assuming that Conjecture 1.1 holds, one should ask which explicit functions may "enjoy" such lower bounds. Two obviously bad choices are

- $F_{\mathtt{all}}^{t,n}(x^{(1)},...,x^{(t)}) = \sum_{i_1,...,i_t\in[n]} x_{i_1}^{(1)}\cdots x_{i_t}^{(t)}$, and
- $F_{\mathtt{diag}}^{t,n}(x^{(1)},...,x^{(t)}) = \sum_{i\in[n]} x_{i}^{(1)}\cdots x_{i}^{(t)}$,

since each of them is easily reducible to an n-way parity (the lower bounds for which we wish to surpass).[2] The same holds for any function that corresponds either to a rectangular tensor (i.e., $T = I_1\times\cdots\times I_t$, where $I_1,..,I_t\subseteq[n]$) or to a sparse tensor (e.g., $T\subseteq[n]^t$ such that $|T|=O(n)$). Ditto w.r.t the sum of few such tensors. Indeed, one should seek tensors $T\subseteq[n]^t$ that are far from the sum of few rectangular tensors (i.e., far from any tensor of low rank [30]). On the other hand, it seems good to stick to as "simple" tensors as possible so as to facilitate their analysis (let alone have the corresponding multilinear function be computable in exponential-time (i.e., in $\mathcal{E}$)).[3]

A Less Obviously Bad Choice. Consider the function $F_{\mathtt{leq}}^{t,n} : (\{0,1\}^n)^t \to \{0,1\}$ such that

$$F_{\mathtt{leq}}^{t,n}(x^{(1)},x^{(2)},...,x^{(t)}) = \sum_{1\le i_1\le i_2\le\cdots\le i_t\le n} x_{i_1}^{(1)}x_{i_2}^{(2)}\cdots x_{i_t}^{(t)} \qquad (2)$$

(having the corresponding tensor $T_{\mathtt{leq}}^{t,n} = \{(i_1,...,i_t)\in[n]^t : i_1\le i_2\le\cdots\le i_t\}$). Note that this function is polynomial-time computable (e.g., via dynamic programming),[4] and that $t=1$ corresponds to `Parity`. Unfortunately, for every constant $t\ge 2$, the function $F_{\mathtt{leq}}^{t,n}$ is not harder than parity: It has depth-three

---

[2] Note that $F_{\mathtt{all}}^{t,n}(x^{(1)},...,x^{(t)}) = \prod_{j\in[t]}\sum_{i_j\in[n]} x_{i_j}^{(j)}$, which means that it can be computed by a t-way conjunction of n-way parity circuits, whereas $F_{\mathtt{diag}}^{t,n}$ is obviously an n-way parity of t-way conjunctions of variables.

[3] Thus, these tensors should be constructible within $\exp(tn)$-time. Note that we can move from the tensor to the multilinear function (and vice versa) in $n^t \ll \exp(tn)$ oracle calls.

[4] Note that $F_{\mathtt{leq}}^{t,n}(x^{(1)},...,x^{(t)})$ equals $\sum_{i\in[n]} F_{\mathtt{leq}}^{t-1,i}(x_{[1,i]}^{(1)},...,x_{[1,i]}^{(t-1)})\cdot x_i^{(t)}$, where $x_{[1,i]}^{(j)} = (x_1^{(j)},...,x_i^{(j)})$. So, for every $t'\in[t-1]$, the dynamic program uses the n values $(F_{\mathtt{leq}}^{t',i}(x_{[1,i]}^{(1)},...,x_{[1,i]}^{(t')}))_{i\in[n]}$ in order to compute the n values $(F_{\mathtt{leq}}^{t'+1,i}(x_{[1,i]}^{(1)},..., x_{[1,i]}^{(t'+1)}))_{i\in[n]}$.

circuits of size $\exp(O(\sqrt{n}))$; see Proposition 3.4. Thus, we move to the slightly less simple candidates presented next.

Specific Candidates. We suggest to consider the following t-linear functions, $F_{\mathtt{tet}}^{t,n}$ and $F_{\mathrm{mod}\,p}^{t,n}$ (especially for $p \approx 2^t \approx n$), which are presented next in terms of their corresponding tensors (i.e., $T_{\mathtt{tet}}^{t,n}$ and $T_{\mathrm{mod}\,p}^{t,n}$, resp).

$$T_{\mathtt{tet}}^{t,n} = \left\{ (i_1, ..., i_t) \in [n]^t : \sum_{j \in [n]} |i_j - \lfloor n/2 \rfloor| \leq \lfloor n/2 \rfloor \right\} \tag{3}$$

$$T_{\mathrm{mod}\,p}^{t,n} = \left\{ (i_1, ..., i_t) \in [n]^t : \sum_{j \in [t]} i_j \equiv 0 \pmod p \right\} \tag{4}$$

(The shorthand `tet` was intended to stand for tetrahedon, since the geometric image of one eighth of $T_{\mathtt{tet}}^{3,n}$ resembles a "slanted tetrahedon". Indeed, $T_{\mathtt{tet}}^{3,n}$ as a whole looks more like a regular octahedon.)

Note that the functions $F_{\mathtt{tet}}^{t,n}$ and $F_{\mathrm{mod}\,p}^{t,n}$ are also computable in polynomial-time.[5] For $p < n$, it holds that $F_{\mathrm{mod}\,p}^{t,n}(x^{(1)}, ..., x^{(t)})$ equals $F_{\mathrm{mod}\,p}^{t,p}(y^{(1)}, ..., y^{(t)})$, where $y_r^{(j)} = \sum_{i \in [n]: i \equiv r \pmod p} x_i^{(j)}$ for every $j \in [t]$ and $r \in [p]$. This reduction may have a forbidding "size cost" in the context of circuits of a specific depth (especially if $p \ll n$), but its cost is insignificant if we are willing to double the depth of the circuit (and aim at lower bounds that are larger than those that hold for parity). Thus, in the latter cases, we may assume that $p = \Omega(n)$, but of course $p < tn$ must always hold.

We note that none of the bilinear versions of the foregoing functions can serve for beating the $\exp(\sqrt{n})$ lower bound. Specifically, the failure of $F_{\mathrm{mod}\,p}^{2,n}$ is related to the aforementioned reduction, whereas the failure of $F_{\mathtt{tet}}^{2,n}$ is due to the fact that $T_{\mathtt{tet}}^{2,n}$ is very similar to $T_{\mathtt{leq}}^{2,n}$ (i.e., each fourth of $T_{\mathtt{tet}}^{2,n}$ is isomorphic to $T_{\mathtt{leq}}^{2,n}$ (under rotation and scaling)). But these weaknesses do not seem to propagate to the trilinear versions (e.g., the eighthes of the tensor $T_{\mathtt{tet}}^{3,n}$ are not isomorphic to $T_{\mathtt{leq}}^{3,n}$).

What's Next? In an attempt to study the viability of our suggestions and conjectures, we defined two restricted classes of depth-three circuits and tried to prove lower bounds on the sizes of circuits (from these classes) that compute the foregoing functions. Our success in proving lower bounds was very partial,

[5] Again, we use dynamic programming, but here we apply it to generalizations of these functions. Specifically, let $T_{\mathtt{tet}}^{t,n,d} = \{(i_1, ..., i_t) \in [n]^t : \sum_{j \in [n]} |i_j - \lfloor n/2 \rfloor| \leq d\}$ and note that the associated function satisfies $F_{\mathtt{tet}}^{t,n,d}(x^{(1)}, ..., x^{(t)}) = \sum_{i \in [n]} F_{\mathtt{tet}}^{t-1,n,d-i}(x^{(1)}, ..., x^{(t-1)}) \cdot x_i^{(t)}$. Likewise, consider the tensor $T_{\mathrm{mod}\,p}^{t,n,r} = \left\{ (i_1, ..., i_t) \in [n]^t : \sum_{j \in [t]} i_j \equiv r \pmod p \right\}$ and note that the associated function satisfies $F_{\mathrm{mod}\,p}^{t,n,r}(x^{(1)}, ..., x^{(t)}) = \sum_{i \in [n]} F_{\mathrm{mod}\,p}^{t-1,n,r-i}(x^{(1)}, ..., x^{(t-1)}) \cdot x_i^{(t)}$.

and will be discussed next—as part of the discussion of these two classes (in Sects. 1.3 and 1.4). Subsequent work [9] was more successful in that regard.

1.3 Design by Direct Composition: The D-Canonical Model

What is a natural way of designing depth-three Boolean circuits that compute multilinear functions?

Let us take our cue from the linear case (i.e., $t = 1$). The standard way of obtaining a depth-three circuit of size $\exp(\sqrt{n})$ for n-way parity is to express this linear function as the $\sqrt{n}$-way sum of $\sqrt{n}$-ary functions that are linear in disjoint sets of variables. The final (depth-three) circuit is obtained by combing the depth-two circuit for the outer sum with the depth-two circuits computing the $\sqrt{n}$ internal sums.

Hence, a natural design strategy is to express the target multilinear function (denoted F) as a polynomial (denoted H) in some auxiliary multilinear functions (i.e., F_i's), and combine depth-two circuits that compute the auxiliary multilinear functions with a depth-two circuit that computes the main polynomial (i.e., H). That is, we "decompose" the multilinear function on the algebraic level, expressing it as a polynomial in auxiliary multilinear functions (i.e., $F = H(F_1, ..., F_s)$), and implement this decomposition on the Boolean level (i.e., each polynomial is implemented by a depth-two Boolean circuit). Specifically, to design a depth-three circuit of size $\exp(O(s))$ for computing a multilinear function F the following steps are taken:

1. Select s arbitrary multilinear functions, $F_1, ..., F_s$, each depending on at most s input bits;
2. Express F as a polynomial H in the F_i's;
3. Obtain a depth-three circuit by combining depth-two circuits for computing H and the F_i's.

Furthermore, we mandate that $H(F_1, ..., F_s)$ is a *syntactically multilinear function*; that is, the monomials of H do not multiply two F_i's that depend on the same block of variables. The size of the resulting circuit is *defined* to be $\exp(\Theta(s))$: The upper bound is justified by the construction, and the lower bound by the assumption that (low degree) polynomials that depend on s variables require depth-two circuits of $\exp(s)$ size. (The latter assumption is further discussed in Sect. 2.2.)[6]

Circuits that are obtained by following this framework are called D-canonical, where "D" stands for *direct* (or *deterministic*, for reasons that will become apparent in Sect. 1.4). Indeed, D-canonical circuits seem natural in the context of computing multilinear functions by depth-three Boolean circuits.

[6] In brief, when computing t-linear polynomials, a lower bound of $\exp(\Omega(s/2^t))$ on the size of depth-two circuits can be justified (see [11, Apdx C]). Furthermore, for $2^t \ll s$, a lower bound of $\exp(\Omega(s))$ can be justified if the CNFs (or DNFs) used are "canonical" (i.e., use only s-way gates at the second (i.e., F_i's) level).

For example, the standard design, reviewed above, of depth-three circuits (of size $\exp(\sqrt{n})$) for (n-way) parity yields D-canonical circuits. In general, D-canonical circuits for a target multilinear function are obtained by combining depth-two circuits that compute auxiliary multilinear functions with a depth-two circuit that computes the function that expresses the target function in terms of the auxiliary functions. The freedom of the framework (or the circuit designer) is reflected in the choice of auxiliary functions, whereas the restriction is in insisting that the target multilinear function be computed by composition of a polynomial and multilinear functions (and that this composition corresponds to a syntactically multilinear function).

Our main results regarding D-canonical circuits are a generic upper bound on the size of D-canonical circuits computing any t-linear function and a matching lower bound that refers to almost all t-linear functions. That is:

Theorem 3.1: *For every $t \geq 2$, every t-linear function $F: (\{0,1\}^n)^t \to \{0,1\}$ can be computed by D-canonical circuits of size* $\exp(O(tn)^{t/(t+1)})$.

(Corollary to) Theorem 4.1: *For every $t \geq 2$, it holds that almost all t-linear functions F: $(\{0,1\}^n)^t \to \{0,1\}$ require D-canonical circuits of size at least* $\exp(\Omega(tn)^{t/(t+1)})$.

Needless to say, the begging question is what happens with explicit multilinear functions.

Problem 1.2 (main problem regarding D-canonical circuits): *For every fixed $t \geq 2$, prove a $\exp(\Omega(tn)^{t/(t+1)})$ lower bound on the size of D-canonical circuits computing some explicit function. Ditto when t may vary with n, but $t \leq \mathrm{poly}(n)$.*

We mention that subsequent work of Goldreich and Tal [9] proved an $\exp(\widetilde{\Omega}(n^{2/3}))$ lower bound on the size of D-canonical circuits computing some explicit trilinear functions (e.g., $F_{\mathtt{tet}}^{3,n}$). A very recent result of Goldreich [8] asserts that, for every constant $\epsilon > 0$, there exists an explicit $\mathrm{poly}(1/\epsilon)$-linear that requires D-canonical circuits of size at least $\exp(n^{1-\epsilon})$.

1.4 Design by Nested Composition: The ND-Canonical Model

As appealing as D-canonical circuits may appear, it turns out that one can build significantly smaller circuits by employing the "guess and verify" technique (see Theorem 2.3). This allows to express the target function in terms of auxiliary functions, which themselves are expressed in terms of other auxiliary functions, and so on. That is, the "composition depth" is no longer 1—it is even not *a priori* bounded—and yet the resulting Boolean circuit has depth-three.

Assuming we want to use s auxiliary functions of arity s, the basic idea is to use s non-deterministic guesses for the values of these s functions, and to verify each of these guesses based on (some of) the other guesses and at most s bits of the original input. Thus, the verification amounts to the conjunction of s conditions, where each condition depends on at most $2s$ bits (and can thus be verified by a CNF of size $\exp(2s)$). The final depth-three circuit is obtained by replacing the s non-deterministic guesses by a 2^s-way disjunction.

This way of designing depth-three circuits leads to a corresponding framework, and the circuits obtained by it are called ND-canonical, where "ND" stands for *non-determinism.* In this framework depth-three circuits of size $\exp(O(s))$ for computing a multilinear function F are designed by the following three-step process:

1. Select s auxiliary multilinear functions, $F_1, ..., F_s$;
2. Express F as well as each of the other F_i's as a polynomial in the subsequent F_i's and in at most s input bits;
3. Obtain a depth-three circuit by combining depth-two circuits for computing these polynomials, where the combination implements s non-deterministic choices as outlined above.

As in the D-canonical framework, the polynomials used in Step (2) should be such that replacing the functions F_i's in them yields multilinear functions (i.e., this is a syntactic condition). Again, the size of the resulting circuit is *defined* to be $\exp(\Theta(s))$.

Note that, here (i.e., in the case of ND-canonical circuits), the combination performed in Step (3) is not a functional composition (as in the case of the D-canonical circuits). It is rather a verification of the claim that there exists $s+1$ values that fit all $s+1$ expressions (i.e., of F and the F_i's). The implementation of Step (3) calls for taking the conjunction of these $s+1$ depth-two computations as well as taking a 2^{s+1}-way disjunction over all possible values that these computations may yield.

The framework of ND-canonical circuits allows to express F in terms of F_i's that are themselves expressed in terms of F_j's, and so on. (Hence, the composition is "nested".) In contrast, in the D-canonical framework, the F_i's were each expressed in terms of s input bits. A natural question is whether this generalization actually helps. We show that the answer is positive.

Theorem 2.3: There exists bilinear functions $F : (\{0,1\}^n)^2 \to \{0,1\}$ that have ND-circuits of size $\exp(O(\sqrt{n}))$ but no D-circuits of size $\exp(o(n^{2/3}))$.

Turning to our results regarding ND-circuits, the upper bound on D-canonical circuits clearly holds for ND-circuits, whereas our lower bound is actually established for ND-canonical circuits (and the result for D-canonical circuits is a corollary). Thus, we have

(Corollary to) Theorem 3.1: *For every $t \geq 2$, every t-linear function $F : (\{0,1\}^n)^t \to \{0,1\}$ can be computed by ND-canonical circuits of size* $\exp(O(tn)^{t/(t+1)})$.

Theorem 4.1: *For every $t \geq 2$, it holds that almost all t-linear functions $F : (\{0,1\}^n)^t \to \{0,1\}$ require ND-canonical circuits of size at least* $\exp(\Omega(tn)^{t/(t+1)})$.

Again, the real challenge is to obtain such a lower bound for explicit multilinear functions.

Problem 1.3 (main problem regarding ND-canonical circuits): *For every fixed $t \geq 2$, prove a $\exp(\Omega(tn)^{t/(t+1)})$ lower bound on the size of ND-canonical circuits computing some explicit function. Ditto when t may vary with n, but $t \leq \text{poly}(n)$.*

The subsequent work of Goldreich and Tal [9] establishes an $\exp(\widetilde{\Omega}(n^{0.6}))$ lower bound on the size of ND-canonical circuits computing the trilinear function $F_{\mathtt{tet}}^{3,n}$ and an $\exp(\widetilde{\Omega}(n^{2/3}))$ lower bound on the size of ND-canonical circuits computing some explicit 4-linear functions. It does so by following the path suggested in the original version of this work [11], where we wrote:

> For starters, *prove a $\exp(\Omega(tn)^{0.51})$ lower bound on the size of ND-canonical circuits computing some explicit t-linear function.*
> As a possible step towards this goal we reduce the task of proving such a lower bound for $F_{\mathtt{tet}}^{3,n}$ to proving a lower bound on the rigidity of matrices with parameters that were not considered before. In particular, an $\exp(\omega(\sqrt{n}))$ lower bound on the size of ND-canonical circuits computing $F_{\mathtt{tet}}^{3,n}$ will follow from the existence of an n-by-n Toeplitz matrix that has rigidity $\omega(n^{3/2})$ with respect to rank $\omega(n^{1/2})$.

For more details, see Sect. 4.2 (as well as Sect. 4.3).

1.5 The Underlying Models of Arithmetic Circuit and AN-Complexity

Underlying the two models of canonical circuits (discussed in Sects. 1.3 and 1.4) is a new model of arithmetic circuits (for computing multilinear functions). Specifically, the expressions representing the value of (the target and auxiliary) functions in terms of the values of auxiliary functions and original variables correspond to gates in a circuit. These gates can compute arbitrary polynomials (as long as the multilinear condition is satisfied). In the case of D-canonical circuits, the corresponding arithmetic circuits have depth two (i.e., a top gate and at most one layer of intermediate gates), whereas for ND-canonical circuits the corresponding arithmetic circuits have unbounded depth. In both cases, the key complexity measure is the *maximum between the arity of the gates and their number.*

In both cases, a canonical Boolean circuit (for computing a multilinear function F) is obtained by presenting a Boolean circuit that emulates the computation of an arithmetic circuit (computing F). Specifically, a D-canonical circuit is obtained by a straightforward implementation of a depth-two arithmetic circuit that computes F, the arithmetic circuit computes F by applying a function H (in the top gate) to intermediate results computed by the intermediate gates (i.e., $F = H(F_1, ..., F_s)$, where F_i is computed by the ith intermediate gate). The ND-canonical circuits are obtained by a Valiant-like (i.e., akin [32]) decomposition of the computation of (unbounded depth) arithmetic circuits; that is, by guessing and verifying the values of all intermediate gates. In both cases, the size of the resulting Boolean circuit is exponential in the maximum between the *arity of these gates* the *number of gates*. Indeed, this parameter (i.e., the maximum of the two measures) restricts the power of the underlying arithmetic circuits or rather serves as their complexity measure, called *AN-complexity*, where "A" stands for arity and "N" for number (of gates). Let us spell out these two models of arithmetic circuit complexity.

The arithmetic circuits we refer to are directed acyclic graphs that are labeled by arbitrary multilinear functions and variables of the target function (i.e., F). These circuits are restricted to be syntactically multilinear; that is, each gate computes a function that is multilinear in the variables of the target function (i.e., arguments that depend on variables in the same block are not multiplied by such gates). Specifically, a gate that is labelled by a function H_i and is fed by gates computing the auxiliary functions $F_{i_1}, ..., F_{i_{m'}}$ and m'' original variables, denoted $z_1, ..., z_{m''}$ (out of $x^{(1)}, x^{(2)}, ..., x^{(t)}$), computes the function

$$\begin{aligned} &F_i(x^{(1)}, x^{(2)}, ..., x^{(t)}) \\ &= H_i(F_{i_1}(x^{(1)}, x^{(2)}, ..., x^{(t)}), ..., F_{i_{m'}}(x^{(1)}, x^{(2)}, ..., x^{(t)}), z_1, ..., z_{m''}). \end{aligned}$$

This holds also for the top gate that computes $F = F_0$. In case of depth-two circuits, the top gate is the only gate in the circuit that may be fed by intermediate gates (and we may assume, with no loss of generality, that it is not fed by any variable).[7] As we shall see later (see, e.g., Remark 3.5), the benefit of circuits of larger depth is that they may contain gates that are fed both by other gates and by variables. Let us summarize this discussion and introduce some notation.

- Following [23], we say that an arithmetic circuit is multilinear if its input variables are partitioned into blocks and the gates of the circuit compute multilinear functions such that *if two gates have directed paths from the same block of variables, then the results of these two gates are not multiplied together.*
- We say that the direct-composition complexity of F, denoted $\mathtt{AN}_2(F)$, is at most s if *F can be computed by a depth-two multilinear circuit with at most s gates that are each of arity at most s.*
- We say that the nested-composition complexity of F, denoted $\mathtt{AN}(F)$, is at most s if *F can be computed by a multilinear circuit with at most s gates that are each of arity at most s.*

We stress that the multilinear circuits in the foregoing definition employ arbitrary multilinear gates, whereas in the standard arithmetic model the gates correspond to either (unbounded) addition or multiplication. Our complexity measure is related to but different from circuit size: On the one hand, we only count the number of gates (and discard the number of leaves, which in our setting may be larger). On the other hand, our complexity measure also bounds the arity of the gates.

Note that for any *linear* function F, it holds that $\mathtt{AN}_2(F) = \Theta(\mathtt{AN}(F))$, because all intermediate gates can feed directly to the top gate (since, in this case,

[7] Since such directly fed variables can be replaced by dummy gates that are each fed by the corresponding variable.

all gates compute linear functions).[8] Also note that $\mathtt{AN}_2(F)$ equals the square root of the number of variables on which the *linear* function F depends. In general, $\mathtt{AN}(F) \geq \sqrt{tn}$ for any t-linear function F that depends on all its variables, and $\mathtt{AN}(F) \leq \mathtt{AN}_2(F) \leq tn$ for any t-linear function F. Thus, our complexity measures (for non-degenerate t-linear functions) range between $\sqrt{tn}$ and tn.

Clearly, F has a D-canonical (resp., ND-canonical) circuit of size $\exp(\Theta(s))$ if and only if $\mathtt{AN}_2(F) = s$ (resp., $\mathtt{AN}(F) = s$). Thus, all results and open problems presented above (i.e., in Sects. 1.3 and 1.4) in terms of canonical (Boolean) circuits are actually results and open problems regarding the (direct and nested) composition complexity of multilinear circuits (i.e., $\mathtt{AN}_2(\cdot)$ and $\mathtt{AN}(\cdot)$). Furthermore, the results are actually proved by analyzing these complexity measures. Specifically, we have:

Theorem 3.1: For every t-linear function $F : (\{0,1\}^n)^t \to \{0,1\}$, it holds that $\mathtt{AN}(F) \leq \mathtt{AN}_2(F) = O((tn)^{t/(t+1)})$.
Theorem 4.1: For almost all t-linear function $F : (\{0,1\}^n)^t \to \{0,1\}$, it holds that $\mathtt{AN}_2(F) \geq \mathtt{AN}(F) = \Omega((tn)^{t/(t+1)})$.
Theorem 2.3: There exists a bilinear function $F : (\{0,1\}^n)^2 \to \{0,1\}$ such that $\mathtt{AN}(F) = O(\sqrt{n})$ but $\mathtt{AN}_2(F) = \Omega(n^{2/3})$.

We stress that the foregoing lower bounds are existential, whereas we seek $\omega(\sqrt{n})$ lower bounds for explicit multilinear functions. (As noted above, this initial goal was achieved by the subsequent work of Goldreich and Tal [9], which establishes an $\mathtt{AN}(F) = \widetilde{\Omega}(n^{2/3})$ for some explicit 4-linear functions F.)

Summary and Additional Comments. This paper introduces and initiates a study of a new model of arithmetic circuits and accompanying new complexity measures. The new model consists of multilinear circuits *with arbitrary multilinear gates*, rather than the standard multilinear circuits that use only addition and multiplication gates. In light of this generalization, the *arity of gates* becomes of crucial importance and is indeed one of our complexity measures. Our second complexity measure is the *number of gates* in the circuit, which (in our context) is significantly different from the number of wires in the circuit (which is typically used as a measure of size). Our main complexity measure is the maximum of these two measures (i.e., the maximum between the arity of the gates and the number of gates in the circuit). Our initial motivation for the study of this arithmetic model is its close relation to canonical Boolean circuits, and from this perspective depth-two arithmetic circuits have a special appeal.

A natural question is whether our complexity measure (i.e., $\mathtt{AN}$) decreases if one waives the requirement that the arithmetic circuit be a multilinear one (i.e., the gates compute multilinear functions and they never multiply the outcomes

[8] Doing so may increase the arity of the top gate, but this increase is upper-bounded by the number of gates. A more general argument is presented in Remark 2.4, which asserts that if gate G computes a monomial that contains no leaves, then this monomial can be moved up to the parent of G.

of gates that depend on the same block of variables). The answer is that waiving this restriction in the computation of any t-linear function may decrease the complexity by at most a factor of 2^t (see Remark 2.5).

We note that the arithmetic models discuss above make sense with respect to any field. The reader may verify that all results stated for $\mathtt{AN}_2(\cdot)$ and $\mathtt{AN}(\cdot)$ hold for every field, rather than merely for the binary field. Ditto for the open problems.

1.6 Related Work

Multilinear functions were studied in a variety of models, mostly in the context of algebraic and arithmetic complexity. In particular, Nisan and Wigderson [23] initiated a study of *multilinear circuits* as a natural model for the computation of multilinear functions. Furthermore, they obtained an exponential (in t) lower bound on the size of depth-three multilinear circuits that compute a natural t-linear function (i.e., iterated matrix multiplication for 2-by-2 matrices).[9]

The multilinear circuit model was studied in subsequent works (cf., e.g., [25]); but, to the best of our knowledge, the complexity measure introduced in Sect. 1.5 was not studied before. Nevertheless, it may be the case that techniques and ideas developed in the context of the multilinear circuit model will be useful for the study of this new complexity measure (and, equivalently, in the study of canonical circuits). For example, it seems that the latter study requires a good understanding of tensors, which were previously studied with focus at a different type of questions (cf., e.g., [24]).

In the following two paragraphs we contrast our model of multilinear circuits, which refers to arbitrary gates of arity that is reflected in our complexity measure, with the standard model of multilinear circuits [23], which uses only addition and multiplication gates (of unbounded arity). For the sake of clarity, we shall refer to canonical circuits rather than to our model of multilinear circuits, while reminding the reader that the two are closely related.

The difference between the standard model of constant-depth *multilinear circuit* and the model of constant-depth Boolean circuits is rooted in the fact that the (standard) *multilinear circuit* model contains unbounded fan-in addition gates as basic components, whereas unbounded fan-in addition is hard for constant-depth Boolean circuits. Furthermore, the very fact that n-way addition requires $\exp(n)$-size depth-two Boolean circuits is the basis of the approach that we are suggesting here. In contrast, hardness in the multilinear circuit model is related to the total degree of the function to be computed.[10]

---

[9] Thus, $n = 4$ and t is the number of matrices being multiplied.

[10] Concretely, the conjectured hardness of computing a multilinear function by constant-depth Boolean circuits may stem from the number (denoted n) of variables of the same type (i.e., the variables in $x^{(j)}$), even when the arity of multiplication (denoted t) is relatively small (e.g., we even consider bilinear functions), whereas in the multilinear circuits hardness seem to be related to t (cf., indeed, the aforementioned lower bound for iterated matrix multiplication).

The foregoing difference is reflected in the contrast between the following two facts: (1) multilinear functions of low degree have small depth-two *multilinear circuits* (i.e., each t-linear function $F : (\{0,1\}^n)^t \to \{0,1\}$ can be written as the sum of at most n^t products of variables), but (2) almost all such functions require depth-three Boolean circuits of subexponential size (because parity is reducible to them). Furthermore, (2′) almost all t-linear functions require depth-three *canonical* circuits of size at least $\exp(\Omega(tn)^{t/(t+1)})$, see Theorem 4.1. Hence, in the context of low-degree multilinear functions, depth-three Boolean circuits (let alone canonical ones) are weaker than standard (constant-depth) multilinear circuits, and so proving lower bounds for the former may be easier.

Decoupling Arity from the Number of Gates. In a work done independently (but subsequent to our initial posting[11]), Hrubes and Rao studied Boolean circuits with general gates [15]. They decoupled the two parameters (i.e., the number of gates and their arity), and studied the asymmetric case of large arity and a small number of gates. We refrained from decoupling these two parameters here, because, for our application, their maximum is the governing parameter. Lastly, we mention that a different relation between the arity and the number of gates is considered in a subsequent work [10] that extends the notion of canonical circuits to constant depth $d > 3$.

1.7 Subsequent Work

The subsequent works of Goldreich and Tal [9,10] were already mentioned several times in the foregoing. While [10] deals with an extension of the current models, the other work (i.e., [9]) is directly related to the current work; specifically, it resolves many of the specific open problems suggested in this work. As done so far, we shall report of the relevant progress whenever reproducing text (of our original work [11]) that raises such an open problem.

1.8 Various Conventions

As stated up-front, throughout this paper, when we say that a function $f : \mathbb{N} \to \mathbb{N}$ is exponential, we mean that $f(n) = \exp(\Theta(n))$. Actually, $\exp(n)$ often means $\exp(cn)$, for some unspecified constant $c > 0$. Throughout this paper, we restrict ourselves to the field GF(2), and all arithmetic operations are over this field.[12]

Tensors. Recall that any t-linear function $F : (\{0,1\}^n)^t \to \{0,1\}$ is associated with the tensor $T \subseteq [n]^t$ that describes its existing monomials (cf., Eq. (1)). This tensor is mostly viewed as a subset of $[n]^t$, but at times such a tensor is viewed in terms of its corresponding characteristic predicate or the predicate's truth-table; that is, $T \subseteq [n]^t$ is associated with the predicate $\chi_T : [n]^t \to \{0,1\}$ or with the t-dimensional array $(\chi_T(i_1, ..., i_t))_{i_1,...,i_t \in [n]})$ such that $\chi_T(i_1, ..., i_t) = 1$

[11] See ECCC TR13-043, March 2013.

[12] However, as stated in Sect. 1.5, our main results extend to other fields.

if and only if $(i_1, ..., i_t) \in T$. The latter views are actually more popular in the literature, and they also justify our convention of writing $\sum_{k\in[m]} T_k$ instead of the symmetric difference of $T_1, ..., T_m \subseteq [n]^t$ (i.e., $(i_1, ..., i_t) \in \sum_{k\in[m]} T_k$ iff $|\{k \in [m] : (i_1, ..., i_t) \in T_k\}|$ is odd).

In the case of $t = 2$, the tensor (viewed as an array) is a matrix. In that case, we sometimes denote the variable-blocks by x and y (rather than $x^{(1)}$ and $x^{(2)}$).

1.9 Organization and Additional Highlights

The rest of this paper focuses on the study of the direct and nested composition complexity of multilinear functions (and its relation to the two canonical circuit models). This study is conducted in terms of the arithmetic model outlined in Sect. 1.5; that is, of multilinear circuits with general multilinear gates and a complexity measure, termed AN-complexity, that accounts for both the arity of these gates and their number. The basic definitional issues are discussed in Sect. 2, upper bounds are presented in Sect. 3, and lower bounds in Sect. 4. These sections are the core of the current paper.

We now highlight a few aspects that were either not mentioned in the introducion or mentioned too briefly.

On the Connection to Matrix Rigidity. As mentioned in Sect. 1.4, we show a connection between proving lower bounds on the AN-complexity of explicit functions and matrix rigidity. In particular, in Sect. 4.2, we show that $\mathtt{AN}(F^{3,n}_{\mathtt{tet}}) = \Omega(m)$ if there exists an n-by-n Toeplitz matrix that has rigidity m^3 with respect to rank m. This follows from Theorem 4.4, which asserts that *if T is an n-by-n matrix that has rigidity m^3 for rank m, then the corresponding bilinear function F satisfies $\mathtt{AN}(F) > m$.* In Sect. 4.3 we show that the same holds for a relaxed notion of rigidity, which we call *structured rigidity*. We also show that structured rigidity is strictly separated from the standard notion of rigidity. All these connections were used in the subsequent work of Goldreich and Tal [9].

On Further-Restricted Models. In Sect. 5, we consider two restricted models of multilinear circuits, which are obtained by imposing constraints on the models outlined in Sect. 1.5.

1. In Sect. 5.1, we consider circuits that compute functions without relying on cancellations. We show that such circuits are weaker than the multilinear circuits considered in the bulk of the paper. Specifically, we prove a $\Omega(n^{2/3})$ lower bound on the complexity of circuits that compute some explicit functions (i.e., $F^{3,n}_{\mathtt{tet}}$ and $F^{2,n}_{\mathtt{had}}$) without cancellation, whereas one of these functions (i.e., $F^{2,n}_{\mathtt{had}}$) has AN-complexity $\widetilde{O}(\sqrt{n})$.[13]

[13] In fact, $\mathtt{AN}_2(F^{2,n}_{\mathtt{had}}) = \widetilde{O}(\sqrt{n})$. In contrast, by [9], $\mathtt{AN}_2(F^{3,n}_{\mathtt{tet}}) = \widetilde{\Omega}(n^{2/3})$ and $\mathtt{AN}(F^{3,n}_{\mathtt{tet}}) = \widetilde{\Omega}(n^{0.6})$.

2. In Sect. 5.2 we study a restricted multilinear model obtained by allowing only standard addition and multiplication gates (and considering the same complexity measure as above, except for not counting multiplication gates that are fed only by variables). While this model is quite natural, it is quite weak. Nevertheless, this model allows to separate $F_{\mathtt{all}}^{t,n}$ and $F_{\mathtt{diag}}^{t,n}$ from the "harder" $F_{\mathtt{leq}}^{2,n}$, which is shown to have AN-complexity $\Theta(n^{2/3})$ in this restricted model.

Note that in both these restricted models, we are able to prove a non-trivial lower bound on an explicit function.

2 Multilinear Circuits with General Gates

In this section we introduce a new model of arithmetic circuits, where gates may compute arbitrary multilinear functions (rather than either addition or multiplication, as in the standard model). Accompanying this new model is a new complexity measure, which takes into account both the number of gates and their arity. This model (and its restriction to depth-two circuits) is presented in Sect. 2.1 (where we also present a separation between the general model and its depth-two restriction). As is clear from the introduction, the model is motivated by its relation to canonical depth-three Boolean circuits. This relation is discussed in Sect. 2.2.

Recall that we consider t-linear functions of the form $F : (\mathrm{GF}(2)^n)^t \to \mathrm{GF}(2)$, where the tn variables are partitioned into t blocks with n variables in each block, and F is linear in the variables of each block. Specifically, for t and n, we consider the variable blocks $x^{(1)}, x^{(2)}, ..., x^{(t)}$, where $x^{(j)} = (x_1^{(j)}, ..., x_n^{(j)}) \in \mathrm{GF}(2)^n$.

2.1 The Two Complexity Measures

We are interested in multilinear functions that are computed by composition of other multilinear functions, and define a conservative (or syntactic) notion of linearity that refers to the way these functions are composed. Basically, we require that this composition does not result in a polynomial that contains terms that are not multilinear, even if these terms cancel out. Let us first spell out what this means in terms of standard multilinear circuits that use (unbounded) addition and multiplication gates, as defined in [23]. This is done by saying that a function is J-linear whenever it is multilinear (but not necessarily homogeneous) in the variables that belongs to blocks in J, and does not depend on variables of other blocks.

- Each variable in $x^{(j)}$ is a $\{j\}$-linear function.
- If an addition gate computes the sum $\sum_{i\in[m]} F_i$, where F_i is a J_i-linear function computed by its ith child, then this gate computes a $\left(\bigcup_{i\in[m]} J_i\right)$-linear function.

- If a multiplication gate computes the product $\prod_{i\in[m]} F_i$, where F_i is a J_i-linear function computed by its $i^{\rm th}$ child, and the J_i's are pairwise disjoint, then this gate computes a $\left(\bigcup_{i\in[m]} J_i\right)$-linear function.

We stress that *if the J_i's mentioned in the last item are not pairwise disjoint, then their product cannot be taken by a gate in a multilinear circuit.*

We now extend this formalism to arithmetic circuits with arbitrary gates, which compute arbitrary polynomials of the values that feed into them. Basically, we require that when replacing each gate by the corresponding depth-two arithmetic circuit that computes this polynomial as a sum of products (a.k.a monomials), we obtain a standard multilinear circuit. In other words, we require the following.

Definition 2.1 (multilinear circuits with general gates): *An arithmetic circuit with arbitrary gates is called* multilinear *if each of its gates is J-linear for some $J \subset [t]$, where J-linearity is defined recursively as follows. Suppose that a gate computes $H(F_1, ..., F_m)$, where H is a polynomial and F_i is a J_i-linear function computed by the $i^{\rm th}$ child of this gate.*[14] *Then, each monomial in H computes a function that is J-linear, where J is the disjoint union of the sets J_i that define the linearity of the functions multiplied in that monomial; that is, if for some set $I \subseteq [m]$ this monomial multiplies J_i-linear functions for $i \in I$, then these J_i's should be disjoint and their union should equal J* (i.e., $J_{i_1} \cap J_{i_2} = \emptyset$ for all $i_1 \neq i_2$ and $\bigcup_{i\in I} J_i = J$). *The function computed by the gate is J'-linear if J' is the union of all the sets that define the linearity of the functions that correspond to the different monomials in H.*

Alternatively, we may require that if a gate multiplies two of its inputs (in one of the monomials computed by this gate), then the sub-circuits computing these two inputs do not depend on variables from the same block (i.e., the two sets of variables in the directed acyclic graphs rooted at these two gates belong to two sets of blocks with empty intersection).

Definition 2.2 (the AN-complexity of multilinear circuits with general gates): *The* arity *of a multilinear circuit is the maximum arity of its* (general) *gates, and the* number of gates *counts only the general gates and not the leaves* (variables). *The* AN-complexity *of a multilinear circuit is the maximum between its arity and the number of its* (general) *gates.*

- *The* general (or unbounded-depth or nested) AN-complexity *of a multilinear function F, denoted $\mathtt{AN}(F)$, is the minimum AN-complexity of a multilinear circuit with general gates that computes F.*

[14] Clearly, w.l.o.g., H is multilinear in its m inputs, since we are considering multiplication over GF(2). However, what we consider next is not the dependency of H on its own inputs, but rather its dependency on the inputs of the circuits as reflected in the composed function $H(F_1, ..., F_m)$. Furthermore, we do not consider this function *per se*, but rather its syntactic form (before cancellations).

- *The* depth-two (or direct) AN-complexity *of a multilinear function F, denoted* $\mathtt{AN}_2(F)$, *is the minimum AN-complexity of a* depth-two *multilinear circuit with general gates that computes F.*

More generally, for any $d \geq 3$, we may denote by $\mathtt{AN}_d(F)$ the minimum AN-complexity of a depth d multilinear circuit with general gates that computes F.

Clearly, $\mathtt{AN}_2(F) \geq \mathtt{AN}(F)$ for every multilinear function F. For linear functions F, it holds that $\mathtt{AN}_2(F) \leq 2 \cdot \mathtt{AN}(F)$, because in this case all gates are addition gates and so, w.l.o.g., all intermediate gates can feed directly to the top gate (while increasing its arity by at most $\mathtt{AN}(F) - 1$ units). This is no longer the case for bilinear functions; that is, there exists bilinear functions F such that $\mathtt{AN}_2(F) \gg \mathtt{AN}(F)$.

Theorem 2.3 (separating $\mathtt{AN}_2$ from $\mathtt{AN}$): *There exist bilinear functions $F : (\mathrm{GF}(2)^n)^2 \to \mathrm{GF}(2)$ such that $\mathtt{AN}(F) = O(\sqrt{n})$ but $\mathtt{AN}_2(F) = \Omega(n^{2/3})$. Furthermore, the upper bound is established by a depth-three multilinear circuit.*

The furthermore clause is no coincidence: As outlined in Remark 2.4, for every t-linear function F, it holds that $\mathtt{AN}_{t+1}(F) = O(\mathtt{AN}(F))$.

Proof: Consider a generic bilinear function $g : \mathrm{GF}(2)^{n+s} \to \mathrm{GF}(2)$, where g is linear in the first n bits and in the last $s = \sqrt{n}$ bits. Using the fact that g is linear in the first n variables, it will be useful to write $g(x, z)$ as $\sum_{i \in [s]} g_i((x_{(i-1)s+1}, ..., x_{is}), z)$, where each g_i is a bilinear function on $\mathrm{GF}(2)^s \times \mathrm{GF}(2)^s$. Define $f : \mathrm{GF}(2)^{2n} \to \mathrm{GF}(2)$ such that $f(x, y) = g(x, L_1(y), ..., L_s(y))$, where $L_i(y) = \sum_{k=(i-1)s+1}^{si} y_k$. That is, f is obtained from g by replacing each variable z_i (of g) by the linear function $L_i(y)$; in the sequel, we shall refer to this f as being derived from g.

Clearly, $\mathtt{AN}(f) \leq 2s+1$ by virtue of a depth-three multilinear circuit that first computes $v \leftarrow (L_1(y), ..., L_s(y))$ (using s gates each of arity s), then computes $w_i \leftarrow (g_i((x_{(i-1)s+1}, ..., x_{is}), v)$ for $i \in [s]$ (using s gates of arity $2s$), and finally compute the sum $\sum_{i \in [s]} w_i$ (in the top gate). The rest of the proof is devoted to proving that for a random g, with high probability, the corresponding f satisfies $\mathtt{AN}_2(f) = \Omega(n^{2/3})$.

We start with an overview of the proof strategy. We consider all functions $f : \mathrm{GF}(2)^n \times \mathrm{GF}(2)^n \to \mathrm{GF}(2)$ that can be derived from a generic bilinear function $g : \mathrm{GF}(2)^n \times \mathrm{GF}(2)^s \to \mathrm{GF}(2)$ (by letting $f(x, y) = g(x, L_1(y), ..., L_s(y))$). For each such function f, we consider a hypothetical depth-two multilinear circuit of AN-complexity at most $m = 0.9n^{2/3}$ that computes f. Given such a circuit, using a suitable (random) restriction, we obtain a circuit that computes the underlying function g such that the resulting circuit belongs to a set containing at most $2^{0.9sn}$ circuits. But since the number of possible functions g is 2^{sn}, this means that most functions f derived as above from a generic g do not have depth-two multilinear circuit of AN-complexity at most $m = 0.9n^{2/3}$; that is, for almost all such functions f, it holds that $\mathtt{AN}_2(f) > 0.9n^{2/3}$. The actual argument follows.

Consider an arbitrary *depth-two* multilinear circuit of AN-complexity m that computes a generic f (derived as above from a generic g). (We shall assume, w.l.o.g., that the top gate of this circuit is not fed directly by any variable, which can be enforced by replacing such variables with singleton linear functions.)[15] By the multilinear condition, the top gate of this circuit computes a function of the form

$$B(F_1(x), ..., F_{m'}(x), G_1(y), ..., G_{m''}(y)) + \sum_{i \in [m''']} B_i(x, y), \tag{5}$$

where B is a bilinear function (over $\mathrm{GF}(2)^{m'} \times \mathrm{GF}(2)^{m''}$), the F_i's and G_i's are linear functions, the B_i's are bilinear functions, and each of these functions depends on at most m variables. Furthermore, $m' + m'' + m''' < m$. (That is, Eq. (5) corresponds to a generic description of a depth-two multilinear circuit of AN-complexity m that computes a bilinear function. The top gate computes the sum of a bilinear function of $m' + m''$ intermediate linear gates and a sum of m''' intermediate bilinear gates, whereas all intermediate gates are fed by variables only.)

We now consider a random restriction of y that selects at random $i_j \in \{(j-1)s+1, ..., js\}$ for each $j \in [s]$, and sets all other bit locations to zero. Thus, for a selection as above, we get y' such that $y'_i = y_i$ if $i \in \{i_1, ..., i_s\}$ and $y'_i = 0$ otherwise. In this case, $f(x, y')$ equals $g(x, y_{i_1}, ..., y_{i_s})$. We now look at the effect of this random restriction on the expression given in Eq. (5).

The key observation is that the expected number of "live" y' variables (i.e., $y'_i = y_i$) in each B_i is at most m/s; that is, in expectation, $B_i(x, y')$ depends on m/s variables of the y-block. It follows that each $B_i(x, y')$ can be specified by $((m + (m/s)) \log_2 n) + m \cdot (m/s)$ bits (in expectation), because $B_i(x, y')$ is a bilinear form in the surviving y-variables and in at most m variables of x, whereas such a function can be specified by identifying these variables and describing the bilinear form applied to them. Hence, in expectation, the residual $\sum_i B_i(x, y')$ is specified by less than $(2m^2 \log_2 n) + (m^3/s)$ bits, and we may pick a setting (of $i_1, ..., i_s$) that yields such a description length. This means that, no matter from which function g (and f) we start, the number of possible (functionally different) circuits that result from Eq. (5) is at most

$$2^{m^2} \cdot \left(\sum_{k \in [m]} \binom{n}{k} \right)^m \cdot 2^{m^3/s + 2m^2 \log_2 n} \tag{6}$$

where the first factor reflects the number of possible bilinear functions B, the second factor reflects the possible choices of the linear functions $F_1, ..., F_{m'}, G_1, ..., G_{m''}$, and the third factor reflects the number of possible bilinear functions that can be computed by $\sum_i B_i(x, y')$. Note that, for $m \geq n^{\Omega(1)}$,

[15] Actually, this may increase m by one unit. The reason is that if the top gate if fed by i variables, then the number of intermediate gates in the circuit is at most $m - i$. So introducing intermediate singleton gates yields a depth-two circuit with at most $(m - i) + i$ intermediate gates.

the quantity in Eq. (6) is upper-bounded by $2^{m^2+\widetilde{O}(m^2)+(m^3/s+\widetilde{O}(m^2))}$, and for $m > \widetilde{O}(n^{1/2})$ the dominant term in the exponent is m^3/s. In particular, for $m = 0.9n^{2/3}$, the quantity in Eq. (6) is smaller than $2^{1.1m^3/s} < 2^{0.9sn}$, which is much smaller than the number of possible functions g (i.e., 2^{sn}). Hence, for $m = 0.9n^{2/3}$, not every function f can be computed as in Eq. (5), and the theorem follows. ■

Digest. The proof of the lower bound of Theorem 2.3 may be decoupled into two parts pivoted at an artificial complexity class, denoted G, that contains all functions g that have multilinear circuits of a relatively small description (i.e., description length at most $0.9n^{1.5}$). Using the random restriction, we show that if f has depth-two AN-complexity at most $0.9n^{2/3}$, then the underlying g is (always) in G. A counting argument then shows that most g's are not in G. Combining these two facts, we conclude that most functions f (constructed based on a function g as in the proof) have depth-two AN-complexity greater than $0.9n^{2/3}$. (A more appealing abstraction, which requires a slightly more refined proof, is obtained by letting G contains all functions g that have depth-two multilinear circuits of AN-complexity at most $0.9n^{2/3}$ such that each gate is fed by at most $n^{1/6}$ variables from the short block.)[16]

Remark 2.4 (on the depth of multilinear circuits achieving $\mathtt{AN}$): *In light of the above, it is natural to consider the depth of general multilinear circuits* (as in Definition 2.1), *and study the trade-offs between depth and other parameters* (as in Definition 2.2). *While this is not our primary focus here, we make just one observation: If* $\mathtt{AN}(F) = s$ *for any* t*-linear function* F*, then there is a depth* $t+1$ *circuit with arity and size* $O(s)$ *computing* F *as well; that is, for any* t*-linear* F*, it holds that* $\mathtt{AN}_{t+1}(F) = O(\mathtt{AN}(F))$*. This observation is proved in Proposition 4.5.*

Remark 2.5 (waiving the multilinear restriction): *We note that arbitrary arithmetic circuits* (with general gates) *that compute* t*-linear functions can be simulated by multilinear circuits of the same depth, while increasing their AN-complexity measure by a factor of at most* 2^t*. This can be done by replacing each* (intermediate) *gate in the original circuit with* $2^t - 1$ *gates in the multilinear circuit such that the gate associated with* $I \subseteq [t]$ *computes the monomials that are* I*-linear (but not* I'*-linear, for any* $I' \subset I$*). The monomials that are not multilinear are not computed, and this is OK because their influence must cancel*

[16] The point is that this alternative class G does not refer to the "description length" but rather to the complexity measures defined in this section. In this case, we may show that a random restriction of the type used in the original proof leaves m/s live variables in each G_i, in expectation, just as it holds for the B_i's. Using $m = 0.9n^{2/3}$, it holds that, with high probability, none of the gates exceeds this expectation by a factor of $1/0.9$. Next, we upper-bound the size of G, very much as done in the foregoing proof, where here the crucial fact is that each B_i has only $m \cdot n^{1/6}$ live terms, whereas $m^2 \cdot n^{1/6} = 0.81 \cdot n^{3/2}$.

out at the top gate.[17] *Indeed, the top gate performs the $2^t - 1$ computations that corresponds to the different I-linear sums, and sums-up the $2^t - 1$ results.*

2.2 Relation to Canonical Circuits

As outlined in Sect. 1.5, the direct and nested AN-complexity of multilinear functions (i.e., $\mathtt{AN}_2$ and $\mathtt{AN}$) are closely related to the size of D-canonical and ND-canonical circuits computing the functions. Below, we spell out constructions of canonical circuits, which are depth-three Boolean functions, having size that is exponential in the relevant parameter (i.e., D-canonical circuits of size $\exp(\mathtt{AN}_2)$ and ND-canonical circuits of size $\exp(\mathtt{AN})$).

Construction 2.6 (D-canonical circuits of size $\exp(\mathtt{AN}_2)$): *Let $F : (\mathrm{GF}(2)^n)^t \to \mathrm{GF}(2)$ be a t-linear function, and consider a depth-two multilinear circuit that computes F such that the top gate applies an m-ary polynomial H to the results of the m gates that compute $F_1, ..., F_m$, where each F_i is a multilinear function of at most m variables.* (Indeed, we assume, without loss of generality, that the top gate is fed by the second-level gates only, which in turn are fed by variables.)[18] *Then, the following depth-three Boolean circuit computes F.*

1. *Let C_H be a CNF* (resp., DNF) *that computes H.*
2. *For each $i \in [m]$, let C_i be a DNF* (resp., CNF) *that computes F_i, and let C_i' be a DNF* (resp., CNF) *that computes $1 + F_i$.*
3. *Compose C_H with the various C_i's and C_i''s such that a positive occurrence of the ith variable of C_H is replaced by C_i and a negative occurrence is replaced by C_i'.*
   Collapsing the two adjacent levels of OR*-gates* (resp., AND-gates), *yields a depth-three Boolean circuit C.*

The derived circuit C is said to be D-canonical, *and a circuit is said to be D-canonical only if it can be derived as above.*

Clearly, C computes F and has size exponential in m. In particular, we have

Proposition 2.7 (depth-three Boolean circuits of size $\exp(\mathtt{AN}_2)$): *Every multilinear function F has depth-three Boolean circuits of size* $\exp(\mathtt{AN}_2(F))$.

It turns out that the upper bound provided in Proposition 2.7 is not tight; that is, D-canonical circuits do not provide the smallest depth-three Boolean circuits for all multilinear functions. In particular, there exists multilinear functions that have depth-three Boolean circuits of size $\exp(\mathtt{AN}_2(F)^{3/4})$. This follows by combining Theorem 2.3 and Proposition 2.9, where Theorem 2.3 asserts that for some bilinear functions F it holds that $\mathtt{AN}(F) = O(\sqrt{n}) = O(n^{2/3})^{3/4} = O(\mathtt{AN}_2(F))^{3/4}$,

[17] Here, we assume (as is standard in the area) that the cancellations must hold over any extension field of GF(2) (rather than only over GF(2) itself); that is, the polynomial x^i may be cancelled by the polynomial $(2k+1) \cdot x^j$ if and only if $i = j$.

[18] Variables that feed directly into the top gate can be replaced by 1-ary identity gates.

and Proposition 2.9 asserts that every multilinear function F has depth-three Boolean circuits of size $\exp(\mathtt{AN}(F))$. The latter is proved by using ND-canonical circuits, which leads us to their general construction.

Construction 2.8 (ND-canonical circuits of size $\exp(\mathtt{AN})$): *Let $F : (\mathrm{GF}(2)^n)^t \to \mathrm{GF}(2)$ be a t-linear function, and consider a multilinear circuit that computes F such that the each of the m gates applies an m-ary polynomial H_i to the results of prior gates and some variables, where H_1 corresponds to the polynomial applied by the top gate. Consider the following depth-three Boolean circuit that computes F.*

1. *For each $i \in [m]$ and $\sigma \in \mathrm{GF}(2)$, let C_i^σ be a CNF that computes $H_i + 1 + \sigma$. That is, C_i^σ evaluates to 1 iff H_i evaluate to σ.*
2. *For each $\overline{v} \stackrel{\text{def}}{=} (v_1, v_2, ..., v_m) \in \mathrm{GF}(2)^m$, let*

$$C_{\overline{v}}(x^{(1)}, ..., x^{(t)}) = \bigwedge_{i \in [m]} C_i^{v_i}(\Pi_{i,1}(x^{(1)}, ..., x^{(t)}, \overline{v}), ..., \Pi_{i,m}(x^{(1)}, ..., x^{(t)}, \overline{v})),$$

   where the $\Pi_{i,j}$'s are merely the projection functions that describe the routing in the multilinear circuit; that is, $\Pi_{i,j}(x^{(1)}, ..., x^{(t)}, \overline{v})) = v_k$ if the jth input of gate i is fed by gate k and $\Pi_{i,j}(x^{(1)}, ..., x^{(t)}, \overline{v})) = x_k^{(\ell)}$ if the jth input of gate i is fed by the kth variable in the ℓth variable-block (i.e., the variable $x_k^{(\ell)}$). *Indeed, each $C_{\overline{v}}$ is a CNF of size $\widetilde{O}(2^m)$.*
3. *We obtain a depth-three Boolean circuit C by letting*

$$C(x^{(1)}, ..., x^{(t)}) = \bigvee_{(v_2, ..., v_m) \in \mathrm{GF}(2)^{m-1}} C_{(1, v_2, ..., v_m)}(x^{(1)}, ..., x^{(t)})$$

   Hence, C has size $2^{m-1} \cdot \widetilde{O}(2^m)$.

The derived circuit C is said to be ND-canonical, *and a circuit is said to be ND-canonical only if it can be derived as above.*

Note that $C(x^{(1)}, ..., x^{(t)}) = 1$ if and only if there exists $\overline{v} = (v_1, v_2, ..., v_m) \in \mathrm{GF}(2)^m$ such that $v_1 = 1$ and for every $i \in [m]$ it holds that $H_i(z_{i,1}, ..., z_{i,m}) = v_i$, where $z_{i,j}$ is the $\ell_{i,j}^{\text{th}}$ bit in the $(tn + m)$-bit long sequence $(x^{(1)}, ..., x^{(t)}, \overline{v})$ for some predetermined $\ell_{i,j} \in [tn + m]$. For this choice of $\overline{v}$, the v_i's represent the values computed by the gates in the original arithmetic circuit (on an input that evaluates to 1), and it follows that C computes F. Clearly, C has size exponential in m. In particular, we have

Proposition 2.9 (depth-three Boolean circuits of size $\exp(\mathtt{AN})$): *Every multilinear function F has depth-three Boolean circuits of size* $\exp(\mathtt{AN}(F))$.

A key question is whether the upper bound provided in Proposition 2.9 is tight. The answer depends on two questions: The main question is whether smaller depth-three Boolean circuits can be designed by deviation from the construction paradigm presented in Construction 2.8. The second question is whether the upper bound of $\exp(m)$ on the size of the depth-two Boolean circuits used to compute m-ary polynomials (of degree at most t) is tight. In fact, it suffices to consider t-linear polynomials, since only such gates may be used in a multilinear circuit.

The latter question is addressed in [11, Apdx C.1], where it is shown that any t-linear function that depends on m variables requires depth-two Boolean circuits of size at least $\exp(\Omega(\exp(-t)\cdot m))$. (Interestingly, this lower bound is tight; that is, there exist t-linear functions that depends on m variables and have depth-two Boolean circuits of size at most $\exp(O(\exp(-t)\cdot m))$.) Conjecturing that the main question has a negative answer, this leads to the following conjecture.

Conjecture 2.10 (AN yields lower bounds on the size of general depth-three Boolean circuits): *No t-linear function $F : (\mathrm{GF}(2)^n)^t \to \mathrm{GF}(2)$ can be computed by a depth-three Boolean circuit of size smaller than* $\exp(\Omega(\exp(-t)\cdot \mathtt{AN}(F)))/\mathrm{poly}(n)$.

When combined with adequate lower bounds on AN (e.g., Theorem 4.1), Conjecture 2.10 yields size lower bounds of the form $\exp(\Omega(\exp(-t)\cdot n^{t/(t+1)}))$, which yields $\exp(n^{1-o(1)})$ for $t = \sqrt{\log n}$. Furthermore, in some special cases (see [11, Apdx C.3]), multilinear functions that depends on m variables requires depth-two Boolean circuits of size at least $\exp(\Omega(m))$. This suggests making a bolder conjecture, which allows using larger values of t.

Conjecture 2.11 (Conjecture 2.10, stronger form for special cases): *None of the multilinear functions $F \in \{F^{t,n}_{\mathtt{tet}}, F^{t,n}_{\mathrm{mod}\,p} : p \geq 2\}$* (see Eqs. (3) and (4), resp.) *can be computed by a depth-three Boolean circuit of size smaller than* $\exp(\Omega(\mathtt{AN}(F)))/\mathrm{poly}(n)$. *The same holds for almost all t-linear functions.*

When combined with adequate lower bounds on AN (e.g., Theorem 4.1), Conjecture 2.11 yields size lower bounds of the form $\exp(\Omega((tn)^{t/(t+1)}))$, which for $t = \log n$ yields $\exp(\Omega(tn))$.

The authors are in disagreement regarding the validity of Conjecture 2.10 (let alone Conjecture 2.11), but agree that also refutations will be of interest.

3 Upper Bounds

In Sect. 3.1 we present a generic upper bound on the direct AN-complexity of any t-linear function; that is, we show that $\mathtt{AN}_2(F) = O((tn)^{t/(t+1)})$, for every t-linear function F. This bound, which is obtained by a generic construction, is the best possible for almost all multilinear functions (see Theorem 4.1). Obviously, one can do better in some cases, even when this may not be obvious at first glance. In Sect. 3.2, we focus on two such cases (i.e., $F^{t,n}_{\mathtt{leq}}$ and $F^{2,n}_{\mathrm{mod}\,p}$).

3.1 A Generic Upper Bound

The following upper bound on the AN-complexity of multilinear circuits that compute a generic t-linear function is derived by using a depth-two circuit with a top gate that computes addition (i.e., a linear function). This implies that the intermediate gates in this circuit, which are fed by variables only, must all be t-linear gates. While the overall structure of the circuit is oblivious of the t-linear function that it computes, the latter function determines the choice of the t-linear gates.

Theorem 3.1 (an upper bound on $\mathtt{AN}_2(\cdot)$ for any multilinear function): *Every t-linear function $F : (\mathrm{GF}(2)^n)^t \to \mathrm{GF}(2)$ has D-canonical circuits of size $\exp(O(tn)^{t/(t+1)})$; that is, $\mathtt{AN}_2(F) = O((tn)^{t/(t+1)})$.*

Proof: We partition $[n]^t$ into m equal-sized subcubes such that the number of subcubes (i.e., m) equals the number of variables that correspond to each subcube (i.e., $t \cdot \sqrt[t]{n^t/m}$); that is, the side-length of each subcubes is $\ell \stackrel{\mathrm{def}}{=} n/m^{1/t}$ and m is selected such that $m = t \cdot \ell$. We then write the tensor that corresponds to F as a sum of tensors that are each restricted to one of the aforementioned subcubes. Details follow.

We may assume that $t = O(\log n)$, since the claim holds trivially for $t = \Omega(\log n)$. Partition $[n]^t$ into m cubes, each having a side of length $\ell = (n^t/m)^{1/t} = n/m^{1/t}$; that is, for $k_1, ..., k_t \in [n/\ell]$, let $C_{k_1,...,k_t} = I_{k_1} \times \cdots \times I_{k_t}$, where $I_k = \{(k-1)\ell + j : j \in [\ell]\}$. Clearly, $[n]^t$ is covered by this collection of $((n/\ell)^t = m)$ cubes, and the sum of the lengths of each cube is $t\ell$. Let T be the tensor corresponding to F. Then,

$$F(x^{(1)}, ..., x^{(t)}) = \sum_{k_1,...,k_t \in [n/\ell]} F_{k_1,...,k_t}(x^{(1)}, ..., x^{(t)})$$

$$\text{where } F_{k_1,...,k_t}(x^{(1)}, ..., x^{(t)}) = \sum_{(i_1,...,i_t) \in T \cap C_{k_1,...,k_t}} x^{(1)}_{i_1} \cdots x^{(t)}_{i_t}.$$

Each $F_{k_1,...,k_t}$ is computed by a single $[t]$-linear gate of arity $t \cdot \ell$, and it follows that $\mathtt{AN}_2(F) \leq \max(t\ell, m+1)$, since $(n/\ell)^t = m$. Using $m = t\ell$, we get $\mathtt{AN}_2(F) \leq m+1$ and $m = t \cdot n/m^{1/t}$ (equiv., $m^{1+\frac{1}{t}} = t \cdot n$), which yields $\mathtt{AN}_2(F) = O((tn)^{t/(t+1)})$, since $m = (tn)^{\frac{1}{1+(1/t)}}$. ■

3.2 Improved Upper Bounds for Specific Functions (e.g., $F^{t,n}_{\mathtt{leq}}$)

Clearly, the generic upper bound can be improved upon in many special cases. Such cases include various t-linear functions that are easily reducible to linear functions. Examples include (1) $F^{t,n}_{\mathtt{all}}(x^{(1)}, ..., x^{(t)}) = \sum_{i_1,...,i_t \in [n]} x^{(1)}_{i_1} \cdots x^{(t)}_{i_t} = \prod_{j \in [t]} \sum_{i \in [n]} x^{(j)}_i$ and (2) $F^{t,n}_{\mathtt{diag}}(x^{(1)}, ..., x^{(t)}) = \sum_{i \in [n]} x^{(1)}_i \cdots x^{(t)}_i$. Specifically, we can easily get $\mathtt{AN}_2(F^{t,n}_{\mathtt{all}}) \leq t\sqrt{n} + 1$ and $\mathtt{AN}_2(F^{t,n}_{\mathtt{diag}}) \leq t\sqrt{n}$. In both cases, the key observation is that each n-way sum can be written as a sum of $\sqrt{n}$

functions such that each function depends on $\sqrt{n}$ of the original arguments. Furthermore, in both cases, we could derive (depth-three) multilinear formulae of AN-complexity $t\sqrt{n}+1$ that use only ($\sqrt{n}$-way) addition and (t-way) multiplication gates.[19] While such simple multilinear formulae do not exist for $F_{\mathtt{leq}}^{2,n}$ (see Sect. 5.2), the full power of (depth-two) multilinear circuits with *general gates* yields $\mathtt{AN}_2(F_{\mathtt{leq}}^{2,n}) = O(\sqrt{n})$; that is, as in the proof of Theorem 3.1, the following construction also uses general multilinear gates.

Proposition 3.2 (an upper bound on $\mathtt{AN}_2(F_{\mathtt{leq}}^{2,n})$): *The bilinear function* $F_{\mathtt{leq}}^{2,n}$ (of Eq. (2)) *has D-canonical circuits of size* $\exp(O(\sqrt{n}))$*; that is,* $\mathtt{AN}_2(F_{\mathtt{leq}}^{2,n}) = O(\sqrt{n})$.

Recall that the corresponding tensor is $T_{\mathtt{leq}}^{2,n} = \{(i_1, i_2) \in [n]^2 : i_1 \leq i_2\}$.

Proof: Letting $s \stackrel{\text{def}}{=} \sqrt{n}$, we are going to express $F_{\mathtt{leq}}^{2,n}$ as a polynomial in $3s$ functions, where each of these functions depends on at most $2s$ variables. The basic idea is to partition $[n]^2$ into s^2 squares of the form $S_{i,j} = [(i-1)s+1, is] \times [(j-1)s+1, js]$, and note that $\bigcup_{i<j} S_{i,j} \subset T_{\mathtt{leq}}^{2,n} \subset \bigcup_{i\leq j} S_{i,j}$. Thus, $F_{\mathtt{leq}}^{2,n}$ can be computed by computing separately the contribution of the diagonal squares and the contribution of the squares that are off the diagonal. The contribution of the square $S_{i,i}$ can be computed as a function of the $2s$ variables that correspond to it, while the contribution of each off-diagonal square can be computed as the product of the corresponding sum of $x^{(1)}$-variables and the corresponding sum of $x^{(2)}$-variables. Thus, the contribution of each diagonal square will be computed by a designated bilinear gate, whereas the contribution of the off-diagonal squares will be computed by the top gate (which is fed by $2s$ linear gates, each computing the sum of s variables, and computes a suitable bilinear function of these $2s$ sums). Details follow.

- For every $i \in [s]$, let $Q_i(x^{(1)}, x^{(2)}) = \sum_{(j_1,j_2)\in T_{\mathtt{leq}}^{2,s}} x_{(i-1)s+j_1}^{(1)} \cdot x_{(i-1)s+j_2}^{(2)}$, where $T_{\mathtt{leq}}^{2,s} = \{(j_1, j_2) \in [s]^2 : j_1 \leq j_2\}$. This means that $Q_i(x^{(1)}, x^{(2)})$ only depends on $2s$ variables (i.e., $x_{(i-1)s+1}^{(1)}, \ldots, x_{is}^{(1)}$ and $x_{(i-1)s+1}^{(2)}, \ldots, x_{is}^{(2)}$).
  Indeed, $Q_i(x^{(1)}, x^{(2)})$ computes the contribution of the ith diagonal square (i.e., $S_{i,i}$). In contrast, the following linear functions will be used to compute the contribution of the off-diagonal squares.
- For every $i \in [s]$, let $L_i(x^{(1)}) = \sum_{j\in[s]} x_{(i-1)s+j}^{(1)}$, which means that $L_i(x^{(1)})$ only depends on $x_{(i-1)s+1}^{(1)}, \ldots, x_{is}^{(1)}$.
- Likewise, for every $i \in [s]$, let $L_i'(x^{(2)}) = \sum_{j\in[s]} x_{(i-1)s+j}^{(2)}$.

Observing that

$$F_{\mathtt{leq}}^{2,n}(x^{(1)}, x^{(2)}) = \sum_{i\in[s]} Q_i(x^{(1)}, x^{(2)}) + \sum_{1\leq i<j\leq s} L_i(x^{(1)}) \cdot L_j'(x^{(2)}), \quad (7)$$

[19] Depth-two circuits can be derived by combining the t-way multiplication gate with the $\sqrt{n}$-way addition gates feeding it (resp., each $\sqrt{n}$-way addition gate with the t-way multiplication gate feeding it).

the claim follows. Specifically, we use $3s$ intermediate gates that compute the Q_i's, L_i's and L'_j's (and let the top gate compute their combination (per Eq. (7))). ■

The Case of $F^{2,n}_{\bmod p}$. We turn to another bilinear function, the function $F^{2,n}_{\bmod p}$, where $F^{t,n}_{\bmod p}$ is defined in Eq. (4).

Proposition 3.3 (an upper bound on $\mathtt{AN}_2(F^{2,n}_{\bmod p})$): *For every p and n, the bilinear function $F^{2,n}_{\bmod p}$ has D-canonical circuits of size* $\exp(O(\sqrt{n}))$*; that is,* $\mathtt{AN}_2(F^{2,n}_{\bmod p}) = O(\sqrt{n})$.

Recall that $F^{2,n}_{\bmod p}(x^{(1)}, x^{(2)}) = \sum_{i_1,i_2 \in [n]: i_1 \equiv -i_2 \pmod p} x^{(1)}_{i_1} \cdot x^{(2)}_{i_2}$.

Proof: Let $s = \sqrt{n}$, and let's consider first the case $p \leq s$. For every $r \in \mathbb{Z}_p$, consider the functions $L_r(x^{(1)}) = \sum_{i \equiv r \pmod p} x^{(1)}_i$ and $L'_r(x^{(2)}) = \sum_{i \equiv r \pmod p} x^{(2)}_i$. Then,

$$F^{2,n}_{\bmod p}(x^{(1)}, x^{(2)}) = \sum_{r \in \mathbb{Z}_p} L_r(x^{(1)}) \cdot L'_{p-r}(x^{(2)}).$$

Each of the foregoing $p \leq s$ linear functions depend on $\lceil n/p \rceil$ variables, which is fine if $p = \Omega(s)$. Otherwise (i.e., for $p = o(s)$), we replace each linear function by $\lceil n/ps \rceil$ auxiliary functions (in order to perform each $\lceil n/p \rceil$-way summation), which means that in total we have $2p \cdot \lceil n/ps \rceil = O(s)$ functions such that each function depends on $\frac{\lceil n/p \rceil}{\lceil n/ps \rceil} < 2s$ variables. Then, the top gate just computes the suitable (bilinear) combination of these $O(s)$ linear functions.

In the case of $p > s$, we face the opposite problem; that is, we have too many linear functions, but each depends on $n/p < s$ variables. So we group these functions together; that is, for a partition of $\mathbb{Z}_p$ to s equal parts, denoted $P_1, \ldots, P_s$, we introduce s functions of the form

$$Q_i(x^{(1)}, x^{(2)}) = \sum_{r \in P_i} \left(\sum_{j \equiv r \pmod p} x^{(1)}_j \right) \cdot \left(\sum_{j \equiv p-r \pmod p} x^{(2)}_j \right)$$

for every $i \in [s]$. Clearly, $F^{2,n}_{\bmod p}(x^{(1)}, x^{(2)}) = \sum_{i \in [s]} Q_i(x^{(1)}, x^{(2)})$, and each Q_i depends on $2 \cdot |P_i| \cdot \lceil n/p \rceil = O(s)$ variables, since $|P_i| \leq \lceil p/s \rceil$. ■

The Case of $F^{t,n}_{\mathtt{leq}}$ *for* $t > 2$. Finally, we turn to t-linear functions with $t > 2$. Specifically, we consider the t-linear function $F^{t,n}_{\mathtt{leq}}$ (of Eq. (2)), focusing on $t \geq 3$.

Proposition 3.4 (an upper bound on $\mathtt{AN}_2(F^{t,n}_{\mathtt{leq}})$): *For every t, it holds that* $\mathtt{AN}_2(F^{t,n}_{\mathtt{leq}}) = O(\exp(t) \cdot \sqrt{n})$.

Proof: The proof generalizes the proof of Proposition 3.2, and proceeds by induction on t. We (again) let $s \stackrel{\text{def}}{=} \sqrt{n}$ and partition $[n]^t$ into s^t cubes of the form $C_{k_1,...,k_t} = I_{k_1} \times \cdots \times I_{k_t}$, where $I_k = \{(k-1)s + j : j \in [s]\}$. Actually, we prove an inductive claim that refers to the *simultaneously expressibility* of the functions $F_{\mathtt{leq}}^{t,[(k-1)s+1,n]}$ for all $k \in [s]$, where

$$F_{\mathtt{leq}}^{t,[i,n]}(x^{(1)}, ..., x^{(t)}) \stackrel{\text{def}}{=} \sum_{(i_1,...,i_t)\in T_{\mathtt{leq}}^{t,n} : i_1 \geq i} x_{i_1}^{(1)} \cdots x_{i_t}^{(t)}. \quad (8)$$

Indeed, $F_{\mathtt{leq}}^{t,n} = F_{\mathtt{leq}}^{t,[1,n]}$. The *inductive claim*, indexed by $t \in \mathbb{N}$, asserts that the functions $F_{\mathtt{leq}}^{t,[(k-1)s+1,n]}$, for all $k \in [s]$, can be expressed as polynomials in $t2^t \cdot s$ multilinear functions such that each of these functions depends on $t \cdot s$ variables. The base case (of $t = 1$) follows easily by using the s functions $L_i(x^{(1)}) = \sum_{j\in[s]} x_{(i-1)s+j}^{(1)}$.

In the induction step, for every $j \in [t]$, define $T_j \stackrel{\text{def}}{=} \{(k_1, ..., k_t) \in T_{\mathtt{leq}}^{t,s} : k_1 = k_j < k_{j+1}\}$, where $k_{t+1} \stackrel{\text{def}}{=} s + 1$; that is, $(k_1, ..., k_t) \in T_j$ if and only if $k_1 = \cdots = k_j < k_{j+1} \leq \cdots \leq k_t \leq s$ (for $j < t$, whereas $T_t = \{(k, k, ..., k) \in [s]^t : k \in [s]\}$). Note that, for every $k \in [s]$, the elements of $T_{\mathtt{leq}}^{t,[(k-1)s+1,n]}$ are partitioned according to these T_j's; that is, each $(i_1, ..., i_t) \in T_{\mathtt{leq}}^{t,[(k-1)s+1,n]}$ uniquely determines $j \in [t]$ and $k_1 \in [k, n]$ such that $(i_1, ..., i_j) \in I_{k_1} \times \cdots \times I_{k_1}$ and $(i_{j+1}, ..., i_t) \in T_{\mathtt{leq}}^{t-j,[k_1 s+1,n]}$. Thus, for every $k \in [s]$, it holds that

$$\begin{aligned} &F_{\mathtt{leq}}^{t,[(k-1)s+1,n]}(x^{(1)}, ..., x^{(t)}) \\ &= \sum_{j\in[t-1]} \sum_{k_1 \geq k} P_{k_1}^{(j)}(x^{(1)}, ..., x^{(j)}) \cdot F_{\mathtt{leq}}^{t-j,[k_1 s+1,n]}(x^{(j+1)}, ..., x^{(t)}) \\ &\text{where } P_{k_1}^{(j)}(x^{(1)}, ..., x^{(j)}) \stackrel{\text{def}}{=} \sum_{(i_1,...,i_j)\in(T_{\mathtt{leq}}^{j,n} \cap (I_{k_1})^j)} x_{i_1}^{(1)} \cdots x_{i_j}^{(j)}. \end{aligned}$$

It follows that all $F_{\mathtt{leq}}^{t,[(k-1)s+1,n]}$'s are simultaneously expressed in terms of $(t-1) \cdot s$ new functions (i.e., the $P_{k_1}^{(j)}$'s), each depending on at most $t \cdot s$ inputs, and $(t-1) \cdot s$ functions (i.e., the $F_{\mathtt{leq}}^{t-j,[k_1 s+1,n]}$'s) that by the induction hypothesis can be expressed using $\sum_{j\in[t-1]}(t-j)2^{t-j} \cdot s$ multilinear functions (although with different variable names for different j's).[20] So, in total, we expressed all $F_{\mathtt{leq}}^{t,[(k-1)s+1,n]}$'s using less than $ts + \sum_{j\in[t-1]}(t-j)2^{t-j} \cdot s$ functions, each depending on at most ts variables. Noting that $ts + \sum_{j\in[t-1]}(t-j)2^{t-j} \cdot s$

[20] By the induction hypothesis, for every $t' \in [t-1]$, we can express the functions $F_{\mathtt{leq}}^{t-t',[(k-1)s+1,n]}(x^{(1)}, ..., x^{(t-t')})$ for all $k \in [s]$, but here we need the functions $F_{\mathtt{leq}}^{t-t',[(k-1)s+1,n]}(x^{(t'+1)}, ..., x^{(t)})$. Still, these are the same functions, we just need to change the variable names in the expressions.

is upper-bounded by $t2^t s$, the induction claim follows. This establishes that $\mathtt{AN}(F_{\mathtt{leq}}^{t,n}) \leq t2^t \cdot \sqrt{n}$.

In order to prove $\mathtt{AN}_2(F_{\mathtt{leq}}^{t,n}) \leq t2^t \cdot \sqrt{n}$, we take a closer look at the foregoing expressions. Specifically, note that all $F_{\mathtt{leq}}^{t,[(k-1)s+1,n]}$ are expressed in terms of $t2^t s$ functions such that each function is either a polynomial in the input variables or another function of the form $F_{\mathtt{leq}}^{t-j,[k_1 s+1,n]}$. In terms of multilinear circuits, this means that each gate is fed either only by variables or only by other gates (rather than being fed by a mix of both types). It follows that the top gate is a function of all gates that are fed directly by variables only, and so we can obtain a depth-two multilinear circuit with the same (or even slightly smaller) number of gates and the same (up to a factor of 2) gate arity. ■

Remark 3.5 (circuits having no mixed gates yield depth-two circuits): *The last part of the proof of Proposition 3.4 relied on the fact that if no intermediate gate of the circuit is fed by both variables and other gates, then letting all intermediate gates feed directly to the top gate yields a depth-two circuit of AN-complexity that is at most twice the AN-complexity of the original circuit, since this transformation may only increase the arity of the top gate by the number of gates. In contrast, as can be seen in the proof of Theorem 2.3, the benefit of feeding a gate by both intermediate gates and variables is that it may multiply these two types of inputs. Such a* mixed gate, *which may apply an arbitrary multilinear function to its inputs, can be split into two non-mixed gates* only if *it sums a function of the variables and a function of the other gates. It is also* not *feasible to feed the top gate with all variables that are fed to mixed gates, because this may square the AN-complexity.*

4 Lower Bounds

We believe that the generic upper bound established by Theorem 3.1 (i.e., every t-linear function F satisfies $\mathtt{AN}(F) \leq \mathtt{AN}_2(F) = O((tn)^{t/(t+1)})$ is tight for some explicit functions. However, we were only able to show that almost all multilinear functions have a lower bound that meets this upper bound. This result is presented in Sect. 4.1, whereas in Sect. 4.2 we present an approach towards proving such lower bounds for explicit functions.

Before proceeding to these sections, we comment that it is easy to see that the n-way Parity function P_n has AN-complexity at least $\sqrt{n}$; this follows from the fact that the product of the number of gates and their arity must exceed n. Of course, $\mathtt{AN}(P_n) = \Omega(\sqrt{n})$ follows by combining Proposition 2.9 with either [12] or [14], but the foregoing proof is much simpler (*to say the least*) and yields a better constant in the Ω-notation.

4.1 On the AN-Complexity of Almost All Multilinear Functions

Theorem 4.1 (a lower bound on the AN-complexity of almost all t-linear functions): *For all $t = t(n) \geq 2$, almost all t-linear functions* $F : (\mathrm{GF}(2)^n)^t \to \mathrm{GF}(2)$

satisfy $\mathtt{AN}(F) = \Omega(tn^{t/(t+1)})$. *Furthermore, such a t-linear function can be found in* $\exp(n^t)$ *time.*

Combined with Theorem 3.1, it follows that almost all t-linear functions satisfy $\mathtt{AN}(F) = \Theta(tn^{t/(t+1)})$. Here (and elsewhere), we use the fact that $t^{t/(t+1)} = \Theta(t)$.

Proof: For $m > t\sqrt{n}$ to be determined at the end of this proof, we upper bound the fraction of t-linear functions F that satisfy $\mathtt{AN}(F) \leq m$. Each such function F is computed by a multilinear circuit with at most m gates, each of arity at most m. Let us denote by H_i the function computed by the ith gate.

Recall that each of these polynomials (i.e., H_i's) is supposed to compute a $[t]$-linear function. We shall only use the fact that each H_i is t-linear in the original variables and in the other gates of the circuit; that is, we can label each gate with an integer $i \in [t]$ (e.g., i may be an block of variables on which this gate depends) and require that functions having the same label may not be multiplied nor can they be multiplied by variables of the corresponding block.

Thus, each gate specifies (1) a choice of at most m original variables, (2) a t-partition of the m auxiliary functions, and (3) a t-linear function of the m variables and the m auxiliary function. (Indeed, this is an over-specification in many ways.)[21] Thus, the number of such choices is upper-bounded by

$$\binom{tn}{m} \cdot t^m \cdot 2^{((2m/t)+1)^t} \tag{9}$$

where $((2m/t) + 1)^t$ is an upper bound on the number of monomials that may appear in a t-linear function of $2m$ variables, which are partitioned into t blocks.[22] Note that Eq. (9) is upper-bounded by $\exp((m/t)^t + m\log(tn)) = \exp((m/t)^t)$, where the equality is due to $m > t\sqrt{n} > t\log n$ and $t \geq 2$ (as we consider here).

It follows that the number of functions that can be expressed in this way is $\exp((m/t)^t)^m$, which equals $\exp(m^{t+1}/t^t)$. This is a negligible fraction of the number (i.e., 2^{n^t}) of t-linear functions over $(\mathrm{GF}(2)^n)^t$, provided that $m^{t+1}/t^t \ll n^t$, which does hold for $m \leq c \cdot (tn)^{t/(t+1)}$, for some $c > 0$. The main claim follows.

The furthermore claim follows by observing that, as is typically the case in counting arguments, both the class of admissible functions and the class of computable functions (or computing devices) are enumerable in time that is polynomial in the size of the class. Moreover, the counting argument asserts that the class of t-linear functions is the larger one (and it is also larger than 2^{tn}, which represents the number of possible inputs to each such function). ■

[21] For starters, we allowed each gate to be feed by m original variables and m auxiliary functions, whereas the arity bound is m. Furthermore, we allowed each gate to be fed by all other gates, whereas the circuit should be acyclic. Moreover, the choice of the t-partition can be the same for all gates, let alone that the various t-partitions must be consistent among gates and adheres to the multilinearity condition of Definition 2.1.

[22] Denoting by m_j the number of variables and/or gates that belong to the jth block, the number of possible monomials is $\prod_{j\in[t]}(m_j + 1)$, where in our case $\sum_{j\in[t]} m_j \leq 2m$.

Open Problems. The obvious problem that arises is proving similar lower bounds for some explicit multilinear functions. In the original version of this work [11], we suggested the following "modest start":

Problem 4.2 (the first goal regarding lower bounds regarding $\mathtt{AN}$): *Prove that* $\mathtt{AN}(F) = \Omega((tn)^c)$ *for some* $c > 1/2$ *and some explicit multilinear function* $F : (\mathrm{GF}(2)^n)^t \to \mathrm{GF}(2)$.

This challenge was met by Goldreich and Tal [9], who showed that $\mathtt{AN}(F^{3,n}_{\mathtt{tet}}) = \Omega(n^{0.6})$ and that $\mathtt{AN}(F) = \widetilde{\Omega}(n^{2/3})$ holds for some explicit 4-linear F. Referring to Problem 4.2, their work leaves open the case of $t = 2$ (for any $c > 1/2$) as well as obtaining $c > 2/3$ (for any $t > 2$). The more ambitious goal set in [11] remains far from reach, since the techniques of [9] (which are based on the "rigidity connection" made in Sect. 4.2) cannot yield $c > 2/3$.

Problem 4.3 (the ultimate goal regarding lower bounds regarding $\mathtt{AN}$): *For every* $t \geq 2$*, prove that* $\mathtt{AN}(F) = \Omega((tn)^{t/(t+1)})$ *for some explicit* t*-linear function* $F : (\mathrm{GF}(2)^n)^t \to \mathrm{GF}(2)$. *Ditto when* t *may vary with* n*, but* $t \leq \mathrm{poly}(n)$.

Actually, a lower bound of the form $\mathtt{AN}(F) = \Omega((tn)^{\epsilon t/(\epsilon t+1)})$, for some fixed constant $\epsilon > 0$, will also allow to derive exponential lower bounds when setting $t = O(\log n)$.

4.2 The AN-Complexity of Bilinear Functions and Matrix Rigidity

In this section we show that lower bounds on the rigidity of matrices yield lower bounds on the AN-complexity of bilinear functions associated with these matrices. We then show that even lower bounds for non-explicit matrices (e.g., generic Toeplitz matrices) would yield lower bounds for explicit trilinear functions, specifically, for our candidate function $F^{3,n}_{\mathtt{tet}}$ (of Eq. (3)).

Let us first recall the definition of matrix rigidity (as defined by Valiant [31] and surveyed in [19]). We say that a matrix A has rigidity s for target rank r *if every matrix of rank at most* r *disagrees with* A *on more than* s *entries.* Although matrix rigidity problems are notoriously hard, it seems that they were not extensively studied in the range of parameters that we need (i.e., rigidity $\omega(n^{3/2})$ for rank $\omega(n^{1/2})$).[23] Anyhow, here is its basic connection to our model.

Theorem 4.4 (reducing AN-complexity lower bounds to matrix rigidity): *If* T *is an* n*-by-*n *matrix that has rigidity* m^3 *for rank* m*, then the corresponding bilinear function* F *satisfies* $\mathtt{AN}(F) > m$.

[23] Added in Revision: Interestingly, a subsequent work of Dvir and Liu [3,4] shows that no Toeplitz matrix is rigid in the range of parameters sought by Valiant [31]. Specifically, they show that, for any constant $c > 1$, no Toeplitz matrix has rigidity n^c with respect to rank $n/\log n$ (see [4], which builds upon [3]). In contrast, the subsequent work of Goldreich and Tal [9] shows that almost all Toeplitz matrix have rigidity $\widetilde{\Omega}(n^3/r^2)$ with respect to rank $r \in [\sqrt{n}, n/32]$.

In particular, *if there exists an n-by-n Toeplitz matrix that has rigidity m^3 for rank m, then the corresponding bilinear function F satisfies* $\mathtt{AN}(F) > m$.

Proof: As a warm-up, we first prove that $\mathtt{AN}_2(F) > m$; that is, we prove a lower bound referring to depth-two multilinear circuits rather than to general multilinear circuits. Suppose towards the contradiction that $\mathtt{AN}_2(F) \leq m$, and consider the multilinear circuit that guarantees this bound. Without loss of generality,[24] it holds that $F(x,y) = H(F_1(x,y),...,F_{m-1}(x,y))$, where H is computed by the top gate and F_i is computed by its $i^{\rm th}$ child. W.l.o.g, the first m' functions (F_i's) are quadratic functions whereas the others are linear functions (in either x or y). Furthermore, each F_i depends on at most m variables. Since $H(F_1(x,y),...,F_{m-1}(x,y))$ is a syntactically bilinear polynomial (in x and y), it follows that it has the form

$$\sum_{i\in[m']} Q_i(x,y) + \sum_{(j_1,j_2)\in P} L_{j_1}(x)L_{j_2}(y), \tag{10}$$

where $P \subset [m'+1,m''] \times [m''+1,m-1]$ (for some $m'' \in [m'+1,m-2]$) and each Q_i and L_j depends on at most m variables. (Indeed, the same form was used in the proof of Theorem 2.3 (see Eq. (5)).) Furthermore, each of the L_j's is one of the auxiliary functions F_i's, which means that the second sum (in Eq. (10)) depends on at most $m-1$ different (linear) functions.

The key observation is that bilinear functions correspond to matrices; that is, the bilinear function $B : \mathrm{GF}(2)^{n+n} \to \mathrm{GF}(2)$ corresponds to the n-by-n matrix M such that the $(k,\ell)^{\rm th}$ entry of M equals 1 if and only if the monomial $x_k y_\ell$ is included in $B(x,y)$ (i.e., iff $B(0^{k-1}10^{n-k},0^{\ell-1}10^{n-\ell}) = 1$).[25] Now, observe that the matrix that corresponds to the first sum in Eq. (10) has less than m^3 one-entries (since the sum of the Q_i's depends on at most $m' \cdot m^2 < m^3$ variables), whereas the matrix that corresponds to the second sum in Eq. (10) has rank at most $m-1$ (since the sum $\sum_{(j_1,j_2)\in P} L_{j_1} L_{j_2}$, viewed as $\sum_{j_1\in[m-1]} L_{j_1} \cdot \sum_{j_2:(j_1,j_2)\in P} L_{j_2}$, corresponds to the sum of $m-1$ rank-1 matrices).[26] But this contradicts the hypothesis that T has rigidity m^3 for rank m, and so $\mathtt{AN}_2(F) > m$ follows.

Turning to the actual proof (of $\mathtt{AN}(F) > m$), which refers to multilinear circuits of arbitrary depth, we note that in the bilinear case the benefit of depth is very limited. This is so because nested composition is beneficial only when

---

[24] As in Construction 2.6, we may replace variables that feed directly into the top gate by 1-ary identity gates. That is, if $F(x,y) = H(F_1(x,y),...,F_{m'}(x,y), z_{m'+1}...,z_{m-1})$, where each z_i belongs either to x or to y, then we let $F(x,y) = H(F_1(x,y),...,F_{m-1}(x,y))$, where $F_i(x,y) = z_i$ for every $i \in [m'+1,m-1]$.

[25] In terms of Eq. (1), letting T denote the set of one-entries of M, it holds that $B(x,y) = \sum_{(k,\ell)\in T} x_k y_\ell$.

[26] That is, letting $L'_j(y) = \sum_{j_2:(j,j_2)\in P} L_{j_2}(y)$, we consider the sum $\sum_{j_1\in[m-1]} L_{j_1}(x) \cdot L'_{j_1}(y)$, and note that each term corresponds to a rank-1 matrix (i.e., the $(k,\ell)^{\rm th}$ entry of the $j_1^{\rm th}$ matrix equals $L_{j_1}(0^{k-1}10^{n-k}) \cdot L'_{j_1}(0^{\ell-1}10^{n-\ell})$).

it involves occurrence of the original variables (since terms in F_i that are product of auxiliary functions only can be moved from the expression for F_i to the expressions that use F_i; cf., Remark 3.5). In particular, without loss of generality, linear F_i's may be expressed in terms of the original variables only (since a linear F_j that feeds a linear F_i can be moved to feed the gates fed by F_i), whereas quadratic F_i's are expressed in terms of the original variables and possibly linear F_j's (since products of linear F_j's can be moved to the top gate). Thus, the expression for $F(x, y)$ is as in Eq. (10), except that here for every $(j_1, j_2) \in P$ either L_{j_1} or L_{j_2} is one of the auxiliary functions F_i's (whereas the other linear function may be arbitrary).[27] This suffices for completing the argument. Details follow.

Suppose towards the contradiction that $\mathtt{AN}(F) \leq m$, and consider a multilinear circuit that supports this bound. Each of the $m' \leq m$ gates in this circuit computes a bilinear (or linear) function of its feeding-inputs, which are a possible mix of (up to m) original variables and (up to $m-1$) outputs of other gates. This bilinear (or linear) function of the feeding-inputs can be expressed as a sum of monomials of the following three types, where F_i denotes the auxiliary function computed by the ith internal gate (and $F_0 = F$ is the function computed by the top gate).

1. Mixed monomials that consist of the product of a linear auxiliary function (i.e., an F_j) and an original variable. Such monomials cannot exist in the computation of linear functions.
2. Monomials that consist only of auxiliary functions F_j's: Such a monomial may be either a single bilinear (or linear) function or a product of two linear functions.[28]
   Without loss of generality, *such monomials exist only in the computation of the top gate* (and not in the computation for any other gate), because the computation of such monomials can be moved from the current gate to all gates fed by this gate (without effecting the number of variables that feed directly to these gates). Note that the arity of gates in the resulting circuit is at most $m + m$, where one term is due to the number of variables that feed directly into the gate and the other term is due to the total number of gates in the circuit.
   For example, if the monomial $F_k(x)F_\ell(y)$ appears in the expression computed by the jth internal gate (which computes $F_j(x, y)$) that feeds the ith gate (which computes $F_i(x, y)$), where possibly $i = 0$ (i.e., the jth gate feeds the top gate), then we can remove the monomial $F_k F_\ell$ from F_j and add it to F_i, which may require adding F_k and F_ℓ to the list of gates (or rather functions) that feed F_i. Ditto if $F_k(x, y)$ is a monomial of F_j. The process may be repeated till no internal gate contains a monomial that consists only of auxiliary functions.

[27] Actually, we can combine all products that involve F_i, see below.
[28] Since, as argued next, such monomials exist only in the top gate, it follows that (w.l.o.g.) they cannot be a single linear function, because the top gate must compute a homogeneous polynomial of degree 2.

3. **Monomials that contain only original variables.** Each quadratic (resp., linear) function computed by any gate has at most m^2 (resp., m) such monomials.

Hence, we obtain the general form for the computations of the top gate (which computes F) and the intermediate gates (which compute the auxiliary functions F_i's):

$$\begin{aligned} F(x,y) &= \sum_{(k,\ell)\in P_{0,1}} F_k(x)y_\ell + \sum_{(k,\ell)\in P_{0,2}} x_k F_\ell(y) \\ &\quad + \sum_{i\in S} F_i(x,y) + \sum_{(i,j)\in P_3} F_i(x)F_j(y) + \sum_{(i,j)\in P_{0,4}} x_i y_j \\ F_i(x,y) &= \sum_{(k,\ell)\in P_{i,1}} F_k(x)y_\ell + \sum_{(k,\ell)\in P_{i,2}} x_k F_\ell(y) + \sum_{(k,j)\in P_{i,4}} x_k y_j \\ F_i(z) &= \sum_{k\in S_i} z_k \end{aligned}$$

where the P's are subsets of $[m]^2$ (resp., the S's are subsets of $[m]$), and the F_i's (of arity at most $2m$) replace the original F_i's (per the "w.l.o.g."-clause of Item 2). Indeed, as asserted in Item 2, only the top gate contains monomials that are either auxiliary bilinear functions (corresponding to S) or products of auxiliary linear functions (corresponding to P_3).

Summing together all mixed monomials, *regardless of the gate to which they belong*, we obtain at most $m-1$ quadratic forms, where each quadratic form is the product of one of the auxiliary (linear) functions F_i and a linear combination (of an arbitrary number) of the original variables. Let us denote this sum by σ_1; that is,

$$\begin{aligned} \sigma_1 &= \sum_{i\in\{0,1,\ldots,m-1\}} \left(\sum_{(k,\ell)\in P_{i,1}} F_k(x)y_\ell + \sum_{(k,\ell)\in P_{i,2}} x_k F_\ell(y) \right) \\ &= \sum_k F_k(x) \cdot \sum_i \sum_{\ell:(k,\ell)\in P_{i,1}} y_\ell + \sum_\ell F_\ell(y) \cdot \sum_i \sum_{k:(k,\ell)\in P_{i,2}} x_k \end{aligned}$$

Adding to this sum (i.e., σ_1) the sum, denoted σ_2, of all monomials (computed by the top gate) that are a product of two linear F_i's (i.e., $\sigma_2 = \sum_{(i,j)\in P_3} F_i(x)F_j(y)$), we still have at most $m-1$ quadratic forms that are each a product of one of the auxiliary (linear) functions F_i and a linear combination of the original variables. (This uses the fact that $F_i \cdot F_j$ may be viewed as a product of F_i and the linear combination of the original variables given by the expression for F_j.)[29] These sums leave out the monomials that are a product of two original variables (i.e., the sum $\sum_{i\in\{0,1,\ldots,m-1\}} \sum_{(k,j)\in P_{i,4}} x_k y_j$). We stress

[29] Note that $\sigma_1+\sigma_2 = \sum_k F_k(x)\cdot L_k(y) + \sum_\ell L_\ell(x)\cdot F_\ell(y)$, where the L_i's are arbitrary linear functions (which may depend on an arbitrary number of variables in either x or y).

that sum $\sum_{i \in S} F_i(x, y)$ is not included here, since the monomials computed by these F_i's are already accounted by one of the foregoing three types (i.e., they either appear in the sum $\sigma_1 + \sigma_2$ or were left out as products of two variables).

Let T' denote matrix that corresponds to the $F' = \sigma_1 + \sigma_2$. Note that T' has rank at most $m - 1$ (since it is the sum of at most $m - 1$ rank-1 matrices, which correspond to the products of the different linear F_i's with arbitrary linear functions). Lastly, note that $F - F'$ equals $\sum_{i \in \{0,1,\ldots,m-1\}} \sum_{(k,j) \in P_{i,4}} x_k y_j$, which means that T' differs from T on at most m^3 entries. (Actually, the disagreement is smaller, since $|P_{i,4}| \leq \max_{m' \in [m-1]}\{m' \cdot (m - m') \leq (m/2)^2$.) This implies that $T = T' + (T - T')$ does not have rigidity m^3 for rank m, and the claim follows. ■

A Short Detour. Before proceeding, let us state the following result that is obtained by generalizing one of the observations used in the proof of Theorem 4.4.

Proposition 4.5 (on the depth of multilinear circuits that approximately achieve the AN-complexity): *Let F be ay t-linear function. Then, there exists a depth $t+1$ circuit with arity and size at most* $2 \cdot \texttt{AN}(F)$ *that computes F. That is, for any t-linear F, it holds that* $\texttt{AN}_{t+1}(F) \leq 2 \cdot \texttt{AN}(F)$.

Proof: Generalizing an observation made in the proof of Theorem 4.4, note that monomials in the expression for F_j that contain *only* auxiliary functions can be moved to the expressions of all functions that depend on F_j, while at most doubling the AN-complexity of the circuit (i.e., the arity of each gate grows by at most the number of gates).[30] Thus, without loss of generality, each auxiliary function F_j (computed by a internal gate) can be expressed in terms of input variables and auxiliary functions that are of smaller degree (than the degree of F_j). Hence, using induction on $i \geq 0$, it holds that gates that are at distance i from the top gate are fed by auxiliary functions of degree at most $t - i$. It follows that gates at distance t from the top gate are only fed by variables. Thus, the depth of multilinear circuits computing a t-linear function F and having AN-complexity $2 \cdot \texttt{AN}(F)$ needs not exceed $t + 1$. ■

Implications of the "Rigidity Connection" on $\texttt{AN}(F^{3,n}_{\texttt{tet}})$. In the original version of this work [11], we suggested to try to obtain an improved lower bound on the AN-complexity of the trilinear function $F^{3,n}_{\texttt{tet}}$ (see Eq. (3)) via a reduction to proving a rigidity lower bound for a *random* (or actually *any*) Toeplitz matrix. Recall that a Toeplitz matrix is a matrix $(t_{i,j})_{i,j \in [n]}$ such that $t_{i+1,j+1} = t_{i,j}$. The reduction, which is presented next, actually reduces proving lower bounds on $\texttt{AN}(F^{3,n}_{\texttt{tet}})$ to proving lower bounds on the AN-complexity of any bilinear function that corresponds to a Toeplitz matrix.

---

[30] This generalizes the claim made in Remark 3.5. Furthermore, as stated there, mixed gates are potentially beneficial. The observation made here is that this benefit (i.e., a mixed monomial) comes at the "cost" of using auxiliary functions of lower degree.

Proposition 4.6 (from $F_{\mathtt{tet}}^{3,n}$ to Toeplitz matrices): *If there exists an n-by-n Toeplitz matrix such that the corresponding bilinear function F satisfies* $\mathtt{AN}(F) \geq m$, *then* $\mathtt{AN}(F_{\mathtt{tet}}^{3,n}) = \Omega(m)$.

Indeed, a striking feature of this reduction is that a lower bound on an explicit function follows from a lower bound on *any* function in a natural class that contained exponentially many different functions.

Proof: For simplicity, assume that $n = 2n' + 1$ is odd, and consider the trilinear function $F_3 : (\mathrm{GF}(2)^{n'+1})^3 \to \mathrm{GF}(2)$ associated with the tensor $T_3 = \{(i_1, i_2, i_3) \in [[n']]^3 : \sum_j i_j \leq n'\}$, where $[[n']] \stackrel{\rm def}{=} \{0, 1, ..., n'\}$ (and $n' = \lfloor n/2 \rfloor$). Indeed, T_3 is a lightly padded version of one eighth of $T_{\mathtt{tet}}^{3,n}$. Observe that multilinear circuits for $F_{\mathtt{tet}}^{3,n}$ yield circuits of similar AN-complexity for F_3: For $y_{[[n']]}^{(j)} = (y_0^{(j)}, y_1^{(j)}, ..., y_{n'}^{(j)})$, the value of $F_3(y_{[[n']]}^{(1)}, y_{[[n']]}^{(2)}, y_{[[n']]}^{(3)})$ equals $F_{\mathtt{tet}}^{3,n}(0^{n'} y_{[[n']]}^{(1)}, 0^{n'} y_{[[n']]}^{(2)}, 0^{n'} y_{[[n']]}^{(3)})$, since $(i_1, i_2, i_3) \in T_3$ if and only if $(n' + i_1, n' + i_2, n' + i_3) \in T_{\mathtt{tet}}^{3,n}$. This means that we may modify each of the expressions used for $F_{\mathtt{tet}}^{3,n}$ by replacing the first n' variables in each variable-block with the value 0 (i.e., omit the corresponding monomials).[31]

The main observation is that *if* $F_3(x, y, z) = \sum_{(i,j,k) \in T_3} x_i y_j z_k$ *satisfies* $\mathtt{AN}(F_3) \leq m$, *then the same upper bound holds for any bilinear function that is associated with an* $(n' + 1)$-*by*-$(n' + 1)$ *triangular Toeplitz matrix* (i.e., $(t_{j,k})_{j,k \in [[n']]}$ such that $t_{j+1,k+1} = t_{j,k}$ and $t_{j,k} = 0$ if $j < k$). This is actually easier to see for their transpose—triangular Hankel matrices (i.e., $(h_{j,k})_{j,k \in [[n']]}$ such that $h_{j+1,k} = h_{j,k+1}$ and $h_{j,k} = 0$ if $j + k > n'$). The foregoing holds because any linear combination of the 1-slices of T_3 (i.e., the two-dimensional tensors $T_i' = \{(j, k) : (i, j, k) \in T_3\}$ for every $i \in [[n']]$) yields a triangular Hankel matrix, and all such matrices can be obtained by such a combination; that is, for every $I \subseteq [[n']]$, it holds that the matrix $(t_{j,k})_{j,k \in [[n']]}$ such that $t_{j,k} = |\{i \in I : (i, j, k) \in T_3\}| \bmod 2$ satisfies $t_{j,k+1} = t_{j+1,k}$ and $t_{j,k} = 0$ if $j + k > n'$, and each such matrix can be obtained by a choice of such an I (i.e., given a triangular Hankel matrix $(h_{j,k})_{j,k \in [[n']]}$, set I such that $|\{i \in I : (i, j, k) \in T_3\}| \equiv h_{j,k} \pmod 2$ holds).[32]

Finally, note that multilinear circuits for any bilinear function that is associated with a *triangular* Toeplitz matrix yields circuits of similar AN-complexity for *general* Toeplitz matrix. This holds because each Toeplitz matrix can be written as the sum of two triangular Toeplitz matrices (i.e., an upper-triangular one and a lower-triangular one). ■

[31] The opposite direction is equally simple: Just note that $F_{\mathtt{tet}}^{3,n}$ can be expressed as a sum of the values in the eight directions corresponding to $\{\pm 1\}^3$.

[32] Equivalently, we wish to set I such that $|\{i \in I : \chi(i, j, k) = 1\}| \equiv h_{0,j+k} \pmod 2$ holds, where $\chi(i, j, k) = 1$ if $(i, j, k) \in T_3$ (equiv., $i + j + k \leq n'$). Letting $\zeta_i \in \mathrm{GF}(2)$ represent whether $i \in I$, we solve the linear system $\sum_{i \in [[n']]} \chi(i, 0, j + k)\zeta_i \equiv h_{0,j+k}$ (mod 2) for $j + k \in [[n']]$. Note that the matrix corresponding to this linear system has full rank.

Hence, establishing an $\Omega(n^c)$ lower bound on $\mathtt{AN}(F_{\mathtt{tet}}^{3,n})$ reduces to establishing this bound for some Toeplitz matrix. This gives rise to the following open problems posed in [11] and resolved in [9].

Problem 4.7 (on the AN-complexity of Toeplitz matrices): *Prove that there exists an n-by-n Toeplitz matrix such that the corresponding bilinear function F satisfies $\mathtt{AN}(F) \geq n^c$, for some $c > 1/2$.*

(This was proved for $c = 0.6$ in [9, Cor. 1.4].) As we saw, Problem 4.7 would be resolved by

Problem 4.8 (on the rigidity of Toeplitz matrices): *For some $c > 1/2$, prove that there exists an n-by-n Toeplitz matrix T that has rigidity n^{3c} for rank n^c.*

(This was proved for $c = 0.6 - o(1)$ in [9, Thm. 1.2], whereas the improved bound for $c = 0.6$ (in [9, Cor. 1.4]) was established via "structured rigidity" as defined next.)

4.3 On Structured Rigidity

The proof of Theorem 4.4 shows that if a bilinear function F has AN-complexity at most m, then the corresponding matrix T can be written as a sum of a rank $m-1$ matrix T' and a matrix that has at most m^3 one-entries. However, even a superficial glance at the proof reveals that the matrix $T - T'$ is structured: It consists of the sum of m matrices such that the one-entries of each matrix are confined to some m-by-m rectangle. This leads us to the following definition.

Definition 4.9 (structured rigidity): *We say that a matrix T has* structured rigidity (m_1, m_2, m_3) for rank r *if for every matrix R of rank at most r and for every $I_1, ..., I_{m_1}, J_1, ..., J_{m_1} \subseteq [n]$ such that $|I_1| = \cdots = |I_{m_1}| = m_2$ and $|J_1| = \cdots = |J_{m_1}| = m_3$ it holds that $T - R \not\subseteq \bigcup_{k=1}^{m_1}(I_k \times J_k)$, where $M \subseteq S$ means that all non-zero entries of the matrix M reside in the set $S \subseteq [n] \times [n]$. We say that a matrix T has* structured rigidity m^3 for rank r *if T has structured rigidity (m, m, m) for rank r.*

Clearly, rigidity is a lower bound on structured rigidity (i.e., if T has rigidity m^3 for rank r, then T has structured rigidity m^3 for rank r), but (as shown below) this lower bound is not tight. Before proving the latter claim, we apply the notion of structured rigidity to our study.

Theorem 4.10 (reducing AN-complexity lower bounds to structured rigidity): *If T is an n-by-n matrix that has structured rigidity m^3 for rank m, then the corresponding bilinear function F satisfies $\mathtt{AN}(F) \geq m$.*

(As stated above, Theorem 4.10 follows by the very proof of Theorem 4.4.) In particular, *if there exists an n-by-n Toeplitz matrix that has structured rigidity m^3 for rank m, then the corresponding bilinear function F satisfies* $\mathtt{AN}(F) \geq m$. Hence, Problem 4.7 would be resolved by

Problem 4.11 (on the structured rigidity of Toeplitz matrices): *For some $c > 1/2$, prove that there exists an n-by-n Toeplitz matrix T that has structured rigidity n^{3c} for rank n^c.*

Indeed, the lower bound of $\Omega(n^{0.6})$ on the AN-complexity of (the bilinear functions that correspond to) most n-by-n Toeplitz matrices has been proved in [9] by establishing an analogous lower bound on the *structured rigidity* of these matrices, improving over a lower bound of $\widetilde{\Omega}(n^{0.6})$ established in [9] via an analogous lower bound on the standard notion of rigidity (see [9, Thm. 1.2] versus [9, Thm. 1.3]). This provides some weak empirical evidence for the speculation, made in the original version of this work [11], by which Problem 4.11 may be easier than Problem 4.8. This speculation was supported in [11] by the following separation result.

Theorem 4.12 (rigidity versus structured rigidity): *For any $m \in [n^{0.501}, n^{0.666}]$, consider a uniformly selected n-by-n Boolean matrix M with exactly $3mn$ ones. Then, with very high probability, M has structured rigidity m^3 for rank m.*

Note that M does not have rigidity $3nm \ll m^3$ for rank zero, let alone for rank m. Hence, the gap between structured rigidity and standard rigidity (for rank m) is a factor of at least $\frac{m^3}{3nm} = \Omega(m^2/n)$.

Proof: For each sequence $R, S_1, ..., S_m$ such that R has rank m and each $S_i \subseteq [n] \times [n]$ is an m-by-m square (generalized) submatrix (i.e., has the form $I_i \times J_i$ such that $|I_i|, |J_i| \leq m$), we shall show that

$$\Pr_{M \in \mathrm{GF}(2)^{n \times n}:|M|=3mn}\left[M - R \subseteq \bigcup_{i \in [m]} S_i\right] \leq 2^{-3nm}, \quad (11)$$

where M is a uniformly selected n-by-n matrix with exactly $3mn$ ones (and $M - R \subseteq S$ means that all non-zero entries of the matrix $M - R$ reside in the set $S \subseteq [n] \times [n]$). The theorem follows since the number of such sequences (i.e., a rank m matrix R and small submatrices $S_1, ..., S_m$) is smaller than $(2^{2n})^m \cdot \binom{n}{m}^{2m} \ll 2^{2nm+2m^2 \log n}$, where we specify a rank-m matrix by a sequence of m rank-1 matrices (equiv., pairs of subsets of $[n]$). Using $m^2 \log n < nm/4$ (equiv., $m = o(n/\log n)$), the foregoing quantity is upper-bounded by $2^{2.5nm}$. We shall also use $m \leq n^{2/3}/2$, which implies $m^3 \leq n^2/8$ and $3nm = o(n^2)$. In order to prove Eq. (11), we consider two cases

Case 1: R has at least $n^2/3$ one-entries. Since $3nm = o(n^2)$, it follows that $M - R$ has at least $n^2/4$ non-zero entries, but these cannot be covered by the $\bigcup_i S_i$, since the latter has at most $m^3 \leq n^2/8$ elements. Hence, $M - R \subseteq \bigcup_{i \in [m]} S_i$ never holds in this case, which means that the l.h.s. of Eq. (11) is zero.

Case 2: R has at most $n^2/3$ one-entries. In this case the union of the one-entries of R and $\bigcup_i S_i$, denoted U, covers at most half of a generic n-by-n matrix. Now, selecting $3nm$ random entries in the matrix, the probability that all entries reside in U at most $(1/2)^{3nm}$. But if some one-entry of M does not reside in U, then this entry is non-zero in $M - R$ but does not reside in $\bigcup_i S_i$. In this case, $M - R \not\subseteq \bigcup_{i\in[m]} S_i$ holds. Hence, Eq. (11) holds.

To rec-cap: Having established Eq. (11), and recalling the upper bound on the number of $(R, S_1, ..., S_m)$-sequences, we conclude that with probability at least $1 - 2^{2.5nm} \cdot 2^{-3nm} = 1 - 2^{-nm/2}$, the matrix M has structural rigidity (m, m, m) for rank m. ■

5 On Two Restricted Models

Focusing on our arithmetic circuit model, we consider two restricted versions of it: The first restricted model is of computation without cancellation, and the second is of computation that use only addition and multiplication gates while parametrizing their arity.

5.1 On Computing Without Cancellation

A natural model in the context of arithmetic computation is that of computing *without cancellations.*[33] We note that all our upper bounds (of Sect. 3) were obtained by computations that use no cancellations. Nevertheless, as one may expect, computations that use cancellation may be more efficient than computations that do not use it. In fact, obtaining such a separation result is quite easy. A striking example is provided by the bilinear function $F_{\text{had}}^{2,n}$ that corresponds to the Hadamard matrix $T_{\text{had}}^{2,n}$ (i.e., $T_{\text{had}}^{2,n} = \{(i,j) \in [n]^2 : \mathtt{ip}_2(i,j)\}$, where $n = 2^\ell$ and $\mathtt{ip}_2(i,j)$ is the inner product (mod 2) of the ℓ-bit binary expansions of $i-1$ and $j-1$).

Proposition 5.1 (computing $F_{\text{had}}^{2,n}$ without cancellation): *Computing* $F_{\text{had}}^{2,n}$ without cancellations *requires a circuit of AN-complexity $\Omega(n^{2/3})$, where the AN-complexity of circuits is as defined in Definition 2.2. In contrast, $F_{\text{had}}^{2,n}$ can be computed by a circuit of AN-complexity $\widetilde{O}(\sqrt{n})$* with cancellation*; actually,* $\mathtt{AN}_2(F_{\text{had}}^{2,n}) = O(\sqrt{n \log n})$.

Proof: We first prove the lower bound. Suppose that $F_{\text{had}}^{2,n}$ can be computed by a circuit of AN-complexity m that uses no cancellation. Following the argument in the proof of Theorem 4.4, we conclude that $T_{\text{had}}^{2,n}$ is a sum of at most m matrices that have m^2 one-entries each and at most m matrices of rank 1 (see

[33] This means that one considers the syntactic polynomial computed by the circuit (over a generic field) and requires that it equals the target polynomial when the field remains unspecified.

Footnote 29). Specifically, assuming that the first $m' < m$ auxiliary functions (i.e., F_i's) are bilinear functions, we observe that

$$F_{\text{had}}^{2,n}(x,y) = F_0(x,y) = \sum_{i=0}^{m'} Q_i(x,y) + \sum_{i=m'+1}^{m-1} L_i(x,y)F_i(x,y), \qquad (12)$$

where Q_i is a sum of the products of pairs of variables that appear in F_i and the L_i's are arbitrary linear functions (which may depend on an arbitrary number of variables in either x or y).[34] Hence, each Q_i corresponds to a tensor (or matrix) with at most m^2 one-entries, whereas each L_iF_i corresponds to a rectangular tensor.

The punchline is that, by the non-cancellation hypothesis, these rectangles (i.e., the L_iF_i's) must be pairwise disjoint and their one-entries must be one-entries also in $T_{\text{had}}^{2,n}$ (since they cannot be cancelled). But by Lindsey's Lemma (cf., e.g., [5, p. 88]) rectangles of area greater than n must contain zero-entries of $T_{\text{had}}^{2,n}$, which implies that each rectangle may have area at most n. It follows that the total area covered by all m tensors is at most $(m'+1) \cdot m^2 + (m-m') \cdot n < m^3 + mn$, whereas $T_{\text{had}}^{2,n}$ has $n^2/2$ one-entries. The main claim (i.e., $m = \Omega(n^{2/3})$) follows.

The secondary claim (i.e., $\mathtt{AN}(F_{\text{had}}^{2,n}) = \widetilde{O}(\sqrt{n})$) follows by the fact that $T_{\text{had}}^{2,n}$ has rank $\ell = \log_2 n$. The point is that any bilinear function F that corresponds to a rank r matrix can be computed as the sum of r functions that correspond to rectangular tensors, where each of these r functions can be computed as the product of two linear functions, and each linear function can be computed as the sum of $\sqrt{n/2r}$ functions that compute the sum of at most $\sqrt{2rn}$ variables. All in all, we use $1 + 2r \cdot \sqrt{n/2r}$ gates, which are each of arity $\sqrt{2rn}$. This yields a depth-two circuit of AN-complexity $\sqrt{2rn}+1$, where the top gate is a quadratic expression in $\sqrt{2rn}$ linear functions. ■

Computing $F_{\text{tet}}^{3,n}$ Without Cancellation. While we were unable to prove that $\mathtt{AN}(F_{\text{tet}}^{3,n}) = \omega(\sqrt{n})$, it is quite easy to prove such a lower bound for circuits that compute $F_{\text{tet}}^{3,n}$ without cancellation.

Proposition 5.2 (computing $F_{\text{tet}}^{3,n}$ without cancellation): *Computing $F_{\text{tet}}^{3,n}$ without cancellations requires a circuit of AN-complexity $\Omega(n^{2/3})$.*

(Again, recall that the AN-complexity of circuits is defined exactly as in Definition 2.2.)

[34] Recall that, w.l.o.g., gates that compute quadratic F_i's (for $i \in [m']$) may only feed into the top gate. Ditto for gates computing products of two linear F_i's (for $i \in [m'+1, m-1]$). Thus, $F_0 = Q_0 + \sum_{i\in[m']} F_i + \sum_{i=m'+1}^{m-1} L_{0,i}F_i$, where Q_0 is a sum of the products of pairs of variables that appear in F_0, the $L_{0,i}$'s are arbitrary linear functions, and for $i > m'$ the linear function F_i is computed by an internal gate. Furthermore, for every $i \in [m']$, it holds that $F_i = Q_i + \sum_{j=m'+1}^{m-1} L_{i,j}F_j$, where Q_i is a sum of the products of pairs of variables that appear in F_i, the $L_{i,j}$'s are arbitrary linear functions, and for $j > m'$ the linear function F_j is computed by an internal gate. Letting $L_j = \sum_{i=0}^{m'} L_{i,j}$, we get Eq. (12).

Proof: Proceeding as in the proof of Proposition 5.1, we consider the top gate of a circuit (with m gates) that computes $F_{\texttt{tet}}^{3,n}$ without cancellations. Here, we can write $F_{\texttt{tet}}^{3,n}$ as

$$F_0 = \sum_{i=0}^{m'} C_i + \sum_{i=m'+1}^{m'+m''} L_i F_i + \sum_{i=m'+m''+1}^{m'+m''+m'''} Q_i F_i\,, \tag{13}$$

where $m' + m'' + m''' \leq m - 1$, the cubic function C_i is a sum of the products of triples of variables that appear in the cubic function F_i (for $i \in [0, m']$), the L_i's (resp., Q_i's) are arbitrary linear (resp., quadratic) functions (which may depend on an arbitrary number of variables (from adequate variable-blocks)), and the other F_i's are either quadratic (for $i \in [m'+1, m'+m'']$) or linear (for $i \in [m'+m''+1, m'+m''+m''']$).[35] Combining the two last summations in Eq. (13), we obtain

$$F_0 = \sum_{i=0}^{m'} C_i + \sum_{i=m'+1}^{m-1} L'_i Q'_i \tag{14}$$

where the C_i's are as in Eq. (13), and the L'_i's (resp., Q'_i's) are arbitrary linear (resp., quadratic) functions (which may depend on an arbitrary number of variables (from adequate variable-blocks)). Note that C_i corresponds to a tensor with one-entries that are confined to a m-by-m-by-m box, and each $L'_i Q'_i$ corresponds to a tensor that is the outer product of a subset of $[n]$ and a subset of $[n]^2$. By the non-cancellation condition, *all these tensors are disjoint, and none may contain a zero-entry of* $T_{\texttt{tet}}^{3,n}$.

We consider the boundary of the tensor $T_{\texttt{tet}}^{3,n}$ (i.e., the set of one-entries that neighbor zero-entries), and consider the contributions of the aforementioned tensors to covering this boundary (without covering zero-entries of $F_{\texttt{tet}}^{3,n}$). We will upper bound this contribution by $m^3 + mn$, and the claim will follow since the size of the boundary is $\Omega(n^2)$.

Actually, we shall consider covering the upper-boundary of $T_{\texttt{tet}}^{3,n}$, defined as the part of the boundary that resides in $[n/2, n]^3$. In other words, the upper-boundary consists of all points $(i_1, i_2, i_3) \in [n/2, n]$ such that $i_1 + i_2 + i_3 = 2n$, and it has size $\Omega(n^2)$.

---

[35] Recall that, w.l.o.g., gates that compute cubic F_i's (for $i \in [m']$) may only feed into the top gate. Ditto for gates computing products of linear F_i's and quadratic F_i's (for $i \in [m'+1, m-1]$). Thus, $F_0 = C_0 + \sum_{i\in[m']} F_i + \sum_{i=m'+1}^{m'+m''} L_{0,i} F_i + \sum_{i=m'+m''+1}^{m'+m''+m''} Q_{0,i} F_i$, where C_0 is a sum of the products of triples of variables that appear in F_0, the $L_{0,i}$'s (resp., $Q_{0,i}$'s) are arbitrary linear (resp., quadratic) functions, and for $i > m'$ the quadratic (resp., linear) function F_i is computed by an internal gate. Furthermore, for every $i \in [m']$, it holds that $F_i = C_i + \sum_{j=m'+1}^{m'+m''} L_{i,j} F_j + \sum_{j=m'+m''+1}^{m'+m''+m''} Q_{i,j} F_j$, where C_i is a sum of the products of triples of variables that appear in F_i, the $L_{i,j}$'s (resp., $Q_{i,j}$'s) are arbitrary linear (resp., quadratic) functions, and for $j > m'$ the quadratic (resp., linear) function F_j is computed by an internal gate. Letting $L_j = \sum_{i=0}^{m'} L_{i,j}$ and $Q_j = \sum_{i=0}^{m'} Q_{i,j}$, we get Eq. (13).

We first observe that the tensor corresponding to each C_j can cover at most m^2 points of the upper-boundary, because this tensor is confined to an m-by-m-by-m box $I'_j \times I''_j \times I'''_j$ and for each $(i_1, i_2) \in I'_j \times I''_j$ there exists at most one i_3 such that (i_1, i_2, i_3) resides in the upper-boundary. Hence, the contribution of $\sum_{j=0}^{m'} C_j$ to the cover is at most m^3.

Turning to the tensors that correspond to the L_jQ_j's, we note that (w.l.o.g.) each such tensor has the form $I'_j \times I''_j$, where $I'_j \subseteq [n]$ and $I''_j \subseteq [n]^2$. We first observe that only the largest $i_1 \in I'_j$ can participate in (a point that resides in) the upper-boundary, because if $(i_1, i_2, i_3) \in I'_j \times I''_j$ participates in the upper-boundary and $i'_1 > i_1$, then (i'_1, i_2, i_3) must be a zero-entry of $T^{3,n}_{\mathtt{tet}}$ (and contradiction is reached in case $i'_1 \in I'_j$, since then $(i'_1, i_2, i_3) \in I'_j \times I''_j$). Next, fixing the largest $i_1 \in I'_j$, we observe that the upper-boundary contains at most n points of the form $(i_1, \cdot, \cdot)$. Hence, the contribution of $\sum_{j=m'+1}^{m-1} L_jQ_j$ to the cover is at most mn.

Having shown that the union of the aforementioned tensors can cover at most $m^3 + mn$ points in the upper-boundary, the claim follows since the size of the upper-boundary is $\Omega(n^2)$. ■

5.2 Addition and Multiplication Gates of Parameterized Arity

In continuation to Definition 2.2, we consider a restricted complexity measure that refers only to multilinear circuits that use standard addition and multiplication gates. Needless to say, the multiplication gates in a multilinear circuit computing a t-linear function have arity at most t, whereas the arity of the addition gates is accounted for in our complexity measure. Furthermore, in our restricted complexity measure we do *not* count multiplication gates that are *fed by variables only*. For sake of clarify, we spell out the straightforward adaptation of Definition 2.2:

Definition 5.3 (the complexity of multilinear circuits with standard gates): *A* standard multilinear circuit *is a multilinear circuit* (as in Definition 2.2) *having only addition and multiplication gates, and its complexity is the maximum between the arity of its gates and the number of its non-trivial gates, where the* trivial gates *are multiplication gates that are* fed by variables only. *The* restricted complexity *of a multilinear function F, denoted* $\mathtt{RC}(F)$, *is the minimum complexity of a standard multilinear circuit that computes F.*

Indeed, we avoided introducing a depth-two version of Definition 5.3, because the model seems restricted enough as is. Note that for every t-linear function F, it holds that $\mathtt{AN}(F) \leq t \cdot \mathtt{RC}(F)$, since trivial multiplication gates can be eliminated by increasing the arity of the circuit (in the general model) by a factor of at most t.[36]

[36] In a gate that is fed by a trivial multiplication-gate, the argument representing the trivial gate's output is replaced by the (up to) t input variables feeding this trivial gate.

5.2.1 The Restricted Model Separates $F_{\mathtt{all}}^{t,n}$ and $F_{\mathtt{diag}}^{t,n}$ from $F_{\mathtt{leq}}^{2,n}$. As stated (implicitly) in Sect. 3.2, it holds that $\mathtt{RC}(F_{\mathtt{all}}^{t,n}) \leq t\sqrt{n}+1$ and $\mathtt{RC}(F_{\mathtt{diag}}^{t,n}) \leq t\sqrt{n}$. We show that this upper bound does not hold for $F_{\mathtt{leq}}^{2,n}$. We start with a general result.

Theorem 5.4 (the restricted complexity of bilinear functions is lower-bounded by the parameters of matrix rigidity): *Let $F : (\mathrm{GF}(2)^n)^2 \to \mathrm{GF}(2)$ be a bilinear function with a corresponding tensor $T \subseteq [n]^2$. If T has rigidity s with respect to rank $r > 1$, then* $\mathtt{RC}(F) \geq \min(r, \sqrt{s})$.

As shown in Proposition 5.5, the tensor $T_{\mathtt{leq}}^{2,n}$ has rigidity $\Omega(n^2/r)$ with respect to rank r, so letting $r = n^{2/3}$, we obtain $\mathtt{RC}(F_{\mathtt{leq}}^{2,n}) = \Omega(n^{2/3})$, since $\sqrt{n^2/r} = n^{(2-(2/3))/2} = r$. Also, since a random n-by-n matrix has rigidity $\Omega(n^2)$ with respect to rank $\Omega(n)$, it follows that for almost all bilinear functions $F : \mathrm{GF}(2)^{n+n} \to \mathrm{GF}(2)$ it holds that $\mathtt{RC}(F) = \Omega(n)$. The latter lower bound is tight, since (for any $t \geq 1$) any t-linear function F satisfies $\mathtt{RC}(F) \leq n^{t/2}$ (via a multilinear formula with $n^{t/2}$ addition gates, each of arity $n^{t/2}$, that sum-up all the relevant monomials).

Proof: Let $m \stackrel{\text{def}}{=} \mathtt{RC}(F)$ and suppose that $m < \sqrt{s}$, since otherwise the claim follows (i.e., $\mathtt{RC}(F) \geq \min(r, \sqrt{s})$). Consider a standard multilinear circuit that computes F with m' addition gates of arity at most m and m'' *non-trivial* multiplication gates, where $m' + m'' \leq m$. Note that the top gate cannot be a multiplication gate, because such a multilinear circuit can only compute bilinear functions that correspond to rank-1 matrices. Also note that there exists exactly one multiplication gate on each path from the top gate to a variable, since F is bilinear, and that this gate is trivial if and only if it is the last gate on this path. Hence, the circuit, which is a directed acyclic graph (DAG) rooted at the top gate, can be decomposed into a top layer that consists of a DAG of addition gates, an intermediate layer of multiplication gates, and a bottom layer that consists of a DAG of addition gates and variables (which feeds linear functions to the multiplication gates). We note that the number of trivial multiplication gates that feed the top DAG is at most m^2, because this DAG has $m' \leq m$ addition gates each of in-degree at most m.

We truncate the foregoing circuit at the trivial multiplication gates (which compute products of variables), obtaining a new circuit that computes a bilinear function F'; that is, $F - F'$ is the sum of the variable-products computed by the trivial multiplication gates. This new circuit has no trivial gates and it has m'' non-trivial multiplication gates (each computing a bilinear function that corresponds to a rank-1 matrix). Hence, the corresponding tensor, denoted T', has rank at most m'' (since it is the sum of m'' rank-1 matrices), whereas $|T + T'| \leq m^2$ (since $T + T'$ corresponds to the function $F - F'$, which is the sum of at most m^2 products of variables). We consider two cases:

1. If $m'' \leq r$, then T' has rank at most r, and we derive a contradiction to the hypothesis that T has rigidity s with respect to rank r, since $|T+T'| \leq m^2 < s$ (where the last inequality uses $m < \sqrt{s}$).

2. Otherwise, $m'' \geq r$, and it follows that $m \geq r$.

The claim follows (i.e., $\mathtt{RC}(F) \geq \min(r, \sqrt{s})$). ∎

Proposition 5.5 (a bound on the rigidity of $T^{2,n}_{\mathtt{leq}}$): *For every $r < n/O(1)$, the tensor $T^{2,n}_{\mathtt{leq}}$ (of Eq. (2)) has rigidity at least $\Omega(n^2/r)$ with respect to rank r.*

The rigidity lower bound is quite tight, since $T^{2,n}_{\mathtt{leq}}$ is $O(1/r)$-close to $\sum_{k\in[r]}(I_k \times J_k)$, where $I_k = \{(k-1)n/r + 1, ..., kn/r\}$ and $J_k = \{kn/r + 1, ..., n\}$, for every $k \in [r]$. (This is the case since $\sum_{k\in[r]}(I_k \times J_k) \subseteq T^{2,n}_{\mathtt{leq}} \subseteq \sum_{k\in[r]}(I_k \times J_{k-1})$, and $\sum_{k\in[r]} |I_k \times (J_{k-1} - J_k)| = n^2/r$.)

Proof: For a constant $c > 1$ to be determined later, we consider any $r < n/c$. We shall prove that any matrix $R = (R_{i,j})_{i,j\in[n]}$ of rank r is $\Omega(1/r)$-far from $T \stackrel{\text{def}}{=} T^{2,n}_{\mathtt{leq}}$; that is, $|R + T| = \Omega(n^2/r)$.

Let R be an arbitrary matrix of rank at most r. We say that $i \in [n]$ is good if $|\{j \in [n] : R_{i,j} \neq T_{i,j}\}| < n/cr$. The claim of the proposition reduces to proving that at least half of $i \in [n]$ are not good, since in this case R disagrees with T on at least $\frac{n}{2} \cdot \frac{n}{cr} = \frac{n^2}{2cr}$ entries. It is thus left to prove the latter claim.

Let G denote the set of good $i \in [n]$, and supposed towards the contradiction that $|G| > n/2$. For $c' \in [1, c/2]$ to be (implicitly) determined later, select $c'r$ indices $i_1, ..., i_{c'r} \in G$ such that for every $k \in [c'r-1]$ it holds that $i_{k+1} > i_k + \frac{n}{2c'r}$. Let us denote the i_k^{th} row of T by v_k, and the i_k^{th} row of R by w_k. Then, for a random non-empty set $K \subseteq [c'r]$, the following two conditions hold:

1. With probability greater than $1-2^{-r}$, the vector $\left(\sum_{k\in K} v_k \bmod 2\right)$ has weight greater than $n/6$.
   The claim follows from the structure of T (i.e., $v_k = 0^{i_k-1}1^{n-i_k+1}$) and the distance between the i_k's. Specifically, for a random K, the weight of the vector $\left(\sum_{k\in K} v_k \bmod 2\right)$ is distributed as $\sum_{j\in[c'r]}(i_{j+1} - i_j) \cdot X_j$, where $i_{c'r+1} \stackrel{\text{def}}{=} n + 1$ and $X_j = X_j(K) \stackrel{\text{def}}{=} \sum_{k\in K} T_{i_k,i_j} \bmod 2$ indicates the parity of the elements selected in column i_j (which equals the parity in all columns in $[i_j, i_{j+1} - 1]$). Thus, $X_j = \left(\sum_{k\leq j} Y_k \bmod 2\right)$, where $Y_k = 1$ if $k \in K$ and $Y_k = 0$ otherwise, which implies that the X_j's are uniformly and indentially distributed in $\{0, 1\}$. For sufficiently large c', we have $\Pr\left[\sum_{j\in[c'r-1]} X_j > c'r/3\right] > 1 - 2^{-r}$, and the claim follows since $\sum_{j\in[c'r]}(i_{j+1} - i_j) \cdot X_j$ is greater than $\frac{n}{2c'r} \cdot \sum_{j\in[c'r]} X_j$ (and $\frac{n}{2c'r} \cdot \frac{c'r}{3} = n/6$).
2. With probability at least 2^{-r}, the vector $\left(\sum_{k\in K} w_k \bmod 2\right)$ has weight 0.
   This follows from the rank of R. Specifically, consider a maximal set of independent vectors among the $w_1, ..., w_{c'r}$, and denote the corresponding set of indices by I. Then, $\Pr_K\left[\sum_{k\in K} w_k = 0\right] = 2^{-|I|} \geq 2^{-r}$, which can be seen by first selecting a random $K' \subseteq ([c'r] \setminus I)$, and then (for any outcome K') selecting a random $K'' \subseteq ([c'r] \cap I)$.

Combining (1) and (2), it follows that there exists non-empty set $K \subseteq [c'r]$ such that the vector $\sum_{k\in K} v_k$ has weight greater than $n/6$ but the vector $\sum_{k\in K} w_k$ has weight 0. But this is impossible because, by the hypothesis that all i_k's are good, the distance between these two vectors is at most $|K| \cdot \frac{n}{cr} \leq c'r \cdot \frac{n}{cr} < n/6$, where the last inequality require selecting $c > 6c'$. The claim (that $|G| \leq n/2$) follows. ■

Corollary 5.6 (lower bound on the restricted complexity of $F^{2,n}_{\mathtt{leq}}$): $\mathtt{RC}(F^{2,n}_{\mathtt{leq}}) = \Omega(n^{2/3})$.

Indeed, Corollary 5.6 follows by combining Theorem 5.4 and Proposition 5.5, while using $r = n^{2/3}$ and $s = \Omega(n^2/r)$. The resulting lower bound is tight:

Proposition 5.7 (upper bound on the restricted complexity of $F^{2,n}_{\mathtt{leq}}$): $\mathtt{RC}(F^{2,n}_{\mathtt{leq}}) = O(n^{2/3})$.

Proof: Consider a partition of $[n]^2$ into $n^{4/3}$ squares, each with side $s = n^{1/3}$: For $i, j \in [n/s]$, let $S_{i,j} = [(i-1)s+1, is] \times [(j-1)s+1, js]$, and note that $\bigcup_{i<j} S_{i,j} \subset T^{2,n}_{\mathtt{leq}} \subset \bigcup_{i\leq j} S_{i,j}$. Thus, $F^{2,n}_{\mathtt{leq}}$ can be computed by computing separately the contribution of the $n/s = n^{2/3}$ diagonal squares and the contribution of the squares that are above the diagonal; that is,

$$F^{2,n}_{\mathtt{leq}}(x,y) = \sum_{i\in[n^{2/3}]} \sum_{(k,\ell)\in S_{i,i}:k\leq\ell} x_k y_\ell + \sum_{i<j} \sum_{(k,\ell)\in S_{i,j}} x_k y_\ell.$$

The contribution of the square $S_{i,i}$ can be computed as the sum of its relevant $r \stackrel{\text{def}}{=} \binom{s}{2} + s < n^{2/3}$ entries, which means that the sum of the contribution of all $n^{2/3}$ diagonal squares consists of less than $n^{4/3}$ monomials. This sum can be computed by $n^{2/3}+1$ addition gates, each of arity $n^{2/3}$. (We also use $n^{2/3} \cdot r < n^{4/3}/2$ trivial multiplication gates, but these are not counted.)

The contribution of the above-diagonal squares can be computed by writing $\bigcup_{i<j} S_{i,j}$ as $\bigcup_{i\in[n/s]} \left(L_i \times \bigcup_{j>i} L_j\right)$, where $L_i = [(i-1)s+1, is]$. Hence, the total contribution of the off-diagonal squares is

$$\begin{aligned}\sum_{i<j} \sum_{(k,\ell)\in S_{i,j}} x_k y_\ell &= \sum_{i\in[n/s]} \left(\sum_{k\in L_i} x_k\right) \cdot \sum_{j>i} \left(\sum_{\ell\in L_j} y_\ell\right) \\ &= \sum_{i\in[n/s]} F_i(x) \cdot \sum_{j>i} G_j(y),\end{aligned}$$

where each of the F_i's and G_j's can be computed by an addition gate of arity $s = n^{1/3}$, whereas $\sum_{i\in[n/s]} F_i(x) \cdot \sum_{j>i} G_j(y)$ can be computed using $n^{2/3}+1$ addition gates of arity $n^{2/3}$ (and $n^{2/3}$ multiplication gates, each of arity 2). Hence, the total contribution of the off-diagonmal squares can be computed by $4 \cdot n^{2/3} + 1$ gates each having arity at most $n^{2/3}$. The claim follows. ■

Digest: The proof of Proposition 5.7 uses two different strategies for computing a generic bilinear form of $m+m$ inputs. The first strategy, employed for each of the diagonal squares, is to compute the sum of the $r \leq m^2$ relevant input-pairs using a single addition gate of arity r. The second strategy, employed for summing-up the total contribution of the non-diagonal squares, is to use $m+1$ addition gates (of arity m) and m multiplication gates (of arity two). Specifically, $\sum_{i,j\in[m]:b_{i,j}=1} u_i v_j$ is computed as $\sum_{i\in[m]} u_i \cdot \sum_{j\in[m]:b_{i,j}=1} v_j$.

Added in Revision: A Lower Bound on the Restricted Complexity of $F_{\mathtt{tet}}^{3,n}$. Combining [9, Thm. 1.2] with Theorem 5.4, we get $\mathtt{RC}(F_{\mathtt{tet}}^{3,n}) = \widetilde{\Omega}(n^{3/4})$. This follows because by [9, Thm. 1.2] almost all n-by-n Toeplitz matrices have rigidity $\widetilde{\Omega}(n^3/r^2)$ with respect to rank $r \in [\sqrt{n}, n/32]$, and (by Theorem 5.4) each corresponding bilinear function F satisfies $\mathtt{RC}(F) \geq \min(r, \widetilde{\Omega}(n^{3/2}/r)) = \widetilde{\Omega}(n^{3/4})$ (using $r = n^{3/4}$). The bound for $F_{\mathtt{tet}}^{3,n}$ follows analogously to Proposition 4.6.

5.2.2 On the Restricted Complexity of Almost All t-Linear Functions

Recall that for every t-linear function F, it holds that $\mathtt{RC}(F) = O(n^{t/2})$, by a circuit that merely adds all relevant monomials. We prove that for almost all t-linear functions this upper bound is tight up to a logarithmic factor.

Proposition 5.8 (a lower bound on the restricted complexity of almost all t-linear functions): *For all* $t = t(n)$, *almost all t-linear functions* $F : (\mathrm{GF}(2)^n)^t \to \mathrm{GF}(2)$ *satisfy* $\mathtt{RC}(F) = \Omega(n^{t/2}/\sqrt{\log(n^t)})$.

Proof: We upper-bound the number of standard multilinear circuits of restricted complexity m. Each such circuit corresponds to a DAG with m vertices, each representing either an addition gate or a (non-trivial) multiplication gate. In addition, each of these non-trivial gates may be fed by some variables or trivial multiplication gates (which are not part of this DAG), but the number of such gate-entries is at most m and each is selected among at most $(n+1)^t$ possibilities (since there are $(n+1)^t$ possible multilinear monomials). Thus, the number of such circuits is at most

$$2^{\binom{m}{2}} \cdot 2^m \cdot \binom{(n+1)^t}{m}^m \tag{15}$$

where $2^{\binom{m}{2}}$ upper bounds the number of m-vertex DAGs, 2^m accounts for choice of the gate types, and $\binom{(n+1)^t}{m}$ accounts for the choice of "DAG-external inputs" to each gate. Clearly, Eq. (15) is upper-bounded by $((n+1)^t)^{m^2} = \exp(tm^2 \log n)$, whereas the number of t-linear functions is 2^{n^t}. The claim follows. ■

Acknowledgments. We are grateful to Or Meir for extremely helpful discussions, and to Avishay Tal for many suggestions for improving the presentation. Research was partially done while O.G. visited the IAS.

References

1. Ajtai, M.: Σ_1^1-formulae on finite structures. Ann. Pure Appl. Logic **24**(1), 1–48 (1983)
2. Babai, L.: Random oracles separate PSPACE from the polynomial-time hierarchy. IPL **26**, 51–53 (1987)
3. Dvir, Z., Liu, A.: Fourier and circulant matrices are not rigid. In: 34th CCC, pp. 17:1–17:23 (2019). arXiv:1902.07334 [math.CO]
4. Dvir, Z., Liu, A.: Fourier and circulant matrices are not rigid. To appear in TOC, special issue of 34th CCC. See also ECCC, TR19-129, September 2019
5. Erdos, P., Spencer, J.: Probabilistic Methods in Combinatorics. Academic Press Inc., New York (1974)
6. Furst, M.L., Saxe, J.B., Sipser, M.: Parity, circuits, and the polynomial-time hierarchy. Math. Syst. Theory **17**(1), 13–27 (1984). Preliminary version in 22nd FOCS (1981)
7. Goldreich, O.: Computational Complexity: A Conceptual Perspective. Cambridge University Press, Cambridge (2008)
8. Goldreich, O.: Improved bounds on the AN-complexity of multilinear functions. In: ECCC, TR19-171 (2019)
9. Goldreich, O., Tal, A.: Matrix rigidity of random Toeplitz matrices. Comput. Complex. **27**(2), 305–350 (2018). Preliminary versions in 48th STOC (2016) and ECCC TR15-079 (2015)
10. Goldreich, O., Tal, A.: On constant-depth canonical Boolean circuits for computing multilinear functions. In: ECCC, TR17-193 (2017)
11. Goldreich, O., Wigderson, A.: On the size of depth-three Boolean circuits for computing multilinear functions. In: ECCC, TR13-043 (2013)
12. Hastad, J.: Almost optimal lower bounds for small depth circuits. In: Micali, S. (ed.) Advances in Computing Research: A Research Annual, (Randomness and Computation), vol. 5, pp. 143–170 (1989). Extended abstract in 18th STOC (1986)
13. Hastad, J.: Computational Limitations for Small Depth Circuits. MIT Press, Cambridge (1987)
14. Hastad, J., Jukna, S., Pudlak, P.: Top-down lower bounds for depth-three circuits. Comput. Complex. **5**(2), 99–112 (1995)
15. Hrubes, P., Rao, A.: Circuits with medium fan-in. In: ECCC, TR14-020 (2014)
16. Jukna, S.: Boolean Function Complexity: Advances and Frontiers. Algorithms and Combinatorics, vol. 27. Springer, Heidelberg (2012). https://doi.org/10.1007/978-3-642-24508-4
17. Karchmer, M., Wigderson, A.: Monotone circuits for connectivity require super-logarithmic depth. SIAM J. Discret. Math. **3**(2), 255–265 (1990)
18. Kushilevitz, E., Nisan, N.: Communication Complexity. Cambridge University Press, Cambridge (1997)
19. Lokam, S.V.: Complexity lower bounds using linear algebra. Found. Trends Theor. Comput. Sci. **4**, 1–155 (2009)
20. van Melkebeek, D.: A survey of lower bounds for satisfiability and related problems. Found. Trends Theor. Comput. Sci. **2**, 197–303 (2007)
21. Nisan, N.: Pseudorandom bits for constant depth circuits. Combinatorica **11**(1), 63–70 (1991)
22. Nisan, N., Wigderson, A.: Hardness vs randomness. J. Comput. Syst. Sci. **49**(2), 149–167 (1994). Preliminary version in 29th FOCS (1988)

23. Nisan, N., Wigderson, A.: Lower bound on arithmetic circuits via partial derivatives. Comput. Complex. **6**, 217–234 (1996)
24. Raz, R.: Tensor-rank and lower bounds for arithmetic formulas. In: Proceeding of the 42nd STOC, pp. 659–666 (2010)
25. Raz, R., Yehudayoff, A.: Lower bounds and separations for constant depth multilinear circuits. In: ECCC, TR08-006 (2008)
26. Razborov, A.: Lower bounds on the size of bounded-depth networks over a complete basis with logical addition. Matematicheskie Zametki **41**(4), 598–607 (1987). (in Russian). English translation in Math. Notes Acad. Sci. USSR **41**(4), 333–338 (1987)
27. Savitch, W.J.: Relationships between nondeterministic and deterministic tape complexities. JCSS **4**(2), 177–192 (1970)
28. Shaltiel, R., Viola, E.: Hardness amplification proofs require majority. SIAM J. Comput. **39**(7), 3122–3154 (2010). Extended abstract in 40th STOC (2008)
29. Smolensky, R.: Algebraic methods in the theory of lower bounds for Boolean circuit complexity. In: 19th STOC, pp. 77–82 (1987)
30. Strassen, V.: Vermeidung von Divisionen. J. Reine Angew. Math. **264**, 182–202 (1973)
31. Valiant, L.G.: Graph-theoretic arguments in low-level complexity. In: Gruska, J. (ed.) MFCS 1977. LNCS, vol. 53, pp. 162–176. Springer, Heidelberg (1977). https://doi.org/10.1007/3-540-08353-7_135
32. Valiant, L.G.: Exponential lower bounds for restricted monotone circuits. In: 15th STOC, pp. 110–117 (1983)
33. Vazirani, U.V.: Efficiency considerations in using semi-random sources. In: 19th STOC, pp. 160–168 (1987)
34. Yao, A.C.: Separating the polynomial-time hierarchy by oracles. In: 26th FOCS, pp. 1–10 (1985)

On the Communication Complexity Methodology for Proving Lower Bounds on the Query Complexity of Property Testing

Oded Goldreich

Abstract. We consider the methodology for proving lower bounds on the query complexity of property testing via communication complexity, which was put forward by Blais, Brody, and Matulef (*Computational Complexity*, 2012). They provided a restricted formulation of their methodology (via "simple combining operators") and also hinted towards a more general formulation, which we spell out in this paper.

A special case of the general formulation proceeds as follows: In order to derive a lower bound on testing the property Π, one presents a mapping F of pairs of inputs $(x, y) \in \{0,1\}^{n+n}$ for a two-party communication problem Ψ to $\ell(n)$-bit long inputs for Π such that $(x, y) \in \Psi$ implies $F(x, y) \in \Pi$ and $(x, y) \notin \Psi$ implies that $F(x, y)$ is far from Π. Let $f_i(x, y)$ be the ith bit of $F(x, y)$, and suppose that B is an upper bound on the (deterministic) communication complexity of each f_i and that C is a lower bound on the randomized communication complexity of Ψ. Then, testing Π requires at least C/B queries.

The foregoing formulation is generalized by considering randomized protocols (with small error) for computing the f_i's. In contrast, the restricted formulation (via "simple combining operators") requires that each $f_i(x, y)$ be a function of x_i and y_i only, and uses $B = 2$ for the straightforward computation of f_i.

We show that the general formulation cannot yield significantly stronger lower bounds than those that can be obtained by the restricted formulation. Nevertheless, we advocate the use of the general formulation, because we believe that it is easier to work with. Following Blais *et al.*, we also describe a version of the methodology for nonadaptive testers and one-way communication complexity.

An early version of this work appeared as TR13-073 of *ECCC*, and served as basis for [11, Sec. 7.3]. The current revision is quite minimal and presents a few expositional improvements, especially in the introduction. Some intuitive clarifications were added in Sect. 7.

1 Introduction

In the last couple of decades, the area of property testing has attracted much attention (see, e.g., [8,21,22]).[1] Loosely speaking, property testing typically

[1] Added in revision: See also the more recent textbook [11]. In fact, the current work served as basis for [11, Sec. 7.3].

O. Goldreich (Ed.): Computational Complexity and Property Testing, LNCS 12050, pp. 87–118, 2020.
https://doi.org/10.1007/978-3-030-43662-9_7

refers to sub-linear time probabilistic algorithms for deciding whether a given object has a predetermined property or is far from any object having this property. Such algorithms, called testers, obtain local views of the object by performing queries; that is, the object is seen as a function and the testers get oracle access to this function (and thus may be expected to work in time that is sub-linear in the size of the object).

In 2011, Blais, Brody, and Matulef enriched the study of property testing by presenting a connection between property testing and communication complexity [4]. Specifically, they presented a methodology for proving lower bounds on the query complexity of property testing by relying on lower bounds on two-party communication complexity problems.

Encountering their work [4], we were quite surprised. Firstly, this connection seems unexpected, since property testing problems have no topology that can be naturally 2-partitioned to fit the two-party setting of communication complexity. Nevertheless, using this methodology, the authors of [4] were able to resolve a fair number of open problems, some of which have escaped our own attempts in the past (cf., e.g., [4, Thms. 1.1-1.3], which resolve open problems in [10]).

While Blais, Brody, and Matulef hint towards the formulation that we will present here (see a few lines before [4, Def. 2.3]), they preferred to present a more restricted formulation, which is pivoted at "simple combining operators" (see [4, Def. 2.3]). Furthermore, it seems that this restricted formulation is the one that has been disseminated in the literature.[2]

1.1 The Current Work

The main purpose of this paper is to explicitly present a more general and flexible formulation of the foregoing methodology, and demonstrate the ease of using it (in comparison to the use of the restricted formulation). A special case of this general formulation reads as follows:

> In order to derive a lower bound on testing the property Π, present a mapping F of pairs of inputs $(x, y) \in \{0,1\}^{n+n}$ for a two-party communication problem Ψ to $\ell(n)$-bit long inputs for Π such that $(x, y) \in \Psi$ implies $F(x, y) \in \Pi$ and $(x, y) \notin \Psi$ implies that $F(x, y)$ is far from Π. Letting $f_i(x, y)$ denote the ith bit of $F(x, y)$, upper-bound the (deterministic) communication complexity of each f_i, and lower-bound the randomized communication complexity of Ψ. Then, the query complexity of testing Π is lower-bounded by the ratio of the lower bound (for Ψ) over the upper bound (for the f_i's).

[2] Added in revision: This sentence reflects the state of affairs at the time this paper was originally written. At that time, we mentioned that "A notable exception has been provided by the recent work of [5]: To streamline their proof, they take a move that is analogous to ours by replacing the simple combining operators of [4, Def. 2.3] with a 'one-bit one-way combining operator' (see Definition 2.4 and Lemma 2.5 in [5])."

More generally, one may use randomized protocols (with small error) for computing the f_i's. In contrast, the restricted formulation of [4] requires that each $f_i(x, y)$ be a function of x_i and y_i only, and uses the straightforward two-bit communication protocol for f_i. Hence, using the restricted formulation of [4] restricts the users of the methodology, since it places tighter constraints on the choice of the two-party communication problem Ψ. *This difficulty is typically overcome by introducing auxiliary communication problems and lower-bounding their complexity by reduction from known communication problems* (cf. [4]).

In fact, we show that the restricted formulation is actually not significantly weaker than the general one. In other words, we show that any lower bound that can be derived by the general formulation, can also be derived (possibly with a small quantitative loss) by the restricted formulation. This is done by introducing auxiliary problems and showing reductions among them. We note that *such tedious maneuvers are exactly what the general methodology spares the user from.*

Indeed, we advocate the use of the general formulation, because we believe that it is easier to work with. This is demonstrated by using it to derive some (known and new) results regarding the hardness of codeword testing for some codes. Furthermore, we believe that *the statement of the general formulation* (of the methodology) *and its proof reveals better what is actually going on.*

1.2 Organization

After recalling some standard definitions (in Sect. 2), we provide a general formulation of the communication complexity methodology for proving query complexity lower bounds on property testing (see Theorem 3.1 in Sect. 3). The (relative) ease of using this methodology is demonstrated in Sect. 4, and further discussed in the conclusion section (Sect. 8). In our opinion, these sections are the most important parts of the current work.

In Sect. 7, we consider the relation between the general formulation and the restricted one (as presented in [4]). Indeed, Theorem 7.1 (i.e., the "emulation theorem") is the main technical contribution of this work. Essentially, it asserts that any lower bound that can be derived by the general formulation, can also be derived by the restricted formulation.

The methodology of *proving query complexity lower bounds on property testing via reductions from communication complexity problems* can be applied in a variety of contexts. While Sect. 3 focuses on the most popular setting of general testers and general two-party communication protocols, versions for nonadaptive testers and one-way (resp., simultaneous) communication complexity are presented in Sect. 5. Other ramifications are discussed in Sect. 6, and a version for multi-party communication complexity is presented in the Appendix.

2 Preliminaries

For sake of simplicity, we focus on problems that correspond to the binary representation (i.e., to objects that are represented as sequences over a binary alphabet). We shall discuss the general case of non-binary alphabets at a later stage (i.e., in Sect. 6.2).[3]

Also, our main presentation refers to finite problems that correspond to bit strings of fixed length. One should think of these lengths as generic (or varying), and interpret the O-notation (as well as similar notions) as hiding universal constants (which do not depend on any parameter of the problems discussed).

We refer to the standard setting of communication complexity, and specifically to randomized two-party protocols in the model of shared randomness (cf. [19, Sec. 3]). We denote by $\langle A(x), B(y)\rangle(r)$ the (joint) output of the two parties, when the first party uses strategy A and gets input x, the second party uses strategy B and gets input y, and both parties have free access to the shared randomness r. Since many of the known reductions that use the methodology of [4] actually reduce from promise problems, we present communication problems in this more general setting. The standard case of decision problems is obtained by using a trivial promise (i.e., $P = \{0,1\}^{2n}$).

Definition 2.1 (two-party communication complexity): *Let* $\Psi = (P, S)$ *such that* $P, S \subseteq \{0,1\}^{2n}$, *and* $\eta \geq 0$. *A two-party* protocol that solves Ψ with error at most η *is a pair of strategies* (A, B) *such that the following holds* (w.r.t. some $\rho = \rho(n)$):

1. *If* $(x, y) \in P \cap S$, *then* $\Pr_{r\in\{0,1\}^\rho}[\langle A(x), B(y)\rangle(r)=1] \geq 1 - \eta$.
2. *If* $(x, y) \in P \setminus S$, *then* $\Pr_{r\in\{0,1\}^\rho}[\langle A(x), B(y)\rangle(r)=0] \geq 1 - \eta$.

The communication complexity of this protocol *is the maximum number of bits exchanged between the parties when the maximization is over all* $x, y \in \{0,1\}^n$ *and* $r \in \{0,1\}^\rho$. *The* η-error communication complexity of Ψ, *denoted* $\mathsf{CC}_\eta(\Psi)$, *is the minimum communication complexity of all protocols that solve* Ψ *with error at most* η.

For a Boolean function $f : \{0,1\}^{2n} \to \{0,1\}$, the two-party communication problem of computing f is the promise problem $\Psi_f \stackrel{\text{def}}{=} (\{0,1\}^{2n}, f^{-1}(1))$. Abusing notation, we let $\mathsf{CC}_\eta(f)$ denote $\mathsf{CC}_\eta(\Psi_f)$.

Note that randomized complexity with zero error (i.e., $\eta = 0$) collapses to deterministic complexity.[4] This is one reason that we kept η as a free parameter rather than setting it to a small constant (e.g., $\eta = 1/3$), as is the standard. Another reason for our choice is to allow greater flexibility in our presentation. For the same reason, we take the rather unusual choice of making the error

[3] Jumping ahead, we note that, with respect to the general formulation, little is lost by considering only the binary representation.

[4] Note that $\mathsf{CC}_0(\cdot)$ is different from the *standard* notion of zero-error randomized communication complexity, since in the latter one considers the *expected* number of bits exchanged on the worst-case pair of inputs (whereas we considered the worst-case over both the shared randomness and the pair of inputs). Note that the difference between the expected complexity and the worst-case complexity is not very significant in the case of $\Theta(1)$-error communication complexity, but it is crucial in the case of zero-error.

probability explicit also in the context of property testing (where we also denote it by η). In the next definition, as in most work on *lower bounds* in property testing (cf. [8,12,13,22]), we fix the proximity parameter (denoted ϵ).

Definition 2.2 (property testing): *Let $\Pi \subseteq \{0,1\}^\ell$, and $\epsilon, \eta > 0$. An ϵ-tester* with error η for Π *is a randomized oracle machine T that satisfies the following two conditions.*

1. *If $z \in \Pi$, then $\Pr[T^z(\ell)=1] \geq 1-\eta$.*
2. *If $z \in \{0,1\}^\ell$ is ϵ-far from Π, then $\Pr[T^z(\ell)=0] \geq 1-\eta$, where the distance between z and Π is $\min_{z' \in \Pi}\{|\{i \in [\ell] : z_i \neq z'_i\}|/\ell\}$.*

The query complexity *of T is the maximum number of queries that T makes, when the maximization is over all $z \in \{0,1\}^\ell$ and the coin tosses of T. The* η-error query complexity of ϵ-testing Π, *denoted $\mathsf{Q}_\eta(\epsilon, \Pi)$, is the minimum query complexity of all ϵ-testers with error η for Π.*

For any property Π and any $\eta > 0$, it holds that $\mathsf{Q}_\eta(\epsilon, \Pi) = O(\mathsf{Q}_{1/3}(\epsilon, \Pi))$, where the O-notation hides a $\log(1/\eta)$ factor. Thus, establishing a lower bound on the ϵ-testing query complexity of Π for any constant error, yields the same asymptotic lower bound for the (standard) error level of $1/3$. In light of this fact, we may omit the constant error from our discussion; that is, when we say the query complexity of ϵ-testing Π we mean the $1/3$-error query complexity of ϵ-testing Π. Likewise, $\mathsf{Q}(\epsilon, \Pi) = \mathsf{Q}_{1/3}(\epsilon, \Pi)$.

3 The General Formulation of the Methodology

With the above preliminaries in place, we are ready to state the main result, which is proved by a straightforward adaptation of the ideas of [4].

Theorem 3.1 (property testing lower bounds via communication complexity): *Let $\Psi = (P, S)$ be a promise problem such that $P, S \subseteq \{0,1\}^{2n}$, and let $\Pi \subseteq \{0,1\}^\ell$ be a property, and $\epsilon, \eta > 0$. Suppose that the mapping $F : \{0,1\}^{2n} \to \{0,1\}^\ell$ satisfies the following two conditions:*

1. *For every $(x, y) \in P \cap S$, it holds that $F(x, y) \in \Pi$.*
2. *For every $(x, y) \in P \setminus S$, it holds that $F(x, y)$ is ϵ-far from Π.*

Then, $\mathsf{Q}_\eta(\epsilon, \Pi) \geq \mathsf{CC}_{2\eta}(\Psi)/B$, where $B = \max_{i \in [\ell]}\{\mathsf{CC}_{\eta/n}(f_i)\}$ and $f_i(x, y)$ is the ith bit of $F(x, y)$. Furthermore, if $B = \max_{i \in [\ell]}\{\mathsf{CC}_0(f_i)\}$, then $\mathsf{Q}_\eta(\epsilon, \Pi) \geq \mathsf{CC}_\eta(\Psi)/B$.

The main result in [4] refers to a mapping F such that each $f_i(x, y)$ is a function of the ith bit of x and the ith bit of y (i.e., x_i and y_i). Indeed, in that case, $\ell = n$ and $B = 2$ (by the straightforward protocol in which the two parties exchange the relevant bits (i.e., x_i and y_i)).

Proof: Given an ϵ-tester with error η for Π and communication protocols for the f_i's, we present a two-party protocol for solving Ψ. The key idea is that, using their shared randomness, the two parties (holding x and y, respectively) can emulate the execution of the ϵ-tester, while providing it with virtual access to $F(x, y)$. Specifically, when the tester queries the ith bit of the oracle, the parties provide it with the value of $f_i(x, y)$ by first executing the corresponding communication protocol. Details follow.

The Protocol for Ψ: On local input x (resp., y) and shared randomness $r = (r_0, r_1, ..., r_\ell) \in (\{0,1\}^*)^{\ell+1}$, the first (resp., second) party invokes the ϵ-tester on randomness r_0, and answers the tester's queries by interacting with the other party. That is, each of the two parties invokes a local copy of the tester's program, but both copies are invoked on the same randomness, and are fed with identical answers to their (identical) queries. When the tester issues a query $i \in [\ell]$, the parties compute the value of $f_i(x, y)$ by using the corresponding communication protocol, with r_i as its shared randomness, and feed $f_i(x, y)$ to (their local copy of) the tester. Specifically, denoting the latter protocol (i.e., pair of strategies) by (A_i, B_i), the parties answer with $\langle A_i(x), B_i(y)\rangle(r_i)$. When the tester halts, each party outputs the output it has obtained from (its local copy of) the tester.

Turning to the analysis of this protocol, we note that the two local executions of the tester are identical, since they are fed with the same randomness and the same answers (to the same queries). The total number of bits exchanged by the two parties is at most B times the query complexity of ϵ-tester; that is, the communication complexity of this protocol is at most $B \cdot q$, where q denotes the query complexity of the ϵ-tester.

Let us consider first the furthermore clause; that is, suppose that B equals $\max_{i\in[\ell]}\{\mathsf{CC}_0(f_i)\}$. In this case, the parties always provide the ϵ-tester, denoted T, with the correct answers to its queries. Now, if $(x, y) \in P \cap S$, then $F(x, y) \in \Pi$, which implies that $\Pr[T^{F(x,y)}(\ell) = 1] \geq 1 - \eta$, which in turn implies that the parties output 1 with probability at least $1 - \eta$. On the other hand, if $(x, y) \in P \setminus S$, then $F(x, y)$ is ϵ-far from Π, which implies that $\Pr[T^{F(x,y)}(\ell) = 0] \geq 1 - \eta$, which in turn implies that the parties output 0 with probability at least $1 - \eta$. Hence, in this case (and assuming that T has query complexity $\mathsf{Q}_\eta(\epsilon, \Pi)$), we get $\mathsf{CC}_\eta(\Psi) \leq B \cdot \mathsf{Q}_\eta(\epsilon, \Pi)$.

Turning to the main claim, we may assume that $q \leq n$, since otherwise we can just use the trivial communication protocol for Ψ (which has complexity n). Recall that if $(x, y) \in P \cap S$, then $\Pr[T^{F(x,y)}(\ell) = 1] \geq 1 - \eta$. However, the emulation of T is given access to bits that are each correct only with probability $1 - \frac{\eta}{n}$, but the possible error probabilities add-up to at most η; that is, the probability that the protocol outputs 1 is at least $1 - \eta - q \cdot \frac{\eta}{n} \geq 1 - 2\eta$. On the other hand, if $(x, y) \in P \setminus S$, then $\Pr[T^{F(x,y)}(\ell) = 0] \geq 1 - \eta$. Taking account of the errors in computing the f_i's, we conclude that the probability that the protocol outputs 0 in this case is at least $1 - 2\eta$. The claim follows. ∎

Corollary 3.2 (a special case of Theorem *3.1*): *Let $\Psi = (P, S)$, Π, $\epsilon, \eta > 0$, F, and the f_i's be as in Theorem 3.1. Suppose that each $f_i(x, y)$ either depends* *on at most one bit of* x (and possibly some bits of y) *or depends on at most one bit of* y (and possibly some bits of x). *Then,* $\mathsf{Q}_\eta(\epsilon, \Pi) \geq \mathsf{CC}_\eta(\Psi)/2$.

Proof: In this case $\mathsf{CC}_0(f_i) \leq 2$ for each $i \in [\ell]$, by letting the holder of the missing bit communicate it to the other party, who responses with the value of f_i. The claim follows by the furthermore clause of Theorem 3.1. ■

Corollary 3.3 (the special case of "simple combining operator" [4]): *Let* $\Psi = (P, S)$, Π, $\epsilon, \eta > 0$, F, *and the* f_i*'s be as in Corollary 3.2. Suppose that each* $f_i(x, y)$ *depends only on the* i^{th} *bit of* x *and the* i^{th} *bit of* y. *Then,* $\mathsf{Q}_\eta(\epsilon, \Pi) \geq \mathsf{CC}_\eta(\Psi)/2$.

Corollary 3.3, which is a special case of Corollary 3.2, is stated merely for sake of reference. We note that the methodology as presented in [4] is slightly more general than Corollary 3.3, since it refers to sequences over an arbitrary alphabet Σ (rather than to bit strings).[5] For further discussion, see Sect. 6.2.

4 Application to Codeword Testing

The applications presented in this section are (of course) negative ones: They are families of codes for which codeword testing is extremely hard. Such families were known before (cf., e.g., [3]).[6] The following results can also be proved by using the restricted methodology as presented in [4] and Corollary 3.3 (see discussion following the proof of Theorem 4.1), but we believe that deriving them via the general methodology (i.e., using either Corollary 3.2 or Theorem 3.1) is simpler. Recall that the rate of a code $C : \{0,1\}^n \to \{0,1\}^\ell$ is n/ℓ, and its relative distance is d/ℓ such that every two different codewords differ on at least d positions (i.e., for every $x, y \in \{0,1\}^n$ such that $x \neq y$. it holds that $C(x)$ and $C(y)$ disagree on at least d positions).

Theorem 4.1 (on the hardness of testing codewords in some codes): *Let* $\{\Psi_n = (P_n, S_n)\}_{n \in \mathbb{N}}$ *be a family of communication problems such that* $P_n, S_n \subseteq \{0,1\}^{2n}$ *and for some constant* $\eta > 0$ *it holds that* $\mathsf{CC}_\eta(\Psi_n) = \Omega(n)$. *Let* $\{C_n : \{0,1\}^n \to \{0,1\}^{\ell(n)}\}_{n \in \mathbb{N}}$ *be a family of codes of constant relative distance. Then, for some constant* $\epsilon > 0$, *the query complexity of* ϵ*-testing the property* $\Pi = \{\Pi_n\}_{n \in \mathbb{N}}$, *where*

$$\Pi_n \stackrel{\text{def}}{=} \{C_n(x)C_n(y) : (x, y) \in P_n \cap S_n\}, \tag{1}$$

is $\Omega(n)$. *That is,* $\mathsf{Q}(\epsilon, \Pi_n) = \Omega(n)$.

The property Π is quite artificial; it consists of the encoding of the input pairs (in $P_n \cap S_n$) under a code (of constant relative distance) that is applied to each input separately. The reduction merely creates distance between pairs in

[5] In that case, $f_i : \Sigma^{2n} \to \Sigma$, and simple combining operators correspond to the case that each $f_i(x, y)$ depends only on the ith symbol of x and the ith symbol of y. The assertion then is that $\mathsf{Q}_\eta(\epsilon, \Pi) \geq \mathsf{CC}_\eta(\Psi)/2\lceil \log_2 |\Sigma| \rceil$.

[6] In contrast, for locally testable codes (cf., e.g., [9,14]), codeword testing is very easy.

$P_n \cap S_n$ and pairs in $P_n \setminus S_n$. Indeed, the elements of Π are codewords of a code C' that has constant relative distance; that is, each $(x,y) \in P_n \cap S_n$ is encoded by $C'_n(xy) = C_n(x)C_n(y)$. Also note that if C has constant rate, then so does C', because $\log_2|P_n \cap S_n| \geq \mathtt{CC}_\eta(\Psi_n) = \Omega(n)$. In any case, Theorem 4.1 asserts that, for the code C', codeword testing requires $\Omega(n)$ queries. Recall that such codes were known before (cf., e.g., [3]), but the code C' is definitely different (alas not necessarily more appealing).[7] We believe that most readers will find Theorem 4.2 more appealing than Theorem 4.1.

Proof: We invoke Corollary 3.2 while using $F(x,y) = C_n(x)C_n(y)$ and noting that each bit in $F(x,y)$ either depends only on bits of x or depends only on bits of y. By Eq. (1), for every $(x,y) \in P_n \cap S_n$ it holds that $F(x,y)$ is in Π_n. On the other hand, if $(x,y) \in P_n \setminus S_n$, then by the distance of C it holds that $F(x,y)$ is $\Omega(1)$-far from Π. Specifically, if the relative distance of C is δ, then $F(x,y)$ must be $\delta/2$-far from Π_n (since at least one of the two codewords in $F(x,y)$ must be replaced). Indeed, Corollary 3.2 ("only") implies $\mathtt{Q}_\eta(\delta/2, \Pi_n) = \Omega(n)$, but using $\mathtt{Q}(\delta/2, \Pi_n) = \mathtt{Q}_{1/3}(\delta/2, \Pi_n) = \Omega(\mathtt{Q}_\eta(\delta/2, \Pi_n)/\log(1/\eta))$, we are done. ■

An Alternative Proof of Theorem 4.1. As stated up-front, Theorem 4.1 can be proved by applying the communication complexity methodology as formulated in [4] (cf. Corollary 3.3). In order to do this, we need to introduce an auxiliary communication complexity problem, which is related to Ψ. Specifically, let $\Psi'_n = (P'_n, S'_n)$ be such that

$$P'_n \stackrel{\text{def}}{=} \{(C_n(x)0^{\ell(n)}, 0^{\ell(n)}C_n(y)) : (x,y) \in P_n\}$$
$$S'_n \stackrel{\text{def}}{=} \{(C_n(x)0^{\ell(n)}, 0^{\ell(n)}C_n(y)) : (x,y) \in S_n\}.$$

(That is, x is replaced by $C_n(x)0^\ell$, whereas y is replaced by $0^\ell C_n(y)$.) First, note that $\mathtt{CC}_\eta(\Psi'_n) \geq \mathtt{CC}_\eta(\Psi_n)$, since a communication protocol for Ψ is obtained by a straightforward emulation of any communication protocol for Ψ'. Next, we shall reduce the communication problem Ψ'_n to $\delta/2$-testing Π, by using $F'(u,v) \stackrel{\text{def}}{=} u \oplus v$, where $\oplus$ denotes the bit-by-bit XOR of strings (which indeed is a simple combining operator). Indeed, for every $(u,v) \in P'_n$ it holds that $u = C_n(x)0^{\ell(n)}$ and $v = 0^{\ell(n)}C_n(y)$ for some $x,y \in \{0,1\}^n$, and so $F'(u,v) = u \oplus v = C_n(x)C_n(y)$, which equals the value of $F(x,y)$ as defined in the proof of Theorem 4.1. Since F' falls within the restricted framework of [4] (cf. Corollary 3.3), by their result $\mathtt{Q}_\eta(\delta/2, \Pi_n) \geq \mathtt{CC}_\eta(\Psi'_n)/2$. A similar comment applies also to the following result.

[7] The non-appealing aspect of C' is that it encodes each of the two parts of the message separately. The appealing aspect is that these two parts can be nicely related; for example, consider $(P_n, S_n) = (\{0,1\}^{2n}, f^{-1}(1))$ such that $f : \{0,1\}^{2n} \to \{0,1\}$ is the inner product (mod 2) function (and recall that the communication complexity of (P_n, S_n) (equiv., of computing f) is $\Omega(n)$ [7]).

In contrast to Theorem 4.1, which refers to certain sets of the codewords in somewhat artificial codes (i.e., $C'_n(xy) = C_n(y)C_n(x)$), the following Theorem 4.2 refers to sets of codewords of any linear code. (Both results require the code to have constant relative distance.)

Theorem 4.2 (more on the hardness of testing codewords in some codes): *Let* $\{C_n : \{0,1\}^n \to \{0,1\}^{\ell(n)}\}_{n\in\mathbb{N}}$ *be a family of* linear *codes* (i.e., $C_n(x \oplus y) = C_n(x) \oplus C_n(y)$) *of constant relative distance. Let* $\mathrm{wt}(z)$ *denote the Hamming weight of* z*; that is,* $\mathrm{wt}(z) = |\{i \in [|z|] : z_i = 1\}|$. *Then, for some constant* $\epsilon > 0$ *and any function* $k : \mathbb{N} \to \mathbb{N}$ *such that* $k(n)$ *is even and* $k(n) < n/2$, *the query complexity of* ϵ*-testing the property*

$$\Pi_n \stackrel{\text{def}}{=} \{C_n(z) : z \in \{0,1\}^n \wedge \mathrm{wt}(z) = k(n)\} \qquad (2)$$

is $\Omega(k(n))$. *That is,* $\mathtt{Q}(\epsilon, \Pi_n) = \Omega(k(n))$.

Note that Π_n is a code; actually, it is a sub-code of the (linear) code C. In the special case that C is the Hadamard code, the property Π_n is $k(n)$-linearity; that is, the codewords of the Hadamard code corresponds to linear functions (from $\mathrm{GF}(2)^n$ to $\mathrm{GF}(2)$) and the codewords of Π_n are $k(n)$-linear (i.e., they are linear functions that depend on exactly $k(n)$ variables). This special case of Theorem 4.2 was proved in [4]. Interestingly, Theorem 4.2 reveals that the underlying phenomenon is that Π_n consists of the encoding, under a linear code, of all strings of a fixed Hamming weight. (The linearity of the code C_n implies that the ith bit of $C_n(x \oplus y)$ is determined by the ith bits of $C_n(x)$ and $C_n(y)$.)

Proof: We reduce from the communication problem SET DISJOINTNESS, while using Theorem 3.1. Specifically, we consider the $k/2$-disjointness problem, denoted $\{\mathrm{DISJ}_n^{(k)} = (P_n, S_n)\}_{n\in\mathbb{N}}$, where $P_n, S_n \subseteq \{0,1\}^{2n}$ such that $(x,y) \in P_n$ if $\mathrm{wt}(x) = \mathrm{wt}(y) = k(n)/2$, and $(x,y) \in S_n$ if (the "intersection" set) $I(x,y) \stackrel{\text{def}}{=} \{i \in [n] : x_i = y_i = 1\}$ is empty. Thus, for every $(x,y) \in P_n$, it holds that $\mathrm{wt}(x \oplus y) = k(n) - 2 \cdot |I(x,y)|$. We use $F(x,y) = C_n(x \oplus y)$ and note that (by the linearity of C) the ith bit of $C_n(x \oplus y) = C_n(x) \oplus C_n(y)$ can be computed by exchanging the ith bits of $C_n(x)$ and $C_n(y)$.

Note that if $(x,y) \in P_n \cap S_n$ then $F(x,y) \in \Pi_n$, since $\mathrm{wt}(x \oplus y) = k(n)$, whereas if $(x,y) \in P_n \setminus S_n$ then $F(x,y)$ is $\Omega(1)$-far from Π_n, since $\mathrm{wt}(x \oplus y) < k(n)$ (and C has constant relative distance). The claimed lower bound follows by combining the celebrated result $\mathtt{CC}_{1/3}(\mathrm{DISJ}_n^{(k)}) = \Omega(k(n))$, which is implicit in [16] (see also [4, Lem. 2.6]), with Theorem 3.1. ■

Digest. While both Theorems 4.1 and 4.2 can be proved by applying the restricted methodology of [4] (cf. Corollary 3.3) after introducing suitable auxiliary communication complexity problems, our proofs avoid the introduction of such auxiliary problems. Instead, our proofs are based on the existence of simple protocols for exchanging bits in the encoding of the inputs under error correcting codes. Although these bits may depend on a linear number of bits in the original input, each party can compute the relevant bit by itself. Indeed, exactly the same computations take place when using the restricted methodology of [4] (cf. Corollary 3.3), but there these computations take place in the reduction of the original communication problem to an auxiliary one (which must be introduced in order to use the restricted methodology). When using the general methodology, the foregoing computation take place in the communication protocols that demonstrate that each bit in $F(x,y)$ has low communication complexity, and this demonstration is performed without introducing any auxiliary problem. We stress that the issue is not with these simple computations, but rather with whether the lower bound proof requires the (explicit) introduction of auxiliary communication problems.

5 Nonadaptive Testers and One-Way Communication

Following [4], we also present a version of the methodology that relates the complexity of nonadaptive testers to the communication complexity of one-way protocols.

One-Way Communication Complexity. In one-way communication protocols the first party sends a single message to the second party, who is the only party that produces an output. Thus, it is natural to denote the outcome of such a protocol by $B(y,r,A(x,r))$, where A and B are the algorithms employed by the two parties (and x,y,r are the private inputs and the shared randomness, respectively, as in Definition 2.1). For $\Psi = (P,S)$ as in Definition 2.1, the η-error one-way communication complexity of Ψ, denoted $\overrightarrow{\mathtt{CC}}_\eta(\Psi)$, is the minimum communication complexity of all *one-way* protocols that solve Ψ with error at most η.

Nonadaptive Testers. A nonadaptive oracle machine is one that determines all its queries based solely on its explicit input and its internal coin tosses, as opposed to a general (adaptive) oracle machine that may select its queries based also on the answers to prior queries. The η-error nonadaptive query complexity of ϵ-testing Π, denoted $\mathtt{Q}_\eta^{\rm na}(\epsilon,\Pi)$, is the minimum query complexity of all *nonadaptive* ϵ-testers with error η for Π.

Theorem 5.1 (Theorem 3.1, revised for nonadaptive testers vs one-way communication): *Let $\Psi = (P,S)$, Π, $\epsilon,\eta > 0$, F, and the f_i's be as in Theorem 3.1. Then, $\mathtt{Q}_\eta^{\rm na}(\epsilon,\Pi) \geq \overrightarrow{\mathtt{CC}}_{2\eta}(\Psi)/B$, where $B = \max_{i\in[\ell]}\{\overrightarrow{\mathtt{CC}}_{\eta/n}(f_i)\}$. Furthermore, if $B = \max_{i\in[\ell]}\{\overrightarrow{\mathtt{CC}}_0(f_i)\}$, then $\mathtt{Q}_\eta^{\rm na}(\epsilon,\Pi) \geq \overrightarrow{\mathtt{CC}}_\eta(\Psi)/B$.*

Again, the analogous result in [4] uses a mapping F such that each $f_i(x,y)$ is a function of the $i^{\rm th}$ bits of x and y. Indeed, in that case, $\ell = n$ and $B = 1$, by the straightforward one-way protocol in which the first party sends the relevant bit (i.e., x_i) to the second party. (The recent work of [5] refers to the general case of $B = 1$.)

Proof: We merely adapt the proof of Theorem 3.1: Given a nonadaptive ϵ-tester with error η for Π and one-way communication protocols for the f_i's, we present a one-way protocol for solving Ψ. Using their shared randomness, each of the two parties determines the (nonadaptive) queries of the tester, and the first party communicates to the second party the information it needs in order to determine the oracle's answers to the tester. Specifically, if position i is included in the set of nonadaptive queries, then the first party employs the one-way communication protocol for f_i, which results in sending a message that allows the second party to determine $f_i(x, y)$. Using these answers, the second party obtains the verdict of the tester, and outputs it. (Indeed, only the second party invokes the decision-making module of the tester and obtains its verdict.)[8] The rest of the analysis proceeds as in the proof of Theorem 3.1, under the obvious modifications. ■

Nonadaptive Testers and Simultaneous Communication Complexity. As suggested by David Woodruff, Theorem 5.1 remains valid when replacing one-way communication by simultaneous communication. The model of simultaneous communication protocols consists of three parties such that only two parties obtain inputs, whereas (only) the third party (called the referee) produces the output. Communication is unidirectional from each of the two input-holding parties to the referee: Based on its own local input (and the shared randomness), each party sends a (single) message to the referee, who then produces the output (based also on the joint randomness).[9] Thus, it is natural to denote the outcome of such a protocol by $R(r, A(x, r), B(y, r))$, where A and B are the algorithms employed by the two input-holding parties and R is the algorithm employed by the referee. For $\Psi = (P, S)$ as in Definition 2.1, the η-error simultaneous communication complexity of Ψ, denoted $\ddot{\mathsf{CC}}_\eta(\Psi)$, is the minimum communication complexity of all *simultaneous* protocols that solve Ψ with error at most η. Note that $\overrightarrow{\mathsf{CC}}_\eta(\Psi) \leq \ddot{\mathsf{CC}}_\eta(\Psi)$, since the second party (in the one-way communication model) can emulate the referee (of the simultaneous communication model).

Theorem 5.2 (Theorem 3.1, revised for nonadaptive testers vs simultaneous communication): *Let $\Psi = (P, S)$, Π, $\epsilon, \eta > 0$, F, and the f_i's be as in Theorem 3.1. Then, $\mathsf{Q}^{\mathrm{na}}_\eta(\epsilon, \Pi) \geq \ddot{\mathsf{CC}}_{2\eta}(\Psi)/B$, where $B = \max_{i\in[\ell]}\{\ddot{\mathsf{CC}}_{\eta/n}(f_i)\}$. Furthermore, if $B = \max_{i\in[\ell]}\{\ddot{\mathsf{CC}}_0(f_i)\}$, then $\mathsf{Q}^{\mathrm{na}}_\eta(\epsilon, \Pi) \geq \ddot{\mathsf{CC}}_\eta(\Psi)/B$.*

---

[8] Formally, a nonadaptive tester T consists of a pair of algorithms, Q and D, which use the same randomness, such that Q determines the tester's query and D its decision given the corresponding answers; that is, $T^z(r) = D(r, z_{i_1}, ..., z_{i_q})$, where $(i_1, ..., i_q) = Q(r)$. In our one-way communication protocol each of the two parties locally determines $(i_1, ..., i_q) = Q(r_0)$, then, for each $j \in [q]$, the first party sends to the second party the message required for the computation of $f_{i_j}(x, y)$, and finally (after computing all the $f_{i_j}(x, y)$'s) the second party invokes D and outputs $D(r_0, f_{i_1}(x, y), ..., f_{i_q}(x, y))$.

[9] It is crucial that the two input-holding parties have access to the same shared randomness, since they cannot communicate with one another. In contrast, it is less essential that the referee also has access to this shared randomness, since one of the input-holding parties can send it along while relying on the fact that the randomness can be made logarithmic in the input length (cf. [19, Thm. 3.14]).

Strictly speaking, Theorem 5.2 is not stronger than Theorem 5.1, but we do expect it to be more useful (since the possibility of $\ddot{\mathsf{CC}}_\eta(\Psi) \gg \overrightarrow{\mathsf{CC}}_\eta(\Psi)$ seems more promising than the potential cost of $\ddot{\mathsf{CC}}_\eta(f_i) \geq \overrightarrow{\mathsf{CC}}_\eta(f_i)$).

Proof: We merely adapt the proof of Theorem 5.1, replacing one-way protocols by simultaneous ones. Given a nonadaptive ϵ-tester with error η for Π and simultaneous communication protocols for the f_i's, we present a simultaneous protocol for solving Ψ. Again, using their shared randomness, each of the three parties determines the (nonadaptive) queries of the tester, and each of the two input-holding parties communicates to the referee the information it needs in order to determine the oracle's answers to the tester. Specifically, if position i is included in the set of nonadaptive queries, then each input-holding party employs the simultaneous communication protocol for f_i, which results in sending a message to the referee, who upon receiving these two messages (and having determined i) determines the value of $f_i(x, y)$. (Indeed, only the referee invokes the decision-making module of the tester, feeds it with all the answers it has determined, and obtains its verdict.) The rest of the analysis proceeds as in the proofs of Theorems 3.1 and 5.1, under the obvious modifications. ■

6 On One-Sided Error and Non-binary Alphabets

In this section, we briefly comment on two ramifications, which have appeared in [4].

6.1 One-Sided Error Versions

One-sided error testers are testers that are allowed no error when the object has the property; that is, if T is a one-sided error tester for Π, then for every $z \in \Pi$ it holds that $\Pr[T^z(\ell)=1] = 1$. (Error probability is only allowed in the case that $z \notin \Pi$.) Deriving lower bounds for such testers via the communication complexity methodology requires referring to the corresponding one-sided error version of communication complexity. That is, we shall consider communication protocols for (P, S) such that for every $(x, y) \in P \cap S$ it holds that $\Pr_{r\in\{0,1\}^\rho}[\langle A(x), B(y)\rangle(r) = 1] = 1$. In this case we may only use *zero-error* communication protocols for computing the f_i's.[10] For sake of clarity, we state the (main) corresponding result.

Theorem 6.1 (Theorem 3.1, revised for one-sided error): *Let $\Psi = (P, S)$, Π, $\epsilon, \eta > 0$, F, and the f_i's be as in Theorem 3.1. Then, the one-sided η-error query complexity of ϵ-testing Π is at least $1/B$ times the one-sided η-error communication complexity of Ψ, where $B = \max_{i\in[\ell]}\{\mathsf{CC}_0(f_i)\}$.*

By one-sided η-error query (resp., communication) complexity, we mean the complexity of one-sided error testers (resp., protocols) that have error probability at most η on the "no-instances".

[10] The point is that in this case we cannot afford any error, regardless of the value of $f_i(x, y)$.

6.2 Non-binary Alphabets

So far, our treatment of the subject-matter referred to computational problems over binary strings. Clearly, any computational problem over other alphabets can be restated via binary alphabets, but sometimes the former formulation is more appealing. This holds, in particular, for property testing problems. Examples include testing low-degree polynomials (cf. [23]), testing graph properties in the bounded-degree model (cf. [13]), and testing monotonicity over general range (cf. [6]). Thus, we may consider properties and communication problems that refer to sequences over some alphabet Σ, rather that over a binary alphabet. The problem, however, is that the communication protocols themselves need not respect the "integrity of the alphabet" (i.e., messages are arbitrary functions of the input, regardless of the alphabet in which the latter is encoded). (Things are, of course, different in the context of property testing: The tester's queries must respect the input format.)

Given this state of affairs, it seems that we gain little by providing a treatment of the general methodology for non-binary alphabets. Instead, when studying a property that refer to object that are encoded as sequences over Σ, we may consider their encoding as binary strings (which means that we lose a factor of $\log_2 |\Sigma|$ in the query lower bounds that we derive (see below)). When using the trivial encoding, we also lose in the value of the proximity parameter for which the lower bound hold, but this loss may be reduced to a constant by using encoding via a good error correcting code. Details follow.

Suppose that we wish to establish a lower bound for ϵ-testing a property Π of objects that are encoded as sequences over Σ, and suppose that we have a reduction F from some communication problem to testing $\Pi \subseteq \Sigma^\ell$. Then, we may consider the corresponding binary property Π' (i.e., $\Pi' = \{(C(z_1), ..., C(z_\ell)) : (z_1, ..., z_\ell) \in \Pi\}$, where $C : \Sigma \to \{0,1\}^{O(\log|\Sigma|)}$ is a good code), and the corresponding reduction F' (i.e., $F'(x, y) = (C(f_1(x, y)), ..., C(f_\ell(x, y)))$, where $F(x, y) = (f_1(x, y), ..., f_\ell(x, y)))$. Now, if $F(x, y) \in \Pi$ then $F'(x, y) \in \Pi'$, whereas if $F(x, y)$ is ϵ-far from Π then $F'(x, y)$ is ϵ'-far from Π', where $\epsilon' = \Omega(\epsilon)$. Hence, using F', we derive a lower bound on $\mathtt{Q}(\epsilon', \Pi')$, which yields a lower bound on $\mathtt{Q}(\epsilon', \Pi)$; that is, $\mathtt{Q}(\epsilon', \Pi) \geq \mathtt{Q}(\epsilon', \Pi')/O(\log|\Sigma|)$.

We stress that the foregoing discussion refers to the general formulation as presented in Theorem 3.1. In contrast, in the context of the special case of Corollary 3.3 (as presented in [4]) there is a benefit in directly treating arbitrary alphabet (as indeed done in [4]), since this allows for a less restricted notion of simple combining operators. Recall that the formulation in [4] requires that each symbol in $F(x, y)$ can be computed as a function of the corresponding symbols of x and y. When we view $F(x, y)$ as a binary string, this means that the ith bit of $F(x, y)$ is a function of x_i and y_i only. But if we view $F(x, y)$ as a sequence over Σ (or as binary strings partitioned into $O(\log|\Sigma|)$-bit long blocks), then this means that each bit in the description of the ith symbol (resp., i^{th} block) of $F(x, y)$ may depend on all bits in the description of the ith symbol of x and the ith symbol of y (resp., on all bits in the ith blocks of x and y).

7 Emulating the General Formulation by the Restricted One

In this section we show that the restricted formulation of [4] (via simple combination operators, cf. Corollary 3.3) can emulate the general formulation as captured by Theorem 3.1. The emulations we present come at the cost of some degradation in the parameters obtained, but this degradation is relatively small. Furthermore, as will become apparent throughout our proof, the degradation is even smaller in some special cases. In fact, we find it useful to present the proof by going from special cases to more general cases.

Before stating our most general result (i.e., Theorem 7.1), we make a couple of tedious comments. First, we assume for simplicity that $\eta \geq \epsilon$. While this seems the most relevant case (e.g., typically one considers $\eta = 1/3$), the result generalizes to arbitrary $\eta > 0$ (while replacing some $\log(1/\epsilon)$ factors by $\log(1/\eta)$ factors). Second, we restrict ourselves to $\epsilon' = \widetilde{\Omega}(1/n)$, and note that the case of $\epsilon' \in (0, \widetilde{\Omega}(1/n)]$ is uninteresting (in light of the application of Theorem 7.1).[11] Lastly, we stress that the constants hidden in the $\widetilde{O}$-notation are universal constants, which are independent of all the parameters that appear in the statement of the result.

Theorem 7.1 (emulating Theorem 3.1 via simple combining operators): *Let $\Psi = (P, S)$, $\Pi \subseteq \{0,1\}^\ell$, $\eta \geq \epsilon > 0$, $F : \{0,1\}^{2n} \to \{0,1\}^\ell$, and the f_i's be as in Theorem 3.1.*[12] *Suppose that $B = \max_{i\in[\ell]}\{\mathtt{CC}_{\eta/n}(f_i)\}$, which implies $\mathtt{Q}_\eta(\epsilon, \Pi) \geq \mathtt{CC}_{2\eta}(\Psi)/B$ (by the main part of Theorem 3.1). Then, there exists a communication problem $\Psi' = (P', S')$ such that $\mathtt{CC}_\eta(\Psi') \geq \mathtt{CC}_\eta(\Psi)$ and a property Π' such that*

$$\mathtt{Q}_\eta(\epsilon', \Pi) \geq \frac{\mathtt{Q}_\eta(\epsilon'/2, \Pi') - \widetilde{O}(B/\epsilon')}{\widetilde{O}(B) \cdot \log(\mathtt{Q}_\eta(\epsilon'/2, \Pi')/\eta)}$$

(for every $\epsilon' = \widetilde{\Omega}(1/n)$), *whereas Ψ' and Π' are related as follows:*

1. *For every $(u, v) \in P' \cap S'$, it holds that $u \oplus v \in \Pi'$.*
2. *For every $(u, v) \in P' \setminus S'$, it holds that $u \oplus v$ is $\Omega(\epsilon)$-far from Π'.*

This means that whenever $\mathtt{Q}_\eta(\epsilon, \Pi) \geq \mathtt{CC}_{2\eta}(\Psi)/B$ is established by Theorem 3.1, when using $B = \max_{i\in[\ell]}\{\mathtt{CC}_{\eta/n}(f_i)\}$, we can (roughly) establish $\mathtt{Q}_\eta(\Omega(\epsilon), \Pi) = \widetilde{\Omega}(\mathtt{CC}_\eta(\Psi))/B$ by using the formulation as presented in [4] (cf. Corollary 3.3).

[11] In that application, we derive a lower bound on $\mathtt{Q}_\eta(\Omega(\epsilon), \Pi)$. This lower bound is smaller than $\mathtt{CC}_\eta(\Psi) - \widetilde{O}(1/\epsilon)$, which in turn is negative in the case of $\epsilon \leq \mathrm{poly}(\log n)/n$.

[12] Recall that this means that $\Psi = (P, S)$ is a promise problem such that $P, S \subseteq \{0,1\}^{2n}$, that $\Pi \subseteq \{0,1\}^\ell$ is a property, and that the mapping $F : \{0,1\}^{2n} \to \{0,1\}^\ell$ satisfies the following two conditions:
(1) For every $(x, y) \in P \cap S$, it holds that $F(x, y) \in \Pi$.
(2) For every $(x, y) \in P \setminus S$, it holds that $F(x, y)$ is ϵ-far from Π.
Lastly, $f_i(x, y)$ denotes the ith bit of $F(x, y)$.

This alternative derivation uses the mapping $F'(u,v) = u \oplus v$, and proceeded as follows:

$$\begin{aligned}
\mathtt{Q}_\eta(\Omega(\epsilon),\Pi) &\geq \frac{\mathtt{Q}_\eta(\Omega(\epsilon),\Pi') - \widetilde{O}(B/\epsilon)}{\widetilde{O}(B)\cdot\log(\mathtt{Q}_\eta(\Omega(\epsilon),\Pi')/\eta)} \\
&\geq \frac{\mathtt{CC}_\eta(\Psi') - \widetilde{O}(B/\epsilon)}{\widetilde{O}(B)\cdot\log(\mathtt{CC}_\eta(\Psi')/\eta)} \\
&\geq \frac{\mathtt{CC}_\eta(\Psi) - \widetilde{O}(B/\epsilon)}{\widetilde{O}(B)\cdot\log(\mathtt{CC}_\eta(\Psi)/\eta)}
\end{aligned}$$

where the first and third inequalities use Theorem 7.1, the second inequality is by [4] (cf. Corollary 3.3).[13] The lower bound derived this way is quantitatively inferior to the one derived by Theorem 3.1. In particular, in the denominator B is replaced by $\widetilde{O}(B)\cdot\log\mathtt{CC}_\eta(\Psi)$, and the lower bound refers to a smaller value of the proximity parameter (i.e., $\Omega(\epsilon)$ rather than ϵ). However, when we aim at large values of $\mathtt{CC}_{2\eta}(\Psi)/B$, the loss of factors of the form of $\log\mathtt{CC}_{2\eta}(\Psi)$ (and $\log B$) seems relatively small. In any case, the additive loss of $\widetilde{O}(B/\epsilon)$ in the numerator is typically insignificant, since we typically aim at much higher lower bounds.

Organization of the Proof. Theorem 7.1 is proved in three steps. We shall start with the special case in which each f_i can be expressed as a function of $g_i(x)$ and $h_i(y)$ such that $|g_i(x)|, |h_i(y)| \leq B/2$. Indeed, in this case $\mathtt{CC}_0(f_i) \leq B$, via the straightforward protocol in which the parties exchange $g_i(x)$ and $h_i(y)$.[14] We shall then move to the special case where $\mathtt{CC}_0(f_i) \leq B$ (i.e., the case of arbitrary deterministic protocols), and end with the general case (i.e., $\mathtt{CC}_{\eta/n}(f_i) \leq B$).

In each step, we shall introduce an auxiliary communication problem Ψ' and an auxiliary property Π', and establish three relations of the type asserted in the theorem: (1) a relation between the communication complexity problems (i.e., Ψ' and Ψ), (2) a relation between the property testing problems (i.e., Π' and Π), and (3) a relation between the auxiliary problems (i.e., Ψ' and Π').

7.1 Step 1: A Syntactic Special Case

We start by considering the special case in which for every $i \in [\ell]$ it holds that $f_i(x,y) = d_i(g_i(x), h_i(y))$, where $|g_i(x)|, |h_i(y)| \leq B/2$ and $d_i : \bigcup_{j,k\leq B/2}\{0,1\}^{j+k} \to \{0,1\}$, which implies $\mathtt{Q}_\eta(\epsilon,\Pi) \geq \mathtt{CC}_\eta(\Psi)/B$ (by the furthermore clause of Theorem 3.1). This special case will provide a good warm-up to the general case. In this case we shall prove

[13] We also use the fact that $x \geq y$ implies $\frac{x-\beta}{\log(x/\eta)} \geq \frac{y-\beta}{O(\log(y/\eta))}$, provided that $x \geq (1+\Omega(1))\cdot\beta$, which we may assume (since otherwise the bound is quite useless).

[14] Actually, $\mathtt{CC}_0(f_i) \leq (B/2)+1$, via the straightforward protocol in which the first party sends $v \leftarrow g_i(x)$ to the second party, who replies with the value of $f_i(x,y)$ that is computed based on v and $h_i(y)$.

Proposition 7.2 (warm-up):[15] *Let $\Psi = (P, S)$, $\Pi \subseteq \{0,1\}^\ell$, $\eta \geq \epsilon > 0$, $F : \{0,1\}^{2n} \to \{0,1\}^\ell$, and the f_i's be as in Theorem 3.1, and suppose hat $f_i(x, y) = d_i(g_i(x), h_i(y))$, where $|g_i(x)|, |h_i(y)| \leq B/2$. Then, there exists a communication problem $\Psi' = (P', S')$ such that $\mathsf{CC}_\eta(\Psi') \geq \mathsf{CC}_\eta(\Psi)$ and a property Π' such that $\mathsf{Q}_\eta(\epsilon', \Pi) \geq \mathsf{Q}_\eta(\epsilon', \Pi')/B$* (for every $\epsilon' > 0$), *whereas Ψ' and Π' are related as follows:*

1. *For every $(u, v) \in P' \cap S'$, it holds that $u \oplus v \in \Pi'$.*
2. *For every $(u, v) \in P' \setminus S'$, it holds that $u \oplus v$ is ϵ/B-far from Π'.*

This means that whenever $\mathsf{Q}_\eta(\epsilon, \Pi) \geq \mathsf{CC}_\eta(\Psi)/B$ is established by Theorem 3.1, when using f_i's of the above form, we can establish $\mathsf{Q}_\eta(\epsilon/B, \Pi) \geq \mathsf{CC}_\eta(\Psi)/2B$ by using the restricted formulation (with simple combining operators as presented in [4] (cf. Corollary 3.3)). Note that there is some degradation in the parameters also in this special case: The main issue is not that B is replaced by $2B$, but rather than the lower bound refers to a smaller value of the proximity parameter (i.e., ϵ/B rather than ϵ). We shall address this issue when discussing the general case (in Sect. 7.3).

Proof: We start with a few simplifying assumptions, which hold without loss of generality up to some insignificant degradation in some parameters.

1. F is non-shrinking (i.e., $\ell \geq 2n$); actually, $\ell \geq n$ suffices for our purposes.
   Otherwise, for $m \stackrel{\text{def}}{=} \lceil n/\ell \rceil$, consider the property $\Pi^{(m)} \stackrel{\text{def}}{=} \{z^m : z \in \Pi\}$ and the mapping $(x, y) \mapsto F(x, y)^m$, which satisfy the conditions of the theorem (and of the current case). Lower bounds on the query complexity of testing $\Pi^{(m)}$ imply similar bounds for Π, because $\mathsf{Q}_\eta(2\epsilon, \Pi^{(m)}) \leq \mathsf{Q}_\eta(\epsilon, \Pi) + O(\epsilon^{-1} \log(1/\eta))$, since we can test $\Pi^{(m)}$ by combining a tester for Π and a repetition test.[16]
2. The mappings $x \mapsto (g_1(x), ..., g_\ell(x))$ and $y \mapsto (h_1(y), ..., h_\ell(y))$ are one-to-one.
   Replying on the first assumption (i.e., $n \leq \ell$), for each $i \in [n]$, we append x_i to $g_i(x)$; that is, we redefine $g_i(x) \leftarrow g_i(x)x_i$. Ditto for y_i and $h_i(y)$.
3. For each $i \in [\ell]$, it holds that $|g_i(x)| = |h_i(y)| = B/2$. Furthermore, not all $B/2$-bit long strings are in the image of g_i, and ditto for h_i.
   We use a standard encoding of $\bigcup_{j \in [B']} \{0,1\}^i$ by $(B'+1)$-bit long strings (e.g., encoding the string s by $s10^{B'-|s|}$). Again, this means redefining g_i's and h_i's.

[15] The following statement holds under some simplifying assumptions that are listed at the beginning of the proof. Enforcing these assumptions causes an insignificant deterioration in some parameters (i.e., $B \leftarrow B + 3$ and $\mathsf{Q}_\eta(\epsilon', \Pi)$ is replaced by $\mathsf{Q}_\eta(\epsilon', \Pi) + \widetilde{O}(1/\epsilon')$.

[16] Note that if a string is ϵ-far from $\Pi^{(m)}$, then either the first block is ϵ-far from Π, or the other blocks are ϵ-far from a repetition of the first block. In the first case the original tester rejects (w.p. at least $1 - \eta$), and in the second case the repetition test rejects (w.p. at least $1 - \eta$).

4. For each $i \in [\ell]$, the predicate $d_i : \{0,1\}^B \to \{0,1\}$ is onto.
   Using the assumption that not all $B/2$-bit long strings are in the image of g_i and ditto for h_i, we can modify d_i on a pair that is *not* in the image of (h_i, g_i) without affecting the conditions of the theorem (and of the current case).

We now turn to the construction of Ψ' and Π'. First, we define $\Psi' = (P', S')$ such that

$$P' \stackrel{\text{def}}{=} \{(g_1(x)\cdots g_\ell(x)0^{\ell B/2}, 0^{\ell B/2}h_1(y)\cdots h_\ell(y)) : (x,y) \in P\} \tag{3}$$

$$S' \stackrel{\text{def}}{=} \{(g_1(x)\cdots g_\ell(x)0^{\ell B/2}, 0^{\ell B/2}h_1(y)\cdots h_\ell(y)) : (x,y) \in S\}. \tag{4}$$

Note that $\mathtt{CC}_\eta(\Psi) \leq \mathtt{CC}_\eta(\Psi')$, since a protocol for Ψ can proceed by emulating the protocol for Ψ'. Specifically, on input x the first party computes $g_1(x)\cdots g_\ell(x)0^{\ell B/2}$, and likewise the second party computes $0^{\ell B/2}h_1(y)\cdots h_\ell(y)$. By the one-to-one feature of these mappings (i.e., Assumption 2), the answer obtained for Ψ' is valid for Ψ.

Next, we introduce the property $\Pi' \subseteq \{0,1\}^{B\ell}$. For every $a_1, ..., a_\ell, b_1, ..., b_\ell \in \{0,1\}^{B/2}$ the $2\ell \cdot B/2$-bit long string $a_1\cdots a_\ell b_1\cdots b_\ell$ is in Π' if and only if it holds that the ℓ-bit string $d_1(a_1,b_1)\cdots d_\ell(a_\ell,b_\ell)$ is in Π. That is:

$$\Pi' \stackrel{\text{def}}{=} \{a_1\cdots a_\ell b_1\cdots b_\ell : d_1(a_1,b_1)\cdots d_\ell(a_\ell,b_\ell) \in \Pi\}. \tag{5}$$

Claim 7.2.1 (relating Π' to Π): *Let Π' be as in* Eq. (5). *Then,* $\mathtt{Q}_\eta(\epsilon', \Pi') \leq B \cdot \mathtt{Q}_\eta(\epsilon', \Pi)$.

Proof: Basically, an ϵ'-tester for Π' can just emulate the execution of an ϵ'-tester for Π while answering each query $i \in [\ell]$ by reading the two corresponding $B/2$-bit long blocks in its oracle. Specifically, using an ϵ'-tester T for Π, we construct an tester for Π' that emulates the virtual oracle $d_1(a_1,b_1)\cdots d_\ell(a_\ell,b_\ell)$ for T by accessing its own oracle $a_1\cdots a_\ell b_1\cdots b_\ell$ (i.e., query $i \in [\ell]$ in answered by reading a_i and b_i). Hence, each query of T is answered by making $2\cdot(B/2)$ oracle queries. Now, if $a_1\cdots a_\ell b_1\cdots b_\ell \in \Pi'$, then it must be that $d_1(a_1,b_1)\cdots d_\ell(a_\ell,b_\ell) \in \Pi$, and so our tester accepts (with probability at least $1-\eta$). On the other hand, if $a_1\cdots a_\ell b_1\cdots b_\ell$ is ϵ'-far from Π', then $d_1(a_1,b_1)\cdots d_\ell(a_\ell,b_\ell)$ is ϵ'-far from Π, because otherwise it suffices to change less than $\epsilon\ell$ of the (a_i,b_i)-pairs in order to obtain a string in Π' (where here we use the hypothesis that d_i is onto (i.e. Assumption 4)). The claim follows. □

Finally, consider $F' : \{0,1\}^{2\cdot B\ell} \to \{0,1\}^{B\ell}$ such that $F'(u,v) = u \oplus v$.

Claim 7.2.2 (relating Ψ' to Π'): *Let Ψ' be as in Eqs. (3) & (4) and Π' be as in* Eq. (5). *Then:*

1. *For every $(u,v) \in P' \cap S'$ it holds that $F'(u,v) \in \Pi'$.*
2. *For every $(u,v) \in P' \setminus S'$ it holds that $F'(u,v)$ is ϵ/B-far from Π'.*

Proof: The key observation is that for every $(u,v) \in P'$, it holds that $u = u'0^{\ell B/2}$ and $v = 0^{\ell B/2}v'$, and so $F'(u,v) = u \oplus v = u'v'$. Furthermore, in that case there exists $(x,y) \in P$ such that $u' = g_1(x) \cdots g_\ell(x)$ and $v' = h_1(y) \cdots h_\ell(y)$. Using the hypothesis that the mapping $(x,y) \to (u,v)$ is one-to-one (i.e., Assumption 2), we infer that this (x,y) is unique.

Now, if $(u,v) \in P' \cap S'$, then the aforementioned (x,y) must be in $P \cap S$, and it follows that $F(x,y) \in \Pi$ (by the hypothesis of Theorem 3.1 regarding F (reproduced in Footnote 12)). It follows that $F'(u,v) = u'v' \in \Pi'$, because $u'v' = g_1(x) \cdots g_\ell(x)h_1(y) \cdots h_\ell(y)$, whereas for each $i \in [\ell]$ it holds that $d_i(g_i(x), h_i(y))$ is the ith bit in $F(x,y)$.

Having established Item 1, we turn to Item 2: We observe that if $(u,v) \in P' \setminus S'$, then the aforementioned (x,y) must be in $P \setminus S$, and it follows that $F(x,y)$ is ϵ-far from Π (by the theorem's hypothesis regarding F). In this case, $F'(u,v) = g_1(x) \cdots g_\ell(x)h_1(y) \cdots h_\ell(y)$ such that for each $i \in [\ell]$ it holds that $d_i(g_i(x), h_i(y))$ is the ith bit in $F(x,y)$. Given that $F(x,y)$ is ϵ-far from Π (i.e., $F(x,y)$ differs in at least $\epsilon\ell$ positions from any ℓ-bit string in Π), it follows that for at least $\epsilon\ell$ of the $i \in [\ell]$ at least one of the corresponding strings (i.e., $g_i(x)$ and $h_i(y)$) must be modified to place the $B\ell$-bit long string in Π'. Hence, $F'(u,v)$ is $(\epsilon\ell/B\ell)$-far from Π'. □

Combining Claims 7.2.1 and 7.2.2, this completes the proof of the current proposition (i.e., the special case in which the f_i's are of the form $d_i(g_i(x), h_i(y))$). ■

Digest. We have established Theorem 7.1 for the special case of f_i's of the form $d_i(g_i(x), h_i(y))$. This was done by introducing an auxiliary communication problem Ψ' in which the input pair (x,y) is replaced by the (redundant and padded) pair $(x',y') \stackrel{\text{def}}{=} (g_1(x) \cdots g_\ell(x)0^{\ell B/2}, 0^{\ell B/2}h_1(y) \cdots h_\ell(y))$, where the padding was done so that the bit-by-bit XOR of the two inputs results in $F'(x',y') = g_1(x) \cdots g_\ell(x)h_1(y) \cdots h_\ell(y)$, which is in the auxiliary property Π' if and only if $F(x,y) = d_1(g_1(x), h_1(y)) \cdots d_\ell(g_\ell(x), h_\ell(y))$ is in Π. Hence, each bit in $F'(x',y')$ is a function of the corresponding bits in the input strings $x' = g_1(x) \cdots g_\ell(x)0^{\ell B/2}$ and $y' = 0^{\ell B/2}h_1(y) \cdots h_\ell(y)$, which means that a simple combining operator allows for reducing Ψ' to ϵ/B-testing Π', where the testing feature follows from the correspondence between $F'(x',y')$ and $F(x,y)$. Specifically, each bit in $F(x,y)$ is encoded by B bits of $F'(x',y')$ (i.e., the ith bit of $F(x,y)$ is encoded by $g_i(x)$ and $h_i(y)$, which are $B/2$-bit blocks in $F'(x',y')$). The main overhead of the argument is that the proximity parameter (for which a testing lower bound could be inferred) was cut by a factor of B.

In the general case, we shall replace the $g_i(x)$'s (resp., $h_i(y)$'s) by descriptions of *residual strategies* for the first (resp., second) party in the guaranteed low complexity protocols for computing the f_i's, where the residual strategies refer to a fixed value of x (resp., y). In the case of deterministic protocols of complexity B (to be treated in Sect. 7.2), such strategies will have length 2^B, and the overhead will be accordingly. Things will become even worse when handling randomized strategies (in Sect. 7.3), but using locally testable and decodable codes will resolve the issue. Actually, when dealing with the foregoing special case (of $f_i(x,y) = d_i(g_i(x), h_i(y))$), one may use any code $C : \{0,1\}^{B/2} \to \{0,1\}^{O(B)}$ of constant relative distance δ_C and define $F'(x',y') = C(g_1(x)) \cdots C(g_\ell(x))C(h_1(y)) \cdots C(h_\ell(y))$; in this case, the proximity parameter (for which a testing lower bound could be inferred) will be decrease by a factor of $\delta_C/2$ (whereas $\mathsf{Q}_\eta(\epsilon', \Pi) \geq \mathsf{Q}_\eta(\epsilon', \Pi')/O(B)$ rather than $\mathsf{Q}_\eta(\epsilon', \Pi) \geq \mathsf{Q}_\eta(\epsilon', \Pi')/B$).

7.2 Step 2: The Case of Deterministic Protocols

We now turn to the more general case of Theorem 7.1 in which the bound B is guaranteed by arbitrary deterministic protocols (i.e., $B = \max_{i\in[\ell]}\{\mathsf{CC}_0(f_i)\}$).

Proposition 7.3 (the deterministic case): *Let $\Psi = (P, S)$, $\Pi \subseteq \{0,1\}^\ell$, $\eta \geq \epsilon > 0$, $F : \{0,1\}^{2n} \to \{0,1\}^\ell$, and the f_i's be as in Theorem 3.1, and suppose that $B = \max_{i\in[\ell]}\{\mathsf{CC}_0(f_i)\}$. Then, there exists a communication problem $\Psi' = (P', S')$ such that $\mathsf{CC}_\eta(\Psi') \geq \mathsf{CC}_\eta(\Psi)$ and a property Π' such that $\mathsf{Q}_\eta(\epsilon', \Pi) \geq \mathsf{Q}_\eta(\epsilon', \Pi')/B$, whereas Ψ' and Π' are related as follows:*

1. *For every $(u,v) \in P' \cap S'$, it holds that $u \oplus v \in \Pi'$.*
2. *For every $(u,v) \in P' \setminus S'$, it holds that $u \oplus v$ is $\epsilon/2^B$-far from Π'.*

This means that whenever $\mathsf{Q}_\eta(\epsilon, \Pi) \geq \mathsf{CC}_\eta(\Psi)/B$ is established by Theorem 3.1, when using $B = \max_{i\in[\ell]}\{\mathsf{CC}_0(f_i)\}$, we can establish $\mathsf{Q}_\eta(\epsilon/2^B, \Pi) \geq \mathsf{CC}_\eta(\Psi)/B$ by using the formulation as presented in [4] (cf. Corollary 3.3). Again, proximity parameter of the derived lower bound is weaker; that is, the lower bound refers to a smaller value of the proximity parameter (i.e., $\epsilon/2^B$ rather than ϵ). The decrease in the value of the proximity parameter is far more acute than in Sect. 7.1. In particular, when $B > \log_2 \ell$, the alternative derivation only yields a result that refers to the query complexity of exact decision (since the value of the proximity parameter is smaller than $1/\ell$, where ℓ is the input length). This deficiency can be fixed by an idea that is presented in the treatment of the general case, which will follow (see Sect. 7.3).[17] But here we focus on the construction of Ψ' and Π' that satisfy the foregoing (somewhat deficient) claim.

Proof: By the hypothesis, for every $i \in [\ell]$, there exists a deterministic two-party protocol of communication complexity at most B for computing f_i. Let A_i and B_i denote the corresponding strategies of the two parties, and let $A_i^x = A_i(x)$ and $B_i^y = B_i(y)$ denote the **residual strategies** for local inputs x and y, respectively. That is, $A_i^x(\gamma)$ denotes the answer of the first party, holding input x, to a message-sequence γ sent by the second party (ditto for B_i^y).[18]

[17] Specifically, we refer to the use of the encoding of the parties' strategies by suitable error correcting codes.

[18] It is standard to assume that the parties interact by sending single-bit messages and that the first party starts. In such a case, A_i^x will be defined for strings of length at most $(B-1)/2$, including the empty string, while B_i^y will be defined for $\bigcup_{j\in[B/2]}\{0,1\}^j$. In general, the situation may be more complex, but in all cases the length of the description of each of the two strategies is at most 2^{B-1}.

We make the simplifying assumption that the mappings $x \mapsto (A_1^x, ..., A_\ell^x)$ and $y \mapsto (B_1^y, ..., B_\ell^y)$ are one-to-one, where the justification is that (for every $i \in [n]$) the strategy A_i^x may start by sending x_i (and ditto for B_i^y, with $\ell \geq n$ justified as in Sect. 7.1). Let $\langle A_i^x \rangle$ (resp., $\langle B_i^y \rangle$) denote a canonical 2^{B-1}-bit long description of the strategy A_i^x (resp., B_i^y) such that the value of $A_i^x(\gamma)$ (resp., $B_i^y(\gamma)$) appears in a specific bit location in $\langle A_i^x \rangle$ (resp., $\langle B_i^y \rangle$), where this location only depends on γ. Now, define $\Psi' = (P', S')$ such that

$$P' \stackrel{\text{def}}{=} \{(\langle A_1^x \rangle \cdots \langle A_\ell^x \rangle 0^{2^{B-1}\ell}, 0^{2^{B-1}\ell} \langle B_1^y \rangle \cdots \langle B_\ell^y \rangle) : (x, y) \in P\} \tag{6}$$

$$S' \stackrel{\text{def}}{=} \{(\langle A_1^x \rangle \cdots \langle A_\ell^x \rangle 0^{2^{B-1}\ell}, 0^{2^{B-1}\ell} \langle B_1^y \rangle \cdots \langle B_\ell^y \rangle) : (x, y) \in S\}. \tag{7}$$

Note that $\mathtt{CC}_\eta(\Psi) \leq \mathtt{CC}_\eta(\Psi')$, since a protocol for Ψ can proceed by emulating the protocol for Ψ'. Specifically, on input x the first party computes $\langle A_1^x \rangle \cdots \langle A_\ell^x \rangle 0^{2^{B-1}\ell}$, and likewise the second party computes $\langle B_1^y \rangle \cdots \langle B_\ell^y \rangle 0^{2^{B-1}\ell}$. By the one-to-one feature of these mappings (see above), the answer obtained for Ψ' is valid for Ψ.

Let (α, β) be a pair of residual strategies (as considered above) for a two-party communication protocol. We say that (α, β) produce the bit σ if emulating the interaction between these strategies yields the (joint) outcome σ. The emulation proceeds by determining the first message sent according to α, then determining the response according to β, and so on.

Next, we introduce the property $\Pi' \subseteq \{0,1\}^{2^B \ell}$. For every $a_1, ..., a_\ell, b_1, ..., b_\ell \in \{0,1\}^{2^{B-1}}$ the string $a_1 \cdots a_\ell b_1 \cdots b_\ell$ is in Π' if and only if for every $i \in [\ell]$ it holds that (a_i, b_i) describes a pair of strategies that produce the output bit w_i and $w = w_1 \cdots w_\ell \in \Pi$. Denoting the bit produced by these descriptions by $\mathtt{P}(a_i, b_i)$, we have

$$\Pi' \stackrel{\text{def}}{=} \{a_1 \cdots a_\ell b_1 \cdots b_\ell : \mathtt{P}(a_1, b_1) \cdots \mathtt{P}(a_\ell, b_\ell) \in \Pi\}. \tag{8}$$

(Note the similarity to Eq. (5). Likewise, the following two claims are similar to the claims made in Sect. 7.1, and their proofs amount to natural extensions of the arguments made there.)

Claim 7.3.1 (relating Π' to Π): *Let Π' be as in* Eq. (8). *Then,* $\mathtt{Q}_\eta(\epsilon', \Pi') \leq B \cdot \mathtt{Q}_\eta(\epsilon', \Pi)$.

Proof: Using an ϵ'-tester T for Π, we construct an ϵ'-tester for Π' by emulating the execution of T. Specifically, if T makes the query $i \in [\ell]$, then we access the ith pair of strategies included in our own oracle, denoted z (i.e., for $z = a_1 \cdots a_\ell b_1 \cdots b_\ell$, this means accessing a_i and b_i). By making B queries to these strategies, we emulate the computation of the ith bit in a virtual ℓ-bit string tested by T (i.e., the string $\mathtt{P}(a_1, b_1) \cdots \mathtt{P}(a_\ell, b_\ell)$). Specifically, we need only determine the value of the B bits that are exchanged in the interaction between the strategies a_i and b_i, rather than the full description of a_i and b_i. (Recall that each of these communicated bits appears as an explicit bit in the corresponding full description of the strategy.)

Note that when given oracle access to $z = a_1 \cdots a_\ell b_1 \cdots b_\ell$, we emulate a computation of T by providing it with oracle access to the virtual string $\mathtt{P}(a_1, b_1) \cdots \mathtt{P}(a_\ell, b_\ell)$. Now, if $z \in \Pi'$, then (by definition) the corresponding virtual string is in Π. On the other hand, if z is ϵ'-far from Π', then the virtual string must be ϵ'-far from Π, because otherwise it suffices to modify less than $\epsilon'\ell$ pairs of strategies in order to produce a string in Π (which contradicts the hypothesis that z is ϵ'-far from Π').[19] □

Finally, consider $F' : \{0,1\}^{2^{B+1}\ell} \to \{0,1\}^{2^B\ell}$ such that $F'(u, v) = u \oplus v$.

Claim 7.3.2 (relating Ψ' to Π'): *Let Ψ' be as in Eqs. (6) & (7) and Π' be as in* Eq. (8). *Then:*

1. *For every $(u, v) \in P' \cap S'$ it holds that $F'(u, v) \in \Pi'$.*
2. *For every $(u, v) \in P' \setminus S'$ it holds that $F'(u, v)$ is $\epsilon/2^B$-far from Π'.*

Proof: As in the proof of Claim 7.2.2, for every $(u, v) \in P'$ it holds that $u = u'0^{2^{B-1}\ell}$ and $v = 0^{2^{B-1}\ell}v'$, and so $F'(u, v) = u \oplus v = u'v'$. Also, in this case, there exists a unique $(x, y) \in P$ such that $u' = \langle A_1^x \rangle \cdots \langle A_\ell^x \rangle$ and $v' = \langle B_1^y \rangle \cdots \langle B_\ell^y \rangle$.

If $(u, v) \in P' \cap S'$, then the aforementioned (x, y) must be in $P \cap S$, and it follows that $F(x, y) \in \Pi$ (by the hypothesis of Theorem 3.1 regarding F (reproduced in Footnote 12)). It follows that $F'(u, v) = u'v' \in \Pi'$, because $u' = \langle A_1^x \rangle \cdots \langle A_\ell^x \rangle$ and $v' = \langle B_1^y \rangle \cdots \langle B_\ell^y \rangle$, whereas for each $i \in [\ell]$ it holds that A_i^x and B_i^y produce the ith bit in $F(x, y) \in \Pi$.

Having established Item 1, we turn to Item 2: If $(u, v) \in P' \setminus S'$, then the aforementioned (x, y) must be in $P \setminus S$, and it follows that $F(x, y)$ is ϵ-far from Π (by the theorem's hypothesis regarding F). In this case $F'(u, v) = \langle A_1^x \rangle \cdots \langle A_\ell^x \rangle \langle B_1^y \rangle \cdots \langle B_\ell^y \rangle$, where for each $i \in [\ell]$ it holds that A_i^x and B_i^y produce the ith bit in $F(x, y)$. Given that $F(x, y)$ is ϵ-far from Π (i.e., $F(x, y)$ differs in at least $\epsilon\ell$ positions from any ℓ-bit string in Π), it follows that for at least $\epsilon\ell$ of the $i \in [\ell]$ at least one of the corresponding strategies (i.e., A_i^x and B_i^y) must be modified to place the $2^B\ell$-bit long string in Π'. Hence, $F'(u, v)$ is $(\epsilon\ell/2^B\ell)$-far from Π'. □

Combining Claims 7.3.1 and 7.3.2, this completes the proof of the current proposition. ■

Digest. The auxiliary problems introduced in the current case are even more imposing. Specifically, in the auxiliary communication problem the input pair (x, y) is replaced by sequences that for every $i \in [\ell]$ contain the description of residual strategies that allow parties that hold x and y respectively to produce the value $f_i(x, y)$ (i.e., the ith bit in $F(x, y)$). This format allows for reducing Ψ' to $\epsilon/2^B$-testing Π' by using a simple combining operator, but this comes at the cost of using a complex auxiliary communication problem. Things will become

[19] This uses the assumption that $\mathtt{P} : \{0,1\}^{2^B} \to \{0,1\}$ is onto, which can be justified as in Sect. 7.1. Specifically, we can modify A_i^x so that it always starts by sending the bit 1, and let $\mathtt{P}(a, b) = \sigma$ if $a(\lambda) = 0$ and $b(1) = \sigma$.

even more complex in Sect. 7.3, where we handle randomized communication protocols for computing the f_i's.

7.3 Step 3: The General Case

Finally, we turn to the general case in which the bound B is guaranteed by arbitrary (randomized) protocols. That is, here we are only guaranteed that $B = \max_{i\in[\ell]}\{\mathtt{CC}_{\eta/n}(f_i)\}$, which means that we have to deal with randomized protocols (of error probability at most η/n).

The basic idea is to proceed as in Sect. 7.2, while using descriptions of residual randomized strategies, where a description of a residual randomized strategy consists of a sequence of descriptions of the corresponding residual deterministic strategies. This raises a difficulty, because not all possible descriptions (i.e., sequences) correspond to legitimate residual randomized strategies (since the descriptions may correspond to strategies that have higher (than η/n) error probability).[20] Hence, some additional tests will be required when reducing the ϵ'-testing of the (modified) auxiliary property Π' to the ϵ'-testing of the (original) property Π. Specifically, we shall test that at least a $1-\eta/n$ fraction of the pairs in the sequence produce the same bit.

Given the fact that additional tests are used, we seize the opportunity to also address a deficiency we have neglected in Sects. 7.1 and 7.2: the fact that we derived lower bounds for testing Π with smaller proximity parameters (i.e., ϵ/B and $\epsilon/2^B$, respectively). Our solution is to encode the aforementioned descriptions using an error correcting code that is *locally testable* (cf. [14, Def. 2.2]) and *locally decodable* (cf. [17]).

- Local decodability (i.e., decoding each bit in the message based on a constant number of queries to the possibly corrupted codeword) is essential for the emulation of the tester of the original property by a tester for the auxiliary property, because the original property refers to strings that appear in encoded form in the auxiliary property.
- Local testability (i.e., codeword testing) is essential for the testing of the modified Π', because this property contains certain sequences of codewords.

Lastly, it is important that this code has constant relative distance, but its rate does not matter, and so we may just use the Hadamard code. We shall denote this code by C, and denote its relative distance by δ_C. With these preliminaries in place, we are ready to prove the general case of Theorem 7.1.

Proposition 7.4 (Theorem 7.1, restated): *Let $\Psi = (P, S)$, $\Pi \subseteq \{0,1\}^\ell$, $\eta \geq \epsilon > 0$, $F : \{0,1\}^{2n} \to \{0,1\}^\ell$, and the f_i's be as in Theorem 3.1, and suppose that $B = \max_{i\in[\ell]}\{\mathtt{CC}_{\eta/n}(f_i)\}$. Then, there exists a communication problem $\Psi' = (P', S')$ such that $\mathtt{CC}_\eta(\Psi') \geq \mathtt{CC}_\eta(\Psi)$ and a property Π' such that $\mathtt{Q}_\eta(\epsilon', \Pi) \geq$*

[20] The issue is not the specific low level of error, but rather that we have to bound the error away from 1/2 so that we can effectively determine what bit is produced (with probability higher than 1/2) by a pair of residual randomized strategies.

$\frac{\mathrm{Q}_\eta(\epsilon'/2,\Pi')-\widetilde{O}(B/\epsilon')}{\widetilde{O}(B)\cdot\log(\mathrm{Q}_\eta(\epsilon'/2,\Pi')/\eta)}$ (for every $\epsilon' = \widetilde{\Omega}(1/n)$), *whereas* Ψ' *and* Π' *are related as follows:*

1. *For every* $(u,v) \in P' \cap S'$, *it holds that* $u \oplus v \in \Pi'$.
2. *For every* $(u,v) \in P' \setminus S'$, *it holds that* $u \oplus v$ *is* $\Omega(\epsilon)$*-far from* Π'.

Proof: For starters, by the hypothesis, for every $i \in [\ell]$, there exists a randomized two-party protocol of communication complexity at most B for computing f_i (with error probability at most η/n). This protocol is in the shared randomness model, and we denote by ρ the length of the random string in use.[21] Let A_i and B_i denote the corresponding strategies of the two parties, and let $A^x_{i,r} = A_i(x;r)$ and $B^y_{i,r} = B_i(y;r)$ denote the residual strategies for local inputs x and y and shared randomness $r \in \{0,1\}^\rho$. That is, $A^x_{i,r}(\gamma)$ denotes the answer of the first party, holding input x and viewing the shared randomness r, to a message-sequence γ sent by the second party (ditto for $B^y_{i,r}$).

Let $\langle A^x_{i,r}\rangle$ (resp., $\langle B^y_{i,r}\rangle$) denote a canonical 2^{B-1}-bit long description of the strategy $A^x_{i,r}$ (resp., $B^y_{i,r}$), and let $\langle A^x_i\rangle$ (resp., $\langle B^y_i\rangle$) denote the 2^ρ-long sequence of corresponding codewords (under the code C); that is, $\langle A^x_i\rangle \stackrel{\mathrm{def}}{=} (C(\langle A^x_{i,0^\rho}\rangle),...,C(\langle A^x_{i,1^\rho}\rangle))$ and $\langle B^y_i\rangle \stackrel{\mathrm{def}}{=} (C(\langle B^y_{i,0^\rho}\rangle),...,C(\langle B^y_{i,1^\rho}\rangle))$. Hence $L \stackrel{\mathrm{def}}{=} |\langle A^x_i\rangle| = 2^\rho \cdot n_C$, where $n_C = |C(1^{2^{B-1}})|$ denotes the length of the codewords in C. We make the simplifying assumption (with justifications as in Sect. 7.2) that the mappings $x \mapsto (\langle A^x_1\rangle,...,\langle A^x_\ell\rangle)$ and $y \mapsto (\langle B^y_1\rangle,...,\langle B^y_\ell\rangle)$ are one-to-one (and onto). We define $\Psi' = (P', S')$ such that

$$P' \stackrel{\mathrm{def}}{=} \{(\langle A^x_1\rangle \cdots \langle A^x_\ell\rangle 0^{\ell\cdot L}, 0^{\ell\cdot L}\langle B^y_1\rangle \cdots \langle B^y_\ell\rangle) : (x,y) \in P\} \qquad (9)$$

$$S' \stackrel{\mathrm{def}}{=} \{(\langle A^x_1\rangle \cdots \langle A^x_\ell\rangle 0^{\ell\cdot L}, 0^{\ell\cdot L}\langle B^y_1\rangle \cdots \langle B^y_\ell\rangle) : (x,y) \in S\}. \qquad (10)$$

Note that $\mathrm{CC}_\eta(\Psi) \le \mathrm{CC}_\eta(\Psi')$, since a protocol for Ψ can proceed by emulating the protocol for Ψ' (very much as in Sect. 7.2).

As in Sect. 7.2, we say that the (residual) deterministic strategies α and β produce the bit $\sigma = \mathrm{P}(\alpha,\beta)$ if emulating the interaction between these strategies yields the (joint) outcome σ. We say that a sequence of such pairs safely produce the bit σ if at least a $1-\eta/n$ fraction of the pairs in the sequence produce this bit; that is, $\mathrm{SP}((\alpha_{0^\rho},\beta_{0^\rho}),...,(\alpha_{1^\rho},\beta_{1^\rho})) = \sigma$ if $|\{r \in \{0,1\}^\rho : \mathrm{P}(\alpha_r,\beta_r) = \sigma\}| \ge (1-\eta/n)\cdot 2^\rho$.

Next, we introduce the property $\Pi' \subseteq \{0,1\}^{2\ell L}$. Loosely speaking, Π' will contain sequences of C-codewords that each encode ℓ sequences of pairs such that the ith sequence *safely* produces the ith bit of an ℓ-bit string in Π. Namely, for every sequence $(a_1,...,a_\ell,b_1,...,b_\ell)$ such that $a_i = (a_{i,0^\rho},...,a_{i,1^\rho}) \in \{0,1\}^{2^\rho\cdot 2^{B-1}}$ and $b_i = (b_{i,0^\rho},...,b_{i,1^\rho}) \in \{0,1\}^{2^\rho\cdot 2^{B-1}}$, the corresponding $2\ell\cdot 2^\rho\cdot n_C$-bit long string $C(a_{1,0^\rho})\cdots C(a_{\ell,1^\rho})C(b_{1,0^\rho})\cdots C(b_{\ell,1^\rho})$ is in Π' if and only if for every

[21] Indeed, we may assume (w.l.o.g., cf. [19, Thm. 3.14]) that $\rho \stackrel{\mathrm{def}}{=} O(\log(n/\eta))$, but this is not needed for our argument.

$i \in [\ell]$ the sequence of pairs $(a_{i,r}, b_{i,r})_{r\in\{0,1\}^\rho}$ safely produce a bit w_i such that $w = w_1 \cdots w_\ell \in \Pi$. That is:

$$\Pi' \stackrel{\rm def}{=} \left\{ c_{1,0^\rho} \cdots c_{2\ell,1^\rho} : \begin{array}{l} \exists a_{1,0^\rho}, ..., a_{\ell,1^\rho}, b_{1,0^\rho}, ..., b_{\ell,1^\rho} \in \{0,1\}^{2^{B-1}} \\ \text{such that} \\ (1)\ \forall i \in [\ell]\ \forall r \in \{0,1\}^\rho \\ \quad C(a_{i,r}) = c_{i,r} \wedge C(b_{i,r}) = c_{\ell+i,r} \\ (2)\ \exists w_1 \cdots w_\ell \in \Pi\ \forall i \in [\ell] \\ \quad \mathtt{SP}((a_{i,0^\rho}, b_{i,0^\rho}), ..., (a_{i,1^\rho}, b_{i,1^\rho})) = w_i \end{array} \right\}, \tag{11}$$

Claim 7.4.1 (relating Π' to Π): *Let Π' be as in* Eq. (11). *For every $\eta \geq \epsilon' = \widetilde{\Omega}(1/n)$ it holds that $\mathtt{Q}_\eta(\epsilon', \Pi') = \widetilde{O}(B/\epsilon') + \widetilde{O}(B \cdot \mathtt{Q}_\eta(\epsilon'/2, \Pi))$, where the polylogarithmic factor hidden in the second $\widetilde{O}$-notation is $O(\log(B \cdot \mathtt{Q}_\eta(\epsilon'/2, \Pi)/\eta)) \cdot \log B)$.*

Proof: Unlike the proofs of Claims 7.2.1 and 7.3.1, here ϵ'-testing Π' does not reduce to merely emulating an ϵ'-tester for Π, because here strings in Π' have additional structure – they are sequences of codewords that encode pairs (of residual randomized strategies) that *safely* produce some bits. Thus, in addition to emulating an $\epsilon'/2$-tester for Π, we would also perform codeword tests and consistency (i.e., "safe production") tests. We start by describing these new testing activities, while recalling that $n_C = |C(1^{2^{B-1}})|$ denote the length of the codewords in C.

On input $z = (z_1, ..., z_{2\ell \cdot 2^\rho})$, with each $z_i \in \{0,1\}^{n_C}$, we first check that this sequence is $\epsilon'/4$-close to a sequence of codewords of C. This can be done at a cost of $\widetilde{O}(1/\epsilon')$ queries, by selecting, for each $j \in [\lceil \log_2(8/\epsilon') \rceil]$, a random sample of $O(2^j \log(1/\epsilon'))$ indices $I \subseteq [2\ell \cdot 2^\rho]$ and performing an $2^{j-3}\epsilon'$-test (with error probability $\mathrm{poly}(\epsilon')$) on z_i for each $i \in I$. (Note that Levin's Economical Work Investment [11, Sec. 8.2.4] is employed here and below in order to obtain query complexity $\widetilde{O}(1/\epsilon')$ rather than $\widetilde{O}(1/\epsilon')^2$.)[22] The (strong) local testability of the code C asserts that ϵ''-testing its codewords with error probability 2^{-k} can be done by using $O(k/\epsilon'')$ queries.

Let $a_{1,0^\rho}, ..., a_{\ell,1^\rho}, b_{1,0^\rho}, ..., b_{\ell,1^\rho} \in \{0,1\}^{2^{B-1}}$ be such that the concatenation of the corresponding codewords (i.e., $C(a_{1,0^\rho}) \cdots C(a_{\ell,1^\rho}) C(b_{1,0^\rho}) \cdots C(b_{\ell,1^\rho})$) is closest to z. We now check that the sequence of $a_{i,r}$'s and $b_{i,r}$'s is $\epsilon'/4$-close to a sequence that safely produces ℓ bits (i.e., one bit per each value of $i \in [\ell]$), by selecting a sample of i's, taking a sample of r's for each i, and checking that the pairs $(a_{i,r}, b_{i,r})$ produce the same value for each such i. (That is, for each $j \in [\lceil \log_2(8/\epsilon') \rceil]$, we select a random sample of $O(2^j \log(1/\epsilon'))$ indices $I \subseteq [\ell]$ and take a sample of $O(1/(2^j \epsilon'))$ choices of $r \in \{0,1\}^\rho$ for each $i \in I$.)[23]

[22] Indeed, the straightforward method is to select a random sample of $O(1/\epsilon')$ indices $I \subseteq [2\ell \cdot 2^\rho]$ and performing an $(\epsilon'/4)$-test (with error probability $\mathrm{poly}(\epsilon')$) on z_i for each $i \in I$.

[23] Since $\epsilon' = \widetilde{\Omega}(1/n)$, we do not expect to see pairs that produces the opposite value, which is quite rare (i.e., appears in at most a η/n fraction of the pairs).

The aforementioned checking is performed while employing local decodability of the relevant bits (in the description of the strategy). We use local decodability with error probability $\mathrm{poly}(\epsilon'/B)$ (which is guaranteed to work up to relative distance $\delta_C/3$, where δ_C denotes the relative distance of the code C). Furthermore, each of these invocations of the local decodability procedure will also run an $(\delta_C/3)$-tester for C-codewords (again, with error probability $\mathrm{poly}(\epsilon'/B)$), and the tester (for Π') will reject whenever any invocation of the codeword tester rejects. Hence, each pair (i,r) that we check generates $O(B\log(B/\epsilon'))$ queries, whereas we check $\widetilde{O}(1/\epsilon')$ such pairs.

Finally, we get to emulate the execution of the $\epsilon'/2$-tester for Π, denoted T. Specifically, if T makes the query $i \in [\ell]$, then we access the ith pair of sequences (which is typically close to $(C(a_{i,r}))_{r\in\{0,1\}^\rho}$ and $(C(b_{i,r}))_{r\in\{0,1\}^\rho}$), and try to recover the answer by self-correction with error probability $\eta/O(Bq)$, where q is the query complexity of T. This self-correction procedure combines a self-correction for the bit produced by the pairs $(a_{i,r}, b_{i,r})_{r\in\{0,1\}^\rho}$, which in turn relies on local decodability of the relevant bits in the descriptions of the sampled $(a_{i,r}, b_{i,r})$-pairs. We also check whether these sequences are 1/4-close to safely produce this answer (bit), and each such check is also performed with error probability $\eta/O(Bq)$. This means that each query of T is emulated by using $O(B \cdot \log(Bq/\eta) \cdot \log B)$ queries, since we use the codeword tester and decoder with error probability $1/O(B)$ (while using constant proximity parameter in the testing).

If $z \in \Pi'$, then (by definition, cf. Eq. (11)) the string z is a concatenation of codewords that encode pairs that safely produce the bits of some $w \in \Pi$. Hence, when using z towards emulating the execution of T, with high probability, all queries made by T are answered by the corresponding bits of this w, and it follows that our tester accepts (with high probability).[24] On the other hand, if z is ϵ'-far from Π', then at least one of the following cases must hold:

1. Either z is $\epsilon'/4$-far from a sequence of C-codewords;
2. or z is C-decodable to a sequence $(a_{1,0^\rho}, ..., a_{\ell,1^\rho}, b_{1,0^\rho}, ..., b_{\ell,1^\rho})$ that is $\epsilon'/4$-far from safely producing bits of some ℓ-bit long;
3. or the string w that the foregoing sequence (safely) produces is $\epsilon'/2$-far from Π.

As argued next, in each of these cases, we reject with high probability. For Case 1 this follows from the various codeword tests that are performed, since in this case there exists an integer $j \in [\lceil \log_2(4/\epsilon') \rceil]$ such that at least a $1/O(2^j \log(1/\epsilon'))$ fraction of the (n_C-bit long) blocks are $2^{j-3}\epsilon'$-far from the code C. Assuming that Case 1 does not hold, we consider the foregoing sequence of C-decodings $(a_{i,r})_{i,r}, (b_{1,r})_{i,r}$, and what happens when Case 2 holds. In this case, with very high probability, we either detect pairs $(a_{i,r}, b_{i,r})$ and $(a_{i,r'}, b_{i,r'})$ that produce

[24] Note that we may also reject, with very small probability, due to encoutering pairs that produce different values (within a sequence of pairs that safely produces a value). But since the fraction of exceptional pairs is at most η/n, this event occurs with very small probability.

different values (via the self-correction) or detect corresponding blocks that are $\delta_C/3$-far from the code C. Finally, assuming that Cases 1 and 2 do not hold, we consider the foregoing ℓ-bit string w that the said sequence produces. In this case, we either detect a problem when emulating T (i.e., indices $i \in [\ell]$ that correspond to bits that are 1/4-far from being safely produced, or blocks that are $\delta_C/3$-far from C-codewords) or we complete an emulation of T^w, which rejects (with high probability). The claim follows. □

Finally, consider $F' : \{0,1\}^{4\ell L} \to \{0,1\}^{2\ell L}$ such that $F'(u,v) = u \oplus v$.

Claim 7.4.2 (relating Ψ' to Π'): *Let Ψ' be as in Eqs. (9) & (10) and Π' be as in* Eq. (11). *Then:*

1. *For every $(u,v) \in P' \cap S'$ it holds that $F'(u,v) \in \Pi'$.*
2. *For every $(u,v) \in P' \setminus S'$ it holds that $F'(u,v)$ is $\Omega(\epsilon)$-far from Π'.*

Proof: As in the proofs of Claims 7.2.2 and 7.3.2, for every $(u,v) \in P'$ it holds that $u = u'0^{\ell L}$ and $v = 0^{\ell L}v'$, and so $F'(u,v) = u \oplus v = u'v'$. Also, in that case there exists a unique $(x,y) \in P$ such that $u' = \langle A_1^x \rangle \cdots \langle A_\ell^x \rangle$ and $v' = \langle B_1^y \rangle \cdots \langle B_\ell^y \rangle$.

If $(u,v) \in P' \cap S'$, then the aforementioned (x,y) must be in $P \cap S$, and it follows that $F(x,y) \in \Pi$ (by the hypothesis of Theorem 3.1 regarding F (reproduced in Footnote 12)). It follows that $F'(u,v) = u'v' \in \Pi'$, because for each $i \in [\ell]$ it holds that $\langle A_i^x \rangle$ and $\langle B_i^y \rangle$ encode a sequence of pairs that safely produce the ith bit in $F(x,y) \in \Pi$.

Having established Item 1, we turn to Item 2: If $(u,v) \in P' \setminus S'$, then the aforementioned (x,y) must be in $P \setminus S$, and it follows that $F(x,y)$ is ϵ-far from Π (by the theorem's hypothesis regarding F). In this case, for each $i \in [\ell]$, it holds that $\langle A_i^x \rangle$ and $\langle B_i^y \rangle$ encode a sequence of pairs that safely produce the ith bit in $F(x,y)$. Given that $F(x,y)$ is ϵ-far from Π (i.e., $F(x,y)$ differs in at least $\epsilon\ell$ positions from any ℓ-bit string in Π), it follows that for at least $\epsilon\ell$ of the $i \in [\ell]$ either $\langle A_i^x \rangle$ or $\langle B_i^y \rangle$ should be modified such that the encoded sequences safely produce a different value for the ith bit. Recalling that each of the above is a sequence of 2^ρ codewords and that a vast majority of the C-decodable pairs produce the current value (of this bit), it follows that we need to change more than half of these codewords. Since the code C has (constant) relative distance δ_C, this means that we need to change more than $2^{\rho-1} \cdot \delta_C n_C = \Omega(L)$ bits per each such i, which implies that $F'(u,v)$ is $\Omega(\epsilon\ell L/2\ell L)$-far from Π'. The claim follows. □

Combining Claims 7.4.1 and 7.4.2, this completes the proof of the current proposition. ■

A Talmudic Comment: The proof of Proposition 7.3 can be carried out for $B = \max_{i\in[\ell]}\{\mathtt{CC}_{1/3}(f_i)\}$, at the cost of an additive overhead of $\widetilde{O}(B/(\epsilon')^2)$ (rather than $\widetilde{O}(B/\epsilon')$) in Claim 7.4.1. In light of this fact, it seems fair to reconsider the comparison made right after stating Theorem 7.1. In this case (i.e., starting with $B = \max_{i\in[\ell]}\{\mathtt{CC}_{1/3}(f_i)\}$), applying Theorem 3.1 requires performing error-reduction first (i.e., use $\mathtt{CC}_{\eta/n}(f_i) = O(\mathtt{CC}_{1/3}(f_i) \cdot \log(n/\eta))$). Actually, for $C =$

$\mathtt{CC}_{2\eta}(\Psi)$, we can use $\mathtt{CC}_{\eta/C}(f_i) = O(\mathtt{CC}_{1/3}(f_i) \cdot \log(C/\eta))$, since the proof of Theorem 3.1 holds also for $B = \max_{i\in[\ell]}\{\mathtt{CC}_{\eta/C}(f_i)\}$. In this case, for every fixed $\eta > 0$, we get $\mathtt{Q}_\eta(\epsilon, \Pi) \geq C/O(B \log C)$ by using the general formulation, which is closer to the *rough bound*[25] of $\mathtt{Q}_\eta(\Omega(\epsilon), \Pi) \geq C/O(B \log BC)$ that we get by the restricted formulation.

8 Conclusions

As demonstrated in Sect. 4, using the general formulation provided in Theorem 3.1 frees the user from the need to introduce *auxiliary* communication complexity problems as a bridge between known communication complexity problems and property testing problems. Recall that these auxiliary problems are needed because it is not clear how to directly reduce the original communication complexity problems (for which lower bounds are known) to the targeted property testing problems when using simple combining operators (as in [4], cf. Corollary 3.3).[26] In contrast, such direct reductions are easy to design when using the general formulation of Theorem 3.1. This phenomenon is not specific to the examples presented Sect. 4: In fact, it seem to arise in all known applications of the communication complexity methodology (starting from [4] itself).

We believe that the simpler it is to apply a methodology, the more useful the methodology becomes. Work should be shifted from the user (of the methodology) to the methodology itself (or rather to the proof of its validity). We believe that this is done by moving from the restricted formulation of [4] (cf. Corollary 3.3) to the general formulation of Theorem 3.1. The shifting of work is evident when trying to emulate results obtained via the general formulation by the restricted one, as done in Sect. 7. Indeed, we believe that the results of Sect. 7 demonstrate that while the general formulation is not much more powerful (as far as the obtainable lower bounds are concerned), it may be far easier to use (e.g., since the emulations that we found are quite imposing).

Acknowledgments. Part of this work is based on joint research with Dana Ron, who refused to co-author it. The constructions presented in Sect. 4 are partially inspired by some constructions in [15]. We are grateful to David Woodruff for suggesting Theorem 5.2 and allowing us to present it here. Ditto with respect to Tom Gur for Theorem A.2. We thank Eric Blais and Sofya Raskhodnikova for calling our attention to [5]. Lastly, we thank Tom Gur and Ron Rothblum for comments on an early draft of this paper.

---

[25] Indeed, this rough bound neglects the aforementioned additive terms, which are insignificant for constant $\epsilon > 0$.

[26] Instead, one reduces the original communication complexity problem to the auxiliary one, and then reduces the auxiliary communication problem to the property testing problem. The first reduction is performed within the setting of communication complexity, whereas the second reduction is the one in which simple combing operators are used.

Appendix: Generalization to Multi-Party Communication

The formulation presented in Sect. 3 generalizes easily to the model of multi-party communication. The treatment is quite oblivious of the details of the model; for example, it does not matter if one considers the standard model of "input on the forehead" or to the more natural model in which each party gets a part of the input (with no overlap). (These variations can be captured by the promise problems that the parties wish to solve.) The exact way in which the parties communicate is also not crucial, at least as long as the number of parties (denoted m) is small. For simplicity, we consider here a broadcast model, where in each communication round there is a single designated sender (determined by the transcript of the communication so far).

In light of the above, we consider m-party communication protocols in which the local input of the jth party is denoted $x^{(j)}$. Let $\langle A^{(1)}(x^{(1)}), ..., A^{(m)}(x^{(m)})\rangle(r)$ denote the (joint) output of the m parties, when the jth party uses strategy $A^{(j)}$ and gets input $x^{(j)}$, and all parties have free access to the shared randomness r. Considering promise problems $\Psi = (P, S)$ such that $P, S \subseteq \{0,1\}^{m \cdot n}$, Definition 2.1 extends naturally; that is, the η-error communication complexity of Ψ, denoted $\mathsf{CC}_\eta(\Psi)$, is the minimum communication complexity of all m-protocols that solve Ψ with error at most η.

Theorem A.1 (Theorem 3.1, generalized to m-party protocols): *Let $\Psi = (P, S)$ be a promise problem such that $P, S \subseteq \{0,1\}^{m \cdot n}$, and let $\Pi \subseteq \{0,1\}^\ell$ be a property, and $\epsilon, \eta > 0$. Suppose that the mapping $F : \{0,1\}^{m \cdot n} \to \{0,1\}^\ell$ satisfies the following two conditions:*

1. *For every $(x^{(1)}, ..., x^{(m)}) \in P \cap S$, it holds that $F(x^{(1)}, ..., x^{(m)}) \in \Pi$.*
2. *For every $(x^{(1)}, ..., x^{(m)}) \in P \setminus S$, it holds that $F(x^{(1)}, ..., x^{(m)})$ is ϵ-far from Π.*

Then, $\mathsf{Q}_\eta(\epsilon, \Pi) \geq \mathsf{CC}_{2\eta}(\Psi)/B$, where $B = \max_{i \in [\ell]}\{\mathsf{CC}_{\eta/n}(f_i)\}$ and $f_i(x^{(1)}, ..., x^{(m)})$ is the ith bit of $F(x^{(1)}, ..., x^{(m)})$. Furthermore, if $B = \max_{i \in [\ell]}\{\mathsf{CC}_0(f_i)\}$, then $\mathsf{Q}_\eta(\epsilon, \Pi) \geq \mathsf{CC}_\eta(\Psi)/B$.

Theorem A.1 is proved by a straightforward generalization of the proof of Theorem 3.1; that is, we merely replace "two" by "m" (and everything goes through). We believe that this generalization further clarifies the ideas underlying the proof of Theorem 3.1 by presenting them in a slightly more abstract form.

Proof: The following description applies to any communication model in which all parties obtain the output produced by the protocol. Given an ϵ-tester with error η for Π and communication protocols for the f_i's, we present a protocol for solving Ψ. The key idea is that, using their shared randomness, the parties (holding the inputs $x^{(1)}, ..., x^{(m)}$, respectively) can emulate the execution of the ϵ-tester, while providing it with virtual access to $F(x^{(1)}, ..., x^{(m)})$. Specifically, when the tester queries the ith bit of the oracle, the parties provide it with the value of $f_i(x^{(1)}, ..., x^{(m)})$ by first executing the corresponding communication protocol. Details follow.

The protocol for Ψ proceeds as follows: On local input $x^{(j)}$ and shared randomness $r = (r_0, r_1, ..., r_\ell) \in (\{0,1\}^*)^{\ell+1}$, the jth party invokes the ϵ-tester on randomness r_0, and answers the tester's queries by interacting with the other parties. That is, each of the parties invokes a local copy of the tester's program, but all copies are invoked on the same randomness, and are fed with identical answers to their (identical) queries. When the tester issues a query $i \in [\ell]$, the parties compute the value of $f_i(x^{(1)}, ..., x^{(m)})$ by using the corresponding communication protocol, and feed $f_i(x^{(1)}, ..., x^{(m)})$ to (their local copy of) the tester. Specifically, denoting the latter protocol (i.e., sequence of strategies) by $(A_i^{(1)}, ..., A_i^{(m)})$, the parties answer with $\langle A_i^{(1)}(x^{(1)}), ..., A_i^{(m)}(x^{(m)})\rangle(r_i)$. When the tester halts, each party outputs the output it has obtained from (its local copy of) the tester.

We stress that the above description is oblivious to the details of the communication model, as long as in this model all parties obtain the output produced by the protocol.[27] Indeed, the description presented in the proof of Theorem 3.1 is merely a special case (which corresponds to the standard model of two-party computation), and the analysis of the general case (omitted here) is identical to the analysis of the special case presented in the proof of Theorem 3.1. ■

On the Potential Usefulness of the Generalization. Tom Gur has pointed out that the generalization to multi-party communication complexity allows additional flexibility for the design of reductions. To illustrate the point, he suggested the proof outlined below, which refers to a multi-party communication complexity model in which parties obtain non-overlapping inputs and communication is by individual point-to-point channels.

Theorem A.2 (a property testing (encoded) version of the frequency moment problem of [2]):[28] *For $k(n) = n/2$ and $\ell(n) = n^{1+o(1)}$, let $\mathcal{F}$ be a finite field of size n, and $C : \mathcal{F}^{k(n)} \to \mathcal{F}^{\ell(n)}$ be a $\mathcal{F}$-linear code of constant relative distance, denoted δ. For any sequence $x = (x_1, ..., x_k) \in \mathcal{F}^k$ and $v \in \mathcal{F}$, let $\#_v(x)$ denote the number of occurrences of v in x; that is, $\#_v(x) = |\{i \in [k] : x_i = v\}|$. For any constant $c > 1$, let*

$$\Pi = \left\{ C(x) : x \in \mathcal{F}^{k(n)} \wedge \sum_{v \in \mathcal{F}} \#_v(x)^c = k(n) \right\} \tag{12}$$

$$\Pi' = \left\{ C(x) : x \in \mathcal{F}^{k(n)} \wedge \sum_{v \in \mathcal{F}} \#_v(x)^c \leq 2k(n) \right\}. \tag{13}$$

[27] If only a designated subset of the parties obtains the output, then we can emulate only nonadaptive testers (as done in Sect. 5).

[28] The following problem differs from the one in [2] in two aspects. Firstly, the computational model is different (i.e., we consider the query complexity of property testing, whereas [2] refers to the space complexity of streaming algorithms). Secondly, the problems are different: We consider an error-correcting encoding (i.e., $C(x)$) of the information (i.e., x) to which the frequency measure is applied. We stress, however, that the lower bound is not due to the complexity of codeword testing, since codeword testing may be easy for $\ell(n) = k(n)^{1+o(1)}$ (cf., e.g., [14]).

Then, distinguishing inputs in Π from inputs that are δ-far from Π' requires $\Omega(\ell(n)^{1-(7/c)})$ queries.

Indeed, it follows that testing Π requires query complexity $\Omega(\ell(n)^{1-(7/c)})$, but this (and, in fact, a stronger $\Omega(n/\log n)$ lower bound) can be proved by reduction from a two-party communication complexity problem (i.e., DISJ).[29] In contrast, Theorem A.2 refers to a doubly-relaxed decision problem, where one level of relaxation is the approximation of the norm (captured by the gap between Π and Π') and the second is the standard property testing relaxation (captured by the gap between Π' and δ-far from Π'). Such doubly-relaxed problems have been often considered in the property testing literature (cf., e.g., [1,20]), starting with [18]. The following proof, which adapts a proof of [2] (which in turn refers to streaming algorithms), relies on a reduction from a multi-party communication problem. As is the case with its streaming original [2], it is not clear whether Theorem A.2 can be proved by reduction from a two-party communication problem.

Proof: We shall use a reduction from the following multi-party communication problem, denoted (m,t)-DISJ$_n$. In this problem, there are m parties, each holding a t-subset of $[n]$, and the problem is to distinguish the case that the subsets are pairwise disjoint from the case that the intersection of all subsets is non-empty. By [2], if $n \geq 2mt - m + 1$, then the communication complexity of (m,t)-DISJ$_n$ (in the point-to-point channels model) is $\Omega(t/m^3)$.[30]

We set $m = n^{1/c}$ and $t = n/2m$ (so that $n = 2mt$), and represent the input of the jth party by a sequence $x^{(j)} \in \mathcal{F}^t$. Recall that $|\mathcal{F}| = n$ and $k(n) = n/2 = mt$. Now, we let $F(x^{(1)},...,x^{(m)}) = C(x^{(1)} \cdots x^{(m)})$, which equals $\sum_{j\in[m]} C(0^{(j-1)t}x^{(j)}0^{(m-j)t})$ by the $\mathcal{F}$-linearity. Hence, each bit of $F(x^{(1)},...,x^{(m)})$ can be computed (in this communication model) by communicating $m^2 \log_2 n$ bits (i.e., each party sends a single field elements to each of the other parties). Note that if $x = (x^{(1)},...,x^{(m)})$ is a YES-instance of (m,t)-DISJ$_n$ then $\sum_{v\in\mathcal{F}} \#_v(x)^c = mt = k(n)$, since each element that occurs in x occurs in it exactly once (i.e., in one of the $x^{(j)}$'s), which means that $F(x^{(1)},...,x^{(m)})$ is in Π. On the other hand, if $(x^{(1)},...,x^{(m)})$ is a NO-instance of (m,t)-DISJ$_n$ then $\sum_{v\in\mathcal{F}} \#_v(x)^c > m^c = n = 2k(n)$, since at least one element occurs m times

[29] Indeed, this follows from the proof of Theorem A.2, when setting $m = 2$, which correspond to the two-party case, and observing that NO-instances are mapped to instances having norm at least $m^c + (t-1)m > tm = k(n)$. Note that the same lower bound can be proved for Π', by padding the inputs to DISJ with an adequate number of repetitions of some fixed symbol. Note that these arguments rely on the fact that testing Π (or Π') requires distinguishing codewords that encode information (i.e., x) with a norm below some threshold from codewords that encode information with norm just above that threshold. In contrast, Theorem A.2 refers to a relaxation that captures an approximation of the corresponding norm, and a straightforward adaptation of the reduction from the two-party case does not seem to work here.

[30] The result of [2] is actually stronger, since it refers to the case that the NO-instances consist of subsets that have pairwise intersections that all equal the same singleton.

(i.e., in all the $x^{(j)}$'s), which means that $F(x^{(1)}, ..., x^{(m)})$ is not in Π', and so it is δ-far from any codeword in Π' (since it is itself a codeword).

Applying Theorem A.1,[31] it follows that the query complexity of the promise problem of distinguishing Π from the set of $\ell(n)$-long sequences that are δ-far from Π' is lower bounded by $\Omega(t/m^3)/(m^2 \log n)$, which equals $\Omega(n/(m^6 \log n)) = \Omega(n^{1-6(1+o(1))/c})$. Using $n = \ell(n)^{1/(1+o(1))}$, the claim follows. ■

References

1. Alon, N., Dar, S., Parnas, M., Ron, D.: Testing of clustering. SIAM J. Discrete Math. **16**(3), 393–417 (2003)
2. Alon, N., Matias, Y., Szegedy, M.: The space complexity of approximating the frequency moments. J. Comput. Syst. Sci. **58**(1), 137–147 (1999)
3. Ben-Sasson, E., Harsha, P., Raskhodnikova, S.: Some 3CNF properties are hard to test. SIAM J. Comput. **35**(1), 1–21 (2005)
4. Blais, E., Brody, J., Matulef, K.: Property testing lower bounds via communication complexity. Comput. Complex. **21**(2), 311–358 (2012). https://doi.org/10.1007/s00037-012-0040-x
5. Blais, E., Raskhodnikova, S., Yaroslavtsev, G.: Lower bounds for testing properties of functions on hypergrid domains. In: ECCC, TR13-036, March 2013
6. Dodis, Y., Goldreich, O., Lehman, E., Raskhodnikova, S., Ron, D., Samorodnitsky, A.: Improved testing algorithms for monotonicity. In: Hochbaum, D.S., Jansen, K., Rolim, J.D.P., Sinclair, A. (eds.) APPROX/RANDOM 1999. LNCS, vol. 1671, pp. 97–108. Springer, Heidelberg (1999). https://doi.org/10.1007/978-3-540-48413-4_10
7. Chor, B., Goldreich, O.: Unbiased bits from sources of weak randomness and probabilistic communication complexity. SIAM J. Comput. **17**(2), 230–261 (1988)
8. Goldreich, O. (ed.): Property Testing – Current Research and Surveys. LNCS, vol. 6390. Springer, Heidelberg (2010). https://doi.org/10.1007/978-3-642-16367-8
9. Goldreich, O.: Short locally testable codes and proofs: a survey in two parts. In: [8]
10. Goldreich, O.: On testing computability by small width OBDDs. In: Serna, M., Shaltiel, R., Jansen, K., Rolim, J. (eds.) APPROX/RANDOM 2010. LNCS, vol. 6302, pp. 574–587. Springer, Heidelberg (2010). https://doi.org/10.1007/978-3-642-15369-3_43
11. Goldreich, O.: Introduction to Property Testing. Cambridge University Press, Cambridge (2017)
12. Goldreich, O., Goldwasser, S., Ron, D.: Property testing and its connection to learning and approximation. J. ACM **45**, 653–750 (1998). (Extended abstract in 37th FOCS, 1996)
13. Goldreich, O., Ron, D.: Property testing in bounded degree graphs. Algorithmica **32**, 302–343 (2002). https://doi.org/10.1007/s00453-001-0078-7
14. Goldreich, O., Sudan, M.: Locally testable codes and PCPs of almost linear length. J. ACM **53**(4), 558–655 (2006)

[31] Actually, we need to generalize Theorem A.1 so that it applies to doubly-relaxed problems. Such a generalization is straightforward.

15. Gur, T., Rothblum, R.: Non-interactive proofs of proximity. In: ECCC, TR13-078, May 2013
16. Kalyanasundaram, B., Schintger, G.: The probabilistic communication complexity of set intersection. SIAM J. Discrete Math. **5**(4), 545–557 (1992)
17. Katz, J., Trevisan, L.: On the efficiency of local decoding procedures for error-correcting codes. In: Proceedings of the 32nd ACM Symposium on the Theory of Computing, pp. 80–86 (2000)
18. Kearns, M., Ron, D.: Testing problems with sub-learning sample complexity. J. Comput. Syst. Sci. **61**(3), 428–456 (2000)
19. Kushilevitz, E., Nisan, N.: Communication Complexity. Cambridge University Press, Cambridge (1997)
20. Parnas, M., Ron, D.: Testing the diameter of graphs. Random Struct. Algorithms **20**(2), 165–183 (2002)
21. Ron, D.: Property testing: a learning theory perspective. Found. Trends Mach. Learn. **1**(3), 307–402 (2008)
22. Ron, D.: Algorithmic and analysis techniques in property testing. Found. Trends TCS **5**(2), 73–205 (2009)
23. Rubinfeld, R., Sudan, M.: Robust characterization of polynomials with applications to program testing. SIAM J. Comput. **25**(2), 252–271 (1996)

Super-Perfect Zero-Knowledge Proofs

Oded Goldreich and Liav Teichner

Abstract. We initiate a study of super-perfect zero-knowledge proof systems. Loosely speaking, these are proof systems for which the interaction can be perfectly simulated in strict probabilistic polynomial-time. In contrast, the standard definition of perfect zero-knowledge only requires that the interaction can be perfectly simulated by a strict probabilistic polynomial-time that is allowed to fail with probability at most one half.

We show that two types of perfect zero-knowledge proof systems can be transformed into super-perfect ones. The first type includes the perfect zero-knowledge interactive proof system for Graph Isomorphism and other systems of the same form, including perfect zero-knowledge arguments for NP. The second type refers to perfect non-interactive zero-knowledge proof systems. We also present a super-perfect non-interactive zero-knowledge proof system for the set of Blum integers.

An early version of this work appeared as TR14-097 of *ECCC*. The current revision is quite minimal.

1 Introduction

A standard exposition of the notion of zero-knowledge proofs may start by presenting the following oversimplified definition:

> An interactive proof system (P, V) for a set S is called zero-knowledge if for every probabilistic polynomial-time strategy V^* there exists a (strict) probabilistic polynomial-time algorithm (called a simulator) A^* such that $A^*(x)$ is distributed identically to the output of V^* after interacting with P on common input x.

(See, e.g., Definition 9.7 in [12, Sec. 9.2.1] and top page 201 in [10, Sec. 4.3.1].)

However (as stated at the bottom of page 201 in [10, Sec. 4.3.1]), the problem with this oversimplified definition is that it is not known to be materializable (for sets outside $\mathcal{BPP}$). Indeed, [12, Def. 9.7] is labeled "oversimplified" and [10, Sec. 4.3.1] avoids presenting it formally. Instead, the standard definition of *perfect* zero-knowledge (cf. [10, Def. 4.3.1]) relaxes the above requirement by allowing the simulator to output a special failure symbol (i.e., $\perp$) with probability at most one half, and requires a perfect simulation conditioned on not failing. We stress that in both cases, the simulator is required to run in *strict* polynomial time.[1]

[1] Note that this definition of perfect zero-knowledge implies that a perfect simulation can be generated in *expected* (probabilistic) polynomial-time, but the latter does not imply the former. Also recall that the issue does not arise for statistical zero-knowledge, since the failure probability can be made exponentially vanishing (by repeated trials), and then absorbed in the statistical deviation of the simulation. Ditto for computational zero-knowledge.

O. Goldreich (Ed.): Computational Complexity and Property Testing, LNCS 12050, pp. 119–140, 2020.
https://doi.org/10.1007/978-3-030-43662-9_8

In this work, we take the "bold" step of turning the oversimplified definition to an actual definition, which we call *super-perfect* zero-knowledge (ZK). We obtain a few positive results regarding this notion, indicating that it is not a vacuous notion; that is, that it can be materializable non-trivially (i.e., for sets outside $\mathcal{BPP}$). Actually, super-perfect zero-knowledge was implicitly considered by Malka [18, Sec. 4.1] (see further discussion below).

The following overview assumes familiarity with the basic definitions and notations, which are reviewed in Sect. 2.

1.1 Our Results

We present several indications that super-perfect ZK exists beyond $\mathcal{BPP}$. Each of these results comes with some limitations (e.g., losing perfect completeness, being applicable only to argument systems or only to perfect NIZK, or holding only for a specific set).

The Case of Verifier-Oblivious Simulation Failure. Our first result presents a sufficient condition for the existence of super-perfect ZK proof systems. It asserts that any perfect ZK proof system in which *all the relevant simulators output* $\perp$ *with probability that may depend on the input but not on the verifier* (whose interaction with the prover is simulated) *can be converted into super-perfect ZK proof system.* This transformation preserves the soundness error but not the completeness error; in particular, it does not preserve perfect completeness. Specifically, in Sect. 3.1, we prove

Theorem 1 (from perfect ZK to super-perfect ZK): *Suppose that* (P, V) *is an interactive proof system for* S *and that there exists a function* $p : S \to [0, 0.5]$ *such that for every probabilistic polynomial-time strategy* V^* *there exists a probabilistic polynomial-time algorithm* A^* *such that for every* $x \in S$ *it holds that* $\mathbf{Pr}[A^*(x) = \perp] = p(x)$ *and* $\mathbf{Pr}[A^*(x) = \gamma \mid A^*(x) \neq \perp] = \mathbf{Pr}[\langle P, V^* \rangle(x) = \gamma]$*, for every* $\gamma \in \{0,1\}^*$*. Then,* S *has a super-perfect zero-knowledge proof system. Furthermore:*

- *The soundness error is preserved and the increase in the completeness error is exponentially vanishing;*
- *black-box simulation is preserved;*
- *the communication complexities* (i.e., number of rounds and length of messages) *are preserved; and*
- *the new prover strategy can be implemented by a probabilistic polynomial-time oracle machine that is given oracle access to the original prover strategy.*

The same holds for computationally-sound proof systems (a.k.a argument systems).

Theorem 1 is proved by observing that the transformation proposed by Malka [18, Sec. 4.1] applies whenever all simulators fail with the same probability (for each fixed input), and not merely when this probability equals one half. We stress that it is not even required that this probability (i.e., the function p) be efficiently computable. (Essentially, the new prover first invokes a simulator (say for V itself) and proceeds with the original proof system if and only if the output is not $\perp$; otherwise, the interaction is suspended and the verifier rejects.) As noted by Malka, one notable example of an interactive proof system that satisfies the foregoing condition (with $p = 1/2$) is the perfect zero-knowledge proof system for Graph Isomorphism of Goldreich, Micali, and Wigderson [13]. The condition holds also for numerous other interactive proofs that have the same form, including the perfect zero-knowledge arguments for $\mathcal{NP}$ of Naor *et al.* [19] (see also [10, Sec. 4.8.3]). Hence, *assuming the existence of one-way permutations, every set in* $\mathcal{NP}$ *has a super-perfect ZK argument system.*

The Case of Perfect ZK Arguments. In contrast to the previous transformation, the following one does preserve perfect completeness. It refers to a *certain class* of perfect ZK *arguments*, and yields super-perfect ZK *arguments* with perfect completeness, assuming the existence of perfectly binding commitment schemes (which can be constructed based on any one-way permutation). The class (see Definition 3.4) includes the aforementioned proof system for Graph Isomorphism and the perfect zero-knowledge arguments for $\mathcal{NP}$ of Naor *et al.* [19].[2] For details see Sect. 3.2. (We mention that super-perfect ZK arguments (of perfect completeness) for $\mathcal{NP}$ are implicit in the work of Pass and Rosen [20,21], were they are based on the existence of claw-free pairs of permutations and established using non-black-box simulators.)

The Case of Perfect NIZKs. Another case in which perfect completeness can be preserved is the case of *non-interactive zero-knowledge (NIZK)* proof systems. Specifically, we refer to perfect NIZK system in which *the probability that the simulator outputs* $\perp$ *is efficiently computable.* (Recall that in setting of NIZK there is only one simulator, and it simulates the distribution $(\omega, P(x,\omega))$ where ω is a uniformly distributed "common reference string" (and $P(x,\omega)$ is the proof provided by the prover for input x under the reference strong ω).)

Theorem 2 (from perfect NIZK to super-perfect NIZK): *Suppose that* (P,V) *is a non-interactive proof system for* S *and that there exist a polynomial-time computable function* $p : S \to [0, 0.5]$ *and a probabilistic polynomial-time algorithm* A *such that for every* $x \in S$ *it holds that* $\mathbf{Pr}[A(x) = \perp] = p(x)$ *and* $\mathbf{Pr}[A(x) = \gamma \mid A(x) \neq \perp] = \mathbf{Pr}_\omega[(\omega, P(x,\omega)) = \gamma]$, *for every* $\gamma \in \{0,1\}^*$. *Then,* S *has a super-perfect non-interactive zero-knowledge proof system. Furthermore:*

- *The completeness error is preserved and the increase in the soundness error is exponentially vanishing;*

[2] This holds only in the non-standard model of PPTs, discussed in Sect. 1.2. Ditto for the result of Pass and Rosen [20,21] (mentioned next).

- *the proof length is preserved; and*
- *the new prover algorithm can be implemented by a probabilistic polynomial-time oracle machine that is given oracle access to the original prover algorithm.*

This presumes a (non-standard) *model of probabilistic polynomial-time machines that are equipped with a special device that when fed with an integer* n, *returns an element uniformly distributed in* $[n] \stackrel{\text{def}}{=} \{1, ..., n\}$. (See discussion in Sect. 1.2.)

(Note that in the standard model, where such a device is not provided, a strict probabilistic polynomial-time can select a uniformly distributed element in $[n]$ if and only if n is a power of 2.) Unfortunately, we are not aware of any perfect NIZK systems to which Theorem 2 can be applied.[3] (We note that Theorem 2 is proved by a transformation akin to the one used in the proof of Theorem 1, except that the simulation error is moved to the soundness error rather than to the completeness error; this can be done because the common reference string can be trusted by both parties.) Using different ideas, we present a super-perfect NIZK system for a set that is widely believed to be outside of $\mathcal{BPP}$.

Theorem 3 (a super-perfect NIZK for Blum Integers): *Let* B *denote the set of all natural numbers that are of the form* $p^e q^d$ *such that* $p \equiv q \equiv 3 \pmod 4$ *are different odd primes and* $e \equiv d \equiv 1 \pmod 2$. *Then,* B *has a super-perfect NIZK.*

We also use the idea underlying the proof of Theorem 3 for presenting a promise problem that is complete for the class of promise problems having super-perfect NIZK of perfect completeness. The yes-instances of this promise problem are circuits that generate uniform distributions and the no-instances are circuits that generate distributions that cover at most half of the relevant range. For details, see Sect. 5.3.

1.2 Models of PPT

As noted above, the standard model of (strict) PPT refers to machines that can only toss fair coins, and such machines cannot generate a uniform distribution over $\{1, 2, 3\}$. In contrast, one may consider non-standard models (of PPT). One such model allows the machine to sample the uniform distribution over $\{1, ..., c\}$ for some fixed (constant) integer $c \geq 2$; indeed, this is a generalization of standard model where $c = 2$. A more powerful model is one in which a PPT machine is equipped with a special device that when fed with an integer n, returns an element uniformly distributed in $[n] \stackrel{\text{def}}{=} \{1, ..., n\}$. Actually, two such models are possible:

1. A model in which the PPT machine provides n in binary, which allows the machine to obtain a uniform distribution over $[n]$ also when n is exponential in the machine's input length. This is the model used in Theorem 2.

[3] We are only aware of the perfect NIZK arguments of Groth *et al.* [16], but these are in a more liberal model that allows the common reference string to be distributed according to any efficiently sampleable distribution.

2. A model in which the PPT machine provides n in unary (i.e., as 1^n), which allows the machine to obtain a uniform distribution over $[n]$ only when n is polynomial in its input length. This is the model used in our reference to [20, 21], where this ability is used to generate a random permutation over $[n]$.

Note that the issue does not arise in case the PPT machine is allowed to fail with bounded probability (as is the case with the PPT simulators underlying the definition of perfect ZK). We note that standard expositions of perfect ZK simulators seem to refer to the non-standard model of PPT, but they can be easily converted to the standard model by implementing the said device by a machine that is allowed to fail with bounded probability.[4]

1.3 Organization

We start (Sect. 2) by recalling the standard definitions underlying this work. Our results regarding interactive proofs and arguments are proved in Sects. 3.1 and 3.2, respectively. In particular, the proof of Theorem 1 appears in Sect. 3.1. Our results regarding non-interactive ZK systems appear in Sects. 4 and 5. In particular, Theorem 2 is proved in Sect. 4 and Theorem 3 is proved in Sect. 5. We conclude with some open problems that arise naturally from this work (Sect. 6).

2 Preliminaries

In this section, we recall the standard definitions underlying this work. For more details, see [10, Chap. 4].

2.1 Interactive Systems

For (randomized) interactive strategies A and B, we denote by $\langle A, B\rangle(x)$ the output of B after interacting with A on common input x. Since A and B are randomized, $\langle A, B\rangle(x)$ is a random variables. We denote by U_ℓ a random variable uniformly distributed in $\{0,1\}^\ell$.

We say that a strategy is probabilistic polynomial-time (PPT) if the total time it spends when it interacts with any other strategy on common input x is $\text{poly}(|x|)$, where the total time accounts for all computations performed at all stages of the interaction (including the final generation of output). We stress that, throughout this work, PPT mean "strict PPT"; that is, there exists a polynomial p such that the running time on any ℓ-bit input is always at most $p(\ell)$, regardless of the outcome of the coin tosses.[5]

[4] One can generate the uniform distribution over $[n]$ by selecting at random a uniformly distributed $r \in [2^{\log_2 \lceil n \rceil}]$, outputting r if $r \in [n]$, and announcing failure otherwise.

[5] We denote the input length by ℓ, rather than by n, in order to avoid confusion with Sect. 5 where n denotes a large integer (which is part of the input).

Definition 2.1 (interactive proof systems, following Goldwasser, Micali and Rackoff [15]): *Let $\epsilon_c, \epsilon_s : \mathbb{N} \to [0,1)$ such that $\epsilon_c(\ell)$ and $\epsilon_s(\ell)$ are computable in* $\text{poly}(\ell)$*-time and $\epsilon_c(\ell) + \epsilon_s(\ell) < 1 - 1/\text{poly}(\ell)$. Let P and V be interactive strategies such that V is PPT. We say that (P, V) is an* interactive proof system for a set S with completeness error ϵ_c and soundness error ϵ_s *if the following two conditions hold:*

Completeness: *For every $x \in S$, it holds that* $\mathbf{Pr}[\langle P, V\rangle(x) = 1] \geq 1 - \epsilon_c(|x|)$.

Soundness: *For every $x \notin S$ and every strategy P^*, it holds that*

$$\mathbf{Pr}[\langle P^*, V\rangle(x) = 1] \leq \epsilon_s(|x|).$$

If $\epsilon_c \equiv 0$, then the system has perfect completeness.

When we talk of interactive proof systems without specifying their errors, the reader may think of any choice (e.g., $\epsilon_c = \epsilon_s = 1/3$ or $\epsilon_c(\ell) = \epsilon_s(\ell) = \exp(-\ell)$). Recall that interactive proof systems with "average error" that is bounded away from one half (i.e., $(\epsilon_c(\ell) + \epsilon_s(\ell))/2 < 0.5 - 1/\text{poly}(\ell)$) can be converted to ones with negligible error by parallel or sequential composition. Lastly, recall that in computationally-sound systems (a.k.a argument systems) the soundness condition is required to hold only with respect to cheating strategies that can be implemented by polynomial-size circuits [8].[6]

Definition 2.2 (perfect and super-perfect zero-knowledge, following Goldwasser, Micali and Rackoff [15]): *Let (P, V) be an interactive proof system for S.*

Super-Perfect ZK: *The system (P, V) is* super-perfect zero-knowledge *if for every probabilistic polynomial-time strategy V^* there exists a* (strict) *probabilistic polynomial-time algorithm A^* such that for every $x \in S$ it holds that $A^*(x)$ is distributed identically to $\langle P, V^*\rangle(x)$.*

Perfect ZK: *The system (P, V) is* perfect zero-knowledge *if for every probabilistic polynomial-time strategy V^* there exists a* (strict) *probabilistic polynomial-time algorithm A^* such that for every $x \in S$ it holds that $\mathbf{Pr}[A^*(x) = \bot] \leq 1/2$ and*

$$\mathbf{Pr}[A^*(x) = \gamma \mid A^*(x) \neq \bot] = \mathbf{Pr}[\langle P, V^*\rangle(x) = \gamma],$$

for every $\gamma \in \{0,1\}^*$.

The same definition applies to argument systems. The honest-verifier *version of these definitions make a requirement only with respect to a strategy V_{hon} that behaves like V except that it outputs its entire view of the interaction* (i.e., its internal coin tosses as well as the sequence of all messages received from P).

Note that the failure probability in the case of perfect ZK (i.e., $\mathbf{Pr}[A^*(x) = \bot]$) can be reduced to $\exp(-\text{poly}(|x|))$ by repeated applications of the original simulator.

[6] Specifically, for any polynomial p, all sufficiently long $x \notin S$, and any strategy P^* that can be implemented by a circuit of size at most $p(|x|)$, it holds that $\mathbf{Pr}[\langle P^*, V\rangle(x) = 1] \leq \epsilon_s(|x|)$.

While Graph Isomorphism (GI) has a perfect ZK proof system [13], it is not *a priori* clear whether GI has an *honest-verifier* super-perfect ZK proof system, let alone a full-fledged super-perfect ZK system. The problem is that the simulator (even just for the honest-verifier case) needs to generate a uniformly distributed permutation of the vertices of a graph, and it is not clear whether a (strict) PPT can do such a thing. This depends on whether a PPT is only allowed to toss fair coins or is also allowed to generate uniform distributions over arbitrary domains of feasible size – see discussion in Sect. 1.2.

Recall that the foregoing issue (i.e., the difficulty of perfectly generating uniform distributions over arbitrary domains) does not arise in case of perfect ZK, since a machine that is allowed to fail with bounded probability can easily generate such distributions. The latter comment refers both to the various simulators (establishing the ZK feature) and to the prescribed verifier itself.

2.2 Non-interactive Systems

In the non-interactive setting both parties, modeled by standard algorithms, have access to a common reference string, which may be thought of as being generated by some trusted third party.

Definition 2.3 (non-interactive zero-knowledge, following Blum, Feldman and Micali [7]): *Let* $\epsilon_c, \epsilon_s : \mathbb{N} \to [0,1)$ *be as in Definition 2.1, and* P *and* V *be algorithms such that* V *is PPT. Let* ρ *be a positive polynomial. We say that* (P,V) *is an* non-interactive proof system for a set S with completeness error ϵ_c and soundness errorϵ_s *if the following two conditions hold.*

Completeness: *For every* $x \in S$*, it holds that*

$$\mathbf{Pr}_{\omega \leftarrow U_{\rho(|x|)}}[V(x,\omega,P(x,\omega))=1] \geq 1-\epsilon_c(|x|).$$

Soundness: *For every* $x \notin S$ *and every function* P^**, it holds that*

$$\mathbf{Pr}_{\omega \leftarrow U_{\rho(|x|)}}[V(x,\omega,P^*(x,\omega))=1] \leq \epsilon_s(|x|).$$

If $\epsilon_c \equiv 0$*, then the system has* perfect completeness.

Super-Perfect ZK: *The system* (P,V) *is* super-perfect zero-knowledge *if there exists a* (strict) *probabilistic polynomial-time algorithm* A *such that for every* $x \in S$ *it holds that* $A(x)$ *is distributed identically to* $(\omega, P(x,\omega))$*, where* $\omega \leftarrow U_{\rho(|x|)}$.

Perfect ZK: *The system* (P,V) *is* perfect zero-knowledge *if there exists a* (strict) *probabilistic polynomial-time algorithm* A *such that for every* $x \in S$ *it holds that* $\mathbf{Pr}[A(x)=\perp] \leq 1/2$ *and*

$$\mathbf{Pr}[A(x)=\gamma \mid A(x)\neq\perp] = \mathbf{Pr}_{\omega \leftarrow U_{\rho(|x|)}}[(\omega,P(x,\omega))=\gamma]$$

for every $\gamma \in \{0,1\}^*$.

Note that in Definition 2.3 the common reference string is uniformly distributed in $\{0,1\}^{\rho(|x|)}$. A popular relaxation, not used here, allows the common reference string to be taken from any efficiently sampleable distribution.

3 From Perfect ZK to Super-Perfect ZK

In Sect. 3.1 we prove Theorem 1, which yields super-perfect zero-knowledge *proofs* with exponentially vanishing completeness error. In Sect. 3.2 we obtain super-perfect zero-knowledge *arguments with perfect completeness*, while assuming the existence of perfectly binding commitment schemes.

3.1 On Super-Perfect ZK Interactive Proofs

In the following transformation we assume, without loss of generality, that $p(x) < 2^{-|x|}$ for any $x \in S$. The transformation amounts to letting the prover perform the original protocol with probability $1 - p(x)$, and abort otherwise. Of course, the verifier will reject in case the prover aborts, and so perfect completeness is lost, but this will allow a super-perfect simulation. Note that this transformation relies on the hypothesis that, on any $x \in S$, all simulators output $\perp$ with the same probability. As stated in the introduction, the transformation is due to Malka [18, Sec. 4.1], although he states it only for the case of $p \equiv 1/2$. (For sake of simplicity, we also assume, w.l.o.g., that the original prover never sends the empty string, denoted λ.)

Construction 3.1 (the transformation used for establishing Theorem 1): *Let (P, V), S and p be as in the hypothesis of Theorem 1, and suppose that A' is a simulator for any fixed PPT strategy V'* (e.g., V' may equal V or V_{hon}) *such that for every $x \in S$ it holds that $\mathbf{Pr}[A'(x) = \perp] = p(x)$. Then, on common input x, the two parties proceed as follows.*

1. *The prover invokes $A'(x)$ and sends the empty message λ if and only if $A'(x) = \perp$. In such a case, the verifier will reject.*
2. *Otherwise, the parties execute (P, V) on the common input x.*

For every PPT strategy V^*, consider the simulator A^* guaranteed by the hypothesis of Theorem 1. On input x, the corresponding new simulator (for V^*) *computes $\gamma \leftarrow A^*(x)$, outputs γ if $\gamma \neq \perp$ and $V^*(x, \lambda)$ otherwise.*

Note that the foregoing protocol preserves the soundness error of V, whereas the completeness error on input $x \in S$ increases by at most $p(x) \leq 2^{-|x|}$ (i.e., from $\epsilon_c(|x|)$ to $p(x) + (1 - p(x)) \cdot \epsilon_c(|x|)$). Indeed, the verifier rejects if the prover got unlucky (i.e., $A'(x)$ yields $\perp$), and so a cheating prover gains nothing by claiming that it got $\perp$. The new simulators establishes the super-perfect ZK feature, and *Theorem* 1 *follows.* Noting that in the case of honest-verifier ZK the condition made in Theorem 1 hold vacuously, we immediate get the following corollary.

Corollary 3.2 (honest-verifier super-perfect ZK): *Every set S that has a honest-verifier perfect ZK proof system has a honest-verifier super-perfect ZK proof system. All additional features asserted in Theorem 1 hold as well.*[7]

[7] But, again, perfect completeness is lost.

More importantly, applying Theorem 1 (or rather Construction 3.1) to the perfect zero-knowledge arguments of Naor *et al.* [19] (see also [10, Sec. 4.8.3]), we obtain:

Corollary 3.3 (super-perfect ZK for $\mathcal{NP}$): *Assuming the existence of* (non-uniformly strong) *one-way permutations, every set in* $\mathcal{NP}$ *has a* (black-box) *super-perfect ZK argument system.*[8]

As stated in the introduction, super-perfect ZK arguments (with perfect completeness) for $\mathcal{NP}$ are implicit in [20] (see [21, Prop. 4.2]), where they are only claimed to be perfect ZK. Their claim, which refers to the non-standard model of PPT (in which a machine can sample $[n]$ uniformly at cost n), is conditioned on a seemingly stronger assumption (i.e., the existence of claw-free pairs of permutations), and is established using non-black-box simulators.[9] Hence, Corollary 3.3 is incomparable to the corresponding results that can be derived from [20,21]. On the one hand, it is stronger, since it uses the standard model of PPT, and provides black-box simulators, while relying on a seemingly weaker assumption. On the other hand, it is weaker, since it does not provide perfect completeness. Obtaining perfect completeness is the focus of Sect. 3.2.

3.2 On Super-Perfect ZK Arguments with Perfect Completeness

Assuming the existence of perfectly binding commitment schemes, we show that certain perfect ZK proof (or argument) systems can be transformed into super-perfect ZK *arguments* with perfect completeness. The transformation refers to perfect ZK proofs (or arguments) that have simulators that can always output a perfectly random prefix of the interaction that misses only the last message (from the prover). See Condition 3 below (whereas Conditions 1 and 2 are (a strong form of) the standard requirement from perfect ZK).[10]

Definition 3.4 (an admissible class of perfect ZK protocols): *Let* (P, V) *be an argument system for* S*, and let* P_0 *denote the strategy derived from* P *by having it abort just before sending the last message. We say that* P *is* admissible *if for every probabilistic polynomial-time strategy* V^* *there exists a* (strict) *probabilistic polynomial-time algorithm* A^* *such that for every* $x \in S$ *the following three conditions hold.*

[8] Again, the derived systems have exponentially vanishing completeness error.

[9] Specifically, the perfect ZK feature of their argument system is demonstrated using Barak's (non-black-box) simulation technique [3,4], whereas such a demonstration actually yields a super-perfect simulator. This is the case because the simulation (constructed according to Barak's technique) amounts to executing the same protocol as the honest prover, while using the verifier's program as a NP-witness to a composed tatement that the honest prover proves by using an NP-witness to the actual input. The need to use the non-standard model of PPT arises because in the known proof systems (e.g., [19]) the honest prover samples uniformly sets that have size that is not a power of 2.

[10] Specifically, Condition 2 requires perfect simulation of the interaction with P in case of non-failure, which is the standard requirement of perfect ZK, whereas Condition 1 requires that failure occurs with probability exactly 1/2 (rather than at most 1/2).

1. $\mathbf{Pr}[A^*(x)=(1,\cdot)] = 1/2$;
2. $\mathbf{Pr}[A^*(x)=(1,\gamma) \mid A^*(x)=(1,\cdot)] = \mathbf{Pr}[\langle P,V^*\rangle(x)=\gamma]$, *for every* $\gamma \in \{0,1\}^*$.
3. $\mathbf{Pr}[A^*(x)=(0,\gamma) \mid A^*(x)=(0,\cdot)] = \mathbf{Pr}[\langle P_0,V^*\rangle(x)=\gamma]$, *for every* $\gamma \in \{0,1\}^*$.

(Indeed, we parse the output of A^* as a pair of the form $(\sigma,\gamma) \in \{0,1\} \times \{0,1\}^*$, where $\sigma = 0$ indicates a failure to simulate interaction with P.)

Note that, in addition to requiring A^* to output $\langle P_0,V^*\rangle(x)$ whenever it fails to output $\langle P,V^*\rangle(x)$ (i.e., Condition 3), we also required the failure probability to be *exactly one half* (rather than at most 1/2). The latter condition can be assumed, without loss of generality, whenever p is efficiently computable, provided that we adopt the non-standard model of PPT machines in which a machine can sample $[n]$ uniformly at cost $\text{poly}(\log n)$ (as discussed in Sect. 1.2). Furthermore, under a weaker non-standard PPT convention, in which a machine can sample $[n]$ uniformly at cost $\text{poly}(n)$, both the perfect zero-knowledge proof system for Graph Isomorphism (of [13]) and the perfect ZK argument for any set in $\mathcal{NP}$ of Naor *et al.* [19] are admissible by Definition 3.4. (In both cases, the convention is required in order to allow a PPT machine to uniformly select a permutation over a set of a size larger than 2.) Actually, the perfect ZK argument for any set in $\mathcal{NP}$ of Naor *et al.* [19] are admissible by Definition 3.4 even when using the weaker non-standard model of PPT where the machine can sample (only) the uniform distribution over $[c]$ for $c = 6$ (equiv., for $c = 3$).[11]

In the following transformation, we shall use a perfectly binding commitment scheme, denoted C. That is, we shall assume that the distributions $C(0)$ and $C(1)$ are computationally indistinguishable (by polynomial-size circuits) although they have disjoint supports.[12] Such commitment schemes can be constructed, assuming the existence of one-way permutations (see [10, Sec. 4.4.1]). We denote the commitment to value v using coins s by $C_s(v)$.

Construction 3.5 (the transformation of admissible protocols): *Let* (P,V) *be an argument system for* S *such that* P *is admissible by Definition 3.4. On common input* x*, the two parties proceeds as follows.*

1. *The parties execute* (P_0,V) *on common input* x*; that is, they invoke the original protocol, except that the prover does not send its last message, denoted* β*.*
2. *The parties performs a standard coin tossing protocol* (see [11, Sec. 7.4.3.1]). *Specifically, the verifier sends a commitment* $c \leftarrow C(v)$ *to a random bit* v*, the prover responds* (in the clear) *with a random bit* u*, and the verifier de-commits to the commitment* (i.e., provides (v,s) such that $c = C_s(v)$).

[11] Indeed, the perfect ZK argument system for $\mathcal{NP}$ based on 3-Colorability requires that the prover and simulator sample a random permutation of 3 elements. Furthermore, the simulator fails with probability exactly 1/3, for every input and every probabilistic polynomial-time strategy V^*.

[12] Note that in some sources (e.g. [10, Sec. 4.4.1]) the perfect binding property of commitment schemes only requires that the supports of $C(1)$ and $C(0)$ intersect on a set of negligible size, while we require that the supports of $C(0)$ and $C(1)$ are totally disjoint.

3. *If the verifier has de-committed improperly* (i.e., $c \neq C_s(v)$), *then the prover sends the empty message, denoted* λ. *Otherwise, if* $u = v$ *then the prover sends* 0, *and otherwise it sends* β (where we assume, w.l.o.g, that $\beta \notin \{0, \lambda\}$).
4. *If* $u = v$ *then the verifier accepts, otherwise* (i.e., $u \neq v$) *it acts as* $V(\alpha, \beta)$, *where* α *denotes the view of* V *in the interaction with* P_0 (as conducted in Step 1).

This transformation preserves the completeness error of (P, V), but the error in the computational-soundness grows from $\epsilon_s(\ell)$ to $(1+\epsilon_s(\ell) + \mu(\ell))/2$, where μ is a negligible function. Indeed, computational-soundness is established by observing that the prover can cause the verifier to accept only if either it guessed correctly the value committed by the verifier (i.e., if $u = v$) or it could have cheated anyhow in the corresponding (P, V) interaction. To reduce the computational-soundness error, one can always use sequential repetitions, whereas using parallel repetitions does not always work (because of issues with both computational-soundness (cf., e.g., [5]) and ZK (cf., e.g., [10, Sec. 4.5.4.1])).

Turning to the super-perfect ZK feature of the resulting protocol (and assuming that V^* always de-commits properly), we rely on the simulator's ability to set $u = v$ whenever it fails in its attempt to produce a full transcript. Hence, we establish the following claim.

Claim 3.6 (super-perfect simulations): *The prover strategy described in Construction 3.5 is super-perfect zero-knowledge.*

Proof: For every potential PPT strategy V^*, let A^* denote the corresponding simulator as guaranteed by Definition 3.4. The new simulator will act as follows.

1. It invokes A^* on input x, obtaining either a full transcript or a partial transcript.
   Recall that each event happens with probability 1/2, and that a full transcript has the form (α, β), where $\beta \notin \{0, \lambda\}$ is the prover's last message. For sake of convenience, set $\beta = 0$ in the case that A^* produced a partial transcript (denoted α).[13]
2. The simulator obtains a commitment c from V^*.
   Note that c determines a unique value, denoted v, such that a proper de-commitment of c yields v. (Here we rely on the perfect binding feature of C; that is, for every c there exist at most one v such that for some s it holds that $C_s(v) = c$.)
3. The simulator obtains the reaction of V^* to both possible $u \in \{0, 1\}$ (that the verifier expects as the prover's response in Step 2); that is, it obtains $d_u \leftarrow V^*(\alpha, u)$ for both $u \in \{0, 1\}$.
4. If in both cases V^* acted improperly (i.e., did not provide a valid de-commitment to c), then the simulator selects u at random in $\{0, 1\}$, and outputs $V^*(\alpha, u, \lambda)$.
   (Obviously, (α, u, λ) is what V^* sees in the real interaction with the prover.)

[13] Indeed, by Definition 3.4, the output of $A^{**}(x)$ has the form $(0, \alpha)$, with probability 1/2, and $(1, \alpha\beta)$ otherwise.

5. If in both cases V^* de-committed properly to the value v, then the simulator outputs $V^*(\alpha, v, 0)$ if $\beta = 0$ and $V^*(\alpha, 1-v, \beta)$ otherwise.
(Here we rely on the fact that $\mathbf{Pr}[\beta = 0] = 1/2$, whereas in the real interaction the prover responds with $u = v$ with probability 1/2. Hence, in the real interaction the prover will respond with $\beta = 0$ if and only if $u = v$ (and otherwise, when $u = 1 - v$, it response will be the actual message $\beta \neq 0$.)
6. If V^* de-committed properly to the value v only when fed with a single value, denoted u^*, then we distinguish two cases.
Case of $\beta = 0$: Output $V^*(\alpha, 1-u^*, \lambda)$.
Case of $\beta \neq 0$: Output $V^*(\alpha, u^*, \beta)$ if $u^* \neq v$ and $V^*(\alpha, u^*, 0)$ otherwise.

It may be more intuitive to restructure the cases in Step 6 as follows:

Case of $u^* = v$ (i.e., proper de-commitment in response to v only): In this case, we output $V^*(\alpha, u^*, 0)$ if $\beta \neq 0$ and $V^*(\alpha, 1-u^*, \lambda)$ otherwise (i.e., $\beta = 0$). Equivalently, output $V^*(\alpha, u^*, 0)$ with probability 1/2 and $V^*(\alpha, 1-u^*, \lambda)$ otherwise.
In this case, in the real interaction, the prover selects $u = u^*$ with probability 1/2 and seeing a proper de-commitment to v, responds with 0. Otherwise (i.e., $u \neq u^*$), the prover sees an improper de-commitment and responds with λ. Hence, the foregoing output distribution matches the transcript of the real interaction.

Case of $u^* \neq v$ (i.e., proper de-commitment in response to $1-v$ only): In this case, we output $V^*(\alpha, u^*, \beta)$ if $\beta \neq 0$ and $V^*(\alpha, 1-u^*, \lambda)$ otherwise.
Similarly, in this case the prover sees a proper de-commitment with probability 1/2 and responds with $\beta \neq 0$ in that case (and otherwise it responds with λ).

Hence, in both cases considered in Step 6, the simulator produces the same distribution as in the real interaction. The same holds also in the situations considered in Steps 4 and 5. ■

4 From Perfect NIZK to Super-Perfect NIZK

While Construction 3.1 is applicable also in the context of NIZK, where the condition regarding p holds vacuously (cf. Corollary 3.2), this construction does not preserve perfect completeness. Our aim here is to preserve perfect completeness, and this can be done by "transferring" the simulation attempt from the prover (who cannot be trusted to perform it at random) to the common reference string (which is uniformly distributed by definition). Specifically, we will establish Theorem 2, which presupposed that the failure probability function $p : S \to [0, 0.5]$ is efficiently computable. Actually, we assume, without loss of generality, that $p(x) < 2^{-|x|}$ for every $x \in \{0,1\}^*$ (and not merely for $x \in S$). (Again, we assume, w.l.o.g., that the original prover never outputs the empty string λ).

Construction 4.1 (the transformation): *Let (P, V), S, A and p be as in the hypothesis of Theorem 2; and let ρ denote the length of the common reference string and ρ' denote the number of coins used by the simulator A. The new NIZK for inputs of length ℓ is as follows.*

Common random string: *An* $(\rho(\ell)+\rho'(\ell))$*-bit string, denoted* (ω,r)*, where* r *is interpreted as an integer in* $\{0,...,2^{\rho'(\ell)}-1\}$.

Prover (on input $x\in\{0,1\}^\ell$): *If* $r<p(x)\cdot 2^{\rho'(\ell)}$*, then the prover outputs the empty message* λ*. Otherwise* (i.e., $r\geq p(x)\cdot 2^{\rho'(\ell)}$), *the prover outputs* $P(x,\omega)$.

Verifier (on input $x\in\{0,1\}^\ell$ and alleged proof y): *If* $r<p(x)\cdot 2^{\rho'(\ell)}$*, then the verifier accepts. Otherwise* (i.e., $r\geq p(x)\cdot 2^{\rho'(\ell)}$), *the verifier decides according to* $V(x,\omega,y)$.

The new simulator invokes $A(x)$ *obtaining the value* v*. If* $v=\bot$*, then the simulator selects uniformly* $\omega\in\{0,1\}^{\rho(\ell)}$ *and* $r\in\{0,...,p(x)\cdot 2^{\rho'(\ell)}-1\}$*, and outputs* $((\omega,r),\lambda)$*. Otherwise* (i.e., $v=(\omega,y)$), *the simulator selects uniformly* $r\in\{p(x)\cdot 2^{\rho'(\ell)},...,2^{\rho'(\ell)}-1\}$*, and outputs* $((\omega,r),y)$.

The completeness error of the new system on input x is upper bounded by $(1-p(x))\cdot\epsilon_c(|x|)\leq\epsilon_c(|x|)$, whereas the soundness error is upper bounded by $p(|x|)+(1-p(x))\cdot\epsilon_s(|x|)\leq\epsilon_s(|x|)+2^{-|x|}$, where ϵ_c and ϵ_s denote the error bounds of (P,V). Note that the distribution of the verifier's view both in the actual system and in its simulation equals $((U_{\rho(\ell)},U_{\rho'(\ell)}),Y)$, where $Y=P(x,\omega)$ if $r\geq p(x)\cdot 2^{\rho'(\ell)}$ and $Y=\lambda$ otherwise.

Recall that the construction of the new simulator relies on the ability to generate uniform distributions on the sets $[p(x)\cdot 2^{\rho'(\ell)}]$ and $[2^{\rho'(\ell)}-p(x)\cdot 2^{\rho'(\ell)}]$, which is possible in the standard PPT model only if $p(x)=1/2$. This was not the case above, since we started by reducing the simulation error to $p(x)\leq 2^{-|x|}$, which is the reason that Theorem 2 holds only in the non-standard PPT model (as stated in it). However, if we start with $p\equiv 1/2$, then Construction 4.1 yields the following.[14]

Corollary 4.2 (super-perfect NIZK in the standard PPT model): *Let* (P,V)*,* S*,* A *and* p *be as in Construction 4.1, and suppose that* $p\equiv 1/2$*. Further suppose that the sum of completeness and soundness error of* (P,V) *are noticeably smaller than* $1/2$*. Then,* S *has a super-perfect non-interactive zero-knowledge proof system, where the simulation is in the standard PPT model, and the completeness and soundness errors are exponentially vanishing. Furthermore, if* (P,V) *has perfect completeness, then so does the resulting system.*

Note that applying Construction 4.1 to (P,V) yields a system with completeness error $\epsilon_c'\stackrel{\rm def}{=}\epsilon_c/2$ and soundness error at most $\epsilon_s'\stackrel{\rm def}{=}(1+\epsilon_s)/2$, where ϵ_c and ϵ_s denote the error bounds of (P,V). Using the assumption $\epsilon_c(\ell)+\epsilon_s(\ell)<0.5-1/\text{poly}(\ell)$, we have $\epsilon_c'(\ell)+\epsilon_s'(\ell)<1-1/\text{poly}(\ell)$, which allows for error reduction that yields the stated (exponentially vanishing) error bounds. Interestingly, in this case, we apply error-reduction on the resulting NIZK system (rather than on the simulator A provided for the original NIZK system).

[14] Indeed, in this case the construction can be simplified. We may use a common reference string of the form $(\omega,\sigma)\in\{0,1\}^{\rho(\ell)+1}$, have the prover output $P(x,\omega)$ if and only if $\sigma=1$, and have the verifier accept if either $\sigma=0$ or $V(x,\omega,y)$, where y denotes the alleged proof.

5 A Super-Perfect NIZK for Blum Integers

We first recall the definition of (generalized) Blum integers.

Definition 5.1 (Blum Integers): *A natural number is called a* (generalize) Blum Integer *if it is of the form $p^e q^d$ such that $p \equiv q \equiv 3 \pmod 4$ are different odd primes and $e \equiv d \equiv 1 \pmod 2$. The set of Blum integers is denoted B.*

The following standard notations will be used extensively. For any natural number n, we let $\mathbb{Z}_n$ denote the additive group modulo n, and $\mathbb{Z}_n^*$ denote the corresponding multiplicative group.

We let $Q_n \subseteq \mathbb{Z}_n^*$ denote the set of quadratic residues modulo n, and recall the definition of the Jacobi symbol modulo n, viewed as a function $\mathtt{JS}_n : \mathbb{Z} \rightarrow \{-1, 0, 1\}$, and a basic fact regarding it: For a prime p, it holds that $\mathtt{JS}_p(r) = 0$ if $r \equiv 0 \pmod p$, whereas $\mathtt{JS}_p(r) = 1$ if $(r \bmod p) \in Q_p$ and $\mathtt{JS}_p(r) = -1$ otherwise (i.e., $(r \bmod p) \in \mathbb{Z}_n^* \setminus Q_p$). For composite $n = n_1 n_2$, it holds that $\mathtt{JS}_n(r) = \mathtt{JS}_{n_1}(r) \cdot \mathtt{JS}_{n_1}(r)$, yet the Jacobi symbol modulo n can be computed efficiently also when not given the factorization of n. Note that $\mathtt{JS}_n(r) = 1$ for every $r \in Q_n$.

Another important set, first utilized in [2], is $S_n \stackrel{\text{def}}{=} \{r \in \{1, ..., \lfloor n/2 \rfloor\} : \mathtt{JS}_n(r) = 1\} \subset \mathbb{Z}_n^*$. For $n \in B$ it holds that $|S_n| = |\mathbb{Z}_n^*|/4$ (see Claims 5.3 and 5.5). We consider the following three functions:

1. The modular squaring function $g_n : \mathbb{Z} \rightarrow Q_n$ defined as $g_n(r) = r^2 \bmod n$.
2. The "first half" function $h_n : \mathbb{Z}_n \rightarrow \mathbb{Z}_{\lfloor n/2 \rfloor}$ defined as $h_n(r) = r$ if $r < n/2$ and $h_n(r) = n - r$ otherwise. Indeed, if $n \in B$, then h_n maps Q_n to S_n.
3. Their composition $f_n = h_n \circ g_n$; that is, $f_n(r) = h_n(g_n(r))$.

Abusing notation, we extend these functions to sets in the obvious manner.

5.1 Well Known Facts

The following well-known facts will be used in our construction and its analysis. The reader may consider skipping this subsection. We start by recalling two computational facts.

1. The set of prime powers is in $\mathcal{P}$.
   (Justification: Try all possible powers $e \in [\lceil \log_2 n \rceil]$, and use the primality tester of [1].)
2. The set $\{(n, r) : r \in S_n\}$ is in $\mathcal{P}$.
   (Justification: Recall that the Jacobi symbol is efficiently computable.)

We next recall a few elementary facts regarding the foregoing sets and functions.

Claim 5.2 (on the size of Q_n and $f_n(S_n)$):

1. *Suppose that $n = \prod_{i \in [k]} p_i^{e_i}$ such that the p_i's are different odd primes. Then, $|Q_n| = 2^{-k} \cdot |Z_n^*|$.*

2. *For every $n \in \mathbb{N}$, it holds that $|f_n(S_n)| \le |Q_n|$.*

Proof: Part 1 holds since $r \in Z_n^*$ is in Q_n if and only if for every $i \in [k]$ it holds that $r \bmod p_i^{e_i}$ is in $Q_{p_i^{e_i}}$, whereas each $s \in Q_{p_i^{e_i}}$ has exactly two modular square root (which sum-up to $p_i^{e_i}$). Part 2 holds since $f_n(S_n) \subseteq f_n(Z_n^*) = h_n(Q_n)$. ■

Claim 5.3 (on $\mathtt{JS}_n(-1)$ and the form of n): *Suppose that $n = \prod_{i\in[k]} p_i^{e_i}$ such that the p_i's are different odd primes, and let $I = \{i \in [k] : p_i \equiv 3 \bmod 4\}$. Then, the following three conditions are equivalent: (1) $n \equiv 1 \pmod 4$; (2) $\mathtt{JS}_n(-1) = 1$; and (3) $\sum_{i\in I} e_i$ is even.*

In particular if n is a Blum integer, then $\mathtt{JS}_n(-1) = 1$.

Proof: Note that $n \equiv \prod_{i\in I} 3^{e_i} \equiv 3^{\sum_{i\in I} e_i} \pmod 4$, which implies that $n \equiv 1 \pmod 4$ if and only if $\sum_{i\in I} e_i$ is even. On the other hand, note that $\mathtt{JS}_n(-1) = \prod_{i\in[k]} \mathtt{JS}_{p_i}(-1)^{e_i} = \prod_{i\in I}(-1)^{e_i} = (-1)^{\sum_{i\in I} e_i}$, where the second equality holds since for every odd prime p it holds that $\mathtt{JS}_p(-1) = 1$ if and only if $p \equiv 1 \pmod 4$. ■

Claim 5.4 (on f_n when n is a Blum integer): *For $n \in B$, the function f_n is a permutation over S_n.*

Proof: Recall that $n \in B$ has the form $p^e q^d$ such that $p \equiv q = 3 \pmod 4$ are distinct odd primes and $e \equiv d \equiv 1 \pmod 2$. First note that g_n is a permutation over Q_n, because $x^2 \equiv y^2 \pmod n$ implies that $x \equiv \pm y \pmod{p^e}$ whereas $|Q_{p^e} \cap \{r, p^e - r\}| = 1$ for every $r \in \mathbb{Z}_n^*$ (since $\mathtt{JS}_{p^e}(-1) = -1$). Ditto for the situation mod q^d. Next note that h_n is a bijection from Q_n to S_n, because $|Q_n \cap \{r, n-r\}| \le 1$ for any $r \in \mathbb{Z}_n^*$. The claim (restated as $f_n(S_n) = S_n$) follows since $f_n(Q_n) = h_n(g_n(Q_n)) = h_n(Q_n) = S_n$ and $f_n(S_n) = f_n(h_n(S_n)) = f_n(h_n(Q_n)) = f_n(Q_n)$, where the last equality holds since $g_n(-1) = 1$ (and so for every $x \in Z_n^*$ it holds that $g_n(h_n(x)) = g_n(x)$). ■

Claim 5.5 (on the size of S_n): *If $\mathtt{JS}_n(-1) = 1$, then $|S_n| \ge |Z_n^*|/4$, where equality holds if n is not of the form $2^e s^2$ for some $e, s \in \mathbb{N}$.*

Proof: Using $\mathtt{JS}_n(-1) = 1$ it follows that the elements of $\mathbb{Z}_n^*$ that have Jacobi symbol 1 are paired such that $\mathtt{JS}_n(s) = 1$ if and only if $\mathtt{JS}_n(n-s) = 1$. This implies that $|S_n| = |\{s \in \mathbb{Z}_n^* : \mathtt{JS}_n(s) = 1\}|/2$. The latter set contains half of $|Z_n^*|$ if there exists $r \in \mathbb{Z}_n^*$ such that $\mathtt{JS}_n(r) = -1$, since in that case $x \mapsto rx$ is a bijection of $\{s \in \mathbb{Z}_n^* : \mathtt{JS}_n(s) = 1\}$ to $\{s \in \mathbb{Z}_n^* : \mathtt{JS}_n(s) = -1\}$. Lastly, note that such r exists if and only if n is not of the form $2^e s^2$ for some $e, s \in \mathbb{N}$, whereas $\mathtt{JS}_{2^e s^2}(r) = \mathtt{JS}_s(r)^2 = 1$ for every $r \in \mathbb{Z}_{2^e s^2}^*$ (and in this case $|S_{2^e s^2}| = |\mathbb{Z}_{2^e s^2}|/2$). ■

5.2 The Proof System

Recall that there exist (deterministic) polynomial-time algorithms for (1) deciding if a number is a prime (ditto for a prime power), and (2) deciding whether

$r \in S_n$ when given n and r. The main observation underlying the proof system is that when $n \in B$ the function f_n is a permutation over S_n, whereas for $n \notin B$ it holds that $|f_n(S_n)| \le |S_n|/2$ (provided that $n \equiv 1 \pmod 4$ and n is not a prime power). (Establishing the claim regarding $n \notin B$ is the core of the proof of Proposition 5.7 (below).) Hence, the proof system amounts to distinguishing the case $f_n(S_n) = S_n$ from the case $|f_n(S_n)| = |S_n|/2$ by *asking the prover to provide a pre-image under f_n of a uniformly distributed $\omega \in S_n$* (where the case $\omega \notin S_n$ is treated separately).[15] The super-perfect simulator can provide such transcripts by uniformly selecting $r \in S_n$, and outputting $(f_n(r), r)$, where $f_n(r)$ represents the common reference string (and r be the prover's message/output).

Construction 5.6 (a non-interactive proof system for B):

Input: *A natural number n. Let $\ell = \lceil \log_2 n \rceil$.*
Common reference string: *An ℓ-bit string, denoted ω, interpreted as an integer in $\mathbb{Z}_{2^\ell}$.*
Prover: *If $\omega \in S_n$ and there exists $r \in S_n$ such that $f_n(r) = \omega$, then the prover outputs r* (otherwise it outputs 0).
(Note that for $n \in B$ and $\omega \in S_n$, there exists a unique $r \in S_n$ such that $f_n(r) = \omega$.)
Verifier: *When receiving an alleged proof r, the verifier proceeds as follows.*
1. Discarding obvious no-instances: *If n is a prime power or $n \not\equiv 1 \pmod 4$, then the verifier halts outputting 0* (indicating rejection).
2. Handling inputs with a small prime factor: *The verifier checks if there exists a prime $p \in \{3, ..., \ell\}$ that divides n and finds the largest e such that p^e divides n. If n/p^e is not a prime power, then the verifier rejects. Otherwise, letting q^d be this prime power* (i.e., $n = p^e q^d$), *the verifier accepts if $p \equiv q \equiv 3 \pmod 4$ and $e \equiv d \equiv 1 \pmod 2$, and rejects otherwise.*
3. *If $\omega \notin S_n$, then the verifier halts outputting 1* (indicating acceptance).
4. *If $r \in S_n$ and $f_n(r) = \omega$, then the verifier outputs 1. Otherwise, it outputs 0.*

Note that the foregoing system has soundness error *at least* 1/2, due to Step 3. However, as shown next, its soundness error is upper-bounded by a constant smaller than 1, and so we can apply straightforward error reduction (since the system has perfect completeness).

Proposition 5.7 (analysis of Construction 5.6): *Construction 5.6 constitutes a super-perfect NIZK for B with perfect completeness and soundness error smaller than* 16/17.

Proof: Suppose that $n = p^e q^d$ such that $p \equiv q \equiv 3 \pmod 4$ are different odd primes and $e \equiv d \equiv 1 \pmod 2$. Then, f_n is a permutation over S_n (see Claim 5.4), and perfect completeness holds (since no step of Construction 5.6 may cause rejection). In such a case, the super-perfect simulation proceeds as follows.

[15] See Step 3. In addition, Steps 1 and 2 take care of other pathological cases. The main action takes place in Step 4.

1. Select uniformly $r \in \mathbb{Z}_{2^\ell}$.
2. If $r \in S_n$ then output $(f_n(r), r)$, else output $(r, 0)$.

Note that the simulator's output is distributed identically to the distribution produced by the prover. In both distributions of pairs, denoted (ω, y), it holds that ω is distributed uniformly in $\mathbb{Z}_{2^\ell}$, whereas y is a function of ω (and n) determined as follows: If $\omega \notin S_n$, then $y = 0$, and otherwise $y \in S_n$ is the unique pre-image of ω under f_n. Hence, it remains to establish the soundness of the system.

Turning to the soundness condition, suppose that $n \notin B$. We may assume that n is not a prime power and that $n \equiv 1 \pmod 4$ (or else Step 1 would have rejected). We may also assume that n has no prime factor smaller than ℓ (or else Step 2 would have rejected).[16] Now, with probability $n/2^\ell > 1/2$, the random string ω is in $\mathbb{Z}_n$. Conditioned on this event, we consider the prime factorization of $n = \prod_{i \in [k]} p_i^{e_i}$ (where the p_i's are different odd primes), and show that $\omega \notin \mathbb{Z}_n^*$ is unlikely, whereas if $\omega \in \mathbb{Z}_n^*$ then the verifier rejects with probability at least 1/8.

First, recall that $|\mathbb{Z}_n^*| = \prod_{i \in [k]}((p_i - 1) \cdot p_i^{e_i - 1})$. Hence

$$\frac{|\mathbb{Z}_n \setminus \mathbb{Z}_n^*|}{|\mathbb{Z}_n|} = 1 - \prod_{i \in [k]} \frac{p_i - 1}{p_i} \le 1 - \left(1 - \frac{1}{\ell}\right)^k$$

where the inequality is due to $p_i \ge \ell$ for every $i \in [k]$. Hence, $\mathbf{Pr}_\omega[\omega \notin \mathbb{Z}_n^* \mid \omega \in \mathbb{Z}_n] \le 1 - (1 - \ell^{-1})^k < k/\ell = o(1)$, since $k \le \log_\ell n = o(\ell)$. Considering the case that $\omega \in \mathbb{Z}_n^*$, and using the fact that $|S_n| \ge |\mathbb{Z}_n^*|/4$ (see Claims 5.3 and 5.5), we infer that $\omega \in S_n$ with probability

$$\frac{|S_n|}{2^\ell} = \frac{n}{2^\ell} \cdot \frac{|\mathbb{Z}_n^*|}{|\mathbb{Z}_n|} \cdot \frac{|S_n|}{|\mathbb{Z}_n|} > \frac{1}{2} \cdot (1 - o(1)) \cdot \frac{1}{4}$$

which is larger than 2/17. Hence, the verifier executes Step 4 with probability greater than 2/17.

Recalling that $|f_n(S_n)| \le 2^{-k} \cdot |\mathbb{Z}_n^*|$ (see Claim 5.2) while $|S_n| = |\mathbb{Z}_n^*|/4$, we infer that if $k \ge 3$, then Step 4 rejects with probability at least half. We are left with the case of $k = 2$, which means that $n = p^e q^d \notin B$ such that p and q are different odd primes (and $e, d \ge 1$). Hence, w.l.o.g., either $p \equiv 1 \pmod 4$ or $e \equiv 0 \pmod 2$, and $p^e \equiv 1 \pmod 4$ follows in both cases. We shall show that in this case $|f_n(S_n)| \le |S_n|/2$, by showing that for each $r \in S_n$ there exists $r' \in S_n$ such that $r' \ne r$ and $f_n(r') = f_n(r)$.

[16] This is the case since if $n = p^e n' \notin B$ for $e \ge 1$ and an odd prime $p \in [\ell]$ that does not divide n', then either n' is not a prime power or the prime factorization of n is found in Step 2 leading the verifier to reject.

For any $r \in S_n$, let $r_1 = r \bmod p^e$ and $r_2 = r \bmod q^d$. Consider the unique $s \in \mathbb{Z}_n^*$ such that $s \equiv -r_1 \pmod{p^e}$ and $s \equiv r_2 \pmod{q^d}$. Then, $s \neq r$ and $s \neq n-r$, whereas $s^2 \equiv r^2 \pmod{n}$, which implies $f_n(s) = f_n(r)$. On the other hand, $\mathtt{JS}_n(n-s) = \mathtt{JS}_n(s) = \mathtt{JS}_{p^e}(-r_1) \cdot \mathtt{JS}_{q^d}(r_2) = \mathtt{JS}_{p^e}(r_1) \cdot \mathtt{JS}_{q^d}(r_2) = \mathtt{JS}_n(r) = 1$, where the first equality uses $\mathtt{JS}_n(-1) = 1$ (which holds by $n \equiv 1 \pmod 4$ and Claim 5.3), the third equality uses $\mathtt{JS}_{p^e}(-1) = 1$ (which holds by $p^e \equiv 1 \pmod 4$ and Claim 5.3), and the last equality uses $r \in S_n$. Hence, either s or $n-s$ is in S_n (since $\mathtt{JS}_n(n-s) = \mathtt{JS}_n(s) = 1$), and it follows that $|\{r, s, n-s\} \cap S_n| \geq 2$. Having shown for each $r \in S_n$ there exists $r' \in S_n$ such that $r' \neq r$ and $f_n(r') = f_n(r)$, we conclude that $|f_n(S_n)| \leq |S_n|/2$.

Let us recap. If $n \notin B$, then the verifier reject with probability at least

$$\mathbf{Pr}_\omega[\omega \in S_n] \cdot \mathbf{Pr}_\omega[\omega \notin f(S_n) | \omega \in S_n] > \frac{2}{17} \cdot \frac{1}{2} = \frac{1}{17}$$

and the proposition follows. ∎

5.3 A Complete Promise Problem

Following Sahai and Vadhan [22], who identified promise problems that are complete for the class of promise problems that has statistical zero-knowledge proof systems, analogous results were obtained for statistical NIZK proof systems (see [14]) and perfect NIZK proof systems (see [18]). Following Malka [18, Sec. 2], we identify a very natural promise problem that is complete for super-perfect NIZK proof systems with perfect completeness. The promise problem is defined next.

Definition 5.8 (the promise problem $(\mathtt{U}_{\mathrm{yes}}, \mathtt{U}_{\mathrm{no}})$):

- *The set* $\mathtt{U}_{\mathrm{yes}}$ *consists of all circuits* $C : \{0,1\}^\ell \to \{0,1\}^m$ *such that* $C(U_\ell)$ *is distributed identically to* U_m.
- *The set* $\mathtt{U}_{\mathrm{no}}$ *consists of all circuits* $C : \{0,1\}^\ell \to \{0,1\}^m$ *such that the support of* $C(U_\ell)$ *has size at most* 2^{m-1}.

We assume that the circuits are given in a format in which it is easy to determine the number of bits in their inputs and in their outputs. We comment that the promise problem considered by Malka [18, Def. 2.2] is related but different (i.e., it required that for a yes-instance C it holds that $C(U_\ell)_{[m-1]} \equiv U_{m-1}$ and $\mathbf{Pr}[C(U_\ell)_m = 1] \geq 2/3$, whereas for a no-instance $\mathbf{Pr}[C(U_\ell)_m = 1] \leq 1/3$).

The definition of super-perfect NIZK proof systems extend naturally to promise problem (cf. [23]). Loosely speaking, a promise problem $(\Pi_{\mathrm{yes}}, \Pi_{\mathrm{no}})$ is a pair of non-intersecting sets, and the soundness condition refers only to inputs in Π_{no} (rather than to inputs in $\{0,1\}^* \setminus \Pi_{\mathrm{yes}}$). (The completeness and zero-knowledge conditions refer to all inputs in Π_{yes}.)

Theorem 5.9 ($(\mathtt{U}_{\mathrm{yes}}, \mathtt{U}_{\mathrm{no}})$ is complete for $\mathcal{SPNIZK}_1$): *Let* $\mathcal{SPNIZK}_1$ *denote the class of promise problems having a super-perfect NIZK proof system of perfect completeness. Then,* $(\mathtt{U}_{\mathrm{yes}}, \mathtt{U}_{\mathrm{no}})$ *is in* $\mathcal{SPNIZK}_1$ *and every problem in* $\mathcal{SPNIZK}_1$ *is Karp-reducible to* $(\mathtt{U}_{\mathrm{yes}}, \mathtt{U}_{\mathrm{no}})$.

Proof: The idea underlying the proof of Theorem 3 (presented in Sect. 5.2) can be used to present a super-perfect NIZK proof system of perfect completeness for $(\mathtt{U}_{\text{yes}}, \mathtt{U}_{\text{no}})$. Specifically, on input a circuit $C : \{0,1\}^\ell \to \{0,1\}^m$ and common reference string $\omega \in \{0,1\}^m$, the prover outputs a uniformly distributed string $r \in C^{-1}(\omega)$, and the verifier accepts if and only if $C(r) = \omega$. Perfect completeness and soundness error of $1/2$ are immediate by the definition of $(\mathtt{U}_{\text{yes}}, \mathtt{U}_{\text{no}})$, whereas super-perfect ZK is demonstrated by a simulator that uniformly selects $r \in \{0,1\}^\ell$ and outputs $(C(r), r)$, where $C(r)$ represents the simulated common reference string and r represents the simulated proof output by the prover on input C (and common reference string $C(r)$).

Assuming that $(\Pi_{\text{yes}}, \Pi_{\text{no}}) \in \mathcal{SPNIZK}_1$, we show a Karp-reduction of $(\Pi_{\text{yes}}, \Pi_{\text{no}})$ to $(\mathtt{U}_{\text{yes}}, \mathtt{U}_{\text{no}})$. Let (P, V) be the non-interactive proof systems of $(\Pi_{\text{yes}}, \Pi_{\text{no}})$, and A be the corresponding simulation. Let $\ell = \ell(|x|)$ denote the number of coin tosses used by A on input x, and m denote the length of the common reference string. (For sake of simplicity, we assume, without loss of generality, that V is deterministic and that the soundness error of (P, V) is at most $0.5 - 2^{-m}$, where the probability is taken over all possible choices of the common reference string.) Now, on input x, the reduction produces the following circuit $C_x : \{0,1\}^\ell \to \{0,1\}^m$.

1. On input $r \in \{0,1\}^\ell$, the circuit C_x invokes A on input x and coins r, and obtains the outcome (ω, y), where ω represents the simulated common reference string and y represents the simulated proof output by P on input x (and common reference string ω).
2. The circuit C_x outputs ω if $V(x, \omega, y) = 1$ and 0^m otherwise.

Observe that if $x \in \Pi_{\text{yes}}$ then $C_x(U_\ell) \equiv U_m$, whereas if $x \in \Pi_{\text{no}}$ then the support of $C_x(U_\ell)$ contains at most $(0.5 - 2^{-m}) \cdot 2^m + 1 = 2^{m-1}$ strings, where the extra unit is due to the possible case that $V(x, 0^m, P(x, 0^m)) \neq 1$. The claim follows. ■

6 Open Problems

All the following problems refer to ZK proof systems (rather than to ZK argument systems). For a wider perspective, we start with a well-known open problem regarding perfect ZK (see, e.g., [23, Chap. 8]).

Open Problem 6.1 (perfect ZK versus statistical ZK): *Let $\mathcal{SZK}$ be the class of sets having statistical* (a.k.a almost-perfect) *zero-knowledge interactive proof system. Prove or disprove, under reasonable assumptions, the conjecture by which not all sets in $\mathcal{SZK}$ have perfect zero-knowledge interactive proof systems. Ditto for having a perfect zero-knowledge proof system with perfect completeness.*

Recall that any set in $\mathcal{SZK}$ has a statistical zero-knowledge proof system with perfect completeness (via the transformation to public-coin systems and the use of Lautemann's technique [17]; see [23, Chap. 5] and [9], resp.). This is not known to be the case for perfect zero-knowledge. In particular, it is even not known whether all sets in $\mathcal{BPP}$ have perfect zero-knowledge proof systems with perfect completeness. (Indeed, all sets in co$\mathcal{RP}$ do have perfect zero-knowledge proof systems with perfect completeness, in which the prover remains silent.) Turning to the subject-matter of this work (i.e., super-perfect ZK), we ask:

Open Problem 6.2 (super-perfect ZK versus perfect ZK): *Let $\mathcal{PZK}$ be the class of sets having perfect zero-knowledge interactive proof system. Prove or disprove, under reasonable assumptions, the conjecture by which not all sets in $\mathcal{PZK}$ have super-perfect zero-knowledge interactive proof systems. Ditto for zero-knowledge proof systems with perfect completeness, where the question may refer both to the standard and non-standard models of PPT machines.*

Recall that the difference between the standard and non-standard models of PPT machines arises only with respect to super-perfect ZK. Indeed, one may ask the following

Open Problem 6.3 (super-perfect ZK: models of PPT): *Let S be a set having a super-perfect zero-knowledge interactive proof systems with perfect completeness under one of the two non-standard models of PPT machines. Does S necessarily has a super-perfect zero-knowledge interactive proof systems with perfect completeness under the standard model of PPT machines. Ditto for zero-knowledge proof system with non-perfect completeness.*

It is tempting to think that the question regarding non-perfect completeness can be resolved by applying Theorem 1, but this presumes that all (super-perfect ZK) simulators use their distribution generating device in the same manner (i.e., with the same n's and for the same number of times).[17] The same questions arise with respect to NIZK.

Open Problem 6.4 (super-perfect NIZK versus perfect and statistical NIZK): *Address the non-interactive zero-knowledge analogues of Problems 6.1 and 6.2. Ditto for the perfect completeness version of Problem 6.3.*

Acknowledgments. We are grateful to Alon Rosen and Amit Sahai for useful discussions. This research was partially supported by the Minerva Foundation with funds from the Federal German Ministry for Education and Research.

References

1. Agrawal, M., Kayal, N., Saxena, N.: PRIMES is in P. Ann. Math. **160**(2), 781–793 (2004)
2. Alexi, W., Chor, B., Goldreich, O., Schnorr, C.P.: RSA/Rabin functions: certain parts are as hard as the whole. SIAM J. Comput. **17**, 194–209 (1988)

[17] This presumption holds trivially when referring either to the honest-verifier version or to the NIZK version.

3. Barak, B.: How to go beyond the black-box simulation barrier. In: 42nd IEEE Symposium on Foundations of Computer Science, pp. 106–115 (2001)
4. Barak, B.: Non-black-box techniques in crypptography. Ph.D. thesis, Weizmann Institute of Science (2004)
5. Bellare, M., Impagliazzo, R., Naor, M.: Does parallel repetition lower the error in computationally sound protocols? In: 38th IEEE Symposium on Foundations of Computer Science, pp. 374–383 (1997)
6. Blum, M., De Santis, A., Micali, S., Persiano, G.: Non-interactive zero-knowledge proof systems. SIAM J. Comput. **20**(6), 1084–1118 (1991). (Considered the journal version of [7].)
7. Blum, M., Feldman, P., Micali, S.: Non-interactive zero-knowledge and its applications. In: 20th ACM Symposium on the Theory of Computing, pp. 103–112 (1988). See [6]
8. Brassard, G., Chaum, D., Crépeau, C.: Minimum disclosure proofs of knowledge. J. Comput. Syst. Sci. **37**(2), 156–189 (1988). Preliminary version by Brassard and Crépeau in 27th FOCS, 1986
9. Fürer, M., Goldreich, O., Mansour, Y., Sipser, M., Zachos, S.: On completeness and soundness in interactive proof systems. In: Micali, S., (ed.) Randomness and Computation. Advances in Computing Research: A Research Annual, vol. 5, pp. 429–442 (1989)
10. Goldreich, O.: Foundation of Cryptography: Basic Tools. Cambridge University Press, Cambridge (2001)
11. Goldreich, O.: Foundation of Cryptography: Basic Applications. Cambridge University Press, Cambridge (2004)
12. Goldreich, O.: Computational Complexity: A Conceptual Perspective. Cambridge University Press, Cambridge (2008)
13. Goldreich, O., Micali, S., Wigderson, A.: Proofs that yield nothing but their validity or all languages in NP have zero-knowledge proof systems. J. ACM **38**(3), 691–729 (1991). Preliminary version in *27th FOCS*, 1986
14. Goldreich, O., Sahai, A., Vadhan, S.: Can statistical zero knowledge be made non-interactive? Or on the relationship of SZK and *NISZK*. In: Wiener, M. (ed.) CRYPTO 1999. LNCS, vol. 1666, pp. 467–484. Springer, Heidelberg (1999). https://doi.org/10.1007/3-540-48405-1_30
15. Goldwasser, S., Micali, S., Rackoff, C.: The knowledge complexity of interactive proof systems. SIAM J. Computi. **18**, 186–208 (1989). Preliminary version in 17th STOC, 1985. Earlier versions date to 1982
16. Groth, J., Ostrovsky, R., Sahai, A.: Perfect non-interactive zero knowledge for NP. In: Vaudenay, S. (ed.) EUROCRYPT 2006. LNCS, vol. 4004, pp. 339–358. Springer, Heidelberg (2006). https://doi.org/10.1007/11761679_21
17. Lautemann, C.: BPP and the polynomial hierarchy. Inf. Process. Lett. **17**, 215–217 (1983)
18. Malka, L.: How to achieve perfect simulation and a complete problem for non-interactive perfect zero-knowledge. In: Canetti, R. (ed.) TCC 2008. LNCS, vol. 4948, pp. 89–106. Springer, Heidelberg (2008). https://doi.org/10.1007/978-3-540-78524-8_6
19. Naor, M., Ostrovsky, R., Venkatesan, R., Yung, M.: Zero-knowledge arguments for NP can be based on general assumptions. J. Cryptol. **11**, 87–108 (1998). Preliminary version in Crypto92
20. Pass, R., Rosen, A.: New and improved constructions of non-malleable cryptographic protocols. SIAM J. Comput. **38**(2), 702–752 (2008)

21. Pass, R., Rosen, A.: Concurrent non-malleable commitments. SIAM J. Comput. **37**(6), 1891–1925 (2008)
22. Sahai, A., Vadhan, S.: A complete promise problem for statistical zero-knowledge. J. ACM **50**(2), 196–249 (2003). Preliminary version in 38th FOCS, 1997
23. Vadhan, S.: A study of statistical zero-knowledge proofs. Ph.D. thesis, Department of Mathematics, MIT (1999). See http://people.seas.harvard.edu/~salil/research/phdthesis.pdf

On the Relation Between the Relative Earth Mover Distance and the Variation Distance (an Exposition)

Oded Goldreich and Dana Ron

Abstract. The "relative earth mover distance" is a technical term introduced by Valiant and Valiant (*43rd STOC*, 2011), and extensively used in their work. They claimed that, for every two distributions, the relative earth mover distance upper-bounds the variation distance up to relabeling, but this claim was not used in their work. The claim appears as a special case of a result proved by Valiant and Valiant in a later work (*48th STOC*, 2016), but we found their proof too terse. The proof presented here is merely an elaboration of (this special case of) their proof.

This is a drastic revision of a text that was posted on the first author's web-site in February 2016.[1]

1 Introduction

The (total) variation distance between (discrete) distributions, defined as half the ℓ_1-norm of their difference, is the most popular notion of distance between distributions (viewed as functions from their domain to $[0,1]$).[2] The natural appeal of this distance measure comes from the fact that it equals the best possible distinguishing probability (i.e., the difference in the verdict of an observer that is given a single sample from one of the two distributions).

In particular, the variation distance is the measure of choice in the context of distribution testing. In that context, given samples from an unknown distribution X the task is to test whether X has some predetermined property (i.e., is in a predetermined set of distributions) or is far from having this property (i.e., is far from any distribution in the class).

Testing label-invariant properties is of special interest, where a property is label-invariant if for every distribution (viewed as a function) $p : D \rightarrow [0,1]$ and every bijection $\pi : D \rightarrow D$ it holds that p has the property if and only if $p \circ \pi$ has the property. In this case, we implicitly care about the variation distance *up to relabeling*, as defined below (see Definition 2.3).

[1] See http://www.wisdom.weizmann.ac.il/~oded/p_remd.html.

[2] Specifically, for two distributions presented by their probability functions $p, q : D \rightarrow [0,1]$, their variation distance equals $0.5 \cdot \sum_{i \in D} |p(i) - q(i)|$, which in turn equals $\min_{S \subseteq D}\{p(S) - q(S)\}$, where $p(S) = \sum_{i \in S} p(i)$. The set S may be viewed as the set of samples on which an observer (as discussed next) outputs the verdict 1.

O. Goldreich (Ed.): Computational Complexity and Property Testing, LNCS 12050, pp. 141–151, 2020.
https://doi.org/10.1007/978-3-030-43662-9_9

One of the most striking results in area of distribution testing asserts that *every label-invariant property of distributions over* $[n]$ *can be tested using* $s(n,\epsilon) = O(\epsilon^{-2} \cdot n/\log n)$ *samples.* This celebrated result of Valiant and Valiant is not stated explicitly (in these terms) in their paper [3], since they find it easier to work with an alternative notion of distance that they define, which upper-bounds the variation distance up to relabeling. The latter claim is made and proved in their subsequent work [4,5]; actually, it appears there as a special case of the more general Fact 1 (i.e., the case of $\tau = 0$), but we found their proof too terse.

Given the importance of the results derived by the foregoing claim, which appears below as Theorem 3.1, we believe that it is important to provide and make accessible a more detailed proof of this claim. In fact, we wrote a sketch of the current document when working on [2], where we relied on this claim. Furthermore, this claim is pivotal to the exposition in [1, Sec. 11.4], which highlights the aforementioned result of Valiant and Valiant [3] as well as its optimality. These results appear as [1, Cor. 11.28] and [1, Thm. 11.29], respectively, and both of them rely on Theorem 3.1.

Unfortunately, the rest of this text is quite technical. We start with the definitions that constitute the alternative notion of distance defined by Valiant and Valiant [3], which is called the *relative earth-mover distance.* This definition (i.e., Definition 2.2) actually refers to the "relative histrograms" of the distributions, which are defined first (see Definition 2.1). The (simpler) non-relative notions play a pivotal role in the proof, so they are defined too (in Sect. 2). (For sake of good order, we also define the variation distance up to relabeling (see Definition 2.3).) With all these definitions in place, the foregoing claim is stated as Theorem 3.1 and proved in Sect. 3. The proof is quite technical and is merely an elaboration of (the special case of) the proof presented by Valiant and Valiant (in [4, Apdx A] and [5, Apdx B]).

2 Definitions

For simplicity, we consider distributions over $[n] \stackrel{\text{def}}{=} \{1, 2, ..., n\}$. We start by introducing two label-invariant representations of distributions.

Definition 2.1 (histograms and relative histograms of distributions): *For a distribution* $p : [n] \to [0,1]$*, the corresponding* histogram*, denoted* $h_p : [0,1] \to \mathbb{N}$*, counts the number of elements that occur with each possible probability; that is,* $h_p(x) \stackrel{\text{def}}{=} |\{i \in [n] : p(i) = x\}|$ *for each* $x \in [0,1]$*. The corresponding* relative histogram*, denoted* $h_p^R : [0,1] \to \mathbb{R}$*, satisfies* $h_p^R(x) = h_p(x) \cdot x$ *for every* $x \in [0,1]$*.*

That is, $h_p(x)$ equals the *number* of elements in p that are assigned probability mass x, whereas $h_p^R(x)$ equals the *total probability mass* assigned to these elements. Hence, $h_p(0)$ may be positive, whereas $h_p^R(0)$ is always zero. Indeed, both functions are label-invariant (i.e., for every $p : [n] \to [0,1]$ and every bijection $\pi : [n] \to [n]$, it holds that $h_p \equiv h_{p\circ\pi}$ and $h_p^R \equiv h_{p\circ\pi}^R$).

For a non-negative function h, let $S(h) \stackrel{\text{def}}{=} \{x : h(x) > 0\}$ denote the support of h. Observe that, for any distribution $p : [n] \to [0,1]$, it holds that $\sum_{x \in S(h_p)} h_p(x) = n$ and $\sum_{x \in S(h_p^R)} h_p^R(x) = 1$. Also note that the support of h_p may contain 0, whereas the support of h_p^R never contains it; furthermore, $S(h_p^R) = S(h_p) \setminus \{0\}$.

The following definitions interpret the distance between non-negative functions h and h' as the cost of transforming h into h' by moving $m(x,y)$ (fractional) units from location x in h to location y in h' (for every $x \in S(h)$ and $y \in S(h')$), where the cost of moving a single unit from x to y is either $|x - y|$ or $|\log(x/y)|$ (depending on the distance).[3] We stress that $m(x,y) \geq 0$ need not be an integer, although in the case of EMD it will be shown to be an integer (w.l.o.g.).

Definition 2.2 (Earth-Mover Distance and Relative Earth-Mover Distance): *For a pair of non-negative functions $h : [n] \to \mathbb{R}$ and $h' : [n] \to \mathbb{R}$ such that $\sum_{x \in S(h)} h(x) = \sum_{x \in S(h')} h'(x)$, the* earth-mover distance *between them, denoted* $\text{EMD}(h, h')$, *is the minimum of*

$$\sum_{x \in S(h)} \sum_{y \in S(h')} m(x,y) \cdot |x - y|, \tag{1}$$

taken over all non-negative functions $m \colon S(h) \times S(h') \to \mathbb{R}$ that satisfy:

- *For every $x \in S(h)$, it holds that $\sum_{y \in S(h)} m(x,y) = h(x)$, and*
- *For every $y \in S(h')$, it holds that $\sum_{x \in S(h')} m(x,y) = h'(y)$.*

The relative earth-mover distance *between h and h', denoted* $\text{REMD}(h, h')$, *is the minimum of*

$$\sum_{x \in S(h)} \sum_{y \in S(h')} m(x,y) \cdot |\log(x/y)|, \tag{2}$$

subject to the same constraints on m as for EMD.

Note that satisfying the constraints may require having $m(x,x) > 0$ for some x's, but the contribution of the corresponding terms to Eq. (1) (resp., Eq. (2)) is zero (since $|x - x| = 0$ (resp., $\log(x/x) = 0$)).

The term *earth-mover* comes from viewing the functions as piles of earth, where for each $x \in S(h)$ there is a pile of size $h(x)$ in location x and similarly for each $y \in S(h')$ there is a pile of size $h'(y)$ in location y. The goal is to transform the piles defined by h so as to obtain the piles defined by h', with minimum "transportation cost". Specifically, $m(x,y)$ captures the (possibly fractional) number of units transferred from pile x in h to pile y in h'. For EMD the cost of transporting a unit from x to y is $|x - y|$, while for REMD it is $|\log(x/y)|$.

[3] Here and in the sequel, the logarithm is to base 2. The proof of Theorem 3.1 as presented in Sect. 3 remains valid for any base $b \in (1, e]$; our only reference to this base is that it (i.e., b) should satisfy $\log_b z > 1 - (1/z)$ for every $z > 1$. It seems that Valiant and Valiant do mean to take $b = 2$ (although other parts of their text suggest $b = e$). Indeed, both $b = 2$ and $b = e$ seems natural choices.

In what follows, for a pair of distributions p and q over $[n]$ we shall apply EMD to the corresponding pair of histograms h_p and h_q, and apply REMD to the corresponding relative histograms h_p^R and h_q^R. For example, Lemma 3.2 asserts that $\mathtt{EMD}(h_p, h_q) \leq 2 \cdot \mathtt{REMD}(h_p^R, h_q^R)$, for every two distributions p and q.

Variation Distance up to Relabeling. As stated in the introduction, variation distance up to relabeling (defined next) is a natural notion in the context of testing properties of distributions (see [1, Sec. 11.1.3]): It arises naturally when the properties are label-invariant (i.e., properties that are invariant under relabeling of the elements of the distribution).

Definition 2.3 (variation distance up to relabeling): *For two distributions p and q over $[n]$, the* variation distance up to relabeling *between p and q, denoted* $\mathtt{VDR}(p, q)$*, is the minimum over all permutations $\pi : [n] \to [n]$ of*

$$\frac{1}{2} \cdot \sum_{i=1}^{n} |p(i) - q(\pi(i))|. \tag{3}$$

In other words, $\mathtt{VDR}(p, q)$ is the variation distance between the distribution p and the set of distributions obtained by a relabeling of q (i.e., the set $\{q \circ \pi : \pi \in \mathrm{Sym}_n\}$). Observing that the latter set equals the set of distributions having histogram h_p, it follows that $\mathtt{VDR}(p, q)$ equals $\min_{q': h_{q'} = h_q} \{0.5 \cdot \sum_{i=1}^{n} |p(i) - q'(i)|\}$.

3 The Claim and Its Proof

Our goal is to present a proof of the following result.

Theorem 3.1 (special case (i.e., $\tau = 0$) of Fact 1 in [4,5]): *For every two distributions p and q over $[n]$, it holds that*

$$\mathtt{VDR}(p, q) \leq \mathtt{REMD}(h_p^R, h_q^R).$$

The proof will consist of two steps (captured by lemmas):

1. $\mathtt{VDR}(p, q) = \frac{1}{2} \cdot \mathtt{EMD}(h_p, h_q)$.
2. $\mathtt{EMD}(h_p, h_q) \leq 2 \cdot \mathtt{REMD}(h_p^R, h_q^R)$.

Actually, we start with the second step.

Lemma 3.2 (upper-bounding EMD in terms of REMD): *For every two distributions p and q over $[n]$,*

$$\mathtt{EMD}(h_p, h_q) \leq 2 \cdot \mathtt{REMD}(h_p^R, h_q^R) .$$

The following proof shows how to construct, for every transportation function m' used for the relative histograms (h_p^R and h_q^R) a corresponding transportation function m for the corresponding histograms (h_p and h_q) such that the EMD-cost of m is at most twice the REMD-cost of m'.

Proof: It will be convenient to consider two distributions, $\widetilde{p}$ and $\widetilde{q}$ that are slight variations of p and q, respectively. They are both defined over $[2n]$, where $\widetilde{p}(i) = p(i)$ and $\widetilde{q}(i) = q(i)$ for every $i \in [n]$, and $\widetilde{p}(i) = \widetilde{q}(i) = 0$ for every $i \in [2n] \setminus [n]$. Since $h^R_{\widetilde{p}} = h^R_p$ and $h^R_{\widetilde{q}} = h^R_q$, we have that $\texttt{REMD}(h^R_{\widetilde{p}}, h^R_{\widetilde{q}}) = \texttt{REMD}(h^R_p, h^R_q)$. As for $h_{\widetilde{p}}$ and $h_{\widetilde{q}}$, they agree with h_p and h_q, respectively, everywhere except on 0, where $h_{\widetilde{p}}(0) = h_p(0) + n$ and $h_{\widetilde{q}}(0) = h_q(0) + n$, so $\texttt{EMD}(h_{\widetilde{p}}, h_{\widetilde{q}}) = \texttt{EMD}(h_p, h_q)$ as well. Therefore, it suffices to show that $\texttt{EMD}(h_{\widetilde{p}}, h_{\widetilde{q}}) \leq 2 \cdot \texttt{REMD}(h^R_{\widetilde{p}}, h^R_{\widetilde{q}})$.

Let m' be a function over $S(h^R_{\widetilde{p}}) \times S(h^R_{\widetilde{q}})$ that satisfies the constraints stated in Definition 2.2 for the pair of relative histograms $h^R_{\widetilde{p}}$ and $h^R_{\widetilde{q}}$. We next show that there exists a non-negative function m over $S(h_{\widetilde{p}}) \times S(h_{\widetilde{q}})$ that satisfies the constraints stated in Definition 2.2 for the pair of histograms $h_{\widetilde{p}}$ and $h_{\widetilde{q}}$, and also satisfies

$$\sum_{x\in S(h_{\widetilde{p}})} \sum_{y\in S(h_{\widetilde{q}})} m(x,y)\cdot|x-y| \leq 2\cdot \sum_{x\in S(h^R_{\widetilde{p}})} \sum_{y\in S(h^R_{\widetilde{q}})} m'(x,y)\cdot|\log(x/y)|. \quad (4)$$

Note that the range of m' is $[0, 1]$, since it is defined over relative histograms, while m is not upper bounded by 1. However, the constraints that Definition 2.2 imposes on the two functions are related, since for every $x \in S(h^R_{\widetilde{p}}) = S(h_{\widetilde{p}}) \setminus \{0\}$ it is required that $\sum_{y\in S(h^R_{\widetilde{q}})} m'(x,y)/x = h_{\widetilde{p}}(x) = \sum_{y\in S(h_{\widetilde{q}})} m(x,y)$ and for every $y \in S(h^R_{\widetilde{q}}) = S(h_{\widetilde{q}}) \setminus \{0\}$ it is required that $\sum_{x\in S(h^R_{\widetilde{p}})} m'(x,y)/y = h_{\widetilde{q}}(y) = \sum_{x\in S(h_{\widetilde{q}})} m(x,y)$. (Indeed, m is also subjected to constraints on $x = 0$ and $y = 0$, whereas m' is not.)

The construction of the transportation function m. Essentially, for every $(x, y) \in S(h^R_{\widetilde{p}}) \times S(h^R_{\widetilde{q}})$ we set $m(x, y) = m'(x, y)/x$ if $x > y$ and $m(x, y) = m'(x, y)/y$ otherwise, and place the excess (arising from the difference between $m'(x, y)/x$ and $m'(x, y)/y$) in either $m(0, y)$ or $m(x, 0)$ according to its sign. This is done by scanning all pairs $(x, y) \in S(h^R_{\widetilde{p}}) \times S(h^R_{\widetilde{q}})$, in arbitrary order, setting $m(x, y)$, and increasing $m(0, y)$ or $m(x, 0)$ appropriately. Specifically, for each $x \in S(h^R_{\widetilde{p}})$, we initialize $m(x, 0)$ to 0, and similarly we initialize $m(0, y)$ to 0 for each $y \in S(h_{\widetilde{q}})$. For every pair $(x, y) \in S(h^R_{\widetilde{p}}) \times S(h^R_{\widetilde{q}})$, if $m'(x, y) = 0$, then we set $m(x, y) = 0$, and otherwise (i.e., $m'(x, y) > 0$) we proceed as follows:

– If $x > y$, then we set $m(x, y)$ to $m'(x, y)/x$ and increase $m(0, y)$ by $m^x(0, y) \stackrel{\text{def}}{=} \frac{m'(x,y)}{y} - \frac{m'(x,y)}{x} > 0$ units.
  Observe that $m(x, y) \cdot (x - y) = m'(x, y) \cdot (x - y)/x = m^x(0, y) \cdot y$. Therefore, the contribution of $m(x, y) \cdot |x - y| + m^x(0, y) \cdot |0 - y|$ to the left-hand-side of Eq. (4) is

$$\begin{aligned} m(x,y)\cdot(x-y) + m^x(0,y)\cdot y &= 2\cdot m'(x,y)\cdot\frac{x-y}{x} \\ &< 2\cdot m'(x,y)\cdot\log(x/y), \end{aligned}$$

where the last inequality uses $1-(y/x)<\log(x/y)$, which is due to the fact that $f(z)=\log z+(1/z)-1\geq \ln z+(1/z)-1$ is positive for all $z>1$.[4]

– If $x<y$, then we set $m(x,y)$ to $m'(x,y)/y$ and increase $m(x,0)$ by $m^y(x,0)\stackrel{\rm def}{=}\frac{m'(x,y)}{x}-\frac{m'(x,y)}{y}>0$ units.
Similarly to the previous case, we have $m(x,y)\cdot(y-x)=m'(x,y)\cdot(y-x)/y=m^y(x,0)\cdot x$, and the contribution to the left-hand-side of Eq. (4) is

$$\begin{aligned} m(x,y)\cdot(y-x)+m^y(x,0)\cdot x &= 2\cdot m'(x,y)\cdot\frac{y-x}{y}\\ &< 2\cdot m'(x,y)\cdot\log(y/x). \end{aligned}$$

– If $x=y$, then we set $m(x,y)=m'(x,y)/x$ (which equals $=m'(x,y)/y$).
In this case $m(x,y)\cdot|x-y|=0$, so there is no contribution to the left-hand-side of Eq. (4). Note that in this case $m'(x,y)\cdot|\log(x/y)|=0$.

Finally, we set $m(0,0)=h_{\widetilde{p}}(0)-\sum_{y\in S(h_{\widetilde{q}}^R)}m(0,y)$. To see that $m(0,0)\geq 0$, observe that $h_{\widetilde{p}}(0)\geq n$ (since $\widetilde{p}(i)=0$ for every $i\in[2n]\setminus[n]$), whereas

$$\begin{aligned} \sum_{y\in S(h_{\widetilde{q}}^R)} m(0,y) &= \sum_{y\in S(h_{\widetilde{q}}^R)}\ \sum_{x\in S(h_{\widetilde{p}}^R)\cap(y,1]} m^x(0,y)\\ &\leq \sum_{y\in S(h_{\widetilde{q}}^R)}\ \sum_{x\in S(h_{\widetilde{p}}^R)\cap(y,1]} \frac{m'(x,y)}{y}\\ &\leq \sum_{y\in S(h_{\widetilde{q}}^R)} h_{\widetilde{q}}(y)\\ &\leq n \end{aligned}$$

(since $\widetilde{q}(i)=0$ for every $i\in[2n]\setminus[n]$). Lastly, combining the contribution of all pairs $(x,y)\in S(h_{\widetilde{p}})\times S(h_{\widetilde{q}})$ to the left-hand-side of Eq. (4), we have

$$\begin{aligned} &\sum_{(x,y)\in S(h_{\widetilde{p}})\times S(h_{\widetilde{q}})} m(x,y)\cdot|x-y|\\ &= \sum_{(x,y)\in S(h_{\widetilde{p}}^R)\times S(h_{\widetilde{q}}^R):x>y} (m(x,y)\cdot(x-y)+m^x(0,y)\cdot y)\\ &\quad+\sum_{(x,y)\in S(h_{\widetilde{p}}^R)\times S(h_{\widetilde{q}}^R):x<y} (m(x,y)\cdot(y-x)+m^y(x,0)\cdot x)\\ &< \sum_{(x,y)\in S(h_{\widetilde{p}}^R)\times S(h_{\widetilde{q}}^R):x>y} 2\cdot m'(x,y)\cdot\log(x/y)\\ &\quad+\sum_{(x,y)\in S(h_{\widetilde{p}}^R)\times S(h_{\widetilde{q}}^R):x<y} 2\cdot m'(x,y)\cdot\log(y/x) \end{aligned}$$

[4] As stated in Footnote 3, we assume that the logarithm is to base $b\in(1,e]$. Indeed, here we use $\log_b z+(1/z)>1$ for all $z>1$, and this is the only place in the proof in which the choice of b matters.

$$= \sum_{(x,y)\in S(h^R_{\tilde{p}})\times S(h^R_{\tilde{q}})} 2\cdot m'(x,y)\cdot|\log(x/y)|,$$

which equals the right-hand-side of Eq. (4).

Verifying that the transportation function m is legal. It remains to verify that m satisfies the constraints in Definition 2.2. For each $x \in S(h_{\tilde{p}})\setminus\{0\}$,

$$\begin{aligned}
\sum_{y\in S(h_{\tilde{q}})} m(x,y) &= m(x,0) + \sum_{y\in S(h_{\tilde{q}})\cap(0,x]} m(x,y) + \sum_{y\in S(h_{\tilde{q}})\cap(x,1]} m(x,y)\\
&= \sum_{y\in S(h^R_{\tilde{q}})\cap(x,1]} m^y(x,0)\\
&\quad + \sum_{y\in S(h^R_{\tilde{q}})\cap(0,x]} m(x,y) + \sum_{y\in S(h^R_{\tilde{q}})\cap(x,1]} m(x,y)\\
&= \sum_{y\in S(h^R_{\tilde{q}})\cap(x,1]} \left(\frac{1}{x}-\frac{1}{y}\right)\cdot m'(x,y)\\
&\quad + \sum_{y\in S(h^R_{\tilde{q}})\cap(0,x]} \frac{m'(x,y)}{x} + \sum_{y\in S(h^R_{\tilde{q}})\cap(x,1]} \frac{m'(x,y)}{y}\\
&= \sum_{y\in S(h^R_{\tilde{q}})} \frac{m'(x,y)}{x}\\
&= h_{\tilde{p}}(x).
\end{aligned}$$

Similarly, for each $y \in S(h_{\tilde{q}})\setminus\{0\}$,

$$\begin{aligned}
\sum_{x\in S(h_{\tilde{p}})} m(x,y) &= m(0,y) + \sum_{x\in S(h_{\tilde{p}})\cap(0,y]} m(x,y) + \sum_{x\in S(h_{\tilde{p}})\cap(y,1]} m(x,y)\\
&= \sum_{x\in S(h^R_{\tilde{q}})\cap(y,1]} m^x(0,y)\\
&\quad + \sum_{x\in S(h^R_{\tilde{q}})\cap(0,y]} m(x,y) + \sum_{x\in S(h^R_{\tilde{q}})\cap(y,1]} m(x,y)\\
&= \sum_{x\in S(h^R_{\tilde{q}})\cap(y,1]} \left(\frac{1}{y}-\frac{1}{x}\right)\cdot m'(x,y)\\
&\quad + \sum_{x\in S(h^R_{\tilde{q}})\cap(0,y]} \frac{m'(x,y)}{y} + \sum_{x\in S(h^R_{\tilde{q}})\cap(y,1]} \frac{m'(x,y)}{x}\\
&= \sum_{x\in S(h^R_{\tilde{p}})} \frac{m'(x,y)}{y}\\
&= h_{\tilde{q}}(y).
\end{aligned}$$

Recall that we defined $m(0,0)$ such that $m(0,0) + \sum_{y\in S(h_{\tilde{q}}^R)} m(0,y) = h_{\tilde{p}}(0)$, and it follows that $\sum_{y\in S(h_{\tilde{q}})} m(0,y) = h_{\tilde{p}}(0)$. Lastly, we observe that

$$\begin{aligned}\sum_{x\in S(h_{\tilde{p}})} m(x,0) &= \sum_{x\in S(h_{\tilde{p}})} \sum_{y\in S(h_{\tilde{q}})} m(x,y) - \sum_{x\in S(h_{\tilde{p}})} \sum_{y\in S(h_{\tilde{q}})\setminus\{0\}} m(x,y)\\ &= \sum_{x\in S(h_{\tilde{p}})} h_{\tilde{p}}(x) - \sum_{y\in S(h_{\tilde{p}})\setminus\{0\}} \sum_{y\in S(h_{\tilde{q}})} m(x,y)\\ &= 2n - \sum_{y\in S(h_{\tilde{q}})\setminus\{0\}} h_{\tilde{q}}(y)\\ &= h_{\tilde{q}}(0).\end{aligned}$$

This completes the verification of all contraints, and the lemma follows. ■

Lemma 3.3 (VDR equals EMD/2): *For every two distributions p and q over $[n]$,*

$$\mathtt{VDR}(p,q) = \frac{1}{2}\cdot \mathtt{EMD}(h_p,h_q).$$

Intuitively, there is a one-to-one correspondence between the cost of the best *integer-valued* transportation functions m satisfying the conditions of Definition 2.2 (w.r.t h_p and h_q) and the differences (between p and $q\circ\pi$) under the best relabeling permutation π used in Definition 2.3. The core of the following proof is showing that integer-value transportation functions m obtain the minimum for EMD.

Proof: Consider a constrained version of the earth-mover distance in which we also require that $m(x,y)$ is an *integer* for every $x\in S(h_p)$ and $y\in S(h_q)$, and denote this distance measure by IEMD. Using the definition of VDR and IEMD, one can verify that $\mathtt{VDR}(q,p) = \frac{1}{2}\cdot\mathtt{IEMD}(h_p,h_q)$. To see this consider the best permutation π used in Definition 2.3, and let $q' = q\circ\pi$. In this case, the ℓ_1-norm of $p-q'$ equals $\sum_{i\in[n]} |p(i)-q'(i)|$, which in turn equals $\sum_{x,y\in[0,1]} m(x,y)\cdot|x-y|$, where $m(x,y) = |\{i\in[n] : p(i)=x \,\&\, q'(i)=y\}|$ is the desired transportation function. Likewise, an integer-valued transportation function m (from h_p to $h_{q'}$) yields an upper bound on the ℓ_1-norm of $p-q'$. Last, noting that $h_{q'} = h_q$ and that the variation distance between distributions equals half the ℓ_1-norm between them, the claim follows.

It remains to prove that $\mathtt{EMD}(h_p,h_q) = \mathtt{IEMD}(h_p,h_q)$; that is, that an integer-valued function m obtains the minimum of the EMD. To this end, we define a specific integer-valued function m (based on a simple iterative assignment procedure), and show that it is optimal.

Constructing the transportation function m. We use an iterative construction. Initially, $m(x,y)=0$ for every $x\in S(h_p)$ and $y\in S(h_q)$. We also initialize $s(x) = h_p(x)$ for every $x\in S(h_p)$, and $d(y) = h_q(y)$ for every $y\in S(h_q)$. Intuitively, $s(x)$ is the supply of h_p at x, and $d(y)$ is the demand of h_q at y. Note that $\sum_{x \in S(h_p)} s(x) = n = \sum_{y \in S(h_q)} d(y)$, and that the supplies and demands are initially integers (and will remain so throughout the iterations).

In each iteration, we consider the smallest $x \in S(h_p)$ for which $s(x) > 0$ and the smallest $y \in S(h_q)$ for which $d(y) > 0$, set $m(x, y) = \min\{s(x), d(y)\}$ and reduce both $s(x)$ and $d(y)$ by $m(x, y)$. Since the supplies and demands are initially integers, the values of the $m(x, y)$'s that we set are integers, and the supplies and demands remain (non-negative) integers. (Equivalently, we may transport one unit in each iteration such that in the ith iteration we transport the ith smallest p-value to the ith smallest q-value.)[5] By its construction, the function m satisfies the constraints of Definition 2.2.

Showing that m is optimal for EMD. Recall that the cost of a transportation function is the value assigned to it by Eq. (1), and our aim is to show that the constructed function m has minimal cost among all transportation functions that satisfy the constraints of Definition 2.2. Note that these constraints form a linear system, and so the notions of minimiality referred to here and below are well-defined. Assuming that m is not optimal, consider a non-negative function ℓ over $S(h_p) \times S(h_q)$ that satisfies the constraints of Definition 2.2 and is minimal in the following sense.

1. The function ℓ has the smallest cost among all non-negative functions that satisfy the constraints of Definition 2.2. (Indeed, this means that $\ell \neq m$.)
2. Among the functions satisfying the foregoing, the function ℓ agrees with m on the longest prefix of pairs (x, y) according to the lexicographical order on pairs.
   Let (x^*, y^*) be the first pair on which ℓ and m differ; that is, $\ell(x^*, y^*) \neq m(x^*, y^*)$ whereas $\ell(x, y) = m(x, y)$ for every $(x, y) < (x^*, y^*)$.
3. Among all functions satisfying the foregoing, the function ℓ attains a minimal value for $|\ell(x^*, y^*) - m(x^*, y^*)|$.

We reach a contradiction to the hypothesis regarding ℓ by first proving that $\ell(x^*, y^*) < m(x^*, y^*)$, and then presenting a function ℓ' that violates the minimiality of ℓ (as defined above (w.r.t all functions that satisfy the constraints of Definition 2.2)).

Proving that $\ell(x^*, y^*) < m(x^*, y^*)$. Towards this end, we consider the supply of x^* and the demand of y^* just before $m(x^*, y^*)$ is reset; that is, $s(x^*) = h_p(x^*) - \sum_{y<y^*} m(x^*, y)$ and $d(y^*) = h_q(y^*) - \sum_{x<x^*} m(x, y^*)$, where we rely on the fact that (by construction of m) before $m(x^*, y^*)$ is reset it holds that $m(x^*, y) > 0$ only if $y < y^*$ and $m(x, y^*) > 0$ only if $x < x^*$. Assuming towards the contradiction that $\ell(x^*, y^*) > m(x^*, y^*)$, and recalling that $m(x^*, y^*) = \min(s(x^*), d(y^*))$, we note that if $m(x^*, y^*) = s(x^*)$, then

[5] That is, letting π_p and π_q be permutations over $[n]$ such that $p(\pi_p(j)) \leq p(\pi_p(j+1))$ and $q(\pi_q(j)) \leq q(\pi_q(j+1))$ for every $j \in [n-1]$, in the ith iteration we transport one unit from location $p(\pi_p(i))$ of h_p to location $q(\pi_q(i))$ of h_q.

$$\sum_{y \leq y^*} \ell(x^*, y) = \sum_{y < y^*} m(x^*, y) + \ell(x^*, y^*) > \sum_{y < y^*} m(x^*, y) + m(x^*, y^*) = h_p(x^*),$$

which means that ℓ violates a constraint of Definition 2.2. A similar contradiction is obtained by assuming that $\ell(x^*, y^*) > m(x^*, y^*) = d(y^*)$, when in this case we get $\sum_{x \leq x^*} \ell(x, y^*) > h_q(x^*)$.

Demonstrating that ℓ is not minimal. Having shown that $\ell(x^*, y^*) < m(x^*, y^*)$, we now derive a function ℓ' that violates the minimality of ℓ. First, using the hypothesis that ℓ satisfies the constraints of Definition 2.2, we observe that $\ell(x^*, y^*) < m(x^*, y^*)$ (combined with $\ell(x, y) = m(x, y)$ for every $(x, y) < (x^*, y^*)$) implies that there exists $x' > x^*$ such that $\ell(x', y^*) > m(x', y^*)$ and $y' > y^*$ such that $\ell(x^*, y') > m(x^*, y')$. Letting

$$c = \min(m(x^*, y^*) - \ell(x^*, y^*), \ell(x', y^*), \ell(x^*, y')) > 0,$$

define ℓ' as equal to ℓ on all pairs except for the following four pairs that satisfy $\ell'(x^*, y^*) = \ell(x^*, y^*) + c$, $\ell'(x', y^*) = \ell(x', y^*) - c$, $\ell'(x^*, y') = \ell(x^*, y') - c$, and $\ell'(x', y') = \ell(x', y') + c$. Then, ℓ' preserves the constraints of Definition 2.2, but[6] (1) its cost is not higher than that of ℓ, (2) its first point of disagreement with m is not before (x^*, y^*), and (3) $|\ell'(x^*, y^*) - m(x^*, y^*)| = |\ell(x^*, y^*) - m(x^*, y^*)| - c$. This contradicts the minimality of ℓ, since $c > 0$.

Having proved that no transportation function that satisfies the constraints of Definition 2.2 has cost that is smaller than the cost of m, we established $\texttt{EMD}(h_p, h_q) = \texttt{IEMD}(h_p, h_q)$, and the lemma follows. ■

Comments. As noted in [3], there exist distributions p and q for which $\texttt{VDR}(h_p, h_q)$ is much smaller than $\texttt{REMD}(h_p^R, h_q^R)$. The source of this phenomenon is the

[6] To see that (1) holds, note that the cost of ℓ' equals the cost of ℓ plus $c \cdot |x^* - y^*| - c \cdot |x' - y^*| - c \cdot |x^* - y'| + c \cdot |x' - y'|$. Hence, we need to verify that the added value is not positive; equivalently, that $|x^* - y^*| - |x^* - y'| \leq |x' - y^*| - |x' - y'|$. Consider the following cases:

1. The diagonal line $y = x$ does not cross the rectangle spanned by (x^*, y^*) (i.e., either $y' \leq x^*$ or $y^* \geq x'$). If $y' \leq x^*$, then $|y^* - x^*| - |y' - x^*| = y' - y^* = |y^* - x'| - |y' - x'|$, and otherwise $|y^* - x^*| - |y' - x^*| = -(y' - y^*) = |y^* - x'| - |y' - x'|$.
2. The diagonal line $y = x$ separates one corner-point of the rectangle from the other three corner-points (e.g., $y' > x^*$ but $y < x$ for $(x, y) \in \{(x^*, y^*), (x', y^*), (x', y')\}$). If $y' > x^*$, then $|y^* - x^*| - |y' - x^*| < y' - y^* = |y^* - x'| - |y' - x'|$, and similarly for the case that (x', y^*) is separated.
3. The diagonal line $y = x$ crosses both horizontal lines of the rectangle (i.e., $y^*, y' \in [x^*, x']$). In this case, $|y^* - x^*| - |y' - x^*| = -(y' - y^*)$ and $|y^* - x'| - |y' - x'| = y' - y^*$.
4. The diagonal line $y = x$ crosses both vertical lines of the rectangle (i.e., $x^*, x' \in [y^*, y']$). In this case $|y^* - x^*| - |y' - x^*| < |y^* - x'| - |y' - x'|$, since $|y^* - x^*| < |y^* - x'|$ and $|y' - x^*| > |y' - x'|$.

To see that (2) holds, recall that $\ell'(x, y) = \ell(x, y) = m(x, y)$ for every $(x, y) < (x^*, y^*)$.

unbounded cost of transportation under the REMD (i.e., transforming a unit of mass from x to y costs $|\log(x/y)|$). For example, for any $\epsilon \in (0, 0.5)$, consider the pair (p, q) such that p is uniform over $[n]$ (i.e., $p(i) = 1/n$ for every $i \in [n]$) and q is extremely concentrated on a single point in the sense that $q(n) = 1 - \epsilon$ and $q(i) = \epsilon/(n-1)$ for every $i \in [n-1]$. Then, the variation distance between p and q is $\frac{n-1}{n} - \epsilon$, but the REMD is greater than $\frac{n-1}{n} \cdot \log(1/\epsilon)$. This phenomenon is reflected in the proof of Lemma 3.2 at the point we used the inequality $1 - (1/z) < \log z$ for $z > 1$. This inequality becomes more crude when z grows.

References

1. Goldreich, O.: Introduction to Property Testing. Cambridge University Press, Cambridge (2017)
2. Goldreich, O., Ron, D.: On sample-based testers. In: 6th Innovations in Theoretical Computer Science, pp. 337–345 (2015)
3. Valiant, G., Valiant, P.: Estimating the unseen: an $n/\log(n)$-sample estimator for entropy and support size, shown optimal via new CLTs. In: 43rd ACM Symposium on the Theory of Computing, pp. 685–694 (2011). See ECCC TR10-180 for the algorithm, and TR10-179 for the lower bound
4. Valiant, G., Valiant, P.: Instance optimal learning. CoRR abs/1504.05321 (2015)
5. Valiant, G., Valiant, P.: Instance optimal learning of discrete distributions. In: 48th ACM Symposium on the Theory of Computing, pp. 142–155 (2016)

The Uniform Distribution Is Complete with Respect to Testing Identity to a Fixed Distribution

Oded Goldreich

Abstract. Inspired by Diakonikolas and Kane (2016), we reduce the class of problems consisting of testing whether an unknown distribution over $[n]$ equals a fixed distribution to the special case in which the fixed distribution is uniform over $[n]$. Our reduction preserves the parameters of the problem, which are n and the proximity parameter $\epsilon > 0$, up to a constant factor.

While this reduction yields no new bounds on the sample complexity of either problems, it provides a simple way of obtaining testers for equality to arbitrary fixed distributions from testers for the uniform distribution. The reduction first reduces the general case to the case of "grained distributions" (in which all probabilities are multiples of $\Omega(1/n)$), and then reduces this case to the case of the uniform distribution. Using grained distributions as a pivot of the exposition, we call attention to this natural class.

An early version of this work appeared as TR16-015 of *ECCC*. The original version reproduced some text from the author's lecture notes on property testing [8], which were later used as a basis for his book [9]. The current revision is quite minimal, except for the correction of various typos and a significant elaboration of the Appendix.

1 Introduction

Inspired by Diakonikolas and Kane [5], we present, for every fixed distribution D over $[n]$, a simple reduction of the problem of testing whether an unknown distribution over $[n]$ equals D to the problem of testing whether an unknown distribution over $[n]$ equals the uniform distribution over $[n]$. Specifically, we reduce ϵ-testing of equality to D to $\epsilon/3$-testing of equality to the uniform distribution over $[6n]$, denoted U_{6n}.

Hence, the sample (resp., time) complexity of testing equality to D, with respect to the proximity parameter ϵ, is at most the sample (resp., time) complexity of testing equality to U_{6n} with respect to the proximity parameter $\epsilon/3$. Since optimal bounds were known for both problems (cf., e.g., [1,2,4,12,14,17]), our reduction yields no new bounds. Still, it provides a simple way of obtaining testers for equality to arbitrary fixed distributions from testers for the uniform distribution.

O. Goldreich (Ed.): Computational Complexity and Property Testing, LNCS 12050, pp. 152–172, 2020.
https://doi.org/10.1007/978-3-030-43662-9_10

The Setting at a Glance. For any fixed distribution D over $[n]$, we consider the problem of ϵ-testing equality to D, where the tester is given samples drawn from an unknown distribution X and is required to distinguish the case that $X \equiv D$ from the case that X is ϵ-far from D, where the distance is the standard statistical distance. The sample complexity of this testing problem, depends on D, and is viewed as a function of n and ϵ. We write $D \subseteq [n]$ to denote that D ranges over $[n]$.

Wishing to present reductions between such problems, we need to spell out what we mean by such a reduction. Confining ourselves to problems of testing equality to fixed distributions, we use a very stringent notion of a reduction, which we call *reduction via a filter*. Specifically, we say that ϵ-testing equality to $D \subseteq [n]$ reduces to ϵ'-testing equality to $D' \subseteq [n']$ if there exists a randomized process F (called a filter) that maps $[n]$ to $[n']$ such that the distribution D is mapped to the distribution D' and any distribution that is ϵ-far from D is mapped to a distribution that is ϵ'-far from D', where we say that F maps the distribution X to the distribution Y if $Y \equiv F(X)$.

Note that the foregoing is a very stringent notion of reduction between distribution testing problems: Under this notion, a tester T for ϵ-testing equality to D is derived by invoking a ϵ'-tester T' for equality to D' and providing T' with the sample $F(i_1), ..., F(i_s)$, where $i_1, ..., i_s$ is the sample provided to T. Still, our main result can be stated as follows.

Theorem 1.1 (completeness of testing equality to U_n): *For every distribution D over $[n]$ and every $\epsilon > 0$, it holds that ϵ-testing equality to D reduces to $\epsilon/3$-testing equality to U_{6n}, where U_m denotes the uniform distribution over $[m]$. Furthermore, the same reduction F can be used for all $\epsilon > 0$.*

Hence, the sample complexity of ϵ-testing equality to D is upper-bounded by the sample complexity of $\epsilon/3$-testing equality to U_{6n}. We mention that in some cases, testing equality to D can be easier than testing equality to U_n; such natural cases contain grained distributions (see below). (A general study of the dependence on D of the complexity of testing equality to D was undertaken in [17].)

Our Reduction at a Glance. We decouple the reduction asserted in Theorem 1.1 into two steps. In the first step, we assume that the distribution D has a probability function q that ranges over multiples of $1/m$, for some parameter $m \in \mathbb{N}$; that is, $m \cdot q(i)$ is a non-negative integer (for every i). We call such a distribution *m-grained*, and reduce testing equality to any fixed m-grained distribution to testing equality to the uniform distribution over $[m]$. This reduction maps i uniformly at random to a set S_i of size $m \cdot q(i)$ such that the S_i's are disjoint. Clearly, this reduction maps the distribution q to the uniform distribution over m fixed elements, and it can be verified that this randomized mapping preserves distances between distributions.

Since every distribution over $[n]$ is $\epsilon/2$-close to a $O(n/\epsilon)$-grained distribution, it is stands to reason that the general case can be reduced to the grained case. This is indeed true, but the reduction is less obvious than the treatment of the grained case. Actually, we shall use a different "graining" procedure (than the one eluded to above), which yields a better result (i.e., a reduction to the case of $O(n)$-grained distributions rather than to the case of $O(n/\epsilon)$-grained distributions). Specifically, we present a reduction of ϵ-testing equality to any distribution $D \subseteq [n]$ to $\epsilon/3$-testing equality to some $6n$-grained distribution D', where D' depends only on D. This reduction is described next.

Letting $q : [n] \to [0,1]$ denote the probability function of D, the reduction maps $i \in [n]$ to itself with probability $\frac{\lfloor 6n \cdot q(i) \rfloor / 6n}{q(i)}$, and otherwise maps i to $n+1$. This description suffices when $q(i) \geq 1/2n$ for every $i \in [n]$, since in this case $\frac{\lfloor 6n \cdot q(i) \rfloor / 6n}{q(i)} \geq \frac{2}{3}$, and in order to guaranteed this condition (i.e., $q(i) \geq 1/2n$ for every $i \in [n]$) we use a preliminary reduction that maps $i \in [n]$ to itself with probability $1/2$ and maps it uniformity to $[n]$ otherwise. This preliminary reduction cuts the distance between distributions by a factor of two, and it can be shown that the main randomized mapping preserves distances between distributions up to a constant factor (of $2/3$).

History, Credits, and an Acknowledgement. The study of testing properties of distributions was initiated by Batu, Fortnow, Rubinfeld, Smith and White [2].[1] Testers of sample complexity $\text{poly}(1/\epsilon) \cdot \sqrt{n}$ for equality to U_n and for equality to an arbitrary distribution D over $[n]$ were presented by Goldreich and Ron [12] and Batu *et al.* [1], respectively, were the presentation in [12] is only implicit.[2] The tight lower and upper bound of $\Theta(\sqrt{n}/\epsilon^2)$ on the sample complexity of both problems were presented in [4,14,17] (see also [5,6]). For a general survey of the area, the interested reader is referred to Canonne [3].

As stated upfront, our reductions are inspired by Diakonikolas and Kane [5], who presented a unified approach for deriving optimal testers for various properties of distributions (and pairs of distributions) via reductions to testing the equality of two unknown distributions that have small $\mathcal{L}_2$-norm. We note that our reduction from testing equality to grained distributions to testing equality to the uniform distribution is implicit in [6].

Lastly, we wish to thank Ilias Diakonikolas for numerous email discussions, which were extremely helpful in many ways.

Organization. In Sect. 2 we recall the basic context and define the restricted notion of a reduction used in this work. The core of this work is presented in Sect. 3, where we prove Theorem 1.1. In Sect. 4 we briefly consider the problem of testing whether an unknown distribution is grained, leaving an open problem. The appendix addresses a side issue that arises in Sect. 4.

[1] As an anecdote, we mention that, in course of their research, Goldreich, Goldwasser, and Ron considered the feasibility of testing properties of distributions, but being in the mindset that focused on complexity that is polylogarithmic in the size of the object (see discussion in [9, Sec. 1.4]), they found no appealing example and did not report of these thoughts in their paper [11].

[2] Testing equality to U_n is implicit in a test of the distribution of the endpoint of a relatively short random walk on a bounded-degree graph.

2 Preliminaries

We consider *discrete* probability distributions. Such distribution have a finite *support*, which we assume to be a subset of $[n]$ for some $n \in \mathbb{N}$, where the support of a distribution is the set of elements assigned positive probability mass. We represent such distributions either by random variables, like X, that are assigned values in $[n]$ (indicated by writing $X \in [n]$), or by probability functions like $p : [n] \to [0, 1]$ that satisfy $\sum_{i\in[n]} p(i) = 1$. These two representation correspond via $p(i) = \mathbf{Pr}[X=i]$. At times, we also refer to distributions as such, and denote them by D. (Distributions over other finite sets can be treated analogously, but in such a case we may provide the tester with a description of the set; indeed, n serves as a concise description of $[n]$.)

Recall that the study of "distribution testing" refers to testing properties of distributions. That is, the object being testing is a distribution, and the property it is tested for is a property of distributions (equiv., a set of distributions). The tester itself is given samples from the distribution and is required to distinguish the case that the distribution has the property from the case that the distribution is far from having the property, where the distance between distributions is defined as the total variation distance between them (a.k.a the statistical difference). That is, X and Y are said to be ϵ-close if

$$\frac{1}{2} \cdot \sum_i \left|\mathbf{Pr}[X=i] - \mathbf{Pr}[Y=i]\right| \leq \epsilon, \qquad (1)$$

and otherwise they are deemed ϵ-far. With this definition in place, we are ready to recall the standard definition of testing distributions.

Definition 2.1 (testing properties of distributions): *Let* $\mathcal{D} = \{\mathcal{D}_n\}_{n\in\mathbb{N}}$ *be a property of distributions and* $s : \mathbb{N} \times (0, 1] \to \mathbb{N}$. *A* tester, *denoted* T, of sample complexity s for the property $\mathcal{D}$ *is a probabilistic machine that, on input parameters* n *and* ϵ, *and a sequence of* $s(n)$ *samples drawn from an unknown distribution* $X \in [n]$, *satisfies the following two conditions.*

1. The tester accepts distributions that belong to $\mathcal{D}$: *If* X *is in* $\mathcal{D}_n$, *then*

$$\mathbf{Pr}_{i_1,...,i_s\sim X}[T(n, \epsilon; i_1, ..., i_s)=1] \geq 2/3,$$

where $s = s(n, \epsilon)$ *and* $i_1, ..., i_s$ *are drawn independently from the distribution* X.

2. The tester rejects distributions that far from $\mathcal{D}$: *If* X *is* ϵ*-far from any distribution in* $\mathcal{D}_n$ (i.e., X is ϵ-far from $\mathcal{D}$), *then*

$$\mathbf{Pr}_{i_1,...,i_s\sim X}[T(n, \epsilon; i_1, ..., i_s)=0] \geq 2/3,$$

where $s = s(n, \epsilon)$ *and* $i_1, ..., i_s$ *are as in the previous item.*

Our focus is on "singleton" properties; that is, the property is $\{D_n\}_{n\in\mathbb{N}}$, where D_n is a fixed distribution over $[n]$. Note that n fully specifies the distribution

D_n, and we do not consider the complexity of obtaining an explicit description of D_n from n. For sake of simplicity, we will consider a generic n and omit it from the notation (i.e., use D rarher than D_n). Furthermore, we refer to ϵ-testers derived by setting the proximity parameter to ϵ. Nevertheless, all testers discussed here are actually uniform with respect to the proximity parameter ϵ (and also with respect to n, assuming that they already derived or obtained an explicit description of D_n).

Confining ourselves to problems of testing equality to distributions, we formally restate the notion of a reduction used in the introduction. In fact, we explicitly refer to the randomized mapping at the heart of the reduction, and also define a stronger (i.e., uniform over ϵ) notion of a reduction that captures the furthermore part of Theorem 1.1.

Definition 2.2 (reductions via filters): *We say that a randomized process F, called a* filter, reduces ϵ-testing equality to $D \subseteq [n]$ to ϵ'-testing equality to $D' \subseteq [n']$ *if the following two conditions hold:*

1. *The filter F maps the distribution D to the distribution D'; that is, $p'(i) = \sum_j p(j) \cdot \mathbf{Pr}[F(j)=i]$, where p and p' denote the probability functions of D and D', respectively.*
2. *The filter F maps any distribution that is ϵ-far from D to a distribution that is ϵ'-far from D'; that is, if q is ϵ-far from D, then $q'(i) \stackrel{\text{def}}{=} \sum_j q(j) \cdot \mathbf{Pr}[F(j)=i]$ is ϵ'-far from D'.*

We say that F reduces testing equality to $D \subseteq [n]$ to testing equality to $D' \subseteq [n']$ *if, for some constant c and every $\epsilon > 0$, it holds that F reduces ϵ-testing equality to D to ϵ/c-testing equality to D'.*

Recall that we say that F maps the distribution X to the distribution Y if Y and $F(X)$ are identically distributed (i.e., $Y \equiv F(X)$), where we view the distributions as random variables. We stress that if F is invoked t times on the same i, then the t outcomes are (identically and) independently distributed. Hence, a sequence of samples drawn independently from a distribution X is mapped to a sequence of samples drawn independently from the distribution $F(X)$.

Note (added in revision): As stated in the introduction, Definition 2.2 captures a natural but stringent notion of a reduction. First, note that this notion extends to reducing testing any set of distributions $\mathcal{D}$ to testing the set $\mathcal{D}'$ (by requiring that F maps any distribution in $\mathcal{D}$ to some distribution in $\mathcal{D}'$ while mapping any distribution that is ϵ-far from $\mathcal{D}$ to a distribution that is ϵ'-far from $\mathcal{D}'$). However, more general definitions may allow the tester of $\mathcal{D}$ to use the sample provided to it in arbitrary ways and invoke the tester of $\mathcal{D}'$ on an arbitrary sample as long as it distinguishes distributions in $\mathcal{D}$ from distributions that are ϵ-far from $\mathcal{D}$. While such general definitions are analogous to Cook-reductions, Definition 2.2 seems analogous to a very restricted (i.e., "local") notion of a Karp-reduction.

3 The Reduction

Recall that testing equality to a fixed distribution D means testing the property $\{D\}$; that is, testing whether an unknown distribution equals the fixed distribution D. For any distribution D over $[n]$, we present a reduction of the task of ϵ-testing $\{D\}$ to the task of $\epsilon/3$-testing the uniform distribution over $[6n]$.

3.1 Overview

We decouple the reduction into two steps. In the first step, we assume that the distribution D has a probability function q that ranges over multiples of $1/m$, for some parameter $m \in \mathbb{N}$; that is, $m \cdot q(i)$ is a non-negative integer (for every i). We call such a distribution *m-grained*, and reduce testing equality to any fixed m-grained distribution to testing uniformity (over $[m]$). Next, in the second step, we reduce testing equality to any distribution over $[n]$ to testing equality to some $6n$-grained distribution.

Definition 3.1 (grained distributions): *A probability distribution over $[n]$ having a probability function $q : [n] \to [0,1]$ is* m-grained *if q ranges over multiples of $1/m$; that is, if for every $i \in [n]$ there exists a non-negative integer m_i such that $q(i) = m_i/m$.*

Note that the uniform distribution over $[n]$ is n-grained, and it is the only n-grained distribution having support $[n]$. Furthermore, if a distribution D results from applying some *deterministic process* to the uniform distribution over $[m]$, then D is m-grained. On the other hand, any m-grained distribution must have support size at most m.

3.2 Testing Equality to a Fixed Grained Distribution

Fixing any m-grained distribution (represented by a probability function) $q : [n] \to \{j/m : j \in \mathbb{N} \cup \{0\}\}$, we consider a randomized transformation (or "filter"), denoted F_q, that maps the support of q to $S = \{\langle i, j \rangle : i \in [n] \wedge j \in [m_i]\}$, where $m_i = m \cdot q(i)$. (We stress that, as with any randomized process, invoking the filter several times on the same input yields independently and identically distributed outcomes.) Specifically, for every i in the support of q, we map i uniformly to $S_i = \{\langle i, j \rangle : j \in [m_i]\}$; that is, $F_q(i)$ is uniformly distributed over S_i. If i is outside the support of q (i.e., $q(i) = 0$), then we map it to $\langle i, 0 \rangle$. Note that $|S| = \sum_{i \in [n]} m_i = \sum_{i \in [n]} m \cdot q(i) = m$. The key observations about this filter are:

1. *The filter F_q maps q to a uniform distribution:* If Y is distributed according to q, then $F_q(Y)$ is distributed uniformly over S; that is, for every $\langle i, j \rangle \in S$, it holds that

$$\mathbf{Pr}[F_q(Y) = \langle i, j \rangle] = \mathbf{Pr}[Y = i] \cdot \mathbf{Pr}[F_q(i) = \langle i, j \rangle]$$

$$= q(i) \cdot \frac{1}{m_i}$$
$$= \frac{m_i}{m} \cdot \frac{1}{m_i}$$

which equals $1/m = 1/|S|$.

2. *The filter preserves the variation distance between distributions*: The total variation distance between $F_q(X)$ and $F_q(X')$ equals the total variation distance between X and X'. This holds since, for $S' = S \cup \{\langle i,0\rangle : i \in [n]\}$, we have

$$\sum_{\langle i,j\rangle \in S'} \left|\mathbf{Pr}[F_q(X) = \langle i,j\rangle] - \mathbf{Pr}[F_q(X') = \langle i,j\rangle]\right|$$
$$= \sum_{\langle i,j\rangle \in S'} \left|\mathbf{Pr}[X = i] \cdot \mathbf{Pr}[F_q(i) = \langle i,j\rangle] - \mathbf{Pr}[X' = i] \cdot \mathbf{Pr}[F_q(i) = \langle i,j\rangle]\right|$$
$$= \sum_{\langle i,j\rangle \in S'} \mathbf{Pr}[F_q(i) = \langle i,j\rangle] \cdot \left|\mathbf{Pr}[X = i] - \mathbf{Pr}[X' = i]\right|$$
$$= \sum_{i \in [n]} \left|\mathbf{Pr}[X = i] - \mathbf{Pr}[X' = i]\right|.$$

Indeed, this is a generic statement that applies to any filter that maps i to a random variable Z_i, which only depends on i, such that the supports of the different Z_i's are disjoint.

Observing that knowledge of q allows to implement F_q as well as to map S to $[m]$, yields the following reduction.

Algorithm 3.2 (reducing testing equality to m-grained distributions to testing uniformity over $[m]$): *Let D be an m-grained distribution with probability function $q : [n] \to \{j/m : j \in \mathbb{N} \cup \{0\}\}$. On input $(n, \epsilon; i_1, ..., i_s)$, where $i_1, ..., i_s \in [n]$ are samples drawn according to an unknown distribution $p : [n] \to [0,1]$, invoke an ϵ-tester for uniformity over $[m]$ by providing it with the input $(m, \epsilon; i'_1, ..., i'_s)$ such that for every $k \in [s]$ the sample i'_k is generated as follows:*

1. *Generate $\langle i_k, j_k\rangle \leftarrow F_q(i_k)$.*
   Recall that if $m_{i_k} \stackrel{\text{def}}{=} m \cdot q(i_k) > 0$, then j_k is selected uniformly in $[m_k]$, and otherwise $j_k \leftarrow 0$. We stress that if F_q is invoked t times on the same i, then the t outcomes are (identically and) *independently distributed. Hence, the s samples drawn independently from p are mapped to s samples drawn independently from p' such that $p'(\langle i,j\rangle) = p(i)/m_i$ if $j \in [m_i]$ and $p'(\langle i,0\rangle) = p(i)$ if $m_i = 0$.*
2. *If $j_k \in [m_{i_k}]$, then $\langle i_k, j_k\rangle \in S$ is mapped to its rank in S* (according to a fixed order of S), *where $S = \{\langle i,j\rangle : i \in [n] \wedge j \in [m_i]\}$, and otherwise $\langle i_k, j_k\rangle \notin S$ is mapped to $m+1$.*

(Alternatively, the reduction may just reject if any of the j_k's equals 0.)[3]

The forgoing description presumes that the tester for uniform distributions over $[m]$ also operates well when given arbitrary distributions (which may have a support that is not a subset of $[m]$).[4] However, any tester for uniformity can be easily extended to do so (see discussion in Sect. 3.4). In any case, we get

Proposition 3.3 (Algorithm 3.2 as a reduction): *The filter F_q used in Algorithm 3.2 reduces ϵ-testing equality to an m-grained distribution D* (over $[n]$) *to ϵ-testing equality to the uniform distribution over $[m]$, where the distributions tested in the latter case are over $[m+1]$. Furthermore, if the support of D equals* $[n]$ (i.e., $q(i) > 0$ for every $i \in [n]$), *which may happen only if $m \geq n$, then the reduction is to testing whether a distribution over $[m]$ is uniform on $[m]$.*

Using any of the known uniformity tests that have sample complexity $O(\sqrt{n}/\epsilon^2)$,[5] we obtain—

Corollary 3.4 (testing equality to m-grained distributions): *For any fixed m-grained distribution D, the property $\{D\}$ can be ϵ-tested in sample complexity $O(\sqrt{m}/\epsilon^2)$.*

The foregoing *tester for equality to grained distributions* seems to be of independent interest, which extends beyond its usage towards testing equality to arbitrary distributions.

3.3 From Arbitrary Distributions to Grained Ones

We now turn to the problem of testing equality to an arbitrary known distribution, represented by $q : [n] \to [0,1]$. The basic idea is to round all probabilities to multiples of γ/n, for an error parameter γ (which will be a small constant). Of course, this rounding should be performed so that the sum of probabilities equals 1. For example, we may use a randomized filter that, on input i, outputs i with probability $\frac{m_i \cdot \gamma/n}{q(i)}$, where $m_i = \lfloor q(i) \cdot n/\gamma \rfloor$, and outputs $n+1$ otherwise. Hence, if i is distributed according to p, then the output of this filter will be i with probability $\frac{\gamma m_i/n}{q(i)} \cdot p(i)$. This works well if $\gamma m_i/n \approx q(i)$, which is the case if $q(i) \gg \gamma/n$ (equiv., $m_i \gg 1$), but may run into trouble otherwise.

For starters, we note that if $q(i) = 0$, then $\frac{\gamma m_i/n}{q(i)}$ is undefined, and replacing it by either 0 or 1 will not do. More generally, suppose that $q(i) \in (0, \gamma/n)$ (e.g., $q(i) = 0.4\gamma/n$). In this case, setting $m_i = 0$ means that the filter is oblivious

[3] See further discussion in Sect. 3.4.

[4] This may happen if and only if the support of $q : [n] \to [0,1]$ is a strict subset of $[n]$ (equiv., if $m_i = 0$ for some $i \in [n]$). Specifically, for every $X \in [n]$, the support of $F_q(X)$ equals $S'' \stackrel{\text{def}}{=} S \cup \{\langle i, 0\rangle : i \in [n] \& q(i) = 0\} \subseteq S'$, whereas $|S''| = m + |\{i \in [n] : q(i)=0\}|$.

[5] Recall that the alternatives include the tests of [14] and [4] or the collision probability test (of [12]), per its improved analysis in [7,10].

of the probability assigned to this i, and does not distinguish distributions that agree on $\{i : q(i) \geq \gamma/n\}$ but greatly differ on $\{i : q(i) < \gamma/n\}$, which means that it does not distinguish the distribution associated with q from some distributions that are 0.1γ-far from it.[6] Hence, we modify the basic idea such to avoid this problem.

Specifically, we first use a filter that averages the input distribution p with the uniform distribution, and so guarantees that all elements occur with probability at least $1/2n$, while preserving distances between different input distributions (up to a factor of two). Only then, do we apply the foregoing proposed filter (which outputs i with probability $\frac{m_i \cdot \gamma/n}{q(i)}$, where $m_i = \lfloor q(i) \cdot n/\gamma \rfloor$, and outputs $n+1$ otherwise). Details follow.

1. We first use a filter F' that, on input $i \in [n]$, outputs i with probability $1/2$, and outputs the uniform distribution (on $[n]$) otherwise. Hence, if i is distributed according to the distribution p, then $F'(i)$ is distributed according to $p' = F'(p)$ such that

$$p'(i) = \frac{1}{2} \cdot p(i) + \frac{1}{2} \cdot \frac{1}{n}. \tag{2}$$

   (Indeed, we denote by $F'(p)$ the probability function of the distribution obtained by selecting i according to the probability function p and outputting $F'(i)$.)
   Let $q' = F'(q)$; that is, $q'(i) = 0.5 \cdot q(i) + (1/2n) \geq 1/2n$.
2. Next, we apply a filter $F''_{q'}$, which is related to the filter F_q used in Algorithm 3.2. Letting $m_i = \lfloor q'(i) \cdot n/\gamma \rfloor$, on input $i \in [n]$, the filter outputs i with probability $\frac{m_i \cdot \gamma/n}{q'(i)}$, and outputs $n+1$ otherwise (i.e., with probability $1 - \frac{m_i \gamma/n}{q'(i)}$).
   Note that $\frac{m_i \gamma/n}{q'(i)} \leq 1$, since $m_i \leq q'(i) \cdot n/\gamma$. On the other hand, recalling that $q'(i) \geq 1/2n$ and observing that $m_i \cdot \gamma/n > ((q'(i) \cdot n/\gamma) - 1) \cdot \gamma/n = q'(n) - (\gamma/n)$, it follows that $\frac{m_i \gamma/n}{q'(i)} > \frac{q'(i) - (\gamma/n)}{q'(i)} \geq 1 - 2\gamma$, since $q'(i) \geq 1/2n$.
   Now, if i is distributed according to the distribution p', then $F''_{q'}(i)$ is distributed according to $p'' : [n+1] \to [0,1]$ such that, for every $i \in [n]$, it holds that

$$p''(i) = p'(i) \cdot \frac{m_i \cdot \gamma/n}{q'(i)} \tag{3}$$

   and $p''(n+1) = 1 - \sum_{i \in [n]} p''(i)$.
   Let q'' denote the probability function related (by $F''_{q'}$) to q'. Then, for every $i \in [n]$, it holds that $q''(i) = q'(i) \cdot \frac{m_i \gamma/n}{q'(i)} = m_i \cdot \gamma/n \in \{j \cdot \gamma/n : j \in \mathbb{N} \cup \{0\}\}$

[6] Consider, for example, the case that $q(i) = 0.4\gamma/n$ if $i \in [0.5n]$ and $q(i) = (2 - 0.4\gamma)/n$ otherwise, and any distribution X such that $\mathbf{Pr}[X = i] < \gamma/n$ if $i \in [0.5n]$ and $\mathbf{Pr}[X = i] = q(i)$ otherwise. Then, each of these possible X's will be mapped by F to the same distribution, although such distributions may be 0.1γ-far from the distribution associated with q.

and $q''(n+1) = 1 - \sum_{i\in[n]} m_i \cdot \gamma/n < \gamma$, since $m \stackrel{\rm def}{=} \sum_{i\in[n]} m_i > \sum_{i\in[n]}((n/\gamma) \cdot q'(i) - 1) = (n/\gamma) - n$.
Note that if n/γ is an integer, then q'' is n/γ-grained. Furthermore, if $m = n/\gamma$, which happens if and only if $q'(i) = m_i \cdot \gamma/n$ for every $i \in [n]$ (i.e., q' is itself n/γ-grained), then q'' has support $[n]$, and otherwise it has support $[n+1]$.

Combining these two filters, we obtain the desired reduction.

Algorithm 3.5 (reducing testing equality to a general distribution to testing equality to a $O(n)$-grained distributions): *Let D be an arbitrary distribution with probability function $q : [n] \to [0,1]$, and T be an ϵ'-tester for m-grained distributions having sample complexity $s(m, \epsilon')$. On input $(n, \epsilon; i_1, ..., i_s)$, where $i_1, ..., i_s \in [n]$ are $s = s(O(n), \epsilon/3)$ samples drawn according to an unknown distribution p, the tester proceeds as follows:*

1. *It produces a s-long sequence $(i''_1, ..., i''_s)$ by applying $F''_{F'(q)} \circ F'$ to $(i_1, ..., i_s)$, where F' and $F''_{q'}$ are as in* Eqs. (2) and (3); *that is, for every $k \in [s]$, it produces $i'_k \leftarrow F'(i_k)$ and $i''_k \leftarrow F''_{F'(q)}(i'_k)$.*
(Recall that $F''_{q'}$ depends on a universal constant γ, which we shall set to $1/6$.)
2. *It invokes the $\epsilon/3$-tester T for equality to q'' while providing it with the sequence $(i''_1, ..., i''_s)$. Note that this is a sequence over $[n+1]$.*

Using the notations as in Eqs. (2) and (3), we first observe that the total variation distance between $p' = F'(p)$ and $q' = F'(q)$ is half the total variation distance between p and q (since $p'(i) = 0.5 \cdot p(i) + (1/2n)$ and ditto for q'). Next, we observe that the total variation distance between $p'' = F''_{q'}(p')$ and $q'' = F''_{q'}(q')$ is lower-bounded by a constant fraction of the total variation distance between p' and q'. To see this, let X and Y be distributed according to p' and q', respectively, and observe that

$$\begin{aligned}\sum_{i\in[n]} \left|\mathbf{Pr}[F''_{q'}(X) = i] - \mathbf{Pr}[F''_{q'}(Y) = i]\right| &= \sum_{i\in[n]} \left|p'(i) \cdot \frac{m_i\gamma/n}{q'(i)} - q'(i) \cdot \frac{m_i\gamma/n}{q'(i)}\right| \\ &= \sum_{i\in[n]} \frac{m_i\gamma/n}{q'(i)} \cdot |p'(i) - q'(i)| \\ &\geq \min_{i\in[n]} \left\{\frac{m_i\gamma/n}{q'(i)}\right\} \cdot \sum_{i\in[n]} |p'(i) - q'(i)|.\end{aligned}$$

As stated above, recalling that $q'(i) \geq 1/2n$ and $m_i = \lfloor (n/\gamma) \cdot q'(i) \rfloor > (n/\gamma) \cdot q'(i) - 1$, it follows that

$$\frac{m_i\gamma/n}{q'(i)} > \frac{((n/\gamma) \cdot q'(i) - 1) \cdot \gamma/n}{q'(i)} = 1 - \frac{\gamma/n}{q'(i)} \geq 1 - \frac{\gamma/n}{1/2n} = 1 - 2\gamma.$$

Hence, if p is ϵ-far from q, then p' is $\epsilon/2$-far from q', and p'' is $\epsilon/3$-far from q'', where we use $\gamma \leq 1/6$. On the other hand, if $p = q$, then $p'' = q''$. Noting that q'' is an n/γ-grained distribution, provided that n/γ is an integer (as is the case for $\gamma = 1/6$), we complete the analysis of the reduction. Hence,

Proposition 3.6 (Algorithm 3.5 as a reduction): *The filter $F''_{F'(q)} \circ F'$ used in Algorithm 3.5 reduces ϵ-testing equality to any fixed distribution D* (over $[n]$)*) to ϵ-testing equality to an $6n$-grained distribution over $[n']$, where $n' \in \{n, n+1\}$ depends on q.*[7] *Furthermore, the support of $F''_{F'(q)} \circ F'(q)$ equals $[n']$.*

Hence, *the sample complexity of ϵ-testing equality to arbitrary distributions over $[n]$ equals the sample complexity of $\epsilon/3$-testing equality to $6n$-grained distributions* (which is essentially a special case).

Digest. One difference between the filter underlying Algorithm 3.2 and the one underlying Algorithm 3.5 is that the former preserves the exact distance between distributions, whereas the later only preserves them up to a constant factor. The difference is reflected in the fact that the first filter maps the different i's to distributions of disjoint support, whereas the second filter (which is composed of the filters of Eqs. (2) and (3)) maps different i's to distributions of non-disjoint support. (Specifically, the filter of Eq. (2) maps every $i \in [n]$ to a distribution that assigns each $i' \in [n]$ probability at least $1/2n$, whereas the filter of Eq. (3) typically maps each $i \in [n]$ to a distribution with a support that contains the element $n+1$.)

3.4 From Arbitrary Distributions to the Uniform One

Combining the reductions stated in Propositions 3.3 and 3.6, we obtain a proof of Theorem 1.1.

Theorem 3.7 (Theorem 1.1, restated): *For every probability function $q : [n] \to [0, 1]$ the filter $F_{q''} \circ F''_{F'(q)} \circ F'$, where $q'' = F''_{F'(q)} \circ F'(q)$ is as in Algorithm 3.5 and $F_{q''}$ is as in Algorithm 3.2, reduces ϵ-testing equality to q to $\epsilon/3$-testing equality to the uniform distribution over $[6n]$.*

Proof: First, setting $\gamma = 1/6$ and using the filter $F''_{F'(q)} \circ F'$, we reduce the problem of ϵ-testing equality to q to the problem of $\epsilon/3$-testing equality to the $6n$-grained distribution q'', while noting that the distribution q'' has support $[n']$, where $n' \in \{n, n+1\}$ (depending on q). Note that the latter assertion relies on the furthermore part of Proposition 3.6. Next, using the furthermore part of Proposition 3.3, we note that $F_{q''}$ reduces $\epsilon/3$-testing equality to q'' to $\epsilon/3$-testing equality to the uniform distribution over $[6n]$. ■

Observe that the proof of Theorem 3.7 avoids the problem discussed right after the presentation of Algorithm 3.2, which refers to the fact that testing equality to an m-grained distribution over $[n]$ is reduced to testing whether distributions over $[m']$ are uniform over $[m]$, where in some cases $m' \in [m, m+n]$ rather than $m' = m$. These bad cases arise when the support of the m-grained distribution is a strict subset of $[n]$ (see Footnote 4), and it was avoided above

[7] Typically, $n' = n + 1$. Recall that $n' = n$ if and only if D itself is $6n$-grained, in which case the reduction is not needed anyhow.

because we applied the filter of Algorithm 3.2 to distributions $q'' : [n'] \to [0,1]$ that have support $[n']$. Nevertheless, it is nice to have a reduction from the general case of "testing uniformity" to the special case, where the general case refers to testing whether distributions over $[n]$ are uniform over $[m]$, for any n and m, and the special case mandates that $m = n$. Such a reduction is provided next.

Theorem 3.8 (testing uniform distributions, a reduction between two versions): *There exists a simple filter that maps U_m to U_{2m}, while mapping any distribution X that is ϵ-far from U_m to a distribution over $[2m]$ that is $\epsilon/2$-far from U_{2m}. We stress that X is not necessarily distributed over $[m]$ and remind the reader that U_n denotes the uniform distribution over $[n]$.*

Thus, for every n and m, this filter reduces ϵ-testing whether distributions over $[n]$ are uniform over $[m]$ to $\epsilon/2$-testing whether distributions over $[2m]$ are uniform over $[2m]$.

Proof: The filter, denoted F, maps $i \in [m]$ uniformly at random to an element in $\{i, m+i\}$, while mapping any $i \notin [m]$ uniformly at random to an element in $[m]$. Observe that any distribution over $[n]$ is mapped to a distribution, and that $F(U_m) \equiv U_{2m}$. Note that F does not necessarily preserve distances between arbitrary distributions over $[n]$ (e.g., both the uniform distribution over $[2m]$ and the uniform distribution over $[m] \cup [2m+1, 3m]$ are mapped to the same distribution), but (as shown next) F preserves distances to the relevant uniform distributions up to a constant factor. Specifically, note that

$$\sum_{i\in[m+1,2m]} \left|\mathbf{Pr}[F(X)=i] - \mathbf{Pr}[U_{2m}=i]\right| = \sum_{i\in[m]} \left|\mathbf{Pr}[X=i]\cdot\frac{1}{2} - \frac{1}{2m}\right|$$

$$= \frac{1}{2}\cdot\sum_{i\in[m]} \left|\mathbf{Pr}[X=i] - \mathbf{Pr}[U_m=i]\right|$$

and

$$\sum_{i\in[m]} \left|\mathbf{Pr}[F(X)=i] - \mathbf{Pr}[U_{2m}=i]\right| \geq \mathbf{Pr}\left[F(X)\in[m]\right] - \mathbf{Pr}\left[U_{2m}\in[m]\right]$$

$$= \left(\mathbf{Pr}[X\in[m]]\cdot\frac{1}{2} + \mathbf{Pr}[X\notin[m]]\right) - \frac{1}{2}$$

$$= \frac{1}{2}\cdot\mathbf{Pr}[X\notin[m]]$$

$$= \frac{1}{2}\cdot\sum_{i\notin[m]} \left|\mathbf{Pr}[X=i] - \mathbf{Pr}[U_m=i]\right|.$$

Hence, the total variation distance between $F(X)$ and U_{2m} is at least half the total variation distance between X and U_m. ■

4 On Testing Whether a Distribution Is Grained

A natural question that arises from the interest in grained distributions refers to the complexity of testing whether an unknown distribution is grained. Specifically, given n and m (and a proximity parameter ϵ), *how many samples are required in order to determine whether an unknown distribution X over $[n]$ is m-grained or ϵ-far from any m-grained distribution?*

This question can be partially answered by invoking the results of Valiant and Valiant [16]. Specifically, for an upper bound we use their "learning up to relabelling" algorithm, which may be viewed as a learner of histograms (which is what it actually does). Recall that the histogram of the probability function p is defined as the multiset $\{p(i) : i \in [n]\}$ (equiv., as the set of pairs $\{(v, m) : m = |\{i \in [n] : p(i) = v\}| > 0\}$).

Theorem 4.1 (learning the histogram [16, Thm. 1]):[8] *There exists an $O(\epsilon^{-2} \cdot n/\log n)$ time algorithm that, on input n, ϵ and $O(\epsilon^{-2} \cdot n/\log n)$ samples drawn from an unknown distribution $p : [n] \to [0,1]$, outputs, with probability $1 - 1/\text{poly}(n)$, a histogram of a distribution that is ϵ-close to p.*

The implication of this result on testing any label-invariant property of distributions is immediate, where a property of distribution $\mathcal{D}$ is called label-invariant if for every distribution $p : [n] \to [0,1]$ in $\mathcal{D}$ and every permutation $\pi : [n] \to [n]$ it holds that $p \circ \pi$ is in $\mathcal{D}$. In our case, the tester consists of employing the algorithm of Theorem 4.1 with proximity parameter $\epsilon/2$ and accepting if and only if the output fits a histogram of a distribution that is $\epsilon/2$-close to being m-grained. The same holds with respect to estimating the distance from the set of m-grained distributions (which can be captured as a special case of label-invariant properties). Hence, we get

Corollary 4.2 (testing whether a distribution is grained): *For every $n, m \in \mathbb{N}$, the set of m-grained distributions over $[n]$ has a tester of sample complexity $O(\epsilon^{-2} \cdot n/\log n)$. Furthermore, the distance of an unknown distribution to the set of m-grained distributions over $[n]$ can be approximated up to an additive error of ϵ using the same number of samples.*

We comment that it seems that, using the techniques of [16], one can reduce the complexity to $O(\epsilon^{-2} \cdot n'/\log n')$, where $n' = \min(n, m)$. (For the case of testing, this is shown in the Appendix, using a reduction.) On the other hand, for $m \in [\Omega(n), O(n)]$, the above distance approximator is optimal, whereas it makes no sense to consider $m > n/\epsilon$ (since any distribution over $[n]$ is ϵ-close to being n/ϵ-grained). The negative result follows from the corresponding result of Valiant and Valiant [16].

---

[8] Valiant and Valiant [16] stated this result for the "relative earthmover distance" (REMD) and commented that the total variation distance up to relabelling is upper-bounded by REMD. This claim appears as a special case of [18, Fact 1] (using $\tau = 0$), and a detailed proof appears in [13].

Theorem 4.3 (optimality of Theorem 4.1 [16, Thm. 2]):[9] *For every sufficiently small $\epsilon > 0$, there exist two distributions $p_1, p_2 : [n] \to [0,1]$ that are indistinguishable by any label-invariant algorithm that takes $O(\epsilon^{-1}n/\log n)$ samples although p_1 is ϵ-close to the uniform distribution over $[n]$ and p_2 is ϵ-close to the uniform distribution over some set of $n/2$ elements.*

Let us spell out that, in the current context, an algorithm A is called label-invariant if for every permutation $\pi : [n] \to [n]$ and every sample $i_1, ..., i_s$, it holds that $A(n, \epsilon; i_1, ..., i_s) \equiv A(n, \epsilon; \pi(i_1), ..., \pi(i_s))$. Indeed, when estimating the distance to a label-invariant property, we may assume (w.l.o.g.) that the algorithm is label-invariant. Combining Theorem 4.3 with the latter fact, we get—

Corollary 4.4 (optimality of Corollary 4.2): *For any $m \in [\Omega(n), O(n)]$, estimating the distance to the set of m-grained distributions over $[n]$ up to a sufficiently small additive constant requires $\Omega(n/\log n)$ samples.*
Similarly, tolerant testing (cf. [15]) in the sense of *distinguishing distributions that are ϵ_1-close to being m-grained from distributions that are ϵ_2-far from being m-grained requires $\Omega(n/\log n)$ samples*, for any constant $\epsilon_2 \in (0, 1/(2 \cdot \lfloor 2m/n \rfloor))$ and $\epsilon_1 \in (0, \epsilon_2)$.

Proof: The case of $m = n/2$ follows by invoking Theorem 4.3, while observing that p_2 is ϵ-close to being m-grained, whereas p_1 is ϵ-close to a distribution (i.e, U_{2m}) that is $(0.499 - \epsilon)$-far from being m-grained,[10] where p_1, p_2 and ϵ are as in Theorem 4.3. Hence, distinguishing the distributions p_2 and p_1 (in a label-invariant manner) is reducible to $(0.499 - 2\epsilon)$-testing the set of distributions that are ϵ-close to be m-grained, which implies that the latter task has sample complexity $\Omega(n/\log n)$.

The case of $m < n/2$ is reduced to the case of $m = n/2$, by resetting $n \leftarrow 2m$. This yields a lower bound of $\Omega(m/\log m)$. Using the hypothesis $m = \Omega(n)$, we derive the desired lower bound of $\Omega(n/\log n)$.

For $m > n/2$ (equiv., $n < 2m$), we show a reduction of the distinguishing task underlying Theorem 4.3 to the testing problem at hand. Specifically, let $t = \lceil 2m/n \rceil$ and $m' \stackrel{\text{def}}{=} \lfloor m/t \rfloor$, and note that $t \in [2, O(1)]$ and $m' \in [\Omega(n), n/2]$ (by the various hypothesis).[11] We assume for simplicity that $m' = m/t$ (equiv., t divides m).[12] Now, consider a randomized filter, denoted $F_{m,t} : [n] \to [n]$, that

[9] Like in Footnote 8, we note that Valiant and Valiant [16] stated this result for the "relative earthmover distance" (REMD) and commented that the total variation distance up to relabelling is upper-bounded by REMD. This claim appears as a special case of [18, Fact 1] (using $\tau = 0$), and a detailed proof appears in [13].

[10] The constant 0.499 stands for an arbitrary large constant that is smaller than 0.5. Recall that the definition of δ-far mandates that the relevant distance be *greater* than δ.

[11] Specifically, $t \geq 2$ since $2m > n$, whereas $t = O(1)$ and $m' = \Omega(n)$ since $m = O(n)$.

[12] Otherwise the following description reduces the problem of Theorem 4.3 to a testing problem regarding $(t \cdot \lfloor m/t \rfloor)$-grained distributions. In this case, we reduce the latter testing problem to one regarding m-grained distributions (e.g., by using a filter that maps each $i \in [n]$ to itself with probability $t \cdot \lfloor m/t \rfloor / m$ and maps it to n otherwise.

maps each $i \in [m']$ to $m'+i$ with probability $1/t$ and otherwise maps it to itself, but always maps $i \in [n] \setminus [m']$ to $i - m'$. Then:

- $F_{m,t}$ maps the uniform distribution over $[m']$ to an m-grained distribution, since $q_2'(j) \stackrel{\text{def}}{=} \mathbf{Pr}[F_{m,t}(U_{m'})=j]$ equals $\mathbf{Pr}[U_{m'}=j] \cdot \frac{t-1}{t} = \frac{1}{m'} \cdot \frac{t-1}{t} = \frac{t-1}{m}$ if $j \in [m']$ and equals $\mathbf{Pr}[U_{m'}=j - m'] \cdot \frac{1}{t} = \frac{1}{m}$ if $j \in [m'+1, 2m']$.
- $F_{m,t}$ maps the uniform distribution over $[2m']$ to a distribution that is $(0.999/2t)$-far from being m-grained, since $q_1'(j) \stackrel{\text{def}}{=} \mathbf{Pr}[F_{m,t}(U_{2m'}) = j]$ equals $\mathbf{Pr}[U_{2m'} = j + m'] + \mathbf{Pr}[U_{2m'} = j] \cdot \frac{t-1}{t} = \frac{t+t-1}{2m't} = \frac{2t-1}{2m}$ if $j \in [m']$ and equals $\mathbf{Pr}[U_{2m'} = j - m'] \cdot \frac{1}{t} = \frac{1}{2m}$ if $j \in [m'+1, 2m']$.

Applying the filter $F_{m,t}$ to the distributions p_1 and p_2 of Theorem 4.3 (while setting $n = 2m' = 2 \cdot m/t$), we obtain distributions p_2' and p_1' such that p_2' is ϵ-close q_2', which is m-grained, whereas p_1' is ϵ-close to q_1', which is $(0.999/2t)$-far from being m-grained, since filters can only decrease the distance between distributions. Hence, distinguishing the distributions p_2 and p_1 (over $[2m/t]$) is reducible to $((0.999/2t) - 2\epsilon)$-testing the set of distributions that are ϵ-close to being m-grained, which implies that the latter task has sample complexity $\Omega((2m/t)/\log(2m/t))$. (The claim follows by recalling that $1/t = \Omega(1)$, since $m = O(n)$.) ■

Open Problems. Note that Corollary 4.4 does not refer to testing, but rather to distance approximation, and there are natural cases in which the complexity of testing a property of distributions is significantly lower than the corresponding distance approximation task (cf. [12] versus [16]). Hence, we ask—

Open Problem 4.5 (the sample complexity of testing whether a distribution is grained): *For any m and n, what is the sample complexity of testing the property that consists of all m-grained distributions over $[n]$.*

This question can be generalized to properties that allow m to reside in some predetermined set M, where the most natural case is that M is an interval, say of the form $[m', 2m']$.

Open Problem 4.6 (Problem 4.5, generalized): *For any finite set $M \subset \mathbb{N}$ and $n \in \mathbb{N}$, what is the sample complexity of testing the property that consists of all distributions over $[n]$ that are each m-grained for some $m \in M$.*

Appendix: Reducing Testing m-Grained Distributions (over $[n]$) to the Case of $n = O(m)$

Recall that Corollary 4.2 asserts that *for every $n, m \in \mathbb{N}$, the set of m-grained distributions over $[n]$ has a tester of sample complexity $O(\epsilon^{-2} \cdot n/\log n)$.* As commented in the main text, we believe that using the techniques of [16] one can reduce the complexity to $O(\epsilon^{-2} \cdot n'/\log n')$, where $n' = \min(n, m)$. Here we show an alternative proof of this result. Specifically, we shall reduce ϵ-testing m-grained distributions over $[n]$ to $\Omega(\epsilon)$-testing m-grained distributions over $[O(m)]$, and apply Corollary 4.2.

The reduction will consist of using a deterministic filter $f : [n] \to [k]$, where $k = O(m)$, that will be selected uniformly at random among all such filters. We stress that this is fundamentally different from the randomized filters F used in the main text. Specifically, when applying F several times to the same input, we obtained outcomes that are independently and identically distributed, whereas when we apply a function f (which is selected at random) several times to the same input we obtain the same output.

Note that applying any function $f : [n] \to [k]$ to any m-grained distribution yields an m-grained distribution. Our main result is that, for any distribution X over $[n]$ that is ϵ-far from being m-grained, for almost all functions $f : [n] \to [O(m)]$, the distribution $f(X)$ is $\Omega(\epsilon)$-far from being m-grained.

Lemma A.1 (relative preservation of distance from m-grained distributions): *For all sufficiently small $c > 0$ and all sufficiently large n and m, the following holds. If a distribution X over $[n]$ is ϵ-far from being m-grained, then, with probability at least $1 - 36c$ over the choice of a function $f : [n] \to [m/c]$, the distribution $f(X)$ is $0.02 \cdot \epsilon$-far from being m-grained.*

Hence, we obtain a randomized reduction of the general problem of testing m-grained distributions (over $[n]$) to the special case of $n = O(m)$, where the reduction consists of selecting at random a function $f : [n] \to [m/c]$ and using it as a (deterministic) filter for reducing the general problem to its special case.

Proof: Let $k = m/c$ and let $p : [n] \to [0, 1]$ denote the probability function that describes X. Define $r : [n] \to [0, 1/m)$ such that $r(i) = p(i) - \lfloor m \cdot p(i) \rfloor / m$. Denoting by $\Delta_G(p)$ the statistical distance between p and the set of m-grained distributions (i.e., half the norm-1 distance), we have

$$2 \cdot \Delta_G(p) \geq \sum_{i \in [n]} \min(r(i), (1/m) - r(i)) \tag{4}$$

$$2 \cdot \Delta_G(p) \leq 2 \cdot \sum_{i \in [n]} \min(r(i), (1/m) - r(i)) \tag{5}$$

where Eq. (4) is due to the need to transform each $p(i)$ to a multiple of $1/m$ and Eq. (5) is justified by a two-step correction process in which we first round each $p(i)$ to the closest multiple of $1/m$, and then we correct the resulting function so that it sums up to 1 (while keeping its values as multiples of $1/m$).[13] Hence, using

[13] Specifically, let $q : [n] \to [0, 2)$ be the function resulting from the first step (i.e., $q(i) = \lfloor m \cdot p(i) \rfloor / m$ if $r(i) \leq 1/2m$ and $q(i) = \lceil m \cdot p(i) \rceil / m$ otherwise). Then, $\delta \stackrel{\text{def}}{=} \sum_{i \in [n]} |q(i) - p(i)| = \sum_{i \in [n]} \min(r(i), (1/m) - r(i))$ and $|1 - \sum_{i \in [n]} q(i)| \leq \delta$, since $\left|\sum_{i \in [n]} q(i) - \sum_{i \in [n]} p(i)\right| \leq \sum_{i \in [n]} |q(i) - p(i)|$.

Eq. (5). the lemma's hypothesis implies that $\sum_{i\in[n]} \min(r(i), (1/m) - r(i)) > \epsilon$. We shall prove the lemma by lower-bounding (w.h.p.) the corresponding sum that refers to the distribution $f(X)$, when f is selected at random. Specifically, letting $p'(j) = \sum_{i:f(i)=j} p(i)$, we shall lower-bound the probability that $\sum_{j\in[k]} \min(r'(j), (1/m) - r'(j)) = \Omega(\epsilon)$, where $r'(j) = p'(j) - \lfloor m \cdot p'(j) \rfloor / m$, and then apply Eq. (4).

Before doing so, we introduce a few additional notations. Firstly, we let $s(i) = \min(r(i), (1/m) - r(i))$, and let $\delta = \sum_{i\in[n]} s(i)$, which is greater than ϵ by the hypothesis. Next, we let $H = \{i \in [n] : p(i) \geq 1/3m\}$ denote the set of "heavy" elements in X. We observe that $|H| \leq 3m$ and that for every $i \in \overline{H} \stackrel{\text{def}}{=} [n] \setminus H$ it holds that $s(i) = r(i) = p(i)$, since $p(i) < 1/2m$ holds for every $i \in \overline{H}$. We consider two cases, according to whether or not the sum $\sum_{i\in\overline{H}} p(i)$ is smaller than $0.5 \cdot \delta$.

Claim A.1.1 (the first case): *Suppose that* $\sum_{i\in\overline{H}} p(i) < 0.5 \cdot \delta$. *Then, with probability at least* $1 - 16c$ *over the choice of* f, *it holds that* $f(X)$ *is* 0.05ϵ*-far from being* m*-grained.*

Proof: In this case $\sum_{i\in H} s(i) > 0.5 \cdot \delta$, and we shall focus on the contribution of $f(H)$ to the distance of $f(X)$ from being m-grained. We shall show that, for almost all functions f, much of this weight is mapped (by f) in a one-to-one manner, and that the elements in $\overline{H}$ do not change by much the weight mapped by f to $f(H)$. Specifically, we consider a uniformly selected function $f : [n] \to [k]$, and the following two good events defined on this probability space.

1. The first (good) event is that the function f maps at least 0.2δ of the $s(i)$-mass of the i's in H to distinct images. Intuitively, this is very likely given that the total $s(i)$-mass of i's in H is greater than 0.5δ and that $|H| \ll k$. Formally, denoting by H_f the (random variable that represents the) set of $i \in H$ that satisfy $f(i) \notin f(H \setminus \{i\})$ (i.e., for every $i \in H_f$ it holds that $f^{-1}(f(i)) \cap H = \{i\}$), we claim that $\mathbf{Pr}_f\left[\sum_{i\in H_f} s(i) > 0.2\delta\right] > 1 - c$.

   To see this, we first note that, for every $i \in H$, conditioned on the values assigned to $H \setminus \{i\}$, the probability that $f(i) \notin f(H \setminus \{i\})$ is at least $\frac{k-(|H|-1)}{k} > 1 - |H|/k \geq 0.9$, where the inequality is due to $|H| \leq 3m = 3c \cdot k \leq 0.1 \cdot k$. Hence, each $i \in H$ contributes $s(i) \leq 1/2m$ to the sum (of $s(i)$'s with $i \in H_f$) with probability at least 0.9, also when conditioned on all other values assigned by f. It follows that $\mathbf{Pr}_f\left[\sum_{i\in H_f} s(i) > 0.2\delta\right] > 1-c$, where the (typical) case of $\delta = \omega(1/m)$ is straightforward.[14]

2. The second (good) event is that the function f does not map much $p(i)$-mass of i's in $\overline{H}$ to the images occupied by H. Again, this is very likely given that $|H| \ll k$. Specifically, observe that $\mathbb{E}_f\left[\sum_{i\in\overline{H}:f(i)\in f(H)} p(i)\right] \leq \frac{|H|}{k}\cdot\sum_{i\in\overline{H}} p(i) < 3c\cdot\delta/2$, since $p(i) = s(i)$ for every $i \in \overline{H}$ (and $|H| \leq 3m$ and $k = m/c$). Letting $S_f = \sum_{i\in\overline{H}:f(i)\in f(H)} p(i)$, we get $\mathbf{Pr}_f[S_f < 0.1\delta] > 1 - \frac{3c\delta/2}{0.1\delta} = 1-15c$.

Assuming that the two good events occur (which happens with probability at least $1-16c$), it follows that at least 0.2δ of the $s(\cdot)$-mass of H is mapped by f to distinct images and at most 0.1δ of the mass of $\overline{H}$ is mapped to these images. Hence, $f(X)$ corresponds to a probability function p' such that $r'(i) \stackrel{\text{def}}{=} p'(i) - \lfloor m\cdot p'(i)\rfloor/m$ satisfies

$$\sum_{i\in H_f} \min(r'(i), (1/m) - r'(i)) \geq \sum_{i\in H_f} s(i) - \sum_{i\in\overline{H}:f(i)\in f(H)} p(i)$$
$$> 0.2\delta - 0.1\delta,$$

where $H_f = \{i \in H : f^{-1}(f(i)) \cap H = \{i\}\}$ (as above). Hence, recalling that $\delta > \epsilon$ and using Eq. (4), with probability at least $1-16c$ over the choice of f, it holds that $f(X)$ is 0.05ϵ-far from being m-grained. □

Claim A.1.2 (the second case): *Suppose that* $\delta' \stackrel{\text{def}}{=} \sum_{i\in\overline{H}} p(i) \geq 0.5\cdot\delta$. *Then, with probability at least* $1-36c$ *over the choice of* f, *it holds that* $f(X)$ *is* 0.02ϵ*-far from being* m*-grained.*

Proof: In this case $\sum_{i\in\overline{H}} s(i) > 0.5\cdot\delta$, and we shall focus on the contribution of $f(\overline{H})$ to the distance of $f(X)$ from being m-grained. We shall show that, for

[14] Specifically, letting $\zeta_i = \zeta_i(f)$ denote the contribution of $i \in H$ to $\sum_{i\in H_f} s(i)$, we have $\mathbb{E}[\zeta_i] \geq 0.9\cdot s(i)$ and $\mathbb{V}[\zeta_i] \leq \mathbb{E}[\zeta_i^2] \leq s(i)^2 \leq s(i)/2m$. Hence, by Chebyshev's Inequality, $\mathbf{Pr}\left[\sum_{i\in H}\zeta_i \leq 0.2\delta\right] < \frac{\delta/2m}{(0.45\delta-0.2\delta)^2}$, since $\mathbb{V}\left[\sum_{i\in H}\zeta_i\right] \leq \delta/2m$ and $\mathbb{E}\left[\sum_{i\in H}\zeta_i\right] \geq 0.9\cdot 0.5\delta$. This suffices for $\delta = \omega(1/m)$. Actually, the same argument holds if $\sum_{i\in H} s(i)^2 = o(\delta^2)$; the argument for the general case follows.
In general (esp., if $\sum_{i\in H} s(i)^2 = \Omega(\delta^2)$), for a sufficiently small $c' > 0$, we define $H' \stackrel{\text{def}}{=} \{i \in H : s(i) \geq c'\cdot\delta\}$, and consider two cases.
(a) If $\sum_{i\in H\setminus H'} s(i) > 0.3\cdot\delta$, then we use $H\setminus H'$ instead of H, while noting that $\mathbf{Pr}\left[\sum_{i\in H\setminus H'}\zeta_i \leq 0.2\delta\right] < \frac{c'\cdot\delta^2}{(0.07\delta)^2} < c$, since $\mathbb{E}\left[\sum_{i\in H\setminus H'}\zeta_i\right] > 0.9\cdot 0.3\delta$ and $\mathbb{V}[\sum_{i\in H\setminus H'}\zeta_i] \leq \sum_{i\in H\setminus H'} s(i)^2 \leq c'\delta\cdot\delta$.
(b) If $\sum_{i\in H'} s(i) > 0.2\cdot\delta$, then we use H' instead of H, while noting that the probability that $|f(H')| < |H'|$ is at most $\binom{|H'|}{2}\cdot(1/k) \leq \binom{1/c'}{2}\cdot(1/k) < c$, where the last inequality holds for sufficiently large k (i.e., $m = c\cdot k > (1/c')^2$ suffices).
(We proceed with H replaced by either H' or $H\setminus H'$.)

almost all functions f, much of this weight is mapped (by f) to $[k]\setminus H$ and that the mass of the elements of $\overline{H}$ is distributed almost uniformly. Specifically, we first show that more than half of the probability mass of $\overline{H}$ is mapped disjointly of H. That is,

$$\mathbf{Pr}_{f:[n]\to[k]}\left[\sum_{i\in\overline{H}:f(i)\notin f(H)} p(i) > 0.5\cdot\delta'\right] \geq 1-6c \tag{6}$$

where the probability is taken uniformly over all possible choices of f. The proof is similar to the analysis of the second event in the proof of Claim A.1.2. Specifically, we consider random variables ζ_i's such that $\zeta_i = p(i)$ if $f(i)\notin f(H)$ and $\zeta_i = 0$ otherwise, and observe that $\mathbb{E}[\zeta_i] \geq \frac{k-|H|}{k}\cdot p(i) \geq (1-3c)\cdot p(i)$ (since $|H|\leq 3m$ and $m = ck$). Thus, $\mathbb{E}\left[\sum_{i\in\overline{H}}\zeta_i\right] \geq (1-3c)\cdot\delta'$ and Eq. (6) follows by Markov Inequality while using $\sum_{i\in\overline{H}}\zeta_i \leq \sum_{i\in\overline{H}} p(i) = \delta'$. This holds also if we fix the values of f on H and condition on it, which is what we do from this point on. Hence, we fix an arbitrary sequence of value for $f(H)$, and consider the uniform distribution of f conditioned on this fixing as well as on the event in Eq. (6).

Actually, we decompose $f:[n]\to[k]$ into three parts, denoted f', f'' and f''', that represents its restriction to the three-way partition of $[n]$ into (H,B,G) such that $B = \{i\in\overline{H} : f(i)\in f(H)\}$ (and $G = \{i\in\overline{H} : f(i)\notin f(H)\}$); indeed, $f':H\to[k]$ is the restriction of f to H, whereas $f'':B\to f(H)$ and $f''':G\to[k]\setminus f(H)$ are its restrictions to the two parts of $\overline{H}$. We fix arbitrary $f':H\to[k]$ and $f'':B\to f'(H)$, where $B = \{i\in\overline{H} : f''(i)\in f'(H)\}$, such that $\sum_{i\in B} p(i) < 0.5\delta'$, while bearing in mind that such fixing (of f' and f'') arise from the choice of a random f with probability at least $1-6c$. Our aim will be to show that, with high probability over the choice of $f''':G\to[k]\setminus f(H)$, it holds that

$$\sum_{i\in G:f'''(i)\in J(f''')} p(i) > 0.4\delta'. \tag{7}$$

where $J(f''') \stackrel{\text{def}}{=} \{j\in[k] : \sum_{i\in G:f'''(i)=j} p(i) \leq 0.8/m\}$. (Recall that for any $i\in G\subseteq\overline{H}$ it holds that $p(i) = r(i) = s(i) < 1/3m$.) This would imply that, with high probability, the distance of $f(X)$ from being m-grained is at least

$$\begin{aligned}\sum_{j\in J(f''')}\min\left(\sum_{i:f'''(i)=j} p(i)\,,\,\frac{1}{m}-\frac{0.8}{m}\right) &\geq \sum_{j\in J(f''')} 0.25\cdot\sum_{i:f'''(i)=j} p(i)\\ &> 0.25\cdot 0.4\delta'\\ &\geq 0.05\delta,\end{aligned}$$

where the first inequality is due to the fact that $p'(j) \stackrel{\text{def}}{=} \mathbf{Pr}[f(X)=j] = \sum_{i\in G:f'''(i)=j} p(i) \leq 0.8/m$ for every $j\in J(f''')$ and so $\min(p'(j), 0.2/m) \geq p'(j)/4$. So all that remains is to show that Eq. (7) holds with high probability over the choice of f'''.

Letting $K' \stackrel{\text{def}}{=} [k] \setminus f(H)$, we start by observing that, for every $i \in G$, it holds that

$$\begin{aligned}
&\mathbf{Pr}_{f''':G \to K'}[f'''(i) \notin J(f''')] \\
&\leq \mathbf{Pr}_{f''':G \to K'}\left[\sum_{\ell \in G \setminus \{i\}: f'''(\ell) = f'''(i)} p(\ell) > \frac{0.8}{m} - \frac{1}{3m}\right] \\
&= \mathbf{Pr}_{f''':G \to K'}\left[f'''(i) \in \left\{j \in K' : \sum_{\ell \in G \setminus \{i\}: f'''(\ell) = j} p(\ell) > \frac{1.4}{3m}\right\}\right] \qquad (8)
\end{aligned}$$

where the equality can be seen by first fixing f'''-values for all elements in $G \setminus \{i\}$ and then selecting $f'''(i)$ uniformly in K'. Upper-bounding the size of the set in Eq. (8) by $(1.4/3m)^{-1}$, and using $m = ck$ and $|K'| \geq k - 3m$, we get

$$\begin{aligned}
\mathbf{Pr}_{f''':G \to K'}[f'''(i) \notin J(f''')] &\leq \frac{3m}{1.4} \cdot \frac{1}{|K'|} \\
&\leq \frac{3ck}{1.4} \cdot \frac{1}{k - 3ck} \\
&\leq 3c,
\end{aligned}$$

where the last inequality presupposes $1.4 \cdot (1 - 3c) \geq 1$ (equiv., $c \leq 2/21$). It follows that

$$\begin{aligned}
\mathbb{E}_{f''':G \to K'}\left[\sum_{i \in G: f'''(i) \notin J(f''')} p(i)\right] &= \sum_{i \in G} \mathbf{Pr}_{f''':G \to K'}[f'''(i) \notin J(f''')] \cdot p(i) \\
&\leq \sum_{i \in G} 3c \cdot p(i) \\
&\leq 3c \cdot \delta',
\end{aligned}$$

since $\sum_{i \in G} p(i) \leq \sum_{i \in \overline{H}} p(i) = \delta'$. Hence,

$$\mathbf{Pr}_{f''':G \to K'}\left[\sum_{i \in G: f'''(i) \notin J(f''')} p(i) \geq 0.1\delta'\right] \leq \frac{3c}{0.1} = 30c.$$

Recalling that $\sum_{i \in B} p(i) < 0.5\delta'$, which implies $\sum_{i \in G} p(i) > 0.5\delta'$, this implies that Eq. (7) holds with probability at least $1 - 30c$ (over the choice of f''').

Lastly, recall that $\sum_{i \in B} p(i) < 0.5\delta'$, where $B = \{i \in \overline{H} : f''(i) \in f'(H)\}$, holds with probability at least $1 - 6c$ (over the choice of f' and f''). The claim follows, since (as argued above) Eq. (7) implies that $\sum_{j \in J(f''')} \min(p'(j), (1/m) - p'(j)) > 0.1\delta'$ (whereas using $\delta' \geq \delta/2 \geq \epsilon/2$ and Eq. (4), it follows that $f(X)$ is 0.02ϵ-far from being m-grained). □

Combining Claims A.1.1 and A.1.2, the lemma follows. ■

References

1. Batu, T., Fischer, E., Fortnow, L., Kumar, R., Rubinfeld, R., White, P.: Testing random variables for independence and identity. In: 42nd FOCS, pp. 442–451 (2001)
2. Batu, T., Fortnow, L., Rubinfeld, R., Smith, W.D., White, P.: Testing that distributions are close. In: 41st FOCS, pp. 259–269 (2000)
3. Canonne, C.L.: A survey on distribution testing: your data is big. But is it blue? In: ECCC, TR015-063 (2015)
4. Chan, S., Diakonikolas, I., Valiant, P., Valiant, G.: Optimal algorithms for testing closeness of discrete distributions. In: 25th ACM-SIAM Symposium on Discrete Algorithms, pp. 1193–1203 (2014)
5. Diakonikolas, I., Kane, D.: A new approach for testing properties of discrete distributions. arXiv:1601.05557 [cs.DS] (2016)
6. Diakonikolas, I., Kane, D., Nikishkin, V.: Testing identity of structured distributions. In: 26th ACM-SIAM Symposium on Discrete Algorithms, pp. 1841–1854 (2015)
7. Diakonikolas, I., Gouleakis, T., Peebles, J., Price, E.: Collision-based testers are optimal for uniformity and closeness. In: ECCC, TR16-178 (2016)
8. Goldreich, O.: Introduction to property testing: lecture notes. Superseded by [9]. Drafts are available from the author's web-page
9. Goldreich, O.: In: Introduction to Property Testing. Cambridge University Press, Cambridge (2017)
10. Goldreich, O.: On the optimal analysis of the collision probability tester (an exposition). This volume
11. Goldreich, O., Goldwasser, S., Ron, D.: Property testing and its connection to learning and approximation. J. ACM **45**, 653–750 (1998). Extended abstract in 37th FOCS, 1996
12. Goldreich, O., Ron, D.: On testing expansion in bounded-degree graphs. In: ECCC, TR00-020, March 2000
13. Goldreich, O., Ron, D.: On the relation between the relative earth mover distance and the variation distance (an exposition). This volume
14. Paninski, L.: A coincidence-based test for uniformity given very sparsely-sampled discrete data. IEEE Trans. Inf. Theory **54**, 4750–4755 (2008)
15. Parnas, M., Ron, D., Rubinfeld, R.: Tolerant property testing and distance approximation. J. Comput. Syst. Sci. **72**(6), 1012–1042 (2006)
16. Valiant, G., Valiant, P.: Estimating the unseen: an n/log(n)-sample estimator for entropy and support size, shown optimal via new CLTs. In: 43rd ACM Symposium on the Theory of Computing, pp. 685–694 (2011)
17. Valiant, G., Valiant, P.: Instance-by-instance optimal identity testing. In: ECCC, TR13-111 (2013)
18. Valiant, G., Valiant, P.: Instance optimal learning. CoRR abs/1504.05321 (2015)

A Note on Tolerant Testing with One-Sided Error

Roei Tell

Abstract. A tolerant tester with *one-sided error* for a property is a tester that accepts every input that is close to the property, with probability 1, and rejects every input that is far from the property, with positive probability. In this note we show that such testers require a linear number of queries.

This work appeared as TR16-032 of *ECCC*. The current revision is quite minimal.

1 Introduction

Property testing studies sub-linear time algorithms that distinguish between objects that have some predetermined property and objects that are far from the property (see a recent textbook [1]). We will specifically be interested in *tolerant testers*, introduced by Parnas, Ron, and Rubinfeld [2]. These are testers that distinguish, with high probability, between inputs that are close to the property, and inputs that are far from the property.

We prove that it is impossible to test a property tolerantly with both a sub-linear number of queries and *one-sided error*. Specifically, for a property Π and two constants $\epsilon > 0$ and $\epsilon' < \epsilon$, an ϵ'-tolerant ϵ-tester with one-sided error for Π accepts inputs that are ϵ'-close to Π *with probability* 1, and rejects inputs that are ϵ-far from Π with positive probability. We show that for essentially any Π, and any pair of constants $\epsilon > 0$ and $\epsilon' < \epsilon$, any ϵ'-tolerant ϵ-tester with one-sided error for Π requires a linear number of queries.

The proof is based on a simple claim that is implicit in the proofs of two similar previous results (see [3, Thm 1.5 and Prop. 3.3]). In Sect. 4 we reproduce the proofs of these two results, as corollaries of the said claim.

2 Preliminaries

For sake of simplicity, we will focus on properties of n-bit long strings, and avoid asymptotic formalism (although we shall provide the tester with n as explicit input for sake of elegancy). The distance between two n-bit strings, denoted by δ, is the relative Hamming distance (i.e., $\delta(x, y) = \frac{|\{i \in [n]: x_i \neq y_i\}|}{n}$). The distance between a string $x \in \{0,1\}^n$ and a non-empty set $S \subseteq \{0,1\}^n$ is $\min_{s \in S} \delta(x, s)$. We say that a string is ϵ-close to a set (resp., ϵ-far from a set), if its distance from the set is at most ϵ (resp., at least ϵ). In the following definition we refer to algorithms that get oracle access to a string $x \in \{0,1\}^n$; by this we mean that for any $i \in [n]$, the algorithm can query for the value of the ith bit of x.

O. Goldreich (Ed.): Computational Complexity and Property Testing, LNCS 12050, pp. 173–177, 2020.
https://doi.org/10.1007/978-3-030-43662-9_11

Definition 1 (tolerant testers with one-sided error): *Let $\Pi_n \subseteq \{0,1\}^n$, and let $\epsilon > 0$ and $\epsilon' < \epsilon$. An ϵ'-tolerant ϵ-tester with one-sided error for Π_n is a probabilistic algorithm T that satisfies the following conditions for every $x \in \{0,1\}^n$:*

1. *If x is ϵ'-close to Π_n, then* $\Pr[T^x(n) = 1] = 1$.
2. *If x is ϵ-far from Π_n, then* $\Pr[T^x(n) = 0] > 0$.

Note that in Condition (2) of Definition 1 we only require rejection of "far" inputs with *positive* probability, rather than with high probability. Indeed, the lower bound holds even for this relaxed definition.

3 The New Result

The following claim is implicit in the proof of [3, Claim 3.3.1].

Claim 2 (technical claim): *For $n \in \mathbb{N}$, let $S \subseteq \{0,1\}^n$ and $\overline{S} = \{0,1\}^n \setminus S$. Assume that there exists a probabilistic algorithm A that queries any n-bit long input string in q locations, and satisfies the following conditions:*

- *For every $s \in S$ it holds that A accepts s with probability* 1.
- *There exists $w \in \overline{S}$ such that A rejects w with positive probability.*

Then, every $x \in \{0,1\}^n$ is (q/n)-close to $\overline{S}$.

Proof: Let r be random coins such that when A queries w with coins r it holds that A rejects w. Denote by $(i_1, ..., i_q)$ the corresponding q locations in w that A queries, given coins r. Now, let $x \in \{0,1\}^n$. Let x' be the string obtained by modifying the q locations $(i_1, ..., i_q)$ in x to the values $(w_{i_1}, ..., w_{i_q})$. Observe that when A queries x' with coins r, it queries locations $(i_1, ..., i_q)$, sees the values $(w_{i_1}, ..., w_{i_q})$, and thus rejects x'. Hence, it cannot be that $x' \in S$ (otherwise A would have to accept x' with probability 1). It follows that $\delta(x, \overline{S}) \leq \delta(x, x') = q/n$. □

Note that Claim 2 also holds if we switch the roles of "accept" and "reject" in it (i.e., if we assume that A rejects every $s \in S$ with probability 1, and accepts some $w \in S$ with positive probability); we will use this fact in Sect. 4. Here we use Claim 2 (as stated) to derive our lower bound on the query complexity of tolerant testers with one-sided error.

Theorem 3 (tolerant testing with one-sided error): *Let $\Pi_n \subseteq \{0,1\}^n$, and let $\epsilon > 0$. Suppose that there exists $p \in \Pi_n$ and $z \in \{0,1\}^n$ such that $\delta(z, \Pi_n) \geq \epsilon$. Then, for every $\epsilon' < \epsilon$, every ϵ'-tolerant ϵ-tester with one-sided error for Π uses more than $\epsilon' \cdot n$ queries.*

Proof: For $\epsilon' < \epsilon$, let T be an ϵ'-tolerant ϵ-tester with one-sided error for Π_n, and denote the query complexity of T by $q = q(n)$. Let S be the set of strings that are ϵ'-close to Π_n, and let z be such that $\delta(z, \Pi_n) \geq \epsilon$. Note that T accepts every $s \in S$ with probability 1, and rejects z with positive probability. Invoking Claim 2 with the tester T, with the set S and with $w = z$, we deduce that every $x \in \{0,1\}^n$ is (q/n)-close to $\overline{S}$. In particular, it follows that $\delta(p, \overline{S}) \leq q/n$. On the other hand, $\delta(p, \overline{S}) > \epsilon'$, since $\delta(y, \Pi_n) \leq \delta(y, p)$ for all y and so $\delta(y, p) \leq \epsilon'$ implies $y \in S$. Hence, $q/n > \epsilon'$. □

Note that the two requirements in Theorem 3 (regarding the existence of a "positive" instance p and a "far" instance z) only exclude "degenerate" cases: If either of the two requirements does not hold, then the testing problem is trivial to begin with.

4 Previous Results as Corollaries of Claim 2

As mentioned in Sect. 1, Claim 2 is implicit in the proofs of two similar results, which we now reproduce.

4.1 Testers for Dual Problems with One-Sided Error

Dual testing problems were introduced in [3], and involve the testing of properties of the form "all inputs that are far from a predetermined set" (e.g., testing the property of graphs that are far from being connected). Specifically, a tester for the dual problem of a set Π accepts, with high probability, every input that is far from Π, and rejects, with high probability, every input that is far from being so; that is, it rejects every input that is far from the set of inputs that are far from Π.

While dual testing problems turn out to be very interesting in general, solving dual problems with *one-sided error* (i.e., accepting inputs that are far from Π with probability 1) requires a linear number of queries. Let us formally state and prove this.

Definition 4 (testers with one-sided error for dual problems): *Let $\Pi_n \subseteq \{0,1\}^n$, and let $\gamma > 0$ and $\epsilon \leq \gamma$. An ϵ-tester with one-sided error for the γ-dual problem of Π_n is a probabilistic algorithm T that satisfies the following conditions for every $x \in \{0,1\}^n$:*

1. *If x is γ-far from Π_n, then* $\Pr[T^x(n) = 1] = 1$.
2. *If x is ϵ-far from the set of strings that are γ-far from Π_n, then* $\Pr[T^x(n) = 0] > 0$.

Theorem 5 (dual problems with one-sided error; see [3, Thm 1.5]): *Let $\Pi_n \subseteq \{0,1\}^n$, and let $\gamma > 0$. Assume that there exists $p \in \Pi_n$ and $z \in \{0,1\}^n$ such that $\delta(z, \Pi_n) \geq 2 \cdot \gamma$. Then, for every $\epsilon \leq \gamma$, every ϵ-tester with one-sided error for the γ-dual problem of Π uses more than $\gamma \cdot n$ queries.*

Proof: For $\epsilon \leq \gamma$, let T be an ϵ-tester with one-sided error for the γ-dual problem of Π, and denote the query complexity of T by $q = q(n)$. Let S be the set of strings that are γ-far from Π_n, and let w be the string $p \in \Pi_n$. Note that T accepts every $s \in S$ with probability 1. Also observe that p is at distance at least $\gamma \geq \epsilon$ from any $z' \in S$ (because $\delta(z', p) \geq \delta(z', \Pi_n) \geq \gamma$), and thus T rejects p with positive probability. We can thus invoke Claim 2 with the tester T, and with S and $w = p$ as above, and deduce that every $x \in \{0,1\}^n$ is (q/n)-close to being at distance less than γ from Π_n; that is, $\delta(x, \Pi_n) < (q/n) + \gamma$. However, by our hypothesis, there exists z such that $\delta(z, \Pi_n) \geq 2 \cdot \gamma$. It follows that $2 \cdot \gamma \leq \delta(z, \Pi_n) < (q/n) + \gamma$, which implies that $q(n) > \gamma \cdot n$. □

4.2 Testers with Perfect Soundness

The standard notion of property testing with one-sided error refers to testers that accepts every input that has the property with probability one, and reject every input that is far from the property with high probability; that is, such testers may err (with small probability) only on inputs that are far from the property. We now consider a dual notion of testers with *perfect soundness*, which are testers that may err (with small probability) only on inputs that have the property. Specifically, a *tester with perfect soundness for Π* is required to accept every input in Π with positive probability, and reject every input that is far from Π with probability 1. It turns out that this task also requires a linear number of queries. The proof of this result, which we now detail, is very similar to the proof of Theorem 5.

Definition 6 (testers with perfect soundness): *Let $\Pi_n \subseteq \{0,1\}^n$, and let $\epsilon > 0$. An ϵ-tester with perfect soundness for Π_n is a probabilistic algorithm T that satisfies the following conditions for every $x \in \{0,1\}^n$:*

1. If $x \in \Pi_n$, then $\Pr[T^x(n) = 1] > 0$.
2. If x is ϵ-far from Π_n, then $\Pr[T^x(n) = 0] = 1$.

Theorem 7 (testing with perfect soundness; see [3, Prop. 3.3]): *Let $\Pi_n \subseteq \{0,1\}^n$, and let $\epsilon > 0$. Assume that there exists $p \in \Pi_n$ and $z \in \{0,1\}^n$ such that $\delta(z, \Pi_n) \geq 2 \cdot \epsilon$. Then, every ϵ-tester with perfect soundness for Π uses more than $\epsilon \cdot n$ queries.*

Proof: Let T be an ϵ-tester with perfect soundness for Π, and denote its query complexity by $q = q(n)$. Let S be the set of strings that are ϵ-far from Π_n, and let w be the string $p \in \Pi_n$. Note that T rejects every $s \in S$ with probability 1, and accepts w with positive probability. Invoking Claim 2 with the tester T, and with these S and w, while switching the roles of accept and reject, we deduce that every $x \in \{0,1\}^n$ is (q/n)-close to being at distance less than ϵ from Π_n. However, by our hypothesis, there exists z such that $\delta(z, \Pi_n) \geq 2 \cdot \epsilon$, which implies that $q(n) > \epsilon \cdot n$, since $2 \cdot \epsilon \leq \delta(z, \Pi_n) < (q/n) + \epsilon$. □

Acknowledgements. The author thanks his advisor, Oded Goldreich, for suggesting the question that this note answers, and for encouraging the author to write this note. This research was partially supported by the Israel Science Foundation (grant No. 671/13).

References

1. Goldreich, O.: Introduction to Property Testing. Cambridge University Press, Cambridge (2017)
2. Parnas, M., Ron, D., Rubinfeld, R.: Tolerant property testing and distance approximation. J. Comput. Syst. Sci. **72**(6), 1012–1042 (2006)
3. Tell, R.: On being far from far and on dual problems in property testing. In: Electronic Colloquium on Computational Complexity (ECCC), vol. 22, p. 72 (Rev. 4) (2015)

On Emulating Interactive Proofs with Public Coins

Oded Goldreich and Maya Leshkowitz

Abstract. The Goldwasser-Sipser emulation of general interactive proof systems by *public-coins* interactive proof systems proceeds by selecting, at each round, a verifier-message such that each message is selected with probability that is at most polynomially larger than its probability in the original protocol. Specifically, the possible messages are essentially clustered according to the probability that they are selected in the original protocol, and the emulation selects a message at random among those that belong to the heaviest cluster.

We consider the natural alternative in which, at each round, if the parties play honestly, then each verifier-message is selected with probability that approximately equals the probability that it is selected in the original (private coins) protocol. This is done by selecting a cluster with probability that is proportional to its weight, and picking a message at random in this cluster. The crux of this paper is showing that, essentially, no matter how the prover behaves, it cannot increase the probability that a message is selected by more than a constant factor (as compared to the original protocol). We also show that such a constant loss is inevitable.

An early version of this work appeared as TR16-066 of *ECCC*. On top of correcting some typos, the current revision simplifies and clarifies the analysis of the emulation (provided in Sect. 3.2).

1 Introduction

The notion of interactive proof systems was introduced by Goldwasser, Micali, and Rackoff [7] in order to capture the most general way in which one party can efficiently verify claims made by another, more powerful party. Interactive proofs generalize and contain as a special case the traditional NP-proof systems. However, we gain a lot from this generalization: the $\mathcal{IP}$ *Characterization Theorem* of Lund, Fortnow, Karloff, Nisan and Shamir [9,10] states that every language in $\mathcal{PSPACE}$ has an interactive proof system.

An interactive proof system is a two-player protocol between a computationally bounded verifier, and a computationally unbounded prover whose goal is to convince the verifier of the validity of some claim. The verifier employs a probabilistic polynomial-time strategy and sends the prover messages, to which the prover responds in order to convince the verifier. It is required that if the claim is true then there exists a prover strategy that causes the verifier to accept with high probability, whereas if the claim is false then the verifier rejects with high

O. Goldreich (Ed.): Computational Complexity and Property Testing, LNCS 12050, pp. 178–198, 2020.
https://doi.org/10.1007/978-3-030-43662-9_12

probability (no matter what strategy the prover employs). A formal definition of an interactive proof system is provided in Sect. 2. The class of sets having an interactive proof system is denoted by $\mathcal{IP}$.

Public Coins Versus Private Coins. A crucial aspect of interactive proofs is the verifier's randomness. Whereas we can assume, without loss of generality, that the prover is deterministic, the verifier must be randomized in order to benefit from the power of interactive proofs. Specifically, without randomness on the verifier's side, interactive proof systems exist only for sets in $\mathcal{NP}$. The verifier's messages in a general interactive proof system are determined based on the input, the interaction preformed so far, and the its internal coin tosses (i.e., the verifier's coin tosses). In that case, we may assume, without loss of generality, that the verifier tosses all coins at the very beginning of the interaction, and it is crucial that (with the exception for the last message) the verifier's messages only reveal partial information about its coins (and keep the rest secret). In contrast, in *public-coin* proof systems, introduced by Babai [1] as *Arthur-Merlin games*, the message sent by the verifier in each round contains (or totally reveals) the outcome of all coin it has tossed at the current round. Thus, these messages reveal the randomness used toward generating them; that is, this randomness becomes public. The class of sets having an interactive *public-coin* proof system is denoted $\mathcal{AM}$.

The relative power of *public-coin* interactive proofs, as compared to general interactive proofs, was first studied by Goldwasser and Sipser [8] who showed that every interactive proof can be emulated using only public coins; hence, $\mathcal{IP} = \mathcal{AM}$. Intuitively, this means that, in order to "test" the prover, the verifier does not need to ask clever questions, which hide some secrets, but it rather suffices to ask random questions (which hide nothing). The fact that $\mathcal{IP} = \mathcal{AM}$ also follows from the *IP characterization theorem* of [9,10], since the proof of this theorem actually establishes $\mathcal{PSPACE} \subseteq \mathcal{AM}$, whereas $\mathcal{IP} \subseteq \mathcal{PSPACE}$.

A finer notion of interactive proofs refers to the number of prover–verifier communication rounds. For an integer function r, the complexity class $\mathcal{IP}(r)$ consists of sets having an interactive proof system in which, on common input x, at most $r(|x|)$ rounds of communication take place. The original proof of Goldwasser and Sipser that $\mathcal{IP} = \mathcal{AM}$ actually provides a *round efficient* emulation of $\mathcal{IP}$ by $\mathcal{AM}$. Specifically, they show that, for any polynomially bounded function $r : \mathbb{N} \to \mathbb{N}$, it holds that $\mathcal{IP}(r) \subseteq \mathcal{AM}(r+1)$.

In addition to being of intrinsic interest, the emulation of general interactive proofs by public-coin interactive coins is instrumental for several fundamental results regarding general interactive proof systems, which are established by reducing them to the analogous results regarding *public-coin* interactive proof systems. Examples include the round-reduction (a.k.a. speed-up) theorem of Babai and Moran asserting that $\mathcal{IP}(2r) \subseteq \mathcal{IP}(r)$, the zero-knowledge emulation asserting that $\mathcal{IP} = \mathcal{ZK}$ (provided that one-way functions exist), and the equivalence between one-sided and two-sided error versions of interactive proof systems. In all three cases, the result is easier to establish for *public-coin* interactive proof systems (see [2,3], and [4], respectively); actually, no "direct proof" that works with arbitrary interactive proof systems is known (and it is even hard to imagine one). We stress that the use of a round-efficient emulation (of general interactive proofs by public coin ones) means that taking this ("via AM") route incurs (almost) no cost in terms of the round complexity of the resulting proof systems.

1.1 The Goldwasser-Sipser Emulation of $\mathcal{IP}$ by $\mathcal{AM}$

The basic idea used in emulating a general interactive proof by a public-coin one is changing the assertion, from proving that *one* (random) interaction using a specific sequence of private coins leads the verifier to accept, to proving that *most* of the sequences of coin tosses lead the verifier to accept. Calling such coin sequences good, the claim that there are many good coin sequences for a potential r-round interaction reduces to showing that the product of the number of verifier-messages (for the first round) times the number of good coin sequences that are consistent with each of these messages (and some prover response to it) is large. Hence, lower-bounding the number of good sequences for the r-round interaction is reduced to lower-bounding the number of good sequences for the remaining $r-1$ rounds.

The foregoing description makes sense when the next verifier message is uniformly distributed in some set, denoted X. In this case, the claim that there are M good coin sequences for the r-round interaction reduces to asserting that there are $|X|$ verifier messages such that each of them yields a $(r-1)$-round interaction with $M/|X|$ good coin sequences. The problem is that the foregoing uniformity condition may not hold in general.

Goldwasser and Sipser [8], who suggested this emulation strategy, resolved the foregoing problem by picking a set of messages that have roughly the same number of good coin sequences. Specifically, they *clustered* the potential messages that the original verifier could have sent (in the next round) into *clusters* according to the (approximate) number of good coin sequences that support each message. A constant-round, public-coin sampling protocol is utilized in order to sample from the cluster of messages that have the largest number of good coin sequences. Hence, the chosen cluster is determined as the "heaviest" one. (We go over the original emulation in more detail in Sect. 2.3.) The emulation succeeds when assuming an initial *gap* between the number of good coin sequences for yes-instances and for no-instances.[1]

Theorem 1.1 (original emulation of $\mathcal{IP}$ by $\mathcal{AM}$, as in [8]): *Suppose that, for $r = r(|x|)$, the set S has an r-round interactive proof system that utilizes $n = n(|x|)$ random coins for an instance x, and a gap of $\Omega(n)^r$ between the number of accepting coins of yes-instances and no-instances. Then, the foregoing emulation* (where the chosen cluster is the heaviest one) *yields an* ($(r+1)$-round) *public-coin interactive system proof for S.*

[1] Such a gap can be created by (sufficiently many) parallel executions of the original interactive proof systems. Indeed, this increases the length of messages but not the number of rounds.

1.2 Our Contribution

We propose an alternative method for preforming a public-coin emulation of $\mathcal{IP}$. Our method is similar to the method of Goldwasser and Sipser [8], but differs in the way the chosen cluster of messages (from which the sampling is preformed) is determined. Whereas in the original emulation the *chosen cluster* is determined as the one with the largest number of coins, in our emulation the *chosen cluster* is selected probabilistically according to its weight (i.e., the number of good coins in the cluster). Intuitively, this method gets closer to sampling from the real distribution of prover-verifier transcripts (see farther discussion in Sect. 1.3). Furthermore, as explained in Sect. 2.3, while the original method looses a factor of $\Theta(n)$ (in the gap between accepting coins of yes- and no-instances) in each round, the new method only looses a constant factor (in each round). Consequently, this method requires a smaller initial gap between the number of accepting coins of yes-instances and no-instances (in order to emulate interactive proofs using public coins).

Theorem 1.2 (new emulation of $\mathcal{IP}$ by $\mathcal{AM}$): *Suppose that, for $r = r(|x|)$, the set S has an r-round interactive proof system for an instance x, and a gap of B^r, for some universal constant $B > 1$, between the number of accepting coins of yes-instances and no-instances. Then, the new emulation* (where the chosen cluster is selected at random according to its weight) *yields an* ($(r+1)$-round) *public-coin interactive proof system for S.*

We present the emulation and the proof of Theorem 1.2 in Sect. 3. We also show that, for the new emulation, the gap that we use is asymptotically tight. Namely, when the initial gap is $O(C^r)$ for some constant $C > 1$, we provide an interactive proof and a prover strategy that fails the new emulation.

Theorem 1.3 (tightness of Theorem 1.2): *For some universal constant $C > 1$, there exists an interactive proof system for a set S that proceeds in $r = r(|x|)$ rounds and has a gap of $\Omega(C^r)$ between the number of accepting coins of yes-instances and no-instances such that emulating this proof system* (as described above) *fails to yield an interactive proof system for S.*

We provide the proof of Theorem 1.3 in Sect. 3.3.

1.3 An Alternative Perspective

As stated in Sect. 1.2, the new emulation can be viewed as an attempt to tightly emulate the original prover-verifier interaction. When choosing a cluster according to its weight, and sampling a message uniformly from this cluster, we are actually selecting a verifier-message with distribution that is quite close to the original, where the deviation is due to approximation that underlies the definition of a cluster (i.e., each cluster contains messages that have approximately, but not necessarily exactly, the same number of coins supporting them). Furthermore, essentially, even malicious behavior of the prover can increase the probability that a specific message is chosen in a specific round by at most a constant factor (as compared to the original interaction).

In contrast, the previous emulation strategy (of Goldwasser and Sipser [8]) selects messages with a distribution that is very far from the original interaction, even in the case that both parties are honest. Recall that this emulation always selects messages from the heaviest cluster, and so it may increase the probability that a message is chosen in a certain round by a factor of $\Theta(n)$. This seems less natural than the emulation we use.

Hence, our contribution is in showing that the natural emulation that emulates the original interaction quite tightly works too, and in fact that it works better. In particular, while the analysis of Goldwasser and Sipser [8] shows that their emulation strategy loses a factor of $O(n)$ in each round, we show that the new emulation strategy loses a constant factor in each round (and that such a factor must be lost).

We comment that choosing clusters according to their weight was also employed by Goldreich, Vadhan, and Wigderson [6], but in their work several such clusters are selected at each round, which makes the analysis of the protocol easier. We cannot afford doing so, since the number of rounds is unbounded here.

2 Preliminaries

We start by recalling the definition of interactive proof systems, and then take a closer look at such systems. In particular, in Sect. 2.2, we define the notion of "accepting coins", which is pivotal to both (the original and our) emulations of general interactive proof systems by public-coin ones. The original emulation (of Goldwasser and Sipser [8]) is reviewed in Sect. 2.3.

2.1 Interactive Proof Systems

We use a general formulation of interactive proof systems in which the completeness and soundness bounds are parameters (rather than fixed constants such as 2/3 and 1/3).

Definition 2.1 (interactive proof systems): *Let $c, s : \mathbb{N} \to [0,1]$ such that $c(n) \geq s(n) + 1/\text{poly}(n)$. An interactive proof system for a set S is a two-party game between a verifier executing a probabilistic polynomial-time strategy, denoted V, and a prover executing a* (computationally unbounded) *strategy satisfying the following two conditions:*

- Completeness with bound c: *For every $x \in S$, with probability at least $c(|x|)$, the verifier V accepts after interacting with the prover P on common input x.*
- Soundness with bound s: *For every $x \notin S$ and every prover strategy P^*, with probability at most $s(|x|)$, the verifier V accepts after interacting with P^* on common input x.*

When c and s are not specified, we mean $c \equiv 2/3$ and $s \equiv 1/3$. We denote by $\mathcal{IP}$ the class of sets having interactive proof systems.

A finer definition of interactive proofs refers to the number of prover-verifier communication rounds (i.e., number of pairs of verifier-message followed by a prover-message). For an integer function r, the complexity class $\mathcal{IP}(r)$ consists of sets having an interactive proof system in which on common input x, at most $r(|x|)$ rounds of communication are executed between the parties.

An interactive proof system is said to be of the public-coin type if the verifier strategy V consists of sending, in each round, the outcome of all coin tosses it made in that round. That is, the verifier makes public all randomness used to generate its current message; it keeps no secrets. The corresponding complexity class, consisting of all sets that have an r-round public-coin interactive proof system, is denoted $\mathcal{AM}(r)$.

The topic of this work is round-efficient transformations of general interactive proof systems into public-coin ones. That is, given an arbitrary interactive proof system for S, the goal is to obtain a public-coin interactive proof system for S, while preserving the number of rounds (up to a constant factor). Furthermore, the public-coin system that we construct will emulate the original system in a round-by-round manner, performing a constant number of rounds per each original round. (The same holds also in the systems constructed by [8].)

2.2 The Notion of Accepting Coins

In order to provide a precise description of the original and new emulations, we formally define the set of *accepting coins* for input x and partial transcript γ. The following definition refers to any fixed pair of deterministic strategies, (P, V), where V is provided with an auxiliary (random) input ρ (which represents the outcomes of coin tosses). When using the following definition in the rest of this paper, we shall always fix V to be the verifier strategy given to us (where the verifier's internal coin tosses are viewed as input to V) and let P be a fixed optimal strategy that maximizes the acceptance probability of V.

Definition 2.2 (accepting coins): *Let (P, V) be a pair of strategies, x be a potential common input, and ρ be a string representing a possible outcome of a sequence of coin tosses.*

- *We denote by $\langle P, V(\rho)\rangle(x)$ the* full transcript *of the interaction of P and V on input x, when V uses coins ρ; that is,*

$$\langle P, V(\rho)\rangle(x) = (\alpha_1, \beta_1, \ldots, \alpha_r, \beta_r, (\sigma, \rho)) \tag{1}$$

where $\sigma = V(x, \rho, \beta_1, \ldots, \beta_r) \in \{0,1\}$ is V's final verdict, *and for every $i = 1, \ldots, r$ it holds that $\alpha_i = V(x, \rho, \beta_1, \ldots, \beta_{i-1})$ and $\beta_i = P(x, \alpha_1, \ldots, \alpha_i)$ are the messages exchanged in round i.*

- *For any partial transcript,* $\gamma = (\alpha_1, \beta_1, \ldots, \alpha_{i-1}, \beta_{i-1})$*, ending with a* P*-message, we denote by* $\mathtt{ACC}_x(\gamma)$ *the set of coin sequences that are consistent with the partial transcript* γ *and lead* V *to accept* x *when interacting with* P*. That is:*

$$\mathtt{ACC}_x(\gamma) = \left\{ \rho \in \{0,1\}^n : \begin{array}{c} \exists \gamma' \in \{0,1\}^{\mathrm{poly}(|x|)} \text{ such that} \\ \langle P, V(\rho)\rangle(x) = (\gamma, \gamma', (1, \rho)) \end{array} \right\} \tag{2}$$

When x *and* γ *are clear from the context we refer to* $\mathtt{ACC}_x(\gamma)$ *as the set of* accepting coins.

Note that we assume, without loss of generality, that the verifier reveals its private coins ρ on the last round, and that its last message also includes its output (or verdict) bit (denoted by σ in Eq. (1)). The set of accepting coins consists of sequences ρ that are consistent with the partial transcript γ (per Eq. (1)) and *may* lead the verifier to accept (per Eq. (2). Specifically, the string γ' represents a possible continuation of the transcript γ that is both consistent with ρ and leads to acceptance of x.

In the sequel, $n = |\rho|$ will serve as our main parameter (rather than $|x|$). Indeed $n \leq \mathrm{poly}(|x|)$, and we may also assume (w.l.o.g.) that $n \geq |x|$.

2.3 The Emulation of Goldwasser and Sipser [8]

The public-coin emulation in [8] is preformed by clustering the possible messages that the verifier may send (at each round) into n clusters according to the approximate number of accepting coins that they each have (i.e., the potential message α is clustered according to $|\mathtt{ACC}_x(\gamma\alpha)|$). Specifically, in [8], the ith cluster contained messages with approximately 2^i accepting coins, but (mainly for clarity) we prefer to use a generic (constant) basis $b > 1$ (while noting that a choice of $b = 2$ is quite good). Thus, we shall use $n' \stackrel{\text{def}}{=} n/\log_2 b = \Theta(n)$ clusters (rather than n clusters). Thus, for the emulation of round r' with partial transcript γ we denote these clusters by $C_0^\gamma, \ldots, C_{n'}^\gamma$, where

$$C_i = C_i^\gamma \stackrel{\text{def}}{=} \left\{ \alpha : b^i \leq |\mathtt{ACC}_x(\gamma\alpha)| < b^{i+1} \right\}. \tag{3}$$

Namely, C_i^γ is the set of messages α that the verifier can send (at round r') that have approximately b^i coins that are consistent with the partial transcript $\gamma\alpha$, and may lead the verifier to accept.

The emulation of [8] proceeds as follows. Denoting by c the completeness parameter of the original r-round interactive proof system, the prover's initial claim is that there are at least $c \cdot 2^n$ accepting coins for x (i.e., that $|\mathtt{ACC}_x(\lambda)| \geq c \cdot 2^n$). The parties then proceed in r iterations that correspond to the r rounds in the original system. Specifically, iteration $r' \in [r]$ refers to the partial transcript generated in the first $r' - 1$ iterations and to a claim about the number of corresponding accepting coins, where initially the partial transcript is empty and the claimed number is $c \cdot 2^n$.

Letting γ denote the partial transcript generated so far and M denote the claimed number of accepting coins, first the prover supplies the verifier with the sizes of the clusters $C_0^\gamma, \ldots, C_{n'}^\gamma$, and the verifier checks that the corresponding number of accepting coins approximately sums-up to the claim; that is, letting $N_0, \ldots, N_{n'}$ denote the claimed sizes, the verifier checks that $\sum_{i=0}^{n'} N_i \cdot b^{i+1} > M$. The verifier then chooses a cluster C_i^γ with the largest number of accepting coins; that is, i is chosen so as to maximize $b^i \cdot N_i$. In order to validate that $|C_i^\gamma| \geq N_i$, and to sample a message α from C_i^γ, the prover and the verifier run a (constant-round) sampling protocol, which utilizes only public coins. Next, the prover supplies its answer β to the sampled message α, and the parties proceed to the next iteration, where the prover claims that there are at least b^i accepting coins that are consistent with the partial transcript $\gamma\alpha\beta$ generated so far.

Hence, in each iteration, the prover clusters the messages according to their weight (i.e., it clusters α according to $|\mathtt{ACC}_x(\gamma\alpha)|$), provides the sizes of the clusters, and the parties sample a message from the heaviest cluster (after the verifier checks that the claimed sizes of the clusters match the claim regarding the number of accepting coins that fit the current partial transcript γ). After the last iteration (corresponding to the last round in the original interaction), the complete prover-verifier transcript is determined, where this transcript also contains the verifier's internal coins tosses. The verifier then checks that the entire transcript is consistent and accepting.

We note that throughout the emulation the verifier does not "challenge" the prover on the number of accepting coins in the clusters other than the selected cluster C_i, and the prover can use this fact in order to safely overstate the total number of accepting coins. For example, even if all of the accepting coins lie in cluster C_i, which has thus weight $|C_i| \cdot b^i$, the prover can claim that there are $|C_i| \cdot b^i - 1$ accepting coins in each other cluster, and get away with this cheating, which allows it to overstate the total number of accepting coins (almost) by a factor of n'. In this way (i.e., by such an unchecked overstating), the gap between the actual number of accepting coins that are consistent with the interaction and the prover's claim regarding this number can be cut by a factor of $\Theta(n)$ in each round. For this reason, the emulation requires an initial gap of $\Theta(n)^r$ between the number of accepting coins in yes-instances and no-instances, where r is the number of rounds of the original interactive proof.

3 The New Emulation

As mentioned in Sect. 2.3, an essential cause for the large initial gap required in the emulation of [8] is the deterministic way in which a cluster of messages is chosen by the verifier. In contrast, a natural and promising approach is to have the verifier choose a cluster with probability proportional to the number of accepting coins that the prover claims are in that cluster. This follows the intuition that we would like to challenge the prover by choosing "heavy" clusters, which contain many accepting coins, with higher probability than "lighter" clusters. The same intuition also underlies [8], but we apply it in a more smooth fashion.

We note that the prover still has a potential of fooling the verifier by sending a message that does not belong to C_i, but rather to some other cluster, when C_i is chosen. Nevertheless, we show that a cheating prover will not be able to gain too much by doing so. This will be prove in Sect. 3.2, but before getting there we spell out the new emulation.

3.1 The Actual Protocols

The original r-round interaction (P, V) is "emulated" in r iterations (each consisting of a constant number of message exchanges). The ith iteration starts with a partial prover-verifier transcript $\gamma_{i-1} = (\alpha_1\beta_1 \ldots \alpha_{i-1}\beta_{i-1})$ and a claimed bound M_{i-1} regarding the size of $\mathtt{ACC}_x(\gamma_{i-1})$. In the first iteration γ_0 is the empty sequence and $M_0 = c \cdot 2^n$, where $c > 0$ is the completeness parameter of the interactive proof system. The ith iteration proceeds as follows.

Construction 3.1 (the ith iteration): *On input γ_{i-1} and M_{i-1}.*

1. Providing the clusters' sizes: *The prover computes the number of messages in each cluster, and sends the corresponding sequence of sizes, $(N_0, \ldots, N_{n'})$, to the verifier, where N_j is the number of messages in cluster $C_j = C_j^{\gamma_{i-1}}$ as defined in* Eq. (3).
Recall that each message in cluster C_j has between b^j and b^{j+1} consistent and accepting coins.
2. Verifier's initial checks: *If $\sum_{j=0}^{n'} N_j \cdot b^{j+1} < M_{i-1}$, then the verifier aborts and rejects.*
3. Verifier's selection of a cluster: *The verifier samples a cluster j according to the probability distribution J that assigns $j \in [n]$ probability proportional to $b^j \cdot N_j$. That is,*
$$\mathbf{Pr}\left[J = j\right] = \frac{N_j \cdot b^j}{\sum_{\ell=0}^{n'} N_\ell \cdot b^\ell} \tag{4}$$
4. Sampling the selected cluster: *The verifier and the prover run a sampling protocol* (as defined below) *to obtain a message α_i that the prover claims to be in cluster C_j. The protocol is invokes with completeness parameter $\epsilon = \frac{1}{3r}$ and soundness parameter $\delta = b$. Specifically, the parties use Construction 3.2. If no output is provided by the sampling protocol, then the verifier rejects.*
5. Completing the current iteration: *Next, the prover determines a message β_i such that $\mathtt{ACC}_x(\gamma_{i-1}, \alpha_i, \beta_i) = \mathtt{ACC}_x(\gamma_{i-1}, \alpha_i)$; that is, the prover selects a message that maximized the number of accepting coins, and sends it to the verifier.*

Toward the next iteration, the parties set $M_i = 2^j$ and $\gamma_i = \gamma_{i-1}\alpha_i\beta_i$.

By our conventions, the last message the verifier sends (in a general interactive proof) contains the outcomes $\rho \in \{0,1\}^n$ of the n coins tossed by the verifier (throughout the entire execution). Thus, this sequence ρ can be easily extracted from the full transcript $\gamma_r = (\alpha_1, \beta_1, \ldots, \alpha_r, \beta_r, (1, \rho))$. Hence, after the last iteration, the new verifier performs final checks and accepts if all of them hold:

(i) *Checking that ρ is accepting for γ_r*; that is, $V(x, \rho, \beta_1, \ldots, \beta_r) = 1$, and for every $i = 1, \ldots, r$ it holds that $\alpha_i = V(x, \rho, \beta_1, \ldots, \beta_{i-1})$.
Note that the verifier needs the full transcipt γ_r (and specifically ρ) in order to verify these conditions, so these checks can only be performed after the last iteration. Also note that if these checks pass, then $|\mathtt{ACC}_x(\gamma_r)| = 1$.
(ii) *Checking that $M_r = 1$*; that is, checking that the prover's last claim was that there is a single sequence of coin tosses (rather than more than one) that is consistent with the full transcript γ_r, which includes ρ.

To complete the description of Construction 3.1, we specify the requirements from the sampling protocol that is used in Step (4), and provide an implementation of it.

The Sampling Protocol: Requirements, Construction, and Analysis. Our protocol utilizes a constant-round, public-coin sampling protocol for sampling in arbitrary sets. The verifier is assisted by a computationally unbounded prover that the verifier does not trust. The prover provides the verifier with an integer N, which is supposed to be a lower bound on the size of the set (in our case the set of messages) denoted $S \subseteq \{0,1\}^\ell$. (We assume for simplicity that the length of the verifier's messages is exactly $\ell = \mathrm{poly}(|x|)$ (which can be justified by padding the messages to be of size ℓ).)

The Requirements from the Sampling Protocol. The sampling protocol with parameters $\epsilon > 0$ and $\delta > 1$, satisfies the following two properties:

Completeness (w.r.t error probability ϵ): If the lower bound on $|S|$ is valid (i.e. $|S| \geq N$), and the prover is honest, then with probability $1 - \epsilon$, the verifier will output an element of S.
Soundness (w.r.t factor δ): For every T such that $|T| < N$, no matter how the prover plays, the probability that verifier will output an element of T is at most $\delta \cdot \frac{|T|}{N}$.

Note that the set T in the soundness condition stands for any set that the prover would like to hit; for example, T may be a small subset of S or S itself (when the lower bound claimed by the prover is wrong). Recall, that our application (i.e., Construction 3.1) uses $\epsilon = 1/3r$ and $\delta = b$.

The Actual Construction. Using any "effective" ensemble of hash functions $\{H_\ell^t\}_{\ell > t}$ such that H_ℓ^t is family of pairwise-independent[2] hash functions mapping $\{0,1\}^\ell$ to $\{0,1\}^t$, the sampling protocol proceeds as follows.

[2] A set H of functions from D to R is called pairwise-independent if for every $x \neq y$ in D and $u, v \in R$ it holds that $\mathbf{Pr}_{h \in H}[h(x) = u \,\&\, h(y) = v] = 1/|R|^2$. Note that such "effective" sets are known; for example, for $D = \{0,1\}^\ell$ and $R = \{0,1\}^t$, the set of all affine transformation will do (e.g., $h_{M,v}(x) = Mx + v$, where M is a t-by-ℓ Boolean matrix and v (resp., x) is a t-dimensional (resp., ℓ-dimensional) Boolean vector). By *effective* we mean that it is easy to select functions in the set and easy to evaluate them on a given input. For more details, see [5, Apdx. D.2].

Construction 3.2 (the sampling protocol): *Using parameters $\epsilon > 0$ and $\delta > 1$, on input ℓ and N, the parties proceed as follows.*

(i) For $t = \lfloor \log_2(\epsilon N) \rfloor - \lceil 2\log_2(\delta/(\delta-1)) \rceil$, the verifier uniformly selects and sends the prover a random hash function $h \in H_\ell^t$, and a random element from the image $y \in \{0,1\}^t$.
(Recall that $h : \{0,1\}^\ell \to \{0,1\}^t$ and note that $2^t < \epsilon \cdot N$.)

(ii) The prover is supposed to answer with $K \stackrel{\text{def}}{=} \lfloor 2^{-t}N/\delta \rfloor$ elements of S that are preimages of y under h; that is, with $x_1, ..., x_K \in S$ such that $h(x_i) = y$ for every $i \in [K]$.

(iii) The verifier checks that the K elements are indeed preimages of y under h. Next, the verifier selects i uniformly in $[K]$, and outputs x_i; that is, it outputs one of these K elements selected uniformly using public randomness.

If less than K elements are provided, or some of the elements are not preimages, then the verifier produces no output.

Intuitively, for any $T \subseteq \{0,1\}^\ell$, we expect $2^{-t} \cdot |T|$ elements of T to be mapped to y under h. Hence, the completeness condition follows by the pairwise independence of H_ℓ^t (since $T = S$ has size at least $N > 2^t/\epsilon$ and $K < 2^{-t} \cdot N$), whereas the soundness condition holds because each element of T is mapped to y with probability 2^{-t} and is chosen by the verifier (if included in the prover's list) with probability $1/K$ (where $2^{-t}/K \approx \delta/N$).

The Actual Analysis. Assuming that the ensemble of hashing function is efficiently sampleable and allows for efficient evaluation, the computational complexity of the protocol for the verifier is polynomial in ℓ/ϵ. This is because $K \le 2^{-t}N/\delta = O_\delta(1/\epsilon)$, and the verifier's actions can be implemented in $\text{poly}(\ell) \cdot K$-time. We now turn to showing that the sampling protocol satisfies the foregoing completeness and soundness conditions.

Lemma 3.3 (analysis of the sampling protocol): *For any constant $\delta > 1$ and all sufficiently small $\epsilon > 0$, the protocol of Construction 3.2 satisfies the foregoing completeness and soundness requirements.*

Proof: We start with the completeness condition. The family of pairwise independent hash functions satisfies an "almost uniform cover" condition (cf. [5, Lem. D.4] and recall that $\delta > 1$); that is, for every $S \subseteq \{0,1\}^\ell$ and every $y \in \{0,1\}^t$, for all but at most a $\frac{2^t}{(1-(1/\delta))^2 \cdot |S|}$ fraction of $h \in H_\ell^t$ it holds that

$$|\{x \in S : h(x) = y\}| > \frac{|S|}{\delta \cdot 2^t}$$

(since the expected size of the set $S \cap h^{-1}(y)$ is $|S|/2^t$). On the other hand, using $|S| \ge N$, we have $K = \lfloor 2^{-t}N/\delta \rfloor \le 2^{-t}|S|/\delta$. Hence, given that the verifier selects $h \in H_\ell^t$ uniformly at random, the prover will fail in sending K

elements of S that are preimages of y (equiv., $|S \cap h^{-1}(y)| < K$) with probability at most

$$\frac{2^t}{(1-(1/\delta))^2 \cdot |S|} \leq \frac{\delta^2 \cdot 2^t}{(\delta-1)^2 \cdot N} \leq \epsilon,$$

where the last inequality is due to $t \leq \log_2(\epsilon N) - 2\log_2(\delta/(\delta-1))$.

Turning to the soundness condition, we consider an arbitrary set $T \subseteq \{0,1\}^\ell$. In this case we fix an arbitrary $h \in H_\ell^t$, and consider the other random choices of the verifier (i.e., the choices of $y \in \{0,1\}^t$ and $i \in [K]$). Let $Y \in \{0,1\}^t$ be a random variable representing an image y chosen by the verifier in Step (i), and $\zeta \in [K]$ denote the index i chosen by the verifier in Step (iii) (indicating a string in the K-long list provided by the prover in Step (ii)). For every $y \in \{0,1\}^t$, denote by T_y the set of preimages of y under h that are in T; that is, $T_y \stackrel{\text{def}}{=} \{\alpha \in T : h(\alpha) = y\}$. Indeed, it holds that $\sum_{y \in \{0,1\}^t} |T_y| = |T|$.

Recall that in Step (ii), the prover provides K preimages (of y under h), denoted $x_1, ..., x_K$, where the x_i's are chosen based on h and y. The prover may try to include in this K-long list as many elements of T (actually of T_y) as possible. In Step (iii), after checking that the list contains only elements of $h^{-1}(y)$, the verifier selects one of the x_i's uniformly at random, as indicated by ζ. Hence, for y with $|T_y|$ preimages in T, the probability that the sampled element resides in T is at most $\frac{|T_y|}{K}$; it may be less if the prover does not provide all the elements in T_y (either because $|T_y| > K$, or because it just acts "foolishly"). Hence, the probability that the output x_ζ is in T equals

$$\begin{aligned} \mathbf{Pr}[x_\zeta \in T] &= \sum_{y \in \{0,1\}^t} \mathbf{Pr}[Y = y \wedge x_\zeta \in T_y] \\ &= \sum_{y \in \{0,1\}^t} \mathbf{Pr}[Y = y] \cdot \mathbf{Pr}[x_\zeta \in T_y] \\ &\leq \sum_{y \in \{0,1\}^t} \frac{1}{2^t} \cdot \frac{|T_y|}{K} \\ &= \frac{|T|}{K \cdot 2^t} \end{aligned}$$

where "$x_\zeta \in T_y$" corresponds to the event "the list provided by the prover in response to (h, y) contains an element of T_y in location ζ". Recalling that $K = \lfloor 2^{-t}N/\delta \rfloor \geq (2^{-t}N/\delta) - 1$, we get

$$\begin{aligned} \mathbf{Pr}[x_\zeta \in T] &\leq \frac{|T|}{((2^{-t} \cdot N/\delta) - 1) \cdot 2^t} \\ &= \delta \cdot \frac{|T|}{N - \delta \cdot 2^t} \\ &< \frac{\delta}{1 - \delta\epsilon} \cdot \frac{|T|}{N} \end{aligned}$$

where the last inequality is due to $N > 2^t/\epsilon$. This means that the claim holds for soundness parameter $\frac{\delta}{1-\delta\epsilon}$, and the original claim follows by replacing δ with $\delta/(1-\delta\epsilon)$ (which is approximately δ in the typical cases where $\epsilon \ll 1/\delta$). ■

The Round Complexity of the Emulation. Combining Constructions 3.1 and 3.2, we obtain a (public-coin) protocol that uses $O(r)$ rounds. Specifically, in Construction 3.1, the prover sends messages in Steps (1), (4) and (5), while the verifier sends messages in Steps (3) and (4), where Step (4) invokes the three-message protocol of Construction 3.2, in which the verifier sends messages in Steps (i) and (iii), and the prover sends a message in Step (ii). Denoting these messages by the sender's initial and the step number, we get the sequence `P1`, `V3`, `V4i`, `P4ii`, `V4iii`, `P5`, which means that two and a half rounds of the public-coin protocol are used in order to emulate a single round of the original proof system.

It is possible to avoid this blowup in the number of rounds by combining the message sent by the prover in Step (ii) of the sampling protocol with its Step (5) message and the Step (1) message of the next iteration in one message. This is possible since the prover can provide the messages that it would have sent for each of the K possible messages of the verifier in Step (iii) of the sampling protocol. Details follow.

Recall that in Step (ii) of the sampling protocol the prover sends K messages allegedly belonging to C_j, and the verifier selects and sends one of these messages, denoted α_i, in Step (iii). The idea is to have the prover provide its response (i.e., β_i) to each of these possible α_i as well as the sizes of the clusters for the next round. All these messages are sent in one new message that the prover sends in a Step (ii) of the modified protocol. So the sequence of messages has the form `V3+V4i, P4ii, V4iii`, where the possible `P5`-messages of the current iteration as well as the possible `P1`-messages of the next iteration are included in the `P4ii`-message. Lastly, the `V4iii`-message of the i-th iteration is combined with the `V3+V4i`-message of the $i+1^{\text{st}}$ iteration. Hence, *an r-round interactive proof system is emulated by an $(r+1)$-rounds public-coin interactive proof system.*

3.2 Analysis of the Emulation

We introduce some notation and terminology that will be useful for the analysis of the proposed emulation. Fixing a generic input x and letting $n = n(|x|)$, we consider an interactive proof system with completeness and soundness parameters $c = c(|x|)$ and $s = s(|x|)$, respectively. Hence if x is yes-instance (resp., a no-intance), then it has at least $c \cdot 2^n$ accepting coins (resp., at most $s \cdot 2^n$ accepting coins). Put differently, there is a *gap* $g_0 \stackrel{\text{def}}{=} \frac{c}{s}$ between the number of accepting coins of yes-instances and no-instances. In the each iteration, the prover's goal is to *lower the gap* regarding the number of accepting coins, where we refer to the following definition.

Definition 3.4 (gaps): *The* gap on the ith iteration, *denoted g_i, is the ratio between the claimed bound regarding the number of accepting coins on the ith*

round (i.e. M_i) *and the number of accepting coins consistent with the partial transcript* γ_i (i.e., $|\mathtt{ACC}_x(\gamma_i)|$). *In case* $|\mathtt{ACC}_x(\gamma_i)| = 0$ *we set* $g_i = \infty$. *That is,*

$$g_i = \begin{cases} \frac{M_i}{|\mathtt{ACC}_x(\gamma_i)|} & \text{if } |\mathtt{ACC}_x(\gamma_i)| > 0 \\ \infty & \text{otherwise} \end{cases} \quad (5)$$

(Indeed, we may assume, without loss of generality that $M_i > 0$, since the verifier rejects whenever $M_i = 0$ for some i). Note that if the prover claims that some no-instance is a yes-instance, then at the beginning of the emulation $M_0 \geq c \cdot 2^n$ whereas $|\mathtt{ACC}_x(\gamma_0)| \leq s \cdot 2^n$, and so $g_0 \geq \frac{c}{s}$. If the verifier accepts the complete emulation, then (in particular) the final checks pass and $M_r = |\mathtt{ACC}_x(\gamma_r)| = 1$ holds, and so $g_r = 1$.

3.2.1 The Effect of a Single Iteration

Recall that we have fixed an arbitrary interactive proof system (P, V), and an input x to it. We consider the public-coin emulation of (P, V) defined in Sect. 3.1, and fix an interaction index $i \in [r]$ as well as the transcript of the first $i - 1$ iterations. Hence, the values γ_{i-1}, g_{i-1} and M_{i-1} are fixed. Denote by G_i the random variable that represents g_i at the end of the ith iteration, which is a function of the public randomness of the emulation protocol (of Construction 3.1 and the sampling protocol of Construction 3.2). Towards proving Theorem 1.2, we analyze the change in the gap caused by the ith iteration, and show that for every $t \in \mathbb{N}$ the gap G_i is reduced by a factor of b^t with probability at most $O(b^{-t})$. It is convenient to prove this claim by letting $j \in \mathbb{N}$ be such that $g_{i-1} \in (b^{j-1}, b^j]$. Hence if $G_i \in (b^{j-t-1}, b^{j-t}]$, then this implies that the gap is reduced by a factor of approximately b^t. The following lemma shows the probability that the gap changed by some factor F can be bounded in a way that is independent of the previous gap, and depends only on the factor F.

Lemma 3.5 (main lemma): *Suppose that* $g_{i-1} \in (b^{j-1}, b^j]$ *and* $t < j$. *Then,*

$$\mathbf{Pr}[G_i \in (b^{j-t-1}, b^{j-t}]] \leq b^{-t+3}.$$

That is, the probability that the gap G_i got reduced by a factor of b^{t-1} is at most $b^{-t+3} = O(b^{-t})$.

Proof: Recall that G_i is defined as the random variable representing the gap g_i, which is the ratio between the number of accepting coins that the prover claims to be consistent with the emulation and the actual number of such accepting coins. The gap G_i is essentially determined by the relation between the cluster that the verifier chooses in Step (3) and the cluster that the message sampled in Step (4) resides in. We are interested in upper-bounding the probability that $G_i \in (2^{j-t-1}, 2^{j-t}]$ for $j > t$. We can write this event as the union of disjoint events regarding to the cluster $C_k = C_k^{\gamma_{i-1}}$ that the verifier chooses in Step (3) of the emulation.

$$\mathbf{Pr}\left[G_i \in (b^{j-t-1}, b^{j-t}]\right] = \sum_{k=0}^{n'} \mathbf{Pr}\left[C_k \text{ is chosen} \wedge G_i \in (b^{j-t-1}, b^{j-t}]\right]. \quad (6)$$

Assume that cluster C_k is chosen by the verifier, which implies that $M_i = b^k$. Recalling that $G_i = \frac{M_i}{|\mathtt{ACC}_x(\gamma_{i-1}\alpha_i)|}$, it holds that if $G_i \in (b^{j-t-1}, b^{j-t}]$, then

$$b^{j-t-1} < \frac{b^k}{|\mathtt{ACC}_x(\gamma_{i-1}\alpha_i)|} \leq b^{j-t}$$

or equivalently

$$b^{k-(j-t)} \leq |\mathtt{ACC}_x(\gamma_{i-1}\alpha_i)| < b^{k-(j-t)+1}.$$

In other words, $G_i \in (b^{j-t-1}, b^{j-t}]$ if and only if the sampled message α_i resides in $C_{k-(j-t)}$ (for $k \geq j-t$). For each $k \in \{0, \ldots, n\}$, we introduce the following Boolean indicator variables:

Y_k: The event that cluster C_k is chosen by the verifier in Step (3).

Z_k: The event that the sampled message in Step (4) resides in cluster C_k

Using the foregoing observation and the notations just introduced, we can rewrite Eq. (6) as

$$\mathbf{Pr}[G_i \in (b^{j-t-1}, b^{j-t}]] = \sum_{k=j-t}^{n'} \mathbf{Pr}[Y_k \wedge Z_{k-(j-t)}]. \tag{7}$$

Next, we upper-bound the probabilities that the events in Eq. (7) occur. We first note that the verifier chooses a cluster (in Step (3)) according to the distribution in Eq. (4), hence

$$\mathbf{Pr}[Y_k] = \frac{N_k \cdot b^k}{\sum_{\ell=0}^{n'} N_\ell \cdot b^\ell} \tag{8}$$

where the N_ℓ's are the sizes claimed by the prover in Step (1). Assume that cluster C_k was chosen by the verifier in Step (3), and recall that the prover claims this cluster to have size N_k. Then, using the soundness property of the sampling protocol (with $T = C_\ell$ and $N = N_k$), we upper-bound the probability that the sampled message resides in C_ℓ as follows.

$$\mathbf{Pr}[Z_\ell \,|\, Y_k] \leq \frac{b \cdot |C_\ell|}{N_k} \tag{9}$$

(since the soundness parameter δ was set to b). Combining Eqs. (8) and (9), we get

$$\begin{aligned}
\mathbf{Pr}[Y_k \wedge Z_{k-(j-t)}] &= \mathbf{Pr}[Y_k] \cdot \mathbf{Pr}[Z_{k-(j-t)} \,|\, Y_k] \\
&\leq \frac{N_k \cdot b^k}{\sum_{\ell=0}^{n'} N_\ell \cdot b^\ell} \cdot \frac{b \cdot |C_{k-(j-t)}|}{N_k} \\
&= \frac{b^{k+1} \cdot |C_{k-(j-t)}|}{\sum_{\ell=0}^{n'} N_\ell \cdot b^\ell} \\
&= \frac{b^{j-t+1} \cdot b^{k-(j-t)} \cdot |C_{k-(j-t)}|}{\sum_{\ell=0}^{n'} N_\ell \cdot b^\ell}
\end{aligned} \tag{10}$$

Note that Eq. (10) does not depend on N_k, which is the purported size of the cluster C_k as claimed by the prover. (This is the case since $\mathbf{Pr}[Y_k]$ is linearly related to N_k, whereas $\mathbf{Pr}[Z_\ell \mid Y_k]$ is linearly related to $1/N_k$.) Moreover, Eq. (10) is proportional to the number of coins in the cluster $C_{k-(j-t)}$, which is approximately $b^{k-(j-t)} \cdot |C_{k-(j-t)}|$. Hence, plugging in the quantity from Eq. (10) in Eq. (7), we get

$$\begin{aligned}
\mathbf{Pr}[G_i \in (b^{j-t-1}, b^{j-t}]] &\leq \sum_{k=j-t}^{n'} \frac{b^{j-t+1} \cdot b^{k-(j-t)} \cdot |C_{k-(j-t)}|}{\sum_{\ell=0}^{n'} N_\ell \cdot b^\ell} \\
&= \frac{b^{j-t+1}}{\sum_{\ell=0}^{n'} N_\ell \cdot b^\ell} \cdot \sum_{k=j-t}^{n'} |C_{k-(j-t)}| \cdot b^{k-(j-t)} \\
&= \frac{b^{j-t+1}}{\sum_{\ell=0}^{n'} N_\ell \cdot b^\ell} \cdot \sum_{\ell=0}^{n'-(j-t)} |C_\ell| \cdot b^\ell
\end{aligned}$$

Thus,

$$\mathbf{Pr}[G_i \in (b^{j-t-1}, b^{j-t}]] \leq b^{j-t+1} \cdot \frac{\sum_{\ell=0}^{n'} |C_\ell| \cdot b^\ell}{\sum_{\ell=0}^{n'} N_\ell \cdot b^\ell} \tag{11}$$

As we shall show the ratio of the two sums that appear in Eq. (11) approximately equals $1/g_{i-1} \approx b^{-j}$, and the lemma will follow (i.e., $\mathbf{Pr}[G_i \in (b^{j-t-1}, b^{j-t}]] = O(b^{-t})$). This is the case since $\sum_{\ell=0}^{n'} |C_\ell| \cdot b^\ell \approx |\mathtt{ACC}_x(\gamma_{i-1})|$ whereas $\sum_{\ell=0}^{n'} N_\ell \cdot b^\ell \approx M_{i-1}$. Details follow.

Recall that the set of accepting coins, $\mathtt{ACC}_x(\gamma_{i-1})$, is partitioned among the clusters $C_0, \ldots, C_{n'}$, where $C_\ell = C_\ell^{\gamma_{i-1}}$. Furthermore, the number of accepting coins in cluster C_ℓ is at least $b^\ell \cdot |C_\ell|$. Thus,

$$\sum_{\ell=0}^{n'} |C_\ell| \cdot b^\ell \leq |\mathtt{ACC}_x(\gamma_{i-1})| \,. \tag{12}$$

On the other hand, recall that passing Step (2) of the emulation protocol mandates that $\sum_{\ell=0}^{n'} N_\ell \cdot b^{\ell+1} \geq M_{i-1}$. Hence,

$$\sum_{\ell=0}^{n'} N_\ell \cdot b^\ell \geq \frac{1}{b} \cdot M_i \tag{13}$$

Combining Eqs. (12) and (13), we upper-bound Eq. (11) as follows

$$\begin{aligned}
\mathbf{Pr}[G_i \in (b^{j-t-1}, b^{j-t}]] &\leq b^{j-t+1} \cdot \frac{|\mathtt{ACC}_x(\gamma_{i-1})|}{\frac{1}{b} \cdot M_i} \\
&= b^{j-t+2} \cdot \frac{1}{g_{i-1}}
\end{aligned}$$

where the last equality is due to $\frac{M_{i-1}}{\text{ACC}_x(\gamma_{i-1})} = g_{i-1}$. Lastly, recalling that $g_{i-1} > b^{j-1}$, we get

$$\begin{aligned}\mathbf{Pr}[G_i \in (b^{j-t-1}, b^{j-t}]] &\leq \frac{b^{j-t+2}}{b^{j-1}} \\ &= b^{-t+3}\end{aligned}$$

which completes the proof. ■

3.2.2 Proof of Theorem 1.2

We shall show that the emulation protocol of Construction 3.1 (combined with the sampling protocol of Construction 3.2) yields a public-coin interactive proof system for any set having r rounds and a gap of at least B^r, where B is some universal constant. Recall that when these two constructions are combined as detailed at the end of Sect. 3.1, the resulting public-coin protocol has $r+1$ rounds. The completeness feature of this protocol is quite straightforward (but will be spelled out next). The soundness feature will be proven later, while relying on Lemma 3.5.

Completeness. We claim that if x is a yes-instance, and the prover is honest, then the verifier accepts with probability greater than $\frac{2}{3}$. We first show that if the sampling goes well—namely, in each iteration, the sampled message reside in the chosen cluster—then the verifier accepts. We then show that the sampling goes well with probability greater than $\frac{2}{3}$.

We prove that if the sampling goes well, then in each iteration i the verifier does not abort and $|\text{ACC}_x(\gamma_i)| \geq M_i$. This is proved by induction on the iteration index, while observing that it holds for $i = 0$. Using the induction hypotheses, it holds that the verifier does not abort up to iteration $i \geq 0$ of the emulation and that $|\text{ACC}_x(\gamma_i)| \geq M_i$. For iteration $i+1$, when the prover sets $N_\ell = |C_\ell|$ as directed by the emulation protocol, the verifier doesn't abort in the Step (2) since the prover is honest and

$$\sum_{\ell=0}^{n'} N_\ell \cdot b^{\ell+1} = \sum_{\ell=0}^{n'} |C_\ell| \cdot b^{\ell+1} > |\text{ACC}_x(\gamma_i)| \geq M_i$$

Now, assume the verifier chooses cluster C_k. When a message α_{i+1} from the chosen cluster C_k is sampled, the prover supplies its response β_{i+1} to the message α_{i+1} so that $|\text{ACC}_x(\gamma_i, \alpha_{i+1}, \beta_{i+1})| \geq b^k = M_{i+1}$. In particular, after the last iteration, γ_r consists of a full transcript that is consistent with verifier's coins ρ and $|\text{ACC}_x(\gamma_r)| = M_r = 1$, so the verifier accepts.

It is left to show that, with probability greater than $\frac{2}{3}$, the sampled messages reside in the chosen cluster in all of the iterations. Recall that we run the sampling protocol with completeness parameter $\frac{1}{3r}$. Since the prover and the verifier follow the sampling protocol, by the properties of the sampling protocol, in each iteration the sampled message resides in the chosen cluster with probability at least $1-\frac{1}{3r}$. Therefore, with probability greater than $\frac{2}{3}$, elements from the chosen clusters are sampled in all the iterations.

Soundness. We show that if x is a no-instance, then for any prover strategy the verifier accepts with probability at most $\frac{1}{3}$. If the verifier accepts after a complete transcript γ_r is sampled, then $M_r = |\mathtt{ACC}_x(\gamma_r)| = 1$ must hold; namely, there is a sequence of accepting coin tosses that is consist with the generated transcript, and this is what the prover claims on the last round. In this case, the "gap" after the last round is 1 (i.e. $g_r = 1$). Therefore, in order to upper-bound the probability the verifier accepts, it suffices to upper-bound the probability that the gap after the last round, denoted g_r, is smaller than or equal to 1. As in the proof of Lemma 3.5, for $i \in \{0, \ldots, r\}$, we denote by G_i the random variable that represent the gap after the ith iteration. We set $G_0 \stackrel{\text{def}}{=} g_0$, where g_0 is the initial gap between the number of accepting coins for yes-instances and no-instances. Hence, it is enough to show that if $g_0 = B^r$, then $\mathbf{Pr}[G_r \leq 1] < \frac{1}{3}$, where B is a constant that will be determined later.

Intuitively, we have a randomized process that starts at a large value (i.e., B^r), and takes few (i.e., r) steps such that each may reduce the value by a factor of F with probability $O(1/F)$, where the latter probability bound is what Lemma 3.5 essentially says. In this case, it is quite unlikely that this process will end with a small value (i.e., the value 1). It may be easier to see this when taking a log scale; that is, we take r steps such that in each step we gain t units with probability that is exponentially vanishing with t. So the expected gain in such an r-step process is $O(r)$, and with probability at least 2/3 we are not going to exceed thrice this expectation. The following detailed analysis merely amounts to this.

To simplify the analysis, we present a "(lower) bounding" process $G'_0, G'_1, ..., G'_r$ such that $G'_0 = b^{\lfloor \log_b G_0 \rfloor}$ and $G'_i = \min(G'_{i-1}, b^{\lfloor \log_b(\max(G_i - 1, 1)) \rfloor})$ for every $i \in [r]$, and focus at upper-bounding $\mathbf{Pr}[G'_r \leq 1]$. Using Lemma 3.5, we infer that for every $t < j$ it holds that $\mathbf{Pr}[G'_i = b^{j-t-1} | G'_{i-1} = b^{j-1}] \leq b^{-t+3}$, which yields

$$\mathbf{Pr}[G'_i = b^{-t} \cdot G'_{i-1}] \leq b^{-t+3}. \tag{14}$$

Next, we let $L_i = \log_b(G'_{i-1}/G'_i) \geq 0$, for $i \in [r]$, and assuming (w.l.o.g.) that B is a power of b, we get

$$\mathbf{Pr}[G_r \leq 1] \leq \mathbf{Pr}[G'_0/G'_r \geq B^r] \tag{15}$$

$$\leq \mathbf{Pr}\left[\sum_{i \in [r]} L_i \geq r \cdot \log_b B\right]. \tag{16}$$

We shall upper-bound the latter probability by upper-bounding the expectation of $\sum_{i \in [r]} L_i$. Using Eq. (14), we get

$$\mu \stackrel{\text{def}}{=} \mathbf{E}\left[\sum_{i\in[r]} L_i\right] \tag{17}$$

$$\le \sum_{i\in[r]}\sum_{t\ge 1} t\cdot \mathbf{Pr}[L_i=t]$$

$$\le \sum_{i\in[r]}\left(3+\sum_{t\ge 4}(t-3)\cdot \mathbf{Pr}[L_i=t]\right)$$

$$\le r\cdot\left(3+\sum_{t\ge 4}(t-3)\cdot b^{-t+3}\right)$$

$$= r\cdot\left(3+\frac{b}{(b-1)^2}\right) \tag{18}$$

where the last equality uses $\sum_{t\ge 1} t\cdot b^{-t} = b/(b-1)^2$. Fixing the constant B such that $\log_b B > 3\cdot(3+(b/(b-1)^2)$ and combining Eq. (15–16) and Eq. (17–18), while recalling that $\mu = \mathbf{E}\left[\sum_{i\in[r]} L_i\right] < \frac{r}{3}\cdot\log_b B$, we get

$$\mathbf{Pr}[G_r \le 1] \le \mathbf{Pr}\left[\sum_{i\in[r]} L_i \ge r\cdot\log_b B\right]$$

$$\le \mathbf{Pr}\left[\sum_{i\in[r]} L_i > 3\cdot\mu\right]$$

which is smaller than 1/3. This establishes the soundness condition, and the theorem follows.

On the Choice of the Base Parameter b. Recall that we essentially set $B = b^{(9(b-1)^2+3b)/(b-1)^2}$, where B^r is the initial gap required by our emulation. Wishing to minimize B calls for minimizing $f(b) = \frac{(9b^2-15b+9)\ln b}{(b-1)^2}$. It turns out that the minimum is obtained at $b \approx 1.79521$, and its value is ≈ 10.2494. This yields $B \approx 28266$, which is not that far from the value $B = 37768$ obtained at $b = 2$. Of course, much better values can be obtained by a more careful analysis (e.g., it can be shown that $\sum_{i\in[r]} L_i$ is concentrated at its mean, provided that r is large enough).[3]

3.3 Lower Bounds

We first observe that for any base parameter $b > 1$, the gap may be reduced by a factor of b in each iteration (of the emulation protocol) due to the mere fact that

[3] The key observations are that the L_i's form a martingale and that with probability at least 3/4 all L_i's reside in $[0, \log_b r + O(1)]$. Hence, $\mathbf{Pr}[\sum_{i\in[r]} L_i > (1+o(1))\cdot\mu] < 1/3$. This allows setting $B = b^{(3(b-1)^2+3b)/(b-1)^2}$, which yields $B \approx (28266)^{1/3} < 31$ at $b \approx 1.79521$.

each element in each C_j is counted as if it has a weight of b^{j+1} whereas its actual weight may be merely b^j. Thus, if b is a constant, then Theorem 1.3 follows (with $C = b$). So we should deal with the case of $b = 1+o(1)$, or, equivalently, establish a bound that is independent of b. Still, in light of the foregoing, we may assume that $b \in (1, 2]$.

The key observation is that the prover can easily reduce the gap when neighboring clusters have similar weight. Specifically, suppose that $|C_j| \cdot b^j = |C_{j+1}| \cdot b^{j+1}$ (and that all messages in C_k have weight exactly b^k). Further suppose that the prover claims that $N_{j+t} = |C_j|$ and $N_{j+t+1} = |C_{j+1}|$, which supports a gap of b^t (since N_k is supposed to equal $|C_k|$). Now, the verifier will select the index $j+t$ with probability half, but the prover can try to let it sample from a set that contains as many elements of C_{j+1} as possible (and use elements of C_j only to fill-up the rest). Indeed, the prover should provided $N_{j+t} = |C_j|$ elements, whereas $|C_{j+1}| = |C_j|/b$. Still, when the prover does so (i.e., makes the verifier sample a set that contains C_{j+1} as well as $|C_j| - |C_{j+1}|$ elements of C_j), the verifier selects an element of C_{j+1} with probability (approximately) $1/b$, and when this happens the parties continue to the next iteration with a gap of $\frac{b^{j+t}}{b^{j+1}} = b^{t-1}$ rather than b^t. These considerations establish the fact that *with probability at least* $1/2b$, *the prover can decrease the gap by a factor of* b. In light of the first paragraph (which established a decrease of b with certainty), this alternative argument seems quite useless, but the point is that the argument can be extended to clusters that are a distance k apart.[4] Specifically, we prove

Claim 3.6 (unavoidable gap decrease): *For any* $k \geq 1$, *with probability at least* $1/2b^k$, *the prover can decrease the gap by a factor of* b^k.

Proof: We mimic the foregoing argument, but use $|C_j| \cdot b^j = |C_{j+k}| \cdot b^{j+k}$ instead. Suppose that the prover claims that $N_{j+t} = |C_j|$ and $N_{j+t+k} = |C_{j+k}|$, which supports a gap of b^t. Now, the verifier will select the index $j+t$ with probability half, and the prover can try to let it sample from a set that contains as many elements of C_{j+k} as possible. When the prover does so (i.e., makes the verifier sample a set that contains C_{j+k} as well as $|C_j| - |C_{j+k}|$ elements of C_j), the verifier selects an element of C_{j+k} with probability (approximately) $1/b^k$, and when this happens the parties continue to the next iteration with a gap of $\frac{b^{t+j}}{b^{j+k}} = b^{t-k}$. ■

Proof of Theorem 1.3. For constants $c, d > 1$ to be determined, we consider two cases. If the base b is greater than c, then Theorem 1.3 follows with $C = b > c$ (by the first argument in this section). Otherwise, we just choose k such that $b^k \in [d, c \cdot d]$, and apply Claim 3.6. It follows that, in each iteration, with probability $1/2b^k > 1/2dc$, the prover can decrease the gap by a factor of at least d. Hence, in this case, Theorem 1.3 follows with $C = d^{1/2dc-o(1)}$, since (for sufficiently large r), with high probability, the prover will be successful in at least

[4] Furthermore, this alternative argument for a decrease of a b-factor does not capitalize on the variance of weights in clusters.

$1/dc - o(1)$ of the iterations. Hence, Theorem 1.3 holds for $C = \min(c, d^{1/2dc})$, which is maximized at $(c, d) \approx (1.170, 2.718)$, yielding $C \approx 1.17 > 7/6$.

References

1. Babai, L.: Trading group theory for randomness. In: 17th STOC, pp. 421–429 (1985)
2. Babai, L., Moran, S.: Arthur-Merlin games: a randomized proof system, and a hierarchy of complexity clusters. J. Comput. Syst. Sci. **36**(2), 254–276 (1988)
3. Ben-Or, M., Goldreich, O., Goldwasser, S., Hastad, J., Kilian, J., Micali, S., Rogaway, P.: Everything provable is provable in zero-knowledge. In: Goldwasser, S. (ed.) CRYPTO 1988. LNCS, vol. 403, pp. 37–56. Springer, New York (1990). https://doi.org/10.1007/0-387-34799-2_4
4. Fürer, M., Goldreich, O., Mansour, Y., Sipser, M., Zachos, S.: On completeness and soundness in interactive proof systems. In: Micali, S. (ed.) Randomness and Computation. Advances in Computing Research: A Research Annual, vol. 5, pp. 429–442 (1989)
5. Goldreich, O.: Computational Complexity: A Conceptual Perspective. Cambridge University Press, Cambridge (2008)
6. Goldreich, O., Vadhan, S., Wigderson, A.: On interactive proofs with a laconic prover. Comput. Complex. **11**(1–2), 1–53 (2002)
7. Goldwasser, S., Micali, S., Rackoff,C.: The knowledge complexity of interactive proof systems. In: 17th STOC, pp. 291–304 (1985)
8. Goldwasser, S., Sipser, M.: Private coins versus public coins in interactive proof systems. In: 18th STOC, pp. 59–68 (1986)
9. Lund, C., Fortnow, L., Karloff, H., Nisan, N.: Algebraic methods for interactive proof systems. J. ACM **39**(4), 859–868 (1992)
10. Shamir, A.: IP = PSPACE. J. ACM **39**(4), 869–877 (1992)

Reducing Testing Affine Spaces to Testing Linearity of Functions

Oded Goldreich

Abstract. For any finite field $\mathcal{F}$ and $k < \ell$, we consider the task of testing whether a function $f : \mathcal{F}^\ell \to \{0,1\}$ is the indicator function of an $(\ell - k)$-dimensional affine space. For the case of $\mathcal{F} = \mathrm{GF}(2)$, an optimal tester for this property was presented by Parnas, Ron, and Samorodnitsky (*SIDMA* 2002), by mimicking the celebrated linearity tester of Blum, Luby and Rubinfeld (*JCSS* 1993) and its analysis. We show that the former task (i.e., testing $(\ell - k)$-dimensional affine spaces) can be efficiently reduced to testing the linearity of a related function $g : \mathcal{F}^\ell \to \mathcal{F}^k$. This reduction yields an almost optimal tester for affine spaces (represented by their indicator function).

Recalling that Parnas, Ron, and Samorodnitsky used testing $(\ell - k)$-dimensional affine spaces as the first step in a two-step procedure for testing k-monomials, we also show that the second step in their procedure can be reduced to testing whether the foregoing function g depends on k of its variables.

A preliminary version of this work was posted in April 2016 as a guest column on the *Property Testing Review*. It was significantly revised and appeared as TR16-080 of *ECCC*. The current version is the result of an even more extensive revision. In particular, the reduction of testing affine spaces to testing linearity (of functions) is simplified and extended to arbitrary finite fields, the second step in the procedure for testing k-monomials is revised, many of the technical justifications are elaborated, and some crucial typos are fixed. In addition, the title has been augmented for clarity, the brief introduction has been expanded, and the high level structure has been re-organized.

1 Introduction

Property Testing is the study of super-fast (randomized) algorithms for approximate decision making. These algorithms are given direct access to items of a huge data set, and determine whether this data set has some predetermined (global) property or is far from having this property, while accessing a small portion of the data set. Thus, property testing is a relaxation of decision problems and it focuses on algorithms, called *testers*, that only read parts of the input. Consequently, the testers are modeled as oracle machines and the inputs are modeled as functions to which the tester has an oracle access.

This paper refers to several basic tasks in property testing, including testing linearity, testing dictatorship, testing (monotone) k-monomials, and testing

O. Goldreich (Ed.): Computational Complexity and Property Testing, LNCS 12050, pp. 199–219, 2020.
https://doi.org/10.1007/978-3-030-43662-9_13

affine spaces. Whereas the first three tasks refer explicitly to an input function (i.e., the object they test is naturally viewed as a function), in the case of testing affine spaces the object is a set of points and representing this set by an indicator function is a natural choice but not an immediate one. In particular, this is a very redundant representation in the case that the set is sparse, but we will consider the case of relatively dense sets. Furthermore, this representation arises naturally in the study of testing k-monomials.

The problem of testing whether a Boolean function is a (monotone) k-monomial was first studied by Parnas, Ron, and Samorodnitsky [9]. The tester that they presented generalizes their tester of dictatorship (i.e., the case $k = 1$), and does so by following the same two-step strategy and using similar arguments at each step. This raises the question of whether the case of general k can be reduced to the special case of $k = 1$. (We mention that this question occurred to us when writing [7, Sec. 5.2.2], and the first version of the current paper was written at that time.)

Specifically, the first step in the strategy of Parnas, Ron, and Samorodnitsky [9] is *testing whether the input function* $f : \{0,1\}^\ell \rightarrow \{0,1\}$ *describes an* $(\ell - k)$*-dimensional affine space*, where the space described by f is $f^{-1}(1)$. In the case of dictatorship (i.e., $k = 1$), this amounts to testing whether f itself is an affine function, but in the case of $k > 1$ a more general task arises (i.e., testing whether $f^{-1}(1)$ is an $(\ell - k)$-dimensional affine space is fundamentally different from testing whether f is affine). In the second step, one tests whether this affine space is of the right form (i.e., is a translation by 1^ℓ of a linear space spanned by axis-parallel vectors). In the case of $k = 1$, the latter task amounts to testing whether the affine function depends on a single variable, but in the case of $k > 1$ another more general task arises.

Both these general tasks were solved by Parnas, Ron, and Samorodnitsky [9], but their solutions mimic the solutions used in the case of $k = 1$ (see Sect. 4.1 for more details). Furthermore, in both cases, the generalization is very cumbersome. We find this state of affairs quite annoying, and believe that it is more appealing to reduce the general case to the special case.

Our Contribution. This paper partially achieves this goal by (1) replacing the first step of [9] with a reduction to the (extensively studied) problem of testing linearity of functions, and (2) replacing the second step of [9] with a reduction to the problem of testing k-linearity of functions. Specifically, we first *reduce the problem of testing affine spaces to the problem of testing the linearity of functions.* This reduction actually hold over any finite field, whereas the application to testing monomials only uses the case of the binary field (which is the case treated in [9]). Next, recalling that k-monomials correspond to axis-parallel $(\ell - k)$-dimensional affine spaces, we reduce the testing of such spaces to testing whether a linear function depends on k of its variables. The complexity of the testers that we derive is only slightly inferior to the complexity of the corresponding optimal testers of [9]; specifically, in the relevant case of $\epsilon = O(2^{-k})$, we get a complexity bound of $\widetilde{O}(\log(1/\epsilon)) \cdot 2^k + O(1/\epsilon) = \widetilde{O}(1/\epsilon)$ rather than $O(1/\epsilon)$. (Indeed, the bounds coincide for $\epsilon < 2^{-k}/\widetilde{O}(k)$.)

Organization. In Sect. 2 we recall the standard definition of property testing and formally define the main properties considered in this paper (i.e., affine spaces and linear functions). The reduction of testing affine spaces to testing linearity of functions is presented in Sect. 3, whereas the problem of testing monomials is considered in Sect. 4. In the Appendix we present a simple and direct procedure for testing whether a linear function depends on a given number of variables.

2 Preliminaries

We assume that the reader is familiar with the basic definition of property testing (see, e.g., [7]), but for sake of good order we reproduce it here. The basic definition refers to functions with domain D and range R.

Definition 2.1 (a tester for property Π): *Let Π be a set of functions of the form $f : D \to R$. A* tester for Π *is a probabilistic oracle machine, denoted T, that, on input a* proximity parameter ϵ *and oracle access to a function $f : D \to R$, outputs a binary verdict that satisfies the following two conditions.*

1. T accepts inputs in Π: *For every $\epsilon > 0$, and for every $f \in \Pi$, it holds that* $\mathbf{Pr}[T^f(\epsilon)=1] \geq 2/3$.
2. T rejects inputs that are ϵ-far from Π: *For every $\epsilon > 0$, and for every function $f : D \to R$ that is ϵ-far from Π it holds that $\mathbf{Pr}[T^f(\epsilon)=0] \geq 2/3$, where f is* ϵ-far from Π *if for every $g \in \Pi$ it holds that $|\{x \in D : f(x) \neq g(x)\}| > \epsilon \cdot |D|$.*

If the first condition holds with probability 1 (i.e., $\mathbf{Pr}[T^f(\epsilon) = 1] = 1$), *then we say that T has* one-sided error; *otherwise, we say that T has* two-sided error.

We focus on the query complexity of such testers, while viewing $|D|$ as an additional parameter. We seek testers of query complexity that is independent of $|D|$, which means that the complexity will be a function of the proximity parameter ϵ and an auxiliary parameter k (of the two properties that we consider).

The properties we shall consider refer to functions over the domain $\mathcal{F}^\ell$, where $\mathcal{F}$ is a finite field. (In the previous versions of this paper, we confined ourselves to the case that $\mathcal{F}$ is the two-element field GF(2), which is the case treated in [9].)

Definition 2.2 (affine spaces): *For fixed $k, \ell \in \mathbb{N}$ and a finite field $\mathcal{F}$, we say that the function $f : \mathcal{F}^\ell \to \{0,1\}$* describes an $(\ell - k)$-dimensional affine space *if $f^{-1}(1) = \{x \in \mathcal{F}^\ell : f(x) = 1\}$ is an $(\ell - k)$-dimensional affine space; that is, $f^{-1}(1) = \{yG + s : y \in \mathcal{F}^{\ell-k}\}$, where $G \in \mathcal{F}^{(\ell-k)\times\ell}$ is an $(\ell - k)$-by-ℓ full-rank matrix and $s \in \mathcal{F}^\ell$. When $s = 0^\ell$, the* described space is linear.

We mention that, for $\mathcal{F} = \mathrm{GF}(2)$, the set of k-monomials (see Definition 4.1) coincides with the set of functions that describe $(\ell - k)$-dimensional affine spaces that are spanned by unit vectors (i.e., the rows of the matrix G are unit vectors).

Definition 2.3 (linear functions): *For fixed $k, \ell \in \mathbb{N}$ and a finite field $\mathcal{F}$, we say that $g : \mathcal{F}^\ell \to \mathcal{F}^k$ is* linear *if $g(x+y) = g(x) + g(y)$ for all $x, y \in \mathcal{F}^\ell$. Equivalently, $g(z) = zT$ for a ℓ-by-k matrix T. We say that f is* affine *if $f(z) = f'(z) + s$ for a linear function f' and some $s \in \mathcal{F}^k$.*

When $k = 1$ and $\mathcal{F} = \mathrm{GF}(2) \equiv \{0,1\}$, it holds that $f : \mathcal{F}^\ell \to \{0,1\}$ describes an $(\ell - k)$-dimensional affine space (resp., linear space) if and only if f is a non-constant affine function (resp., $f+1$ is a non-constant linear function). However, in the other cases, this does not hold; in particular, for other fields a non-constant affine function must range over $\mathcal{F}$ rather than over $\{0,1\}$, whereas for $\mathcal{F} = \mathrm{GF}(2)$ and $k > 1$ the densities do not match (i.e., an $(\ell - k)$-dimensional affine space over GF(2) has density 2^{-k}, but $f^{-1}(1)$ has density $1/2$ for any non-constant affine function $f : \mathrm{GF}(2)^\ell \to \mathrm{GF}(2)$).

Conventions. When writing $\mathbf{Pr}_x[\mathrm{event}(x)]$ we refer to the case that x is selected uniformly in a set that is clear from the context; we sometimes spell out this set by writing $\mathbf{Pr}_{x \in S}[\mathrm{event}(x)]$. For sake of simplicity, we often use the phrase "with high probability" (abbrev., "w.h.p."), which mean that we can obtain arbitrary high constant probability smaller 1 (e.g., 0.99). The image of a function $f : D \to R$ is the set $\{f(e) : e \in D\} \subseteq R$. The symbol $\perp$ denotes a special symbol that is not in $\mathcal{F}^k$.

We view $\mathcal{F}$, ℓ and k as parameters, and when using O-notation we refer to universal constants that are independent of $\mathcal{F}$, ℓ and k. However, when stating time-complexity bounds, we shall assume that basic operations on elements of $\mathcal{F}^\ell$ (e.g., addition, selection of a random element, etc) can be performed at unit cost.

3 The Reduction (of Testing Affine Spaces to Testing Linearity of Functions)

We start by restating the problem. We are given access to a function $h : \mathcal{F}^\ell \to \{0,1\}$ and wish to test whether $h^{-1}(1)$ is an $(\ell - k)$-dimensional affine subspace by reducing this problem to testing linearity (of a function). We present two reductions: The first (and simpler) reduction increases the complexities by a factor of $|\mathcal{F}|^k$, whereas the second reduction only incurs an overhead of $\widetilde{O}(\log(1/\epsilon))$. The first reduction (presented in Sect. 3.2) will be used as a subroutine in the second reduction (presented in Sect. 3.3). Furthermore, the first reduction provides a good warm-up towards the second one. Before describing these reductions, we present and justify some simplifying assumptions.

3.1 Simplifying Assumptions

First, note that we may assume that $\epsilon = O(|\mathcal{F}|^{-k})$, which means that $|\mathcal{F}|^k = O(1/\epsilon)$, since the case of $\epsilon > 4 \cdot |\mathcal{F}|^{-k}$ can be handled by merely estimating the density of $h^{-1}(1)$. Specifically, note that any function that describes an $(\ell - k)$-dimensional subspace (over $\mathcal{F}$) is at distance exactly $|\mathcal{F}|^{-k}$ from the all-zero function. Hence, if $\epsilon > 4 \cdot |\mathcal{F}|^{-k}$ and h is 0.75ϵ-close to the all-zero function, then it is ϵ-close to describing a $(\ell - k)$-dimensional subspace, and it is OK to accept it. On the other hand, if $\epsilon > 4 \cdot |\mathcal{F}|^{-k}$ and h is 0.25ϵ-far from the all-zero function, then it cannot describe a $(\ell - k)$-dimensional subspace, and it is OK to reject it. Hence, we have

Claim 3.1 (reducing to the case of $\epsilon \leq 4 \cdot |\mathcal{F}|^{-k}$): *Testing whether a function $h : \mathcal{F}^\ell \to \{0,1\}$ describes a $(\ell - k)$-dimensional affine subspace can be randomly reduced to the case of $\epsilon \leq 4 \cdot |\mathcal{F}|^{-k}$, where the reduction introduces an additive overhead of $O(|\mathcal{F}|^k)$ queries.*

(An alternative justification boils down to resetting the proximity parameter to $\min(\epsilon, 4 \cdot |\mathcal{F}|^{-k})$; that is, on input proximity parameter ϵ and oracle h, we invoke the given tester on proximity parameter $\min(\epsilon, 4 \cdot |\mathcal{F}|^{-k})$ and provide it with oracle access to h.)

Another simplifying assumption is that we are dealing with linear subspaces rather than with affine ones. Actually, we present a reduction of the general case to this special case.[1]

Claim 3.2 (reducing to the linear case): *Testing whether a function $h : \mathcal{F}^\ell \to \{0,1\}$ describes a $(\ell - k)$-dimensional affine subspace can be randomly reduced to testing whether a function $h' : \mathcal{F}^\ell \to \{0,1\}$ describes a $(\ell - k)$-dimensional linear subspace, where the reduction introduces an additive overhead of $O(|\mathcal{F}|^k)$ queries.*

The foregoing randomized reduction has a two-sided error probability; obtaining an analogous one-sided error reduction is left as an open problem. Yet, since the known testers for linear subspaces (i.e., of [9] and of this paper) have two-sided error probability, our use of Claim 3.2 cause no real loss.

Proof: On input parameter $\epsilon > 0$ and oracle access to h, we proceed as follows.

1. Select uniformly a sample of $O(|\mathcal{F}|^k)$ points in $\mathcal{F}^\ell$. If h evaluates to 0 on all these points, then we reject. Otherwise, let u be a point in this sample such that $h(u) = 1$.
2. Invoke the tester for linear subspaces on input parameter ϵ and oracle access to h' defined by $h'(x) \stackrel{\text{def}}{=} h(x+u)$, and output its verdict. That is, each query x to h' is emulated by making the query $x + u$ to h.

The overhead of the reduction is due to Step 1, whereas in Step 2 we just invoke the tester for the special case.

If h describes an $(\ell - k)$-dimensional affine subspace, then, with high probability, Step 1 finds $u \in h^{-1}(1)$, since $h^{-1}(1)$ has density $|\mathcal{F}|^{-k}$, and we proceed to Step 2. But in this case it holds that $h'(x) = 1$ if and only if $x + u \in h^{-1}(1)$, which means that h' describes the $(\ell - k)$-dimensional linear space $h^{-1}(1) - u$, and so the invoked test accepts (w.h.p.).

(Indeed, if $H = \{yG+s : y \in \mathcal{F}^{\ell-k}\}$ is an affine space (as in Definition 2.2) and $u = (zG+s) \in H$, then $H - u = \{yG+s-u : y \in \mathcal{F}^{\ell-k}\} = \{yG+s-(zG+s) : y \in \mathcal{F}^{\ell-k}\} = \{(y-z)G : y \in \mathcal{F}^{\ell-k}\}$ is a linear space. Likewise, if $H' = \{yG : y \in \mathcal{F}^{\ell-k}\}$ is a linear space, then $H' + u = \{yG + u : y \in \mathcal{F}\}$ is an affine space.)

---

[1] This reduction somewhat simplifies the presentation in Sect. 3.2, and more significantly so in Sect. 3.3.

On the other hand, if h is ϵ-far from being an $(\ell-k)$-dimensional affine subspace, then either Step 1 rejects or else $u \in h^{-1}(1)$. But in this case h' (which was defined by $h'(x) \stackrel{\text{def}}{=} h(x+u)$) must be ϵ-far from describing an $(\ell-k)$-dimensional linear subspace. This is so because if h' is ϵ-close to g' that describes an $(\ell-k)$-dimensional linear subspace (i.e., g' describes the linear space $\{yG : y \in \mathcal{F}^{\ell-k}\}$), then $g(x) \stackrel{\text{def}}{=} g'(x-u)$ (equiv., $g(x+u) = g'(x)$) describes an affine space (i.e., g describes the affine space $\{yG+u : y \in \mathcal{F}^{\ell-k}\}$), whereas h is ϵ-close to g (since $h(x) = h'(x-u)$). Hence, in this case (i.e., h' is ϵ-far from describing an $(\ell-k)$-dimensional linear subspace), Step 2 rejects with high probability. ■

3.2 The First Reduction

The pivotal step in the reduction is the definition of a function $g : \mathcal{F}^\ell \to \mathcal{F}^k \cup \{\perp\}$ such that if $H \stackrel{\text{def}}{=} h^{-1}(1)$ is an $(\ell-k)$-dimensional linear space, then g is linear (with image $\mathcal{F}^k$) and $g^{-1}(0^k) = H$. Furthermore, in that case, $g(x)$ indicates one of the $|\mathcal{F}|^k$ translations of H in which x resides; that is, if $v^{(1)}, ..., v^{(k)} \in \mathcal{F}^\ell$ form a basis for the k-dimensional space that complements H, then $g(x)$ represents coefficients $(c_1, ..., c_k) \in \mathcal{F}^k$ such that $x \in H - \sum_{i\in[k]} c_i v^{(i)}$.

Indeed, the definition of g is based on any fixed sequence of linearly independent vectors $v^{(1)}, ..., v^{(k)} \in \mathcal{F}^\ell$ such that for every non-zero sequence of coefficients $(c_1, ..., c_k) \in \mathcal{F}^k$ it holds that $\sum_{i\in[k]} c_i v^{(i)} \notin H$. Such sequences of vectors exist[2] if H is an $(\ell-k)$-dimensional linear space, and we can find such a sequence in this case (by random sampling and querying h). Failure to find such a sequence will provide good justification for ruling that H is not an $(\ell-k)$-dimensional linear space.

Fixing such a sequence of $v^{(i)}$'s, we define $g : \mathcal{F}^\ell \to \mathcal{F}^k \cup \{\perp\}$ such that $g(x) = (c_1, ..., c_k)$ if $(c_1, ..., c_k) \in \mathcal{F}^k$ is the unique sequence that satisfies $x + \sum_{i\in[k]} c_i v^{(i)} \in H$ and let $g(x) = \perp \notin \mathcal{F}^k$ otherwise. Indeed, a unique sequence $(c_1, ..., c_k) \in \mathcal{F}^k$ exists for each $x \in \mathcal{F}^\ell$ if H is an $(\ell-k)$-dimensional linear space, and in that case $g(x) \in \mathcal{F}^k$ for every $x \in \mathcal{F}^\ell$. But when H is not an $(\ell-k)$-dimensional linear space, it may happen that for some (or even all) x's there is no sequence $(c_1, ..., c_k) \in \mathcal{F}^k$ such that $x + \sum_{i\in[k]} c_i v^{(i)} \in H$; similarly, it may happen that there are several different sequences $(c_1, ..., c_k) \in \mathcal{F}^k$ that satisfy $x + \sum_{i\in[k]} c_i v^{(i)} \in H$. Anyhow, using matrix notation, we restate the foregoing definition next (where the $v^{(i)}$'s are the rows of the matrix V).

Definition 3.3 (the function $g = g_{H,V}$): *Let V be a k-by-ℓ full-rank matrix over $\mathcal{F}$ such that $cV \notin H$ for every $c \in \mathcal{F}^k \setminus \{0^k\}$. Then, $g_{H,V} : \mathcal{F}^\ell \to \mathcal{F}^k \cup \{\perp\}$ is defined such that $g_{H,V}(x) = c$ if $c \in \mathcal{F}^k$ is the unique vector that satisfies $x + cV \in H$, and $g_{H,V}(x) = \perp$ if the number of such k-long vectors is not one.*

Note that $g(x) = c$ implies $x + cV \in H$; hence, in particular, $g_{H,V}(x) = 0^k$ implies $x \in H$; that is, $g_{H,V}^{-1}(0^k) \subseteq H$.

---

[2] Actually, the density of suitable k-long sequences in H is $\prod_{i\in[k]}(1 - |\mathcal{F}|^{-i}) > 1/4$.

3.2.1 Key Observations

The most important observation is that if H is an $(\ell - k)$-dimensional linear space then g is a linear function, whereas if H is far from being an $(\ell - k)$-dimensional linear space then g is far from any linear function. Whenever we say that g is linear, we mean, in particular, that it never assumes the value $\perp$. (Indeed, when emulating g for the linearity tester, we shall reject if we ever encounter the value $\perp$.)

Claim 3.4 (H versus $g_{V,H}$): *Let H, V and $g = g_{H,V}$ be as in Definition 3.3. Then, H is an $(\ell - k)$-dimensional linear space if and only if g is a linear function with image $\mathcal{F}^k$.*

Actually, it turns out that if g is linear, then it has image $\mathcal{F}^k$; a more general statement is proved in Claim 3.6. Furthermore, if g is ϵ-close to being a linear function with image $\mathcal{F}^k$, then $g^{-1}(0^k)$ is ϵ-close to being an $(\ell - k)$-dimensional linear space (i.e., the indicator functions of these sets are ϵ-close). To see this, consider a linear g' that is ϵ-close to g, and note that the $(\ell - k)$-dimensional linear space $H' = \{x \in \mathcal{F}^\ell : g'(x) = 0^k\}$ is ϵ-close to $g^{-1}(0^k)$, since $g'(x) \neq g(x)$ for any x that resides in the symmetric difference of these sets.

Proof: Recall that $g^{-1}(0^k) \subseteq H$ always holds. Furthermore, equality (i.e., $g^{-1}(0^k) = H$) holds if g never assumes the value $\perp$, since in this case $x + cV \in H$ implies that $g(x) = c$ (and so $x \in H$ implies $g(x) = 0^\ell$).

Now, on the one hand, if g is a linear function with image $\mathcal{F}^k$ (i.e., $g(x) = xT$ for some full-rank ℓ-by-k matrix T), then $H = g^{-1}(0^k)$ (i.e., $H = \{x \in \mathcal{F}^\ell : xT = 0^k\}$), which implies that H is an $(\ell - k)$-dimensional linear subspace (since $H = \{yG : y \in \mathcal{F}^{\ell-k}\}$ for any G that is a basis of the space orthogonal to $T^\top$).[3]

On the other hand, if H is an $(\ell - k)$-dimensional linear space, then, for some full-rank $(\ell - k)$-by-ℓ matrix G, it holds that $H = \{yG : y \in \mathcal{F}^{\ell-k}\}$. In this case, for every $x \in \mathcal{F}^\ell$ there exists a *unique* representation of x as $yG - cV$, since V is a basis for a k-dimensional linear space that complements the $(\ell - k)$-dimensional linear space H. Hence, for every $x \in \mathcal{F}^\ell$, there exists a unique $(c, y) \in \mathcal{F}^k \times \mathcal{F}^{\ell-k}$ such that $x + cV = yG \in H$, and $g(x) = c$ follows. We now observe that the image of g equals $\mathcal{F}^k$, since $g(0^\ell - cV) = c$ for every $c \in \mathcal{F}^k$, and that g is linear, since for every $x = yG - cV$ and $x' = y'G - c'V$ in $\mathcal{F}^\ell$, it holds that $g(x) + g(x') = c + c'$ and $c + c' = g(y''G - (c + c')V)$ holds for every $y'' \in \mathcal{F}^{\ell-k}$ (and in particular for $y'' = y + y'$, which implies that $c + c' = g(x + x')$). ■

Claim 3.5 (finding V): *Let $h : \mathcal{F}^\ell \to \{0,1\}$. If $H = h^{-1}(1)$ is an $(\ell - k)$-dimensional linear space, then a matrix V as underlying the definition of $g_{H,V}$ can be found* (w.h.p.) *by making $O(|\mathcal{F}|^k)$ queries to h.*

---

[3] Alternatively, if $g(x + x') = g(x) + g(x')$ for every $x, x' \in \mathcal{F}^\ell$, then $x, x' \in H$ implies $x + x' \in H$ (for every $x, x' \in \mathcal{F}^\ell$), since $g(x) = g(x') = 0^k$ implies $g(x + x') = 0^k$. Hence, $H = g^{-1}(0^k)$ is a linear subspace. Lastly, we note that this subspace has dimension $\ell - k$, since the image of g equals $\mathcal{F}^k$ and $|g^{-1}(0^k)| = |g^{-1}(c)|$ holds for every $c \in \mathcal{F}^k$.

Proof: The matrix V can be found in k iterations as follows. In the $i^{\rm th}$ iteration we try to find a vector $v^{(i)}$ such that $\sum_{j\in[i]} c_j v^{(j)} \notin H$ holds for every $(c_1,...,c_i) \in \mathcal{F}^i \setminus \{0^i\}$. In each trial, we pick $v^{(i)}$ at random, while noting that the probability of success is $1-|\mathcal{F}|^{i-1}\cdot|\mathcal{F}|^{-k} \geq 1/2$, since for every $(c_1,...,c_i) \in \mathcal{F}^i \setminus \{0^i\}$ it holds that $\mathbf{Pr}_{v^{(i)}}[\sum_{j\in[i]} c_j v^{(j)} \in H] = \mathbf{Pr}_{v^{(i)}}[c_i v^{(i)} \in H - \sum_{j\in[i-1]} c_j v^{(j)}] \leq |\mathcal{F}|^{-k}$, where equality holds if $c_i \neq 0$. Lastly, observe that the foregoing condition can be checked by making $|\mathcal{F}|^i - 1$ queries to h. (Actually, $|\mathcal{F}|^{i-1}$ queries suffice for checking in the $i^{\rm th}$ iteration, since it suffices to check the cases in which $c_i = 1$.)[4] ■

Claim 3.6 (on the linear function closest to $g_{V,H}$):[5] *Let H, V and $g = g_{H,V}$ be as in Definition 3.3. If g is 0.499-close to a linear function, then this linear function has image $\mathcal{F}^k$.*

The constant 0.499 can be replaced by any quantity that is smaller than $1 - |\mathcal{F}|^{-1} \geq 1/2$.

Proof: We consider a partition of $\mathcal{F}^\ell$ into $|\mathcal{F}|^{\ell-k}$ equivalence classes such that x and y are in the same class if $x-y$ is spanned by the rows of V; that is, x resides in the class $C_x \stackrel{\rm def}{=} \{x + cV : c \in \mathcal{F}^k\}$ and $C_x = C_{x+c'V}$ for every $c' \in \mathcal{F}^k$. A class is considered good if it contains a single element of H, which happens if and only if $g(x) \in \mathcal{F}^k$. The key observation is that if C_x is good (equiv., $g(x) \in \mathcal{F}^k$), then, for every $c \in \mathcal{F}^k$, it holds that $g(x+cV) = g(x) - c$, since $C_x \cap H = \{x + g(x)V\} = \{(x+cV) + (g(x)-c)V\}$.

Now, let $f : \mathcal{F}^\ell \to \mathcal{F}^k$ be an arbitrary linear function that has an image that is partial to $\mathcal{F}^k$, and note that this image has size at most $|\mathcal{F}|^{k-1}$ (since the image of f must be a linear subspace). Noting that $g(x) = f(x)$ implies $g(x) \in \mathcal{F}^k$, we get

$$\begin{aligned}
\mathbf{Pr}_{x\in\mathcal{F}^\ell}[g(x)=f(x)] &= \mathbf{Pr}_{x\in\mathcal{F}^\ell,c\in\mathcal{F}^k}[g(x)\in\mathcal{F}^k \ \&\ g(x+cV)=f(x+cV)] \\
&\leq \max_{x\in\mathcal{F}^\ell:g(x)\in\mathcal{F}^k} \{\mathbf{Pr}_{c\in\mathcal{F}^k}[g(x+cV)=f(x+cV)]\} \\
&\leq |\mathcal{F}|^{-1},
\end{aligned}$$

since for any such $x \in \mathcal{F}^\ell$ (i.e., $g(x) \in \mathcal{F}^k$) and uniformly distributed $c \in \mathcal{F}^k$ it holds that $g(x+cV) = g(x) - c$ is uniformly distributed over $\mathcal{F}^k$, whereas the image of f contains at most $|\mathcal{F}|^{k-1}$ elements. It follows that g is at distance at least $1 - |\mathcal{F}|^{-1} \geq 1/2$ from any linear function that has image that is partial to $\mathcal{F}^k$. ■

---

[4] First note that there is no need to check the cases in which $c_i = 0$. As for the other cases, by linearity of H, it holds that $\sum_{j\in[i]} c_j v^{(j)} \in H$ if and only if $\sum_{j\in[i]} (c_j/c_i) v^{(j)} \in H$.

[5] This observation was missed in prior versions of this work, leading to unnecessary checks in the original testers.

3.2.2 The Actual Reduction

Combining the above three claims, the desired reduction follows (as detailed next). Note that this reduction has two-sided error, and that the resulting tester has query complexity $O(|\mathcal{F}|^k/\epsilon)$ (rather than $O(1/\epsilon)$, all in case $\epsilon < 4 \cdot |\mathcal{F}|^{-k}$).[6]

Algorithm 3.7 (testing whether H is an $(\ell - k)$-dimensional linear space): *On input a proximity parameter $\epsilon \in (0, 0.5)$ and oracle access to $h : \mathcal{F}^\ell \to \{0,1\}$, specifying $H = h^{-1}(1)$, proceed as follows.*

1. Find an adequate matrix V: *Using $O(|\mathcal{F}|^k)$ queries to h, try to find a k-by-ℓ full-rank matrix V such that for any non-zero $c \in \mathcal{F}^k$ it holds that $cV \notin H$. If such a matrix V is found, then proceed to the next step. Otherwise, reject.*
2. Test whether the function $g = g_{H,V}$ is linear: *Invoke a linearity test with proximity parameter ϵ, while providing it with oracle access to the function $g = g_{H,V}$. When the tester queries g at x, query h on $x+cV$ for all $c \in \mathcal{F}^k$, and answer accordingly; that is, the answer is c if c is the unique vector satisfying $h(x + cV) = 1$, otherwise* (i.e., $g(x) = \perp$) *the execution is suspended and the algorithm rejects. If the execution of the linearity tester is not suspended, then output its verdict.*

Recalling that linearity testing (with proximity parameter ϵ) has complexity $O(1/\epsilon)$, the complexity of the foregoing algorithm is $O(|\mathcal{F}|^k) + O(1/\epsilon) \cdot |\mathcal{F}|^k$, where the two terms correspond to the two steps.

Proposition 3.8 (analysis of Algorithm 3.7): *For any $\epsilon \in (0, 0.5)$, the following holds.*

1. *Algorithm 3.7 accepts every function h that describes an $(\ell - k)$-dimensional linear space with high probability.*
2. *Algorithm 3.7 rejects every function h that is ϵ-far from describing an $(\ell - k)$-dimensional linear space with high probability.*

Hence, Algorithm 3.7 constitutes a tester for $(\ell - k)$-dimensional linear spaces over $\mathcal{F}$ with query complexity $O(|\mathcal{F}|^k/\epsilon)$.

Proof: Suppose that H is an $(\ell - k)$-dimensional linear space. Then, by Claim 3.5, with high probability, a suitable matrix V will be found (in Step 1) and Step 2 will accept, since (by Claim 3.4) the function $g_{H,V}$ is linear. This establishes Part 1.

Turning to Part 2, if h is ϵ-far from describing an $(\ell - k)$-dimensional linear space, then either no suitable matrix V is found in Step 1, and the algorithm rejects, or such V is found. In the latter case, we consider the corresponding function $g_{H,V}$. We shall prove that $g_{H,V}$ is ϵ-far from the set of linear functions, and it follows that Step 2 rejects with high probability.

[6] Needless to say, we would welcome a one-sided error reduction. Recall that the case $\epsilon \geq 4 \cdot |\mathcal{F}|^{-k}$ can be handled by density estimation. A complexity improvement for the main case (of $\epsilon < 4 \cdot |\mathcal{F}|^{-k}$) appears in Sect. 3.3.

Using Claim 3.6, it suffices to show that $g = g_{H,V}$ is ϵ-far from being a linear function with image $\mathcal{F}^k$, since (for $\epsilon < 0.5$) if g is ϵ-close to a linear function then this linear function has image $\mathcal{F}^k$. Now, suppose towards the contradiction that g is ϵ-close to a linear function g' with image $\mathcal{F}^k$. Then, $H' = \{x \in \mathcal{F}^\ell : g'(x) = 0^k\}$ is an $(\ell - k)$-dimensional linear space, since $x, x' \in H'$ implies $x + x' \in H'$ (because $g'(x + x') = g'(x) + g'(x') = 0^k + 0^k = 0^k$), and $|H'| = |\mathcal{F}|^\ell / |\mathcal{F}|^k$ (because each image of g' has the same number of preimages). Next, letting $h' : \mathcal{F}^\ell \to \{0,1\}$ describe H' (i.e., $h'(x) = 1$ iff $x \in H'$), we show that $g'(x) = g(x)$ implies $h'(x) = h(x)$. This is because $g'(x) = g(x) = 0^k$ implies $h'(x) = 1$ and $h(x) = 1$ (since $g^{-1}(0^k) \subseteq h^{-1}(1)$), whereas $g'(x) = g(x) \notin \{0^\ell, \perp\}$ implies $h'(x) = 0$ and $h(x) \neq 1$ (since $h(x) = 1$ implies $g(x) \in \{0^\ell, \perp\}$). It follows that h is ϵ-close to h', which contradicts our hypothesis that h is ϵ-far from describing an $(\ell - k)$-dimensional linear space. ■

Remark 3.9 (extension to affine spaces): *In light of Claim 3.2 there is no real need to extend Algorithm 3.7 to the affine case, but let us outline such an extension nevertheless.*

- *The definition of g will be as in the linear case* (i.e., $g(x) = c$ iff $x + cV \in H$), *except that it will be based on a full-rank k-by-ℓ matrix V such that for some $u \in H$ and every non-zero $c \in \mathcal{F}^k$ it holds that $u + cV \notin H$.* (Indeed, finding such a $u \in H$ is moved from Claim 3.2 to the revised algorithm.)
- *Claim 3.4 is extended to show that H is an $(\ell - k)$-dimensional affine space if and only if g is a affine function with image $\mathcal{F}^k$.*
- *Claim 3.6 is extended to show that if g is 0.499-close to an affine function, then this affine function has image $\mathcal{F}^k$.*

Recall that testing the affinity of $g : \mathcal{F}^\ell \to \mathcal{F}^k$ can be reduced to testing the linearity of the mapping $x \mapsto g(x) - g(0^\ell)$. Alternatively, one can just use the natural extension of the linearity test of [3] that selects $x, y, z \in \mathcal{F}^\ell$ uniformly at random and checks that $g(x + y) - g(y) = g(x + z) - g(z)$.

3.3 The Second Reduction

In Sect. 3.2, for $h : \mathcal{F}^\ell \to \{0,1\}$, we reduced ϵ-testing whether $h^{-1}(1)$ is an $(\ell - k)$-dimensional linear subspace to ϵ-testing the linearity of a function $g : \mathcal{F}^\ell \to \mathcal{F}^k \cup \{\perp\}$, where the value of g at any point can be computed by making $|\mathcal{F}|^k$ queries to h. (Indeed, in order to define the function g, the reduction made $O(|\mathcal{F}|^k)$ additional queries to h.) This yields an ϵ-tester of time complexity $O(|\mathcal{F}|^k / \epsilon)$ for testing $(\ell - k)$-dimensional linear subspaces. In this section we improve this time bound.

Our starting point is the fact that, for every $\epsilon < 1/4$, if g is ϵ-close to being a linear function, then it is ϵ-close to a unique linear function g', which can be computed by self-correction of g (where each invocation of the self-corrector makes two queries to g and is correct with probability at least $1 - 2\epsilon$). Furthermore, if g' has image $\mathcal{F}^k$, then the corresponding Boolean function h' (i.e., $h'(x) = 1$ iff

$g'(x) = 0^k$) describes an $(\ell - k)$-dimensional linear space, whereas if h describes an $(\ell - k)$-dimensional linear space then $g' = g$ and $h' = h$. This suggests a two-step algorithm in which we first invoke Algorithm 3.7 with constant proximity parameter ϵ_0 (e.g., $\epsilon_0 = 0.1$ will do), and then test equality between h and h'.

The key observation is that if h is ϵ-far from describing an $(\ell - k)$-dimensional linear space, then, with high probability, either the first step rejects or it yields a function $g = g_{H,V}$ that is ϵ_0-close to a linear function g' with image $\mathcal{F}^k$. In the latter case, the corresponding h', which describes an $(\ell - k)$-dimensional linear space (i.e., $\{x \in \mathcal{F}^\ell : h'(x) = 1\} = \{x \in \mathcal{F}^\ell : g'(x) = 0^k\}$), must be ϵ-far from h (by the foregoing hypothesis). (On the other hand, if h describes an $(\ell - k)$-dimensional linear space, then $h' = h$.)

High Level Structure of the New Algorithm. Using the foregoing observation, we spell out the resulting algorithm, while leaving a crucial detail to later.

Step I: Invoke Algorithm 3.7 with proximity parameter set to ϵ_0, where $\epsilon_0 \in (0, 0.5)$ is a constant (e.g., $\epsilon_0 = 0.1$). If the said invocation rejects, then reject. Otherwise, let V be the matrix found in Step 1 of that invocation, and let $g = g_{H,V}$ be the corresponding function.

Let g' denote the *linear function closest to* g, and note that g is ϵ_0-close to g' (or else Algorithm 3.7 would have rejected with high probability). Furthermore, by Claim 3.6, the image of g' equals $\mathcal{F}^k$. Defining $h' : \mathcal{F}^\ell \to \{0,1\}$ such that $h'(x) = 1$ if and only if $g'(x) = 0^k$, it follows that h' describes an $(\ell - k)$-dimensional linear subspace. Hence, *ifh is ϵ-far from describing an $(\ell - k)$-dimensional linear subspace, then h is ϵ-far from h'.*

Step II (overview)**:** Test whether h equals h' by using a sample of $O(1/\epsilon)$ points. For each sample point, the value of h is obtained by querying h, whereas the value of h' on all sample points is obtained by obtaining the values of g' on these points (since $h'(x) = 1$ iff $g'(x) = 0^k$), where the values of g' on these points are computed via self-correction of g.

The problem is that each query to g is implemented by making $|\mathcal{F}|^k$ queries to h. Hence a straightforward implementation of the foregoing procedure will result in making $O(|\mathcal{F}|^k/\epsilon)$ queries to h, which is no better than Algorithm 3.7. Instead, we shall use a sample of $O(1/\epsilon)$ *pairwise-independent* points such that their g'-values can be determined by the value of g' at $O(\log(1/\epsilon))$ points, which in turn are computed by self-correction of g that uses $|\mathcal{F}|^k$ queries to h per each point. The details are given in Algorithm 3.10.

Note that if h describes an $(\ell - k)$-dimensional linear subspace, then $g = g'$, and the current step accepts if reached (which happens with high probability). On the other hand, if h is ϵ-far from this property and the current step is reached (which implies that g, g' and h' are well-defined), then h is ϵ-far from h', and a sample of $O(1/\epsilon)$ pairwise-independent points will contain a point of disagreement (w.h.p.).

The key observation here is that Step II can be implemented in complexity $\widetilde{O}(1/\epsilon)$ by taking a sample of $m = O(1/\epsilon)$ *pairwise independent points* in $\mathcal{F}^\ell$ such that evaluating g' on these m points only requires time $O(m + |\mathcal{F}|^k \cdot \widetilde{O}(\log m))$ rather than $O(|\mathcal{F}|^k \cdot m)$ time. This is done as follows.[7]

The Pairwise Independent Sample Points. For $t' = \lceil \log_{|\mathcal{F}|}(m+1) \rceil$, select uniformly $s^{(1)}, ..., s^{(t')} \in \mathcal{F}^\ell$, compute each $g'(s^{(j)})$ via self-correcting g, with error probability $o(1/t')$, and use the sample points $r^{(L)} = L(s^{(1)}, ..., s^{(t')})$ for m non-zero linear function $L : \mathcal{F}^{t'} \to \mathcal{F}$. The key observations are that (1) the $r^{(L)}$'s are pairwise independent, and (2) the values of g' at all $r^{(L)}$'s can be determined based on the values of g' on the $s^{(j)}$'s. This determination is based on the fact that $g'(r^{(L)}) = L(g'(s^{(1)}), ..., g'(s^{(t')}))$, by linearity of g'. Hence, the values of g' on t' random points (i.e., the $s^{(j)}$'s) determines the value of g' on $m \leq |\mathcal{F}|^{t'} - 1$ pairwise independent points (i.e., the $r^{(L)}$'s). This yields the following:

Algorithm 3.10 (implementing Step II): *For* $m = O(1/\epsilon)$ *and* $t' = \lceil \log_{|\mathcal{F}|}(m+1) \rceil$*, set* $t'' = O(\log_2 t')$*, and proceed as follows.*

1. *Select uniformly* $s^{(1)}, ..., s^{(t')} \in \mathcal{F}^\ell$.
2. *For each* $j \in [t']$*, compute the value of* $g'(s^{(j)})$ *by using self-correction on* g*, which in turn queries* h *on* $|\mathcal{F}|^k$ *points per each query to* g*. The self-correction procedure is invoked* t'' *times so that the correct value is obtained with probability* $1 - o(1/t')$.
   Specifically, select uniformly $w^{(1)}, ..., w^{(t'')} \in \mathcal{F}^\ell$*, and set* $\sigma^{(j)}$ *to equal the majority vote of* $g(s^{(j)} + w^{(1)}) - g(w^{(1)}), ..., g(s^{(j)} + w^{(t'')}) - g(w^{(t'')})$*, where the values of* g *at each point* x *is determined according to the value of* h *at the points* $\{x + cV : c \in \mathcal{F}^k\}$.
   Recall that $g(x) = c$ *if* c *is the unique point in* $\mathcal{F}^k$ *such that* $h(x + cV) = 1$*, and is set to* $\perp$ *otherwise. If the value of* g *at any point is set to* $\perp$*, then we can abort this algorithm and reject* h*. Alternatively, we can set* $\sigma^{(j)}$ *to* $\perp$*, and define the* $z + \perp = \perp$ *for every* $z \in \mathcal{F}^k$.
   (Indeed, $\sigma^{(j)}$ is our sound guess for $g'(s^{(j)})$, and this guess is correct with probability $1 - \exp(-\Omega(t'')) = 1 - o(1/t')$).
3. *For each of* m *non-zero linear function* $L : \mathcal{F}^{t'} \to \mathcal{F}$*, let* $r^{(L)} = L(s^{(1)}, ..., s^{(t')})$ *and check whether* $h(r^{(L)})$ *equals our guess for* $h'(r^{(L)})$*, where the later value is set to 1 if and only if* $L(\sigma^{(1)}, ..., \sigma^{(t')}) = 0^k$*. Accept if and only if all checks were successful* (i.e., equality holds in all).
   (Recall that $g'(r^{(L)}) = g'(L(s^{(1)}, ..., s^{(t')})) = L(g'(s^{(1)}), ..., g'(s^{(t')}))$. Hence, $L(\sigma^{(1)}, ..., \sigma^{(t')})$ is our sound guess for $g'(r^{(L)})$, and this guess is correct if all guesses for the $g'(s^{(j)})$'s are correct, which happens with probability $1 - o(1)$).

The time complexity of Algorithm 3.10 is $O(t' \cdot t'' \cdot |\mathcal{F}|^k + m) = \widetilde{O}(\log_{|\mathcal{F}|}(1/\epsilon)) \cdot |\mathcal{F}|^k + O(1/\epsilon)$. Hence, the time complexity of Step II dominates the time complexity of Step I, which is $O(|\mathcal{F}|^k/\epsilon_0) = O(|\mathcal{F}|^k)$. Assuming $\epsilon = O(|\mathcal{F}|^{-k})$, the resulting algorithm has time complexity $\widetilde{O}(\log(1/\epsilon)) \cdot \epsilon^{-1} = \widetilde{O}(1/\epsilon)$. Recall that for $\epsilon > 4 \cdot |\mathcal{F}|^{-k}$ there is an almost trivial tester of complexity $O(1/\epsilon)$, and note

[7] This procedure is inspired by [8] (as presented in [6, Sec. 7.1.3] for $\mathcal{F} = \mathrm{GF}(2)$).

that for $\epsilon < |\mathcal{F}|^{-k-1.01\cdot\log_{|\mathcal{F}|} k}$ it holds that $\widetilde{O}(\log_{|\mathcal{F}|}(1/\epsilon))\cdot|\mathcal{F}|^k = O(1/\epsilon)$. Hence, our complexity bound is (slightly) inferior to the optimal bound of $O(1/\epsilon)$ only for a narrow range of parameters (i.e., for $\epsilon \in [k^{-1.01}\cdot|\mathcal{F}|^{-k}, 4\cdot|\mathcal{F}|^{-k}]$).

Theorem 3.11 (analysis of the foregoing algorithm): *Consider an algorithm that invokes the algorithm captured by the foregoing Steps I and II, where Step II is implemented as detailed in Algorithm 3.10. Then, the resulting algorithm constitutes a tester for $(\ell - k)$-dimensional linear subspaces.*

Note that this tester has two-sided error probability. Obtaining an analogous one-sided error tester is left as an open problem. It is indeed possible that such a tester does not exist.

Proof: If h describes an $(\ell - k)$-dimensional linear subspace, then (w.h.p.) the execution reaches Step II, which always accepts, since in this case $h' = h$. On the other hand, if h is ϵ-far from describing an $(\ell - k)$-dimensional linear subspace, then we consider two cases.

1. If h is ϵ_0-far from describing an $(\ell - k)$-dimensional linear subspace, then Step I rejects (w.h.p).
2. Otherwise, assuming that Step II is reached, we consider the corresponding functions $g = g_{H,V}$ and g'. Recall g' is a linear function that is ϵ_0-close to g, since otherwise Step I rejects with high probability (due to its linearity test), and that the image of g' equals $\mathcal{F}^k$ (by Claim 3.6). Hence, h must be ϵ-far from h', since in this case h' describes an $(\ell-k)$-dimensional linear subspace. In this case, with probability at least $1 - o(1)$, the tester obtains the correct values of g' at all $s^{(j)}$'s and hence determined correctly the values of g' at all the $r^{(L)}$'s. Since these $r^{(L)}$ are uniformly distributed in $\mathcal{F}^\ell$ in a pairwise independent manner, with high probability (i.e., w.p. at least $1 - \frac{m\epsilon}{(m\epsilon)^2}$), the sample contains a point on which h and h' disagree, and Step II rejects.

In conclusion, if h is ϵ-far from the tested property, then the foregoing algorithm rejects with high probability. The theorem follows. ■

Remark 3.12 (extension to affine spaces): *Again, there is no real need to extend the foregoing to the affine case, but we outline such an extension nevertheless. We first note that self-correction of g that is close to an affine g' requires querying g at three random locations rather than at two; specifically, to obtain $g'(s)$, we select uniformly $r, r' \in \mathcal{F}^\ell$ and query g at $s + r + r', r, r'$, while relying on $g'(s) = g'(s + r - r') - g'(r) + g'(r')$. Likewise, the equality $g'(r^{(L)}) = L(g'(s^{(1)}), ..., g'(s^{(t')})$ is replaced by $g'(r^{(L)}) = L(g'(s^{(1)}), ..., g'(s^{(t')}) - L(v, ..., v) + v$, where $v = g'(0^\ell)$ is also obtained by self-correction of g, which calls for making $3\cdot|\mathcal{F}|^k$ additional queries.*

4 On Testing Monotone Monomials

As stated in the introduction, the problem that motivated our study is trying to reduce testing monomials to testing dictatorships. Such a reduction is preferable to an extension of the ideas that underly the tests of (monotone) dictatorship towards testing the set of functions that are (monotone) k-monomials, for any $k \geq 1$. In this section, we first review the said extension, as performed by Parnas, Ron, and Samorodnitsky [9], and then we present our alternative. We start with the definition of the relevant properties.

Definition 4.1 (monomial and monotone monomial): *A Boolean function* $f : \{0,1\}^\ell \to \{0,1\}$ *is called a* k-monomial *if for some* k*-subset* $I \subseteq [\ell]$ *and* $\sigma = \sigma_1 \cdots \sigma_\ell \in \{0,1\}^\ell$ *it holds* $f(x) = \wedge_{i \in I}(x_i \oplus \sigma_i)$. *It is called a* monotone k-monomial *if* $\sigma = 0^\ell$.

Indeed, the definitions of (regular and monotone) 1-monomials coincide with the notions of (regular and monotone) dictatorships. We focus on the task of testing monotone k-monomials, while recalling that the task of testing k-monomials is reducible to it (see [9] or [7, Sec. 5.2.2.1]). We also recall that this testing problem is of interest only when the proximity parameter, denoted ϵ, is small (in relation to 2^{-k}). In contrast, when $\epsilon > 2^{-k+2}$, we may just estimate the density of $f^{-1}(1)$ and accept if and only if the estimate is below $\epsilon/2$.

4.1 The Tester of Parnas, Ron, and Samorodnitsky

We start by interpreting the dictatorship tester of [1,9] in a way that facilitates its generalization. Recall that these works perform a dictatorship test by first testing that the function is linear and then performing a "conjunction check" (i.e., checking that $f(x \wedge y) = f(x) \wedge f(y)$). Now, if f is a monotone dictatorship, then $f^{-1}(1)$ is an $(\ell-1)$-dimensional affine subspace (of the ℓ-dimensional space $\{0,1\}^\ell$), where $\{0,1\}$ is associated with the two-element field GF(2). Specifically, if $f(x) = x_i$, then this subspace is $\{x \in \{0,1\}^\ell : x_i = 1\}$. In this case, the *linearity tester* could be thought of as testing that $f^{-1}(1)$ is an arbitrary $(\ell-1)$-dimensional affine subspace, whereas the "conjunction check" verifies that this subspace is an affine translation by 1^ℓ of a linear space that is spanned by $\ell - 1$ unit vectors (i.e., vectors of Hamming weight 1).[8]

When generalizing the treatment for abitrary k, we observe that if f is a monotone k-monomial, then $f^{-1}(1)$ is an $(\ell - k)$-dimensional affine subspace. So the foregoing two-step procedure generalizes to first testing that $f^{-1}(1)$ is an $(\ell - k)$-dimensional affine subspace, and then testing that it is an affine subspace of the right form (i.e., it has the form $\{x \in \{0,1\}^\ell : (\forall i \in I)\ x_i = 1\}$, for some k-subset I). Following are outlines of the treatment of these two tasks in [9].

[8] That is, we requires that this subspace has the form

$$\left\{1^\ell + \sum_{j \in ([\ell] \setminus \{i\})} c_j e_j : c_1, ..., c_\ell \in \{0,1\}\right\}$$

where $e_1, ..., e_\ell \in \{0,1\}^\ell$ are the ℓ unit vectors (i.e., vectors of Hamming weight 1).

Testing Affine Subspaces. Supposed that the alleged affine subspace H is presented by a Boolean function h such that $h(x) = 1$ if and only if $x \in H$. (Indeed, in our application, $h = f$.) We wish to test that H is indeed an affine subspace. (Actually, we are interested in testing that H has a given dimension, but this extra condition can be checked easily by estimating the density of H in $\{0,1\}^\ell$, since we are willing to have complexity that is inversely proportional to the designated density (i.e., 2^{-k}).)[9]

This task is related to linearity testing and it was indeed solved in [9] using a tester and an analysis that resembles the standard linearity tester of [3]. Specifically, the tester selects uniformly $x, y \in H$ and $z \in \{0,1\}^\ell$ and checks that $h(x - y + z) = h(z)$ (i.e., that $x - y + z \in H$ if and only if $z \in H$). Indeed, we uniformly sample H by repeatedly sampling $\{0,1\}^\ell$ and checking whether the sampled element is in H.

Note that, for co-dimension $k > 1$, the function h is not affine (i.e., $h(x) = h(y) = h(z) = 0$, which means $x, y, z \notin H$, does not determine the value of $h(x - y + z)$ (i.e., whether or not $x - y + z \in H$)).[10] Still, testing whether h describes an affine subspace can be reduced to testing linearity of functions (albeit not of h but rather of a related function), providing an alternative to the tester of [9]. Presenting such a reduction is the core of this paper (see Sect. 3, which handles an arbitrary finite field $\mathcal{F}$, whereas the current application only requires $\mathcal{F} = \mathrm{GF}(2)$).

Testing that an affine subspace is a translation by 1^ℓ of a linear subspace spanned by unit vectors. Suppose that an affine subspace H' is presented by a Boolean function, denoted h', and that we wish to test that H' has the form

$$\left\{ 1^\ell + \sum_{i \in [\ell] \setminus I} c_i e_i : c_1, ..., c_\ell \in \{0,1\} \right\}$$

where $e_1, ..., e_\ell \in \{0,1\}^\ell$ are unit vectors, and $I \subseteq [\ell]$ is arbitrary. That is, we wish to test that $h'(x) = \wedge_{i \in I} x_i$.

This can be done by picking uniformly $x \in H'$ and $y \in \{0,1\}^\ell$, and checking that $h'(x \wedge y) = h'(y)$ (i.e., $x \wedge y \in H'$ if and only if $y \in H'$). Note that if H' has the form $1^\ell + L$, where L is a linear subspace spanned by the unit vectors $\{e_i : i \in [\ell] \setminus I\}$ for some I, then $h'(z) = \wedge_{i \in I} z_i$ holds for all $z \in \{0,1\}^\ell$, and $h'(x \wedge y) = h'(x) \wedge h'(y)$ holds for all $x, y \in \{0,1\}^\ell$. On the other hand, as shown in [9], if H' is an affine subspace that does not have the foregoing form, then the test fails with probability at least 2^{-k-1}.

However, as in the case of $k = 1$, we do not have access to h' but rather to a Boolean function h that is (very) close to h'. So we need to obtain the value of h' at specific points by querying h at uniformly distributed points. Specifically,

[9] Recall that if $\epsilon < 2^{-k+2}$, then $O(2^k) = O(1/\epsilon)$, and otherwise (i.e., for $\epsilon \geq 2^{-k+2}$) testing affinity of H reduces to estimating the density of $h^{-1}(1)$.

[10] Note that $x, y \notin H$ does not determine whether or not $x - y$ is in the linear space $H - H$, let alone that a negative answer does not allow to related $h(x - y + z)$ to $h(z)$. In contrast, $x, y \in H$ implies that $h(x - y + z) = h(z)$ for every $z \in \{0,1\}^\ell$.

the value of h' at z is obtained by uniformly selecting $r, s \in h^{-1}(1)$ and using the value $h(z+r-s)$. In other words, we self-correct h at any desired point z by using the value of h at a point obtained by shifting z by the difference between two random elements of $h^{-1}(1)$, while hoping that these points actually reside in the affine subspace H' (so that their difference is in the linear space $H'-H'$). This hope is likely to materialize when h is $0.01 \cdot 2^{-k}$-close to h'.

The foregoing is indeed related to the conjunction check performed as part of the dictatorship tester of [1,9], and the test and the analysis in [9] (for the case of $k > 1$) resemble the corresponding parts in [1,9] (which handle the case of $k = 1$).

In contrast, building on the main idea of Sect. 3, in Sect. 4.2, we present a simple reduction from the general case (of any $k \geq 1$) to testing that a linear function depends on at most k of its variables. Note that the later problem generalizes the problem that was solved by the conjunction check when $k = 1$.

4.2 An Alternative Tester of Monotone Monomials

As outlined in Sect. 4.1, the function $f : \{0,1\}^\ell \to \{0,1\}$ is a monotone k-monomial if and only if f describes an $(\ell-k)$-dimensional affine space that is a translation by 1^ℓ of an $(\ell-k)$-dimensional axis-parallel linear space; that is, if $f^{-1}(1)$ has the form $\{yG + 1^\ell : y \in \{0,1\}^{\ell-k}\}$, where G is a full-rank $(\ell-k)$-by-ℓ Boolean matrix that contains k all-zero columns. Hence, we may focus on testing that the function $h : \{0,1\}^\ell \to \{0,1\}$ defined by $h(x) \stackrel{\text{def}}{=} f(x+1^\ell)$ describes an $(\ell-k)$-dimensional *axis-parallel* linear space. (Indeed, the reduction of Sect. 3.1 is instantiated here by mandating $u = 1^\ell$.)

Following [9], we first test that the Boolean function h describes an $(\ell-k)$-dimensional linear space, and next test that this linear space has the right form. As detailed Sect. 3, the first task is reduced to testing the linearity of a corresponding function $g : \{0,1\}^\ell \to \{0,1\}^k \cup \{\perp\}$, and if this test passes (w.h.p.) then g must be close to a linear function $g' : \{0,1\}^\ell \to \{0,1\}^k$, which has image $\{0,1\}^k$. In this case, h is closed to the Boolean function h' that describes the $(\ell-k)$-dimensional linear space $\{x \in \{0,1\}^\ell : g'(x) = 1\}$.

The key observation is that h' describes an axis-parallel linear space (i.e., the set $\{x \in \{0,1\}^\ell : h'(x) = 1\}$ equals the linear space $\{yG : y \in \{0,1\}^{\ell-k}\}$ for some full-rank $(\ell-k)$-by-ℓ matrix G with k all-zero columns) if and only if g' depends on k variables.[11] In general:

Claim 4.2 (on axis-parallel linear spaces): *Let $g' : \{0,1\}^\ell \to \{0,1\}^k$ be a linear function with image $\{0,1\}^k$, and $H' = \{x \in \{0,1\}^\ell : g'(x) = 0^k\}$. Then, for every k-subset I, it holds that $H' = \{x \in \{0,1\}^\ell : x_I = 0^k\}$ if and only if g' depends only on the bits in locations I* (equiv., there exists an invertible linear function $T : \{0,1\}^k \to \{0,1\}^k$ such that $g'(x) = T(x_I)$).

---

[11] In some previous versions of this work it was stated that h' describes an axis-parallel linear space if and only if g' is a projection function (i.e., $g'(x) = x_I$ for some k-subset I). As shown next, this statement is wrong.

In case of $k=1$, Claim 4.2 coincides with the assertion that $H' = \{x \in \{0,1\}^\ell : x_i = 0\}$ *if and only if* g' *is the* i^{th} *dictatorship function* (i.e., $g'(x) = x_i$).

Proof: First note that if g' depends only on the bits in locations I, then $H' = \{x : g'(x) = 0^k\}$ is an axis-parallel linear space; specifically, if $g'(x) = T(x_I)$ for some linear function $T : \{0,1\}^k \to \{0,1\}^k$, which must be invertible (since the image of g' equals $\{0,1\}^k$), then $H' = \{x : T(x_I) = 0^k\} = \{x : x_I = 0^k\}$, since $T(z) = 0^k$ if and only if $z = 0^k$.

Turning to the opposite direction, suppose that $H' = \{x : x_I = 0^k\}$, and recall that $H' = \{x : g'(x) = 0^k\}$. Hence, $g'(x) = 0^k$ if and only if $x_I = 0^k$. To see that g' necessarily depends only on variables in locations I, suppose that some bit of $g'(x)$ depends on variable $j \notin I$, and derive a contradiction by considering $e^{(j)} = 0^{j-1}10^{\ell-j}$ (i.e., evidently $e_I^{(j)} = 0^k$, but $g'(e^{(j)}) \neq 0^k$ due to the bit in the value of g' that depends on location j). ■

It seems natural to refer to a linear function that depends only on k of its variables by the term k-linear; this term was used before when referring to Boolean functions (i.e., functions from $\{0,1\}^\ell$ to $\{0,1\}$). Recall that, for constant proximity parameter, testing whether a Boolean function is k-linear can be performed in time $O(k \log k)$ by using either a junta tester (e.g., Blais's [2]) or a general function isomorphism tester (cf. [4]). It seems that these testers extend also to the case of functions from $\{0,1\}^\ell$ to $\{0,1\}^m$, but much simpler testers can be applied when we are guaranteed that the tested function is linear. For sake of self-containment, we present such a tester in the appendix.

However, as in Sect. 3, we do not have access to g', but rather can obtain its values at desired points by applying self-correction to g, which is close to g'. We can afford to compute g at any desired point, since we intend to do so only for $\widetilde{O}(k)$ points. Specifically, each query x of the k-linearity tester is answered by $g(x+w) - g(w)$, where w is selected uniformly in $\{0,1\}^\ell$. To wrap-up, we obtain the following tester, where we assume (for simplicity and w.l.o.g.) that $\epsilon = O(2^{-k}) = o(1/\widetilde{O}(k))$.

Algorithm 4.3 (testing whether f is a monotone k-monomial): *On input a proximity parameter* $\epsilon \in (0, O(2^{-k})]$ *and oracle access to* $f : \{0,1\}^\ell \to \{0,1\}$, *the algorithm proceeds as follows.*

1. *Apply the tester of Sect. 3.3 to test whether* $h : \{0,1\}^\ell \to \{0,1\}$, *defined by* $h(x) \stackrel{\text{def}}{=} f(x + 1^\ell)$, *describes an* $(\ell - k)$*-dimensional linear space over* GF(2). *The said tester is invoked with proximity parameter* ϵ, *and each query* x *is answered by the value* $f(x+1^\ell)$. *If the foregoing tester rejects, then the current algorithm reject.*
2. *Find a matrix* V *as in Step 1 of Algorithm 3.7, and let* $g = g_{h^{-1}(1),V} : \{0,1\}^\ell \to \{0,1\}^k$ *denote the corresponding function.* (Alternatively, we may use the matrix V that is found by the foregoing invocation of Algorithm 3.7.) *Letting* g' *be the linear function closest to* g, *we check whether the function* g' *is* k*-linear.*

(Recall g' is computed by self-reduction of g, whereas the value of g at x is computed by querying h at the points $\{x + cV : c \in \{0,1\}^k\}$.)

The query complexity of Algorithm 4.3 equals $(\widetilde{O}(\log(1/\epsilon)) \cdot 2^k + O(1/\epsilon)) + \widetilde{O}(k) \cdot 2^k$, where the first (resp., second) term is due to Step 1 (resp., Step 2).

Proposition 4.4 (analysis of Algorithm *4.3*): *Assuming that* $\epsilon \in (0, O(2^{-k})]$, *the following holds.*

1. *Algorithm 4.3 accepts any monotone k-monomial with high probability.*
2. *Algorithm 4.3 rejects any function that is* ϵ*-far from being a monotone k-monomial with high probability.*

Hence, Algorithm 4.3 constitutes a tester for monotone k-monomials.

Proof: The small error probability in the case that f is a monotone k-monomial is due to Step 1. Furthermore, in this case h describes an $(\ell - k)$-dimensional axis-parallel linear space, and $g' = g$ depends on k variables, and so Step 2 (if reached) accepts with probability 1.

The analysis of the case of functions that are ϵ-far from being monotone k-monomials reduces to the analysis of Step 2, in which we may assume that g is ϵ-close to a linear function g' with image $\{0,1\}^k$, since otherwise Step 1 rejects (w.h.p.). By Claim 4.2, g' must depend on more than k variables, since otherwise h is ϵ-close to describing an $(\ell - k)$-dimensional axis-parallel linear space. But in this case Step 2 rejects with high probability, because the probability that any invocation of the self-corrector is wrong is at most $\widetilde{O}(k) \cdot 2\epsilon = o(1)$. ■

Comparison to [9]. Algorithm 4.3 differs from the tester of [9] in two aspects. Firstly, Step 1 uses the tester of Sect. 3.3, which is based on a reduction of testing affine spaces to testing linear functions. Specifically, we reduce testing that $f : \{0,1\}^\ell \to \{0,1\}$ describes an affine space to testing that a related $h : \{0,1\}^\ell \to \{0,1\}$ describes a linear space, which in turn is reduced to testing that a related $g : \{0,1\}^\ell \to \{0,1\}^k$ is linear. In contrast, testing affine spaces is performed in [9] by modifying the linearity tester of [3] and mimicking the known analysis of this tester.

Second, Step 2 uses the foregoing function g (of Step 1), and reduces testing that the linear space described by h has the right form to testing that g depends on k of its variables, which generalizes the condition used in the case of $k = 1$. In contrast, as can be seen in Sect. 4.1, the path taken by [9] involves a modification of the procedure and analysis used in the case of $k = 1$.

Acknowledgements. I am grateful to Roei Tell for reading prior versions of this text and pointing out numerous inaccuracies and gaps. I also wish to thank Clement Canonne and Tom Gur for their comments. Lastly, I am grateful to Eric Blais for helpful discussions regarding the problem of testing k-linearity, and for permission to include his ideas in the appendix. This research was partially supported by the Israel Science Foundation (grant No. 671/13).

Appendix: On Testing k-Linearity

For an arbitrary finite field $\mathcal{F}$ and integers $k, \ell, m \in \mathbb{N}$, we say that $f : \mathcal{F}^\ell \to \mathcal{F}^m$ is k-linear if f is a linear function that depends on at most k (out of its ℓ) variables. Assuming that the tested function f is linear, we present a simple test that always accepts k-linear functions and rejects (w.h.p.) linear functions that are not k-linear.

The tester uses the key idea of the k-junta test of Fischer *et al.* [5], which is to randomly partition the ℓ variables to $O(k^2)$ sets, hoping that (in the case of a yes-instance) each set contains at most one variable that influences the function's value, and detecting which of the sets contain such a variable, hereafter called an influential variable. We accept if and only if at most k sets are detected as containing influential variables. To check whether a set $S \subset [\ell]$ contains influential variables, we select an arbitrary (e.g., random) assignment to the variables in $\overline{S} = [\ell] \setminus S$ and two random (and independent) assignments to the variables in S, and check whether the function's value is the same under both assignments.[12]

Given that the tested function is linear, the analysis of this procedure is quite straightforward (and, indeed, much simpler than in [5]): If S contains some influential variables of f, then the function's value changes with probability at least $1 - |\mathcal{F}|^{-1}$ (e.g., it is exactly $1 - |\mathcal{F}|^{-1}$ in case $m = 1$), whereas the value does not change if S contains no influential variable. The complexity of this procedure is $\Omega(k^2)$, but we can improve it by using two levels of partitions. Specifically, we suggest the following procedure (where the stated expectations refer to the case that the function is $O(k)$-linear).[13]

1. Select a random partition of $[\ell]$ into $t = k/\log k$ parts, expecting $O(\log k)$ influential variables in each part.
2. For each part S in this t-partition, perform the following (three-step) trial for $t' = O(\log k)$ times.
   (a) Select a random partition of S to $O(\log k)^2$ sub-parts, expecting at most a single influential variable in each sub-part.
   (b) For each sub-part, check whether it contains influential variables (by selecting two random assignments as described above). Actually, we perform this check $O(\log k)$ times such that a sub-part that contains influential variables is detected to be so with probability at least $1 - o(1/k)$.
   (c) Set the vote of the current trial to equal the number of sub-parts that were detected as containing influential variables.
   (Note that if the sub-partition selected in Step (a) is good (i.e., each sub-part contains at most one influential variable) and all relevant sub-parts

[12] We can set the values of the variables in $\overline{S}$ arbitrarily since we are testing a linear function. In contrast, when testing for k-junta (as in [5]) it is essential to assign the variables in $\overline{S}$ at random, and perform this test of influence many times.

[13] In the general case, these expectations may not be satisfied, but such violation provides statistical evidence for rejection.

were detected as containing influential variables, then the current vote equals the number of influential variables in S.)

For each S, set the verdict for number of influential variables in S to equal the largest vote obtained in the t' corresponding trials, expecting the answer to be correct with probability $1 - o(1/t)$.

3. Accept if and only if the sum of the verdicts is at most k.

The foregoing test has complexity $t \cdot t' \cdot O(\log k)^3 = \widetilde{O}(k)$, but this can be improved using an adaptive procedure. Specifically, the following recursive procedure was suggested to us by Eric Blais.

1. Select a random partition of $[\ell]$ into k parts, expecting $\Omega(k)$ parts with a single variable.
2. For each part S in this k-partition, determine whether it contains exactly one variable by performing the following (two-step) trial for $t' = O(\log k)$ times.
   (a) Select a random 2-partition of S and check whether each sub-part contains influential variables. (We stress that each of these two checks is performed once.)
   (b) Set the vote of the current trial to equal the number of sub-parts that were detected as containing influential variables.

   For each S, declare this part as having a single influential variable if and only if the maximum vote for it equals 1.

   (Note that, with probability $1-o(1/k)$, the declaration regarding S is correct; that is, S is declared as having a single influential variable if and only if this is actually the case.)
3. Let U be the union of the parts declared to have a single influential variable, and k' be their number. If more than k parts were found to contain influential variables (e.g., $k' > k$) then reject; if $k' \leq k$ and no part was found to contain more than one influential variable then accept. Otherwise, recurse with the function restricted to $[\ell] \setminus U$ and $k \leftarrow k - k'$.

Observing that for k-linear functions, with high probability, the number of parts in all recursion calls is $O(k)$, it follows that this procedure has complexity $O(k \log k)$.

References

1. Bellare, M., Goldreich, O., Sudan, M.: Free bits, PCPs and non-approximability - towards tight results. SIAM J. Comput. **27**(3), 804–915 (1998). Extended abstract in 36th FOCS (1995)
2. Blais, E.: Testing juntas nearly optimally. In: 41st ACM Symposium on the Theory of Computing, pp. 151–158 (2009)
3. Blum, M., Luby, M., Rubinfeld, R.: Self-testing/correcting with applications to numerical problems. J. Comput. Syst. Sci. **47**(3), 549–595 (1993). Extended abstract in 22nd STOC (1990)
4. Chakraborty, S., Garcia-Soriano, D., Matsliah, A.: Nearly tight bounds for testing function isomorphism. In: 22nd ACM-SIAM Symposium on Discrete Algorithms, pp. 1683–1702 (2011)

5. Fischer, E., Kindler, G., Ron, D., Safra, S., Samorodnitsky, A.: Testing juntas. J. Comput. Syst. Sci. **68**(4), 753–787 (2004). Extended abstract in 44th FOCS (2002)
6. Goldreich, O.: Computational Complexity: A Conceptual Perspective. Cambridge University Press, Cambridge (2008)
7. Goldreich, O.: Introduction to Property Testing. Cambridge University Press, Cambridge (2017)
8. Goldreich, O., Levin, L.A.: A hard-core predicate for all one-way functions. In: The Proceedings of 21st ACM Symposium on the Theory of Computing, pp. 25–32 (1989)
9. Parnas, M., Ron, D., Samorodnitsky, A.: Testing basic Boolean formulae. SIAM J. Discrete Math. Algorithm **16**(1), 20–46 (2002)

Deconstructing 1-Local Expanders

Oded Goldreich

Abstract. A 1-local $2d$-regular 2^n-vertex graph is represented by d bijections over $\{0,1\}^n$ such that each bit in the output of each bijection is a function of a single bit in the input. An explicit construction of 1-local expanders was presented by Viola and Wigderson (*ECCC*, TR16-129, 2016), and the goal of the current work is to de-construct it; that is, make its underlying ideas more transparent.

Starting from a generic candidate for a 1-local expander (over $\{0,1\}^n$), we first observe that its underlying bijections consists of pairs of ("relocation") permutations over $[n]$ and offsets (which are n-bit long strings). Next, we formulate a natural problem regarding "coordinated random walks" (CRW) on the corresponding (n-vertex) "relocation" graph, and prove the following two facts:

1. Any solution to the CRW problem yields 1-local expanders.
2. Any constant-size expanding set of generators for the symmetric group (over $[n]$) yields a solution to the CRW problem.

This yields an alternative construction and different analysis than the one used by Viola and Wigderson. Furthermore, we show that solving (a relaxed version of) the CRW problem is equivalent to constructing 1-local expanders.

An early version of this work appeared as TR16-152 of *ECCC*; it was written in a rather laconic style and reflected the train of thoughts of the author. The current version was significantly revised, adding various clarifications and elaborations with the aim of better serving a wider readership. In addition, Theorem A.2 was discovered and included at the last stages of the revision.

1 Introduction

Expander graphs are families of regular graphs of fixed degree and constant "expansion" factor (equiv., logarithmic mixing time), where the family consists of graphs for varying number of vertices. Expander graphs exist in abundance (i.e., a random $O(1)$-regular graph is an expander, w.v.h.p.) and have numerous applications in the theory of computation (see, e.g., [3]). Hence, explicit constructions of expanders are of great interest to this field, and the more explicit the construction—the better.

A strong notion of explicitness refers to the computation of the neighborhoods in the expander. Specifically, given the label of a vertex v and an index i, the task is to find the ith neighbor of v. It is also desirable to have $2d$-regular expanders that can be represented by d *simple* permutations of the vertex set, where each permutation corresponds to a collection of disjoint cycles that covers the vertex-set. But, *how simple can these bijections be?*

O. Goldreich (Ed.): Computational Complexity and Property Testing, LNCS 12050, pp. 220–248, 2020.
https://doi.org/10.1007/978-3-030-43662-9_14

It turns out that these bijections can be extremely simple. Specifically, considering graphs over the vertex set $\{0,1\}^n$, Viola and Wigderson [6] showed that each bit in the output of each of the corresponding bijections may depend on a single bit in the input (i.e., the label of the vertex); that is, these bijections are 1-local. Recall that a function $f:\{0,1\}^n \to \{0,1\}^n$ is called t-local if each bit in its output depends on at most t bits in its input. The aforementioned result of Viola and Wigderson [6] asserts:

Theorem 1.1 (a construction of 1-local expanders [6]): *There exists a constant d and a set of d explicit 1-local bijections, $\{f_1, ..., f_d : \{0,1\}^n \to \{0,1\}^n\}_{n\in\mathbb{N}}$, such that the $2d$-regular 2^n-vertex graph that consists of the vertex set $\{0,1\}^n$ and the edge multi-set $\bigcup_{i\in[d]}\{\{x, f_i(x)\} : x \in \{0,1\}^n\}$ is an expander.*

Indeed, by association, we refer to a regular 2^n-vertex graph as in Theorem 1.1 by the term 1-local; that is, a $2d$-regular graph $G = (\{0,1\}^n, E)$ is called 1-local if $E = \bigcup_{i\in[d]}\{\{x, f_i(x)\} : x \in \{0,1\}^n\}$ such that each f_i is a 1-local bijection. (By saying that the foregoing bijections are explicit, we mean that they can be evaluated in poly(n)-time, where throughout the paper we think of n as varying.)

1.1 Initial Observations, Which Should Not Be Skipped

We first observe that each 1-local bijection $f_i : \{0,1\}^n \to \{0,1\}^n$ is determined by a permutation of the bit locations $\pi^{(i)} : [n] \to [n]$, called the relocation (permutation), and an offset $s^{(i)} \in \{0,1\}^n$ such that $f_i(x) = x_{\pi^{(i)}} \oplus s^{(i)}$, where $x_{\pi^{(i)}} = x_{\pi^{(i)}(1)} \cdots x_{\pi^{(i)}(n)}$; that is, $f_i(x)$ is the string obtained by relocating the bits of x according to $\pi^{(i)}$ and offsetting the result by $s^{(i)}$ (equiv., the jth bit of $f_i(x)$ equals the sum of the $\pi^{(i)}(j)^{\text{th}}$ bit of x and the jth bit of $s^{(i)}$).

Obtaining a 1-local expander requires using *both* the offsets (i.e., $s^{(i)}$'s) and the relocation permutations, because without the offsets the f_i's maintain the Hamming weight of the vertex (and so the 2^n-vertex graph is not even connected), whereas without the permutations the 2^n-vertex graph decomposes into even smaller connected components (i.e., each of size at most 2^d). On the other hand, using both offsets and relocations, it is quite easy to obtain 1-local 4-regular graphs with polylogarithmic mixing time (equiv., the rate of convergence is bounded away from 1 by the reciprocal of a polylogarithmic function in the size of the graph (see Sect. 1.3)).

Observation 1.2 (the "shuffle exchange" graph is a 1-local "weak expander"):[1] *Let $f_1(x) = \mathtt{sh}(x)$ and $f_2(x) = x \oplus 0^{n-1}1$, where $\mathtt{sh}(x_1 \cdots x_n) = (x_2 \cdots x_n x_1)$ is a*

[1] A similar result holds for the 4-regular graph that uses the bijections $f_1(x) = \mathtt{sh}(x)$ and $f_2(x) = \mathtt{sh}(x) \oplus 0^{n-1}1$. Note that, when taking an n-step random walk on the 2-regular *directed* graph in which edges are directed from each vertex x to the vertices $\mathtt{sh}(x)$ and $\mathtt{sh}(x) \oplus 0^{n-1}1$, the final vertex is uniformly distributed (regardless of the start vertex). However, there is a fundamental difference between random walks on directed graphs and random walks on the underlying undirected graphs. For further discussion, see Sect. 1.4.

cyclic shift that corresponds to the relocation permutation $\pi(i) = (i \bmod n) + 1$. *Then, the* 4*-regular* 2^n*-vertex 1-local graph that consists of the vertex set* $\{0,1\}^n$ *and the edge multi-set* $\bigcup_{i\in[2]}\{\{x, f_i(x)\} : x \in \{0,1\}^n\}$ *has second* (normalized) *eigenvalue* $1 - \Theta(1/n^2)$.

(Indeed, in this graph, x is connected to $x \oplus 0^{n-1}1$ by two parallel edges, and the other pairs of edges (i.e., $\{x, \mathtt{sh}(x)\}$ and $\{x, \mathtt{sh}^{-1}(x)\}$ for each x) may also be non-distinct.)

Proof Sketch: We claim that taking a random walk of length $t = O(t' \cdot n^2)$ on this graph yields a distribution that is $2^{-t'}$-close to uniform, whereas the rate of convergence (a.k.a. the second eigenvalue) is closely related to the distance from uniformity (see Sect. 1.3). Specifically, it follows that the convergence rate is at least $2^{-(t'-n)/t}$, which equals $2^{1/O(n^2)} = 1 - \Omega(n^{-2})$ for sufficiently large t' (e.g., $t' > 2n$).

The foregoing claim is proved by observing that during a t-step random walk, with probability at least $1 - 2^{-t'}$, each position in the original string appeared at the rightmost position at some time during the walk (and that at the next step the corresponding value is randomized, since at that step f_2 is applied with probability one half).[2] ■

The Relocation Graph. The foregoing argument refers implicitly to a random walk on the n-vertex cycle, which represents the shift relocation permutation $i \mapsto (i \bmod n) + 1$ used in the 1-local 2^n-vertex graph (considered in Observation 1.2). In general, we shall consider the n-vertex graph that corresponds to the relocation permutations of the 1-local 2^n-vertex graph that we wish to analyze. Hence, we shall be discussing two graphs: The 2^n-vertex graph with transitions that are 1-local, and an n-vertex graph that describes the relocation permutations used in the 1-local graph.

Definition 1.3 (a generic 1-local graph and the corresponding relocation graph): *Let* $\pi^{(1)}, ..., \pi^{(d)} : [n] \to [n]$ *be permutations and* $s^{(1)}, ..., s^{(d)} \in \{0,1\}^n$.

1. *The* 1-local graph associated with $\pi^{(1)}, ..., \pi^{(d)}$ and $s^{(1)}, ..., s^{(d)}$ *is the 2d-regular* 2^n*-vertex graph that consists of the vertex set* $\{0,1\}^n$ *and the edge multi-set*

$$\bigcup_{i\in[d]} \left\{ \{x, x_{\pi^{(i)}} \oplus s^{(i)}\} : x \in \{0,1\}^n \right\} \tag{1}$$

where $x_\pi = x_{\pi(1)} \cdots x_{\pi(n)}$.

---

[2] After i steps, the jth bit in the original string (which is originally located at position j) is located at position $(j - 1 + \sum_{k\in[i]} X_k \bmod n) + 1$, where the X_k's are the $\{0, \pm1\}$-indicators of the chosen transitions (i.e., $X_k = 1$ (resp. $X_k = -1$) if the transition $\mathtt{sh}$ (resp., $\mathtt{sh}^{-1}$) was taken in the kth step and $X_k = 0$ otherwise (i.e., if the offset $0^{n-1}1$ was applied)). Note that each block of $t/t' = O(n^2)$ random variables has absolute value of at least $3n$ with probability at least $1/2$. Hence, looking at t' partial sums that correspond to t' such disjoint blocks, we observe that the probability that all these partial sums are in the interval $[-n, n]$ is at most $2^{-t'}$. Finally, note that if any of these partials sums has value outside $[-n, n]$, then in the corresponding $O(n^2)$ steps each original bit position appeared in the rightmost location.

2. *The* relocation graph *associated with* $\pi^{(1)}, ..., \pi^{(d)}$ *is the* $2d$*-regular* n*-vertex graph that consists of the vertex set* $[n]$ *and the edge multi-set* $\bigcup_{i\in[d]}\{\{j, \pi^{(i)}(j)\} : j \in [n]\}$.

The mapping $x \mapsto x_{\pi^{(i)}} \oplus s^{(i)}$ (resp., $j \mapsto \pi^{(i)}(j)$) *is called a* forward transition, *whereas the reverse mapping* $y \mapsto (y \oplus s^{(i)})_{\pi^{(-i)}}$ (resp., $k \mapsto \pi^{(-i)}(k)$) *is called a* reverse transition, *where* $\pi^{(-i)}$ *denotes the inverse of* $\pi^{(i)}$.

Note that $((x_{\pi^{(i)}} \oplus s^{(i)}) \oplus s^{(i)})_{\pi^{(-i)}} = (x_{\pi^{(i)}})_{\pi^{(-i)}} = x$ and $((y \oplus s^{(i)})_{\pi^{(-i)}})_{\pi^{(i)}} \oplus s^{(i)} = (y \oplus s^{(i)}) \oplus s^{(i)} = y$.

The proof of Observation 1.2 is based on the fact that the corresponding relocation graph (i.e., the n-cycle) has cover time $O(n^2)$. Using an n-vertex expander as the relocation graph, and relying on the fact that it has cover time $\widetilde{O}(n)$, we may infer that the corresponding 1-local 2^n-vertex graph has spectral gap $1/\widetilde{O}(n)$ (see Appendix).

It turns out that obtaining a 1-local $O(1)$-regular 2^n-vertex graph with constant spectral gap (i.e., expanders), mandates using an n-vertex relocation graph that has a stronger "mixing" property than a standard expander. (Studying this property is the core of the current work.) In addition, we need to use offsets that are not of Hamming weight 1. In fact, we need to use offsets that have Hamming weight $\Omega(n)$.

Proposition 1.4 (using only light offsets can not yield an expander): *Consider a* $2d$*-regular* 2^n*-vertex graph as in Definition 1.3, and suppose that* $|s^{(i)}| = o(n)$ *for all* $i \in [d]$*. Then, this 1-local* 2^n*-vertex graph is not an expander.*

The proof of Proposition 1.4 appears in the Appendix, where it is also shown that using also offsets of Hamming weight $n - o(n)$ does not help. In view of the above, we must use at least one offset that has Hamming weight in $[\Omega(n), n - \Omega(n)]$.

1.2 Our Main Results

As further discussed in Sect. 1.3, the salient property of expander graphs is that a random walk of logarithmic length ends in a vertex that is almost uniformly distributed on the vertex set, regardless of the start vertex. Our plan is to show that a 2^n-vertex 1-local graph is an expander by showing that a random walk of length $t = \omega(n)$ ends in a vertex that is $\exp(-\Omega(t))$-close to be uniformly distributed in $\{0, 1\}^n$.

As stated above, we will show this by relying on the hypothesis that the corresponding n-vertex relocation graph has a property that is stronger than standard expansion. The stronger property that we shall use refers to "coordinated (t-step) random walks" that start at the n different vertices of the graph, where the t-step random walks are specified by t indices that determine the choices of neighbors at each step. That is, the sequence $(\sigma_1, ..., \sigma_t)$ corresponds to t-step walks that in the ith step move to the σ_i^{th} neighbor of the current vertex. Hence, by *coordinated random walks* of length t on a $2d$-regular n-vertex graph, we mean selecting uniformly one sequence $(\sigma_1, ..., \sigma_t) \in [2d]^t$, and considering the n corresponding walks (such that the jth walk starts at vertex j).

A standard property of a *single random walk* refers to the number of times that the walk hits a set of constant density. In a standard n-vertex expander, the fraction of hits in a sufficiently long random walk (i.e., one of $\Omega(\log n)$ length) closely approximates the density of the set (with probability that is exponential in the length of the walk). The property that we shall consider for *coordinated random walks* of length $t \geq (1 + \Omega(1)) \cdot n$ is that the matrix describing the hitting pattern of the n coordinated walks has full rank with probability $1 - \exp(-\Omega(t - n))$. That is, for a fixed set T (of constant density), we consider a random Boolean t-by-n matrix such that the $(i, j)^{\text{th}}$ entry is 1 if and only if the jth walk hits the set T in the ith step. This property (of coordinated random walks) is the pivot of our results. Specifically, we prove the following two facts:

Graphs satisfying the coordinated random walks property yield 1-local expanders (see Theorem 2.3): *If the n-vertex relocation graph satisfies the foregoing property* (of coordinated random graphs), *then the corresponding 1-local 2^n-vertex graph coupled with adequate offsets is an expander.* (Actually, given a $2d$-regular n-vertex relocation graph, we consider a $8d$-regular 2^n-vertex 1-local graph, where each relocation permutation is coupled with four offsets (and the offsets are easily computed based on the permutation).)

Obtaining graphs that satisfy the coordinated random walks property (see Theorem 3.1): *Any constant-size expanding set of generators for the symmetric group* (over $[n]$) *yields an n-vertex* (relocation) *graph that satisfies the foregoing property* (of coordinated random graphs).

Lastly, a result of Kassabov [4], which was also used in [6], asserts that the symmetric group has an explicit generating set that is expanding and of constant size.

Combining these three results, we obtain an alternative proof of Theorem 1.1. In addition, we show that constructing an n-vertex graph that satisfies (a relaxed version of) the foregoing property is equivalent to constructing 1-local 2^n-vertex expanders.

Organization of the Rest of this Paper. Sections 1.3 and 1.4 provide necessary background and a useful clarification towards the rest of the paper. The main results (i.e., Theorems 2.3 and 3.1) are proved in Sects. 2 and 3, respectively. Specifically, in Sect. 2 we present the coordinated random walks (CRW) property and show that satisfying it yields 1-local expanders, and in Sect. 3 we show that any constant size set of expanding generators for the symmetric group over $[n]$ yields an n-vertex $O(1)$-regular graph that satisfies the CRW property. In Sect. 4 we prove the aforementioned equivalence (between a relaxed version of the CRW property and constructing 1-local expanders), and in Sect. 5 we generalize the main results to non-binary alphabets.

1.3 The Algebraic Definition of Expansion and Convergence Rate

The combinatorial definition of expansion, which refers to the relative size of neighborhoods of sets of vertices (e.g., the number of vertices that neighbor the set but are not in it as a function of the size of the set), has a strong intuitive appeal. The same holds with respect to the algebraic definition, which refers to the (normalized) second eigenvalue of the corresponding adjacency matrix, provided that one realizes that the (normalized) second eigenvalue represents *the rate at which a random walk on a regular graph converges to the uniform distribution.* Indeed, in this work we shall use the term convergence rate when referring to this eigenvalue. In an expander the convergence rate is a constant smaller than 1, whereas in a general (regular and non-bipartite) N-vertex graph the convergence rate is upper-bounded by $1 - \frac{1}{\text{poly}(N)}$.

When trying to estimate the convergence rate of a regular graph, it is useful to consider a sufficiently long random walk and relate the convergence rate to the distance of the distribution of its end-vertex from the uniform distribution over the vertex-set. Specifically, consider an N-vertex regular graph, and let λ denote its convergence rate and $\Delta_t^{(p)}$ denote the distance (in norm L_p) of the uniform distribution from the distribution of the final vertex in a t-step random walk that starts at the worst possible vertex. Then, the following two facts relate λ^t and $\Delta_t^{(p)}$ (up-to a poly(N)-factor slackness):[3]

1. *The distance is upper-bounded in terms of the convergence rate*: $\Delta_t^{(1)} \leq \sqrt{N} \cdot \Delta_t^{(2)} \leq \sqrt{N} \cdot \lambda^t$.
2. *The distance is lower-bounded in terms of the convergence rate*: $N^{-1} \cdot \lambda^t \leq \Delta_t^{(2)} \leq \Delta_t^{(1)}$.

Hence, for sufficiently large t (i.e., $t \gg \log N$), it holds that $\lambda \approx (\Delta_t^{(1)})^{1/t}$.

We shall use $\lambda \leq (N \cdot \Delta_t^{(1)})^{1/t}$ quite extensively. In fact, we have already used it in the proof of Observation 1.2, where, using $N = 2^n$ and $t = O(t' \cdot n^2)$, we showed that $\Delta_t^{(1)} \leq 2^{-t'}$ and (using $t' \geq 2n$) inferred that $\lambda \leq (2^n \cdot 2^{-t'}) = 2^{1/O(n^2)} = 1 - \Omega(n^{-2})$. In the following sections, for sufficiently large t, we shall show that $\Delta_t^{(1)} = \exp(-\Omega(t))$, and infer that $\lambda = \exp(-\Omega(1)) < 1$ (i.e., the convergence rate is upper-bounded by a constant smaller than 1).

1.4 A Technical Source of Complication

As commented in Footnote 1, the proof of Observation 1.2 is less simple than it could have been because we have to account for both forward and reverse

[3] The first inequality (i.e., $\Delta_t^{(2)} \leq \lambda^t$) is well-known and extensively used. It captures the fact that the corresponding linear operator shrinks each vector that is orthogonal to the uniform one. The second inequality (i.e., $\Delta_t^{(2)} \geq \lambda^t/N$) is far less popular. It can be proved by considering a random walk that starts in a probability distribution that is described by the vector $u + v_2$, where $u = (1/N, ..., 1/N)$ is the uniform distribution and v_2 is a vector in the direction of the second eigenvector such that no coordinate in v_2 has value lower than $-1/N$.

transitions. Unfortunately, the same phenomenon occurs in the proofs of our main results. In other words, it would have been simpler to analyze a random walk on the directed graph that corresponds to forward transitions only (and, ditto, of course, for the directed graph of reverse transitions). We may refer to such directed walks in the warm-ups, but we cannot reduce the analysis to them. Such a reduction would require a de-composition result that is not true in general but may be true in some special cases (and in particular in those that are of interest to us here).

Open Problem 1.5 (de-composing random walks on regular graphs): *For a function $f : (0,1) \rightarrow (0,1)$, we say that a class of $2d$-regular graphs defined by d bijections on their vertex-sets is f-*good *if for every graph in the class the following holds: If the convergence rate of the directed graph containing only forward transitions is at most λ, then the convergence rate of the undirected graph is at most $f(\lambda)$. We ask:*

1. *Is the class of 1-local graphs f-good for some f?*
2. *Which natural classes of graphs are f-good for some f?*

Note that Item 1 can only hold with $f(\lambda) \geq 1 - O(1-\lambda)^2$. This is the case because for the 1-local 2^n-vertex graph of Observation 1.2 the convergence rate of the directed graph is $1 - \Theta(1/n)$, whereas the convergence rate of the graph itself is $1 - \Theta(1/n^2)$.[4]

2 A Sufficient Condition: The Coordinated Random Walks Property

As stated upfront, a 1-local $2d$-regular 2^n-vertex graph is associated with d relocation permutations and d offsets, which means that constructing 1-local expander graphs reduces to constructing suitable relocation graphs and offsets. In this section we identify a property of the relocation graph (equiv., of the d relocating permutations) that suffices for showing that the corresponding 1-local graph is an expander. In Sect. 2.1 we present the basic intuition, while referring to a random walk on the directed graph of forward transitions only. The actual analysis is presented in Sect. 2.2, and it is intended to be understood also when skipping Sect. 2.1.

2.1 The Intuition

The general problem we face is of finding relocation graphs and offsets that yield 1-local expanders. For sake of simplicity, we consider the case of using

[4] On the one hand, an $3n$-step random walk on the directed graph yields an almost uniformly distributed vertex, since (w.v.h.p.) such a walk uses the forward shift at least n times. On the other hand, an $o(n^2)$-step random walk on the graph itself that starts at the vertex 0^n is unlikely to reach a vertex that has Hamming weight at least $n/3$.

a single non-zero offset $s \in \{0,1\}^n$, since it turns out that this suffices. As a warm-up towards the actual problem, we consider a generic 1-local ($4d$-regular) graph with d relocation permutations, $\pi^{(1)}, ..., \pi^{(d)} : [n] \to [n]$ and $2d$ forward transitions such that the $2i-1^{\text{st}}$ (resp., the $2i^{\text{th}}$) transition is $x \mapsto x_{\pi^{(i)}} \oplus s$ (resp., $x \mapsto x_{\pi^{(i)}}$). Indeed, we use each relocation permutation both with the offset s and without it; that is, the set of corresponding bijections is $\{f_{(\sigma,b)} : \{0,1\}^n \to \{0,1\}^n\}_{(\sigma,b)\in[d]\times\{0,1\}}$ such that $f_{(\sigma,1)}(x) = x_{\pi^{(\sigma)}} \oplus s$ and $f_{2\sigma}(x) = x_{\pi^{(\sigma)}}$.

We shall consider a t-step random walk (on the 1-local graph) that uses only the forward transitions (and starts at the vertex 0^n). Such a walk is specified by a sequence of pairs, denoted $((\sigma_1, b_1), ..., (\sigma_t, b_t)) \in ([d] \times \{0,1\})^t$, such that at the ith step the $(2\sigma_i - b_i)^{\text{th}}$ forward transition is used (so that the result is moving from the current vertex v to the vertex $v_{\pi^{(\sigma_i)}} \oplus s^{b_i}$, where $s^0 = 0^n$ and $s^1 = s$). Hence, in the ith step, the label of the current vertex is permuted (according to $\pi^{(\sigma_i)}$) and then offset by s if $b_i = 1$ (and is only permuted otherwise).

Actually, it is instructive to say that, in the ith step, the offset $s^{(i)} \stackrel{\text{def}}{=} s_{\pi^{(\sigma_i)}\circ\cdots\circ\pi^{(\sigma_1)}}$ is added to the initial label (i.e., 0^n) if and only if $b_i = 1$. Hence, the label of the final vertex in the walk equals $\sum_{i\in[t]:b_i=1} s^{(i)}$ permuted according to $\pi^{(\sigma_t)} \circ \cdots \circ \pi^{(\sigma_1)}$. Note that, for any fixed sequence $(\sigma_1, ..., \sigma_t) \in [d]^t$, we get a distribution that depends only on the random b_i's, since the $s^{(i)}$'s depend only on the sequence $\sigma_1, ..., \sigma_t$ (and on s). The punch-line is that *this distribution* (i.e., $\sum_{i\in[t]:b_i=1} s^{(i)}$ for random b_i's) *is uniform over* $\{0,1\}^n$ *if and only if the* t*-sequence* $(s^{(1)}, ..., s^{(t)})$ *spans* $\{0,1\}^n$.

Focusing on the question of whether or not the t-sequence $(s^{(1)}, ..., s^{(t)})$ spans $\{0,1\}^n$, we observe that this question refers to a property of the corresponding walk on the relocation graph, since the $s^{(i)}$'s are determined by the sequence $\sigma_1, ..., \sigma_t$ (and s). Specifically, for each $j \in [n]$, we consider a walk that starts at vertex j and proceeds according to the sequence $(\sigma_1, ..., \sigma_t) \in [d]^t$. After i steps, this (j^{th}) walk is at vertex $\pi^{(\sigma_i)} \circ \cdots \circ \pi^{(\sigma_1)}(j)$. Now, the key observation is that the jth bit of $s^{(i)} = s_{\pi^{(\sigma_i)}\circ\cdots\circ\pi^{(\sigma_1)}}$ is 1 if and only if this walk hits the set of vertices $T = \{k : s_k = 1\}$ (i.e., iff $\pi^{(\sigma_i)} \circ \cdots \circ \pi^{(\sigma_1)}(j) \in T$). Hence, $(s^{(1)}, ..., s^{(t)})$ spans $\{0,1\}^n$ if and only if the t-by-n Boolean matrix that describes the pattern of hitting T is full rank.

That is, for every sequence $\overline{\sigma} = (\sigma_1, ..., \sigma_t)$, we consider a t-by-n Boolean matrix $B = B^{(\overline{\sigma})}$ such that the $(i,j)^{\text{th}}$ entry in this matrix is 1 if and only if the walk that starts at j and proceeds according to $\overline{\sigma}$ hits T in the ith step (i.e., if and only if $\pi^{(\sigma_i)} \circ \cdots \circ \pi^{(\sigma_1)}(j) \in T$). The foregoing observation is that, for $s^{(i)} = s_{\pi^{(\sigma_i)}\circ\cdots\circ\pi^{(\sigma_1)}}$, it holds that $(s^{(1)}, ..., s^{(t)})$ spans $\{0,1\}^n$ if and only if $B^{(\overline{\sigma})}$ is full rank.

To summarize, we are looking at coordinated random walks on the relocation graph. These walks start at different vertices $j \in [n]$ but proceed according to a single random sequence $\overline{\sigma} = (\sigma_1, ..., \sigma_t) \in [d]^t$. For a fixed set T, we are interested in the event that the corresponding matrix (i.e., $B^{(\overline{\sigma})}$) has full rank (i.e., its rows span $\{0,1\}^n$). Whenever this happens, the corresponding random walk on the 1-local graph, which is further randomized by the choice of the b_i's, is uniformly distributed over $\{0,1\}^n$. The property that we wish to hold is that, with probability at least $1 - \exp(-\Omega(t))$ over the choice of $\overline{\sigma}$, the matrix $B^{(\overline{\sigma})}$ has full rank.

2.2 The Actual Analysis

It turns out that it suffices to use a single non-zero offset $s \in \{0,1\}^n$ (of Hamming weight approximately $n/2$), along with the offsets that are derived from it when considering also the reverse transitions. That is, for each relocation permutation $\pi : [n] \to [n]$, we consider the four transitions $x \mapsto (x \oplus s^b)_\pi \oplus s^c$, where $b, c \in \{0,1\}$ and $s^0 = 0^n$ (and $s^1 = s$). (Note that such a generic transition can be viewed as $x \mapsto x_\pi \oplus (s_\pi)^b \oplus s^c$, and that the reverse transition has the form $y \mapsto (y \oplus s^c)_{\pi^{-1}} \oplus s^b = y_{\pi^{-1}} \oplus (s_{\pi^{-1}})^c \oplus s^b$.)[5] In other words, referring to Definition 1.3 and assuming that d is a multiple of 4, we postulate that for some $s \in \{0,1\}^n \setminus \{0^n\}$ and every $i \in [d/4]$ and $b, c \in \{0,1\}$ it holds that $\pi^{(4i-2b-c)} = \pi^{(4i)}$ and $s^{(4i-2b-c)} = (s_{\pi^{(4i)}})^b \oplus s^c$.

Note that in this case, for every i, taking at random one of the four corresponding (forward) transitions has the effect of randomizing the vertex label by the offset s (by virtue of the random value of $c \in \{0,1\}$), and the same holds when taking the reverse transition (by virtue of the random value of $b \in \{0,1\}$). When taking a random walk on this graph, we consider only the randomizing effect of this offset (i.e., of the choice of c in a forward move, and the choice of b in a reverse move).[6]

To clarify the above and motivate the following property, suppose that we take $t = \Omega(n)$ random steps on the 1-local graph, and consider the t-by-n Boolean matrix describing the activity status of the location to which each of the initial positions is moved during these t steps, where an initial position is said to be currently active if it currently reside in a location in $\{k : s_k = 1\}$. That is, fixing the choice of relocation permutations (but leaving the choice of the b's and c's undermined), the $(i,j)^{\text{th}}$ entry in the matrix indicate whether or not, in the ith step of the fixed random walk being considered, the jth initial location is mapped to an active location (i.e., a 1-entry in the offset s). Later, when considering the effect of using random b's and c's, this will have the effect of flipping all active locations (together) with probability $1/2$.

---

[5] In contrast, if we were to use only the transitions $x \mapsto x_\pi \oplus s^c$, then the reverse transitions would have had the form $y \mapsto (y \oplus s^c)_{\pi^{-1}} = y_{\pi^{-1}} \oplus (s_{\pi^{-1}})^c$, which would have hindered the argument that follows (i.e., the proof of Theorem 2.3); see also the following paragraph. Of course, the issue would not have arose if we were analyzing random walks on the directed graph of forward transitions only (see Sects. 1.4 and 2.1).

[6] If we are currently at vertex x and take the forward transition associated with (π, b, c), then we move to vertex $x_\pi \oplus (s_\pi)^b \oplus s^c$, and the foregoing randomization effect refers to the addition of the offset s (to $(x \oplus s^b)_\pi$), which occurs if and only if $c = 1$. Likewise, if we are currently at vertex y and take the reverse transition associated with (π, b, c), then we move to vertex $(y \oplus s^c)_{\pi^{-1}} \oplus s^b$, and the foregoing randomization effect refers to the addition of the offset s (to $(y \oplus s^c)_{\pi^{-1}}$), which occurs if and only if $b = 1$.

Note that the foregoing matrix, which is defined based on a fixed random walk on the 1-local 2^n-vertex graph, describes n coordinated walks on the n-vertex relocation graph, each starting at a different vertex of the graph and all proceeding according to the same sequence of (random) choices. (Note that each step on the n-vertex relocation graph, which has degree $2d/4$, only determines $i \in [d/4]$ and the direction of the transition (i.e., forward or backward), while leaving the choice of the corresponding bits b and c unspecified.)

The punch-line is that, *if the foregoing t-by-n matrix has full rank, then the t random choices of whether to apply the offset s* (which are governed by the random choice of the corresponding bits b and c) *correspond to a random linear combination of the t rows of the matrix, which yields a uniformly distributed n-bit long string.* In this case, the corresponding random walk on the 2^n-vertex graph yields a uniform distribution (since the resulting n-bit string is added to the initial vertex in the walk). That is, fixing a random walk on the n-vertex relocation graph, we observe that if the matrix that corresponds to this walk has full rank, then the final vertex in the corresponding random walk on the 1-local 2^n-vertex graph is uniformly distributed in $\{0,1\}^n$, since it is (essentially) a random linear combination of the rows of the matrix (where the randomization is due to the choices of the corresponding b's and c's).

The CRW Property. We are finally ready to define the coordinated random walks (CRW) property. The following definition is actually more general than needed; we are interested in the special case that the g_σ's are the permutations and their inverses (in an d-regular relocation (undirected) graph).[7]

Definition 2.1 (a property of coordinated random walks):[8] *For $d = O(1)$, consider a d-regular n-vertex directed graph such that for every $\sigma \in [d]$ the function $g_\sigma : [n] \to [n]$ that maps each vertex to its σth neighbor is a bijection.*

- *For an integer t, consider a random sequence $\overline{\sigma} = (\sigma_1, ..., \sigma_t) \in [d]^t$ and the n corresponding* coordinate random walks (CRW) *such that the jth walk starts at vertex j and moves in the ith step to the σ_i^{th} neighbor of the current vertex.*
- *For a set $T \subseteq [n]$, consider a t-by-n Boolean matrix $B^{(\overline{\sigma})} = B_T^{(\overline{\sigma})}$ such that its $(i,j)^{\text{th}}$ entry indicates whether the jth walk passed through T in its ith step; that is, the $(i,j)^{\text{th}}$ is 1 if and only if $g_{\sigma_i}(\cdots(g_{\sigma_1}(j))\cdots) \in T$.*

[7] Indeed, Definition 2.1 is stated in more general terms that fit an arbitrary directed graph that is described in terms of d directed cycle covers; that is, each g_σ describes a collection of directed cycles that cover all the graph's vertices, and the formulation refers to random walks in the direction of the edges. The special case we are interested in refers to the case that $g_{2\sigma'}$ is the inverse of $g_{2\sigma'-1}$; in this case, the directed graph consists of anti-parallel edges that correspond to the forward and reverse transitions, and a random walk may take forward and reverse transitions (by picking either $g_{2\sigma'}$ or $g_{2\sigma'-1}$).

[8] An alternative way of defining the matrix $B^{(\overline{\sigma})}$ proceeds by considering a sequence of permutations over $[n]$, denoted $\pi_0, \pi_1, ..., \pi_t$, such that π_0 is the identity permutation, and $\pi_i(j) = g_{\sigma_i}(\pi_{i-1}(j))$. The ith row of $B^{(\overline{\sigma})}$ is then defined as the T-indicator of π_i; that is, the $(i,j)^{\text{th}}$ entry in the matrix is 1 if and only if $\pi_i(j) \in T$.

The desired CRW property *is that, for some* $T \subseteq [n]$ *and* $t \in \mathbb{N}$, *with probability at least* $1 - 2^{-n-\Omega(t)}$ *over the choice of* $\overline{\sigma} \in [d]^t$, *the corresponding matrix* $B_T^{(\overline{\sigma})}$ *has full rank* (over GF(2)).[9] *In this case we say that the* CRW property holds w.r.t T.

Obviously, the CRW property mandates $t \geq n$. Furthermore, the proof of Proposition 1.4 actually shows that the CRW property mandates that the set T must have size in $[\Omega(n), n - \Omega(n)]$. We are definitely not concerned of these restrictions.

Intuitively, the CRW property postulates that, with extremely high probability, the coordinated random walks are "linearly independent" with respect to hitting the set T. The allowed failure probability is $\exp(-\Omega(t))$, which is extremely low given that the probability space is of size $\exp(O(t))$.

Standard Expanders May Not Satisfy the CRW Property. We note that using an arbitrary expander graph and an arbitrary set T of size $\approx n/2$ *will not do*: Indeed, in this case, each column in the matrix corresponding to a random walk has approximately $t/2$ 1-entries (w.v.h.p.), but this matrix may not have full rank. For example, consider an n-vertex expander that consists of two $n/2$-vertex expanders that are connected by a matching, and let T be the set of vertices in one of these two expanders. Then, w.r.t this T, coordinated walks on this graph always yields a Boolean matrix of rank at most two, since all the (coordinated) walks that start at vertices in T (resp., in $[n] \setminus T$) always move together to T or to $[n] \setminus T$. (Nevertheless, it may be that for every n-vertex expander, there exists a set T such that the CRW property holds (w.r.t it).)

So Which Graphs Satisfy the CRW Property? Indeed, one may wonder whether the CRW property can be satisfied at all. As stated in the introduction, combining Theorem 3.1 with Kassabov's result [4], implies that such graphs exists and can even be efficiently constructed. Still, the general question remains open—

Open Problem 2.2 (the CRW problem): *For which graphs and which sets* T*'s does the CRW property* (as in Definition 2.1) *hold?*

A partial answer to this question is postponed to Sect. 3, and is revisited in Sect. 4. But before proceeding there, we establish the usefulness of the CRW property for constructing 1-local expanders.

The CRW Property Implies 1-Local Expanders. As outlined above, any $2d$-regular relocation graph that satisfies the CRW property yields an $8d$-regular 1-local 2^n-vertex expander. Let us formally state and prove this claim.

---

[9] The failure bound is set to $\tau = 2^{-n-\Omega(t)}$ in order to facilitate deriving an upper bound on the convergence rate of the corresponding 1-local graph. Specifically, we shall use $(2^n \cdot \tau)^{1/t} < 1$. An alternative formulation that will support this application is to require error probability at most $\exp(-\Omega(t))$ for some $t = \omega(n)$ (or error probability at most 2^{-ct} for some constant $c > 0$ and some $t \geq \frac{1+c}{c} \cdot n$).

Theorem 2.3 (graphs satisfying the CRW property yield 1-local expanders): *Let $\pi^{(1)}, ..., \pi^{(d)} : [n] \to [n]$ be d permutations and $s \in \{0,1\}^n$. If the $2d$-regular n-vertex graph with the edge multi-set $\bigcup_{i\in[d]}\{\{j, \pi^{(i)}(j)\} : j \in [n]\}$ satisfies the CRW property with respect to the set $\{j \in [n] : s_j = 1\}$, then the $8d$-regular 2^n-vertex 1-local graph with the edge multi-set*

$$\bigcup_{i\in[d],b,c\in\{0,1\}} \{\{x, (x \oplus s^b)_{\pi^{(i)}} \oplus s^c\} : x \in \{0,1\}^n\} \tag{2}$$

is an expander.

Indeed, the $8d$-regular 1-local graph given by Eq. (2) is not the 1-local graph that correspond to the foregoing $2d$-regular relocation graph (per Definition 1.3), but it does correspond to the $8d$-regular relocation graph with the edge multi-set $\bigcup_{i\in[d],b,c\in\{0,1\}}\{\{j, \pi^{(i,b,c)}(j)\} : j \in [n]\}$, where $\pi^{(i,b,c)} = \pi^{(i)}$ for every $i \in [d]$ and $b, c \in \{0,1\}$. Needless to say, the difference between these two relocation graphs is immaterial.

Proof: Recall that a t-step random walk on the $2d$-regular relocation graph is specified by a sequence $(\sigma_1, ..., \sigma_t) \in [2d]^t$, whereas a random walk on the $8d$-regular 1-local graph is specified by a sequence $((\sigma_1, b_1, c_1), ..., (\sigma_t, b_t, c_t)) \in ([2d] \times \{0,1\}^2)^t$. By the hypothesis, for some $t = \Omega(n)$, the t-by-n matrix that corresponds to a random walk on the n-vertex relocation graph has full rank with probability at least $1 - 2^{-n-\Omega(t)}$. Fixing an arbitrary walk $\overline{\sigma} = (\sigma_1, ..., \sigma_t) \in [2d]^t$ on this n-vertex graph such that $B^{(\overline{\sigma})} = B^{(\overline{\sigma})}_{\{j\in[n]:s_j=1\}}$ has full rank, for each $i \in [t]$, we consider the residual random choices of $b_i, c_i \in \{0,1\}$ for the ith step of the corresponding random walk on the 2^n-vertex graph. Specifically, we consider a random process that selects these bits uniformly, in two stages.

- In the first stage, for every $i \in [t]$, if the ith transition is in the forward direction, we select b_i at random, otherwise we select c_i at random.
- In the second stage, we make the remaining choices; that is, for every $i \in [t]$, if the ith transition is in the forward direction, we select c_i at random, otherwise we select b_i at random.

Fixing any sequence of choices for the first stage, the label of the final vertex (in the corresponding random walk on the 2^n-vertex graph) is a random variable that depends only on the random choices made in the second stage. The key observation is that these random choices have the effect of randomizing the vertex-label by adding to it a corresponding random linear combination of the rows of the matrix $B^{(\overline{\sigma})}$. Specifically, row i is taken to this linear combination if and only if the relevant c_i or b_i equals 1 (where for a forward direction c_i determines whether the current label is offset by s, and for the reverse direction this choice is determined by b_i).

Detailed (alas tedious) Analysis: Denoting the initial vertex in the walk on the 1-local graph by v_0, the ith vertex in the walk, denoted v_i, satisfies $v_i = (v_{i-1} \oplus s^{b_i})_\pi \oplus s^{c_i}$ (resp., $v_i = (v_{i-1} \oplus s^{c_i})_{\pi^{-1}} \oplus s^{b_i}$) if σ_i indicates a forward (resp., reverse) transition according to π. Denoting by π_i the relocation permutation applied in the ith step of the walk (i.e., $\pi_i = \pi$ (resp., $\pi_i = \pi^{-1}$) if σ_i indicates a forward (resp., reverse) transition according to π), note that

$$v_i = (v_{i-1})_{\pi_i} \oplus (s_{\pi_i})^{x_i} \oplus s^{y_i},$$

where $(x_i, y_i) = (b_i, c_i)$ if the ith step takes a forward transition and $(x_i, y_i) = (c_i, b_i)$ otherwise. Note that the x_i's were fixed in the first stage, whereas the π_i's were fixed at the onset. In contrast, the y_i's are selected at random in the second stage. In both cases (i.e., regardless if $(x_i, y_i) = (b_i, c_i)$ or $(x_i, y_i) = (c_i, b_i)$), the ith row in the matrix, denoted r_i, equals $s_{(\pi_i \circ \cdots \circ \pi_1)^{-1}}$, where $\pi_i \circ \cdots \circ \pi_1$ is the composition of the relocation permutations applied in the i first steps. Hence,

$$\begin{aligned}
&(v_i)_{(\pi_i \circ \cdots \circ \pi_1)^{-1}} \\
&\quad = (v_{i-1})_{(\pi_{i-1} \circ \cdots \circ \pi_1)^{-1}} \oplus (s_{(\pi_{i-1} \circ \cdots \circ \pi_1)^{-1}})^{x_i} \oplus (s_{(\pi_i \circ \cdots \circ \pi_1)^{-1}})^{y_i} \\
&\quad = (v_{i-1})_{(\pi_{i-1} \circ \cdots \circ \pi_1)^{-1}} \oplus (s_{(\pi_{i-1} \circ \cdots \circ \pi_1)^{-1}})^{x_i} \oplus r_i^{y_i}.
\end{aligned}$$

It follows that $(v_t)_{(\pi_t \circ \cdots \circ \pi_1)^{-1}} = v_0 \oplus w \oplus \bigoplus_{i \in [t]} r_i^{y_i}$, where w denotes $\bigoplus_{i \in [t]} (s_{(\pi_{i-1} \circ \cdots \circ \pi_1)^{-1}})^{x_i}$. This means that the label of the vertex reached by this random walk is the sum of an already-fixed value (i.e., $v_0 \oplus w$) and the random linear combination of the rows of the matrix (i.e., $\bigoplus_{i \in [t]} r_i^{y_i}$, where the y_i's are uniformly distributed).

Hence, in this case (i.e., when $B^{(\overline{\sigma})}$ has full rank), the corresponding random walk on the 2^n-vertex graph yields a uniform distribution (regardless of the start vertex). Using the hypothesis that $B^{(\overline{\sigma})}$ has full rank with probability it least $1 - 2^{-n-\Omega(t)}$, it follows that the distribution of the label of the final vertex in a t-step random walk on the 2^n-vertex graph is $2^{-n-\Omega(t)}$-close to the uniform distribution over $\{0,1\}^n$, which implies that the convergence rate of the 2^n-vertex graph is bounded away from 1 (i.e., it is at most $(2^n \cdot 2^{-n-\Omega(t)})^{1/t} = 2^{-\Omega(1)}$), which means that this 1-local graph is an expander. ■

3 Constructions that Satisfy the CRW Property

For the benefit of the reader, we distinguish between the main result of this section (presented in Sect. 3.1) and two secondary comments that follow its presentation (in Sect. 3.2).

3.1 The Main Result of This Section

Recall that Kassabov's result [4], which was also used in [6], asserts that the symmetric group has an explicit generating set that is expanding and of constant size.[10] We shall show that using this set of permutations (i.e., as our set of relocating permutations) along with the set $T = [n']$ such that $n' \approx n/2$ is odd (e.g., odd $n' \in \{\lfloor n/2 \rfloor, \lfloor n/2 \rfloor + 1\}$) yields an n-vertex graph that satisfies the coordinated random walks property (of Definition 2.1). Combined with Theorem 2.3, this yields an alternative proof of Theorem 1.1. In fact, our result, presented next, is more general (i.e., it refers to any generating set that is expanding and has constant size, and to any $T \subset [n]$ of odd size $n' \approx n/2$).

Theorem 3.1 (graphs satisfying the CRW property): *For $d = O(1)$, let $\Pi = \{\pi^{(i)} : i \in [d]\}$ be a generating set of the symmetric group of n elements and suppose that Π is expanding.*[11] *Then, the n-vertex graph that consists of the vertex set $[n]$ and the edge multi-set $\bigcup_{i \in [d]} \{\{j, \pi^{(i)}(j)\} : j \in [n]\}$ satisfies the coordinated random walks property of Definition 2.1 with respect to any set of odd size $n' \approx n/2$.*

Proof: For a sufficiently large t and any set T of size n', consider a random t-by-n Boolean matrix that corresponds to coordinated random walks (from all possible start vertices) on the n-vertex graph; that is, for any $\overline{\sigma} \in [2d]^t$, we consider the matrix $B = B_T^{(\overline{\sigma})}$. We shall show that, for every non-empty set $J \subseteq [n]$, with probability at least $1 - \exp(-\Omega(t) + O(n \log n))$, the sum of columns of B in positions J is non-zero. (This establishes CRW property for any sufficiently large $t = \Omega(n \log n)$.)[12]

Claim 3.1.1 (the distribution of a specific linear combination of the columns): *For every non-empty set $J \subseteq [n]$, with probability at least $1 - \exp(-\Omega(t) + O(n \log n))$ over the choice of a t-step random walk on the n-vertex graph, the sum* (mod 2) *of columns J in the corresponding Boolean matrix is non-zero.*

Proof: For $J = [n]$ this follows from the fact that n' is odd. Otherwise (i.e., for $J \subset [n]$), we shall prove the claim by using the correspondence between (t-step) random walks on the n-vertex graph and (t-step) random walks on the set of generators Π (of the n-element symmetric group), where in a random step on Π the current permutation is composed with the selected generator (or its inverse).[13] That is, for $\sigma \in [2d]$, selecting the σth neighbor in the random walk

[10] Indeed, this refers to a third graph, which is the corresponding Cayley graph with $n!$ vertices (i.e., the vertices are all the possible permutations over $[n]$). To reduce confusion, in the main text (unlike in footnotes), we shall not refer explicitly to this graph, but rather refer to the generating set of the symmetric group, and refer to its vertices as to states.

[11] That is, letting $\mathtt{Sym}_n$ denote the symmetric group of n elements, we consider the Cayley graph consisting of the vertex set $\mathtt{Sym}_n$ and the edge multi-set $\bigcup_{i \in [d]} \{\{\pi, \pi^{(i)} \circ \pi\} : \pi \in \mathtt{Sym}_n\}$, where $\circ$ denote composition of permutations. The hypothesis postulates that this Cayley graph is an expander.

[12] We comment that the CRW property can be established for any sufficiently large $t = \Omega(n)$; see Claim 3.1.2.

[13] That is, we use the correspondence between (coordinated) random walks on the n-vertex graph and random walks on the $n!$-vertex Cayley graph.

on the n-vertex graph, a choice that determines a transition (i.e., $\lceil\sigma/2\rceil \in [d]$) as well as the direction (i.e., forward or reverse) in which the transition is applied, corresponds to selecting the $\lceil\sigma/2\rceil^{\text{th}}$ generating permutation and moving by composing it or its inverse (according to the value of σ mod 2). Hence, the result of a t-step walk on Π is the composition of the t selected permutations, which means that we effectively consider random walks (on Π) that start at the identity permutation.

In our argument, we shall refer to a set of permutations over $[n]$, denoted $\mathtt{Sym}_n$, and consider the set of permutation, denoted W, consisting of permutations having an J-image that contains an odd number of elements of T; that is, $\pi \in W$ if and only if $|\{j \in J : \pi(j) \in T\}|$ is odd. The claim will follow by proving the following two facts:

Fact 1: $|W| \approx |\mathtt{Sym}_n|/2$.

Fact 2: For every $\overline{\sigma} \in [2d]^t$, the sum of columns J in the matrix $B_T^{(\overline{\sigma})}$ equals the all-zero vector if and only if the corresponding random walk on Π does not visit W.

We first show that W has density approximately half within the set of all $n!$ permutations over $[n]$. This can be shown by considering, w.l.o.g., the case of $|J| \leq n/2$ (or else consider $[n] \setminus J$). The easy case is when $|J| = o(n^{1/2})$. In this case, for a uniformly selected $\pi \in \mathtt{Sym}_n$, the set $\pi(J) \stackrel{\text{def}}{=} \{\pi(j) : j \in J\}$ is close to a sample of $|J|$ independent points in $[n]$, and so the probability that $|\pi(J) \cap T|$ is odd is approximately 1/2 (since $|T| = n' \approx n/2$). Turning to the case of $|J| \geq 2n^{1/3}$, we let J' be an $n^{1/3}$-subset of J and $J'' = J \setminus J'$. Using $|J| \leq n/2$, it follows that, with very high probability, $\frac{|\pi(J'')\cap T|}{|J''|} \approx \frac{|T|}{n} \approx 1/2$ holds, which implies that $\frac{|T\setminus\pi(J'')|}{|[n]\setminus\pi(J'')|} \approx 1/2$. It follows that conditioned on the values of $\pi(J'')$, the set $\pi(J')$ is close to a sample of $|J'|$ independent points in $[n] \setminus \pi(J'')$, and so the probability that $|\pi(J') \cap T|$ is odd is approximately 1/2, assuming the foregoing (extremely likely) event (of $\frac{|T\setminus\pi(J'')|}{|[n]\setminus\pi(J'')|} \approx 1/2$). This establishes Fact 1.

We now turn to the proof of Fact 2. Let $\overline{\sigma} = (\sigma_1, ..., \sigma_t) \in [2d]^t$ describe coordinated random walks on the relocation graph, and recall that, in the ith step, the jth walk hits T if and only if the $(i,j)^{\text{th}}$ entry in $B_T^{(\overline{\sigma})}$ is 1. Now, consider the sequence of permutation, denoted $\pi_1, ..., \pi_t \in \mathtt{Sym}_n$, that correspond to this coordinated random walks, and observe that it corresponds to a sequence of generators or inverses that are selected in a random walk on Π (i.e., π_i is a generator or its inverse as determined by σ_i).[14] Then, in the ith step of the coordinated walks, the jth walk is in position $\pi_i(\cdots(\pi_1(j))\cdots)$, whereas the $(i,j)^{\text{th}}$ entry in $B_T^{(\overline{\sigma})}$ is 1 if and only if $\pi_i(\cdots(\pi_1(j))\cdots) \in T$. Hence, for every $i \in [t]$, the sum of the entries in row i and columns J (of $B_T^{(\overline{\sigma})}$) equals 1 (mod 2) if and only if $\pi_i \circ \cdots \circ \pi_1 \in W$. This establishes Fact 2.

[14] That is, $\pi_i = (\pi^{(\lceil\sigma_i/2\rceil)})^{d_i}$, where $d_i = (-1)^{\sigma_i \bmod 2}$ is the direction in which the transition is applied.

Having established both facts, we now establish the claim by upper-bounding the probability that a t-step random walk on Π does not pass through states in W. By the expansion property of the generating set Π (for the symmetric group), the probability that a t-step random walk does not pass through a fixed set of constant density is at most $\exp(-\Omega(t-O(n\log n)))$, where the first $O(\log(n!)) = O(n\log n)$ steps are taken for convergence to the uniform distribution (when starting at the identity permutation) and the remaining steps are used for hitting attempts. □

Using a union bound (over all non-empty sets $J \subset [n]$), we conclude that, with probability at least $1-(2^n-2)\cdot\exp(-\Omega(t)+O(n\log n))$, the corresponding t-by-n Boolean matrix has full rank. Taking a sufficiently large $t = \Omega(n\log n)$, the theorem follows. ■

Conclusion: Indeed, as stated upfront, applying Theorem 2.3 to the n-vertex graph (and set) analyzed in Theorem 3.1 implies that any constant-size generating set for the symmetric group that is expanding yields a 1-local expander. Using Kassabov's construction of such a set [4], yields an alternative proof of Theorem 1.1.

3.2 Secondary Comments

In this section, we make two comments about Theorem 3.1. The first comment refers to its proof, and the second comment refers to its converse (i.e., we show that a converse of a weak form of Theorem 3.1 fails).

For Sake of Elegancy: Tightening the Proof of Theorem 3.1. As noted in Footnote 12, the probability bound of Claim 3.1.1 can be tightened. Specifically, the error bound of $\exp(-\Omega(t)+O(n\log n))$ can be improved to $\exp(-\Omega(t)+O(n))$, which is optimal.

Claim 3.1.2 (the distribution of linear combinations of the columns, revisited): *Let T be an odd set of size $n' \approx n/2$. Then, with probability at least $1-\exp(-\Omega(t)+O(n))$ over the choice of a t-step random walk $\overline{\sigma}$ on the n-vertex graph, for every non-empty set $J \subseteq [n]$, the sum* (mod 2) *of the columns J of the matrix $B_T^{\overline{\sigma})}$ is non-zero.*

Proof Sketch: We proceed as in the proof of Claim 3.1.1, but consider random walks (on the set Π (which generates $\mathtt{Sym}_n$)) that start at a state that is uniformly distributed in a specific set S (rather than start at the identity permutation). The set S is the set of all permutations such that each location in T holds an element of T; that is, $\pi \in S$ if and only if $\{i \in T : \pi(i)\} = T$. Using $|T| = n' \approx n - |T|$, observe that S has density approximately $\frac{(n'!)^2}{n!}$, which is approximately 2^{-n}.

The key observation is that the Boolean matrix that represents coordinated random walks on the n-vertex graph equals (up to a permutation of its columns) the matrix that represents the same walks on any isomorphic copy of that graph that leaves T invariant (i.e., rather than walking on an n-vertex graph $G =$

$([n], E)$, we walk on its isomorphic copy $\phi(G) = ([n], \{\{\phi(i), \phi(j)\} : \{i,j\} \in E\})$, where $\phi : [n] \to [n]$ is a permutation such that $\phi(j) \in T$ for every $j \in T$). That is, *if the matrix B represents coordinated random walks on the original graph and $\phi : [n] \to [n]$ is a permutation that leaves T invariant, then the matrix obtained by permuting the columns of B according to ϕ represents coordinated random walks on the isomorphic copy of the original graph obtained by relabeling its vertices according to ϕ.* (This is the case because the jth column in B indicates whether the walk on G that starts at vertex j hits T in each of the t steps, but this column also indicates whether the same walk on $\phi(G)$ that starts at $\phi(j)$ hits $\phi(T) = T$ in each of the t steps.)

Now, since B is full rank if and only if permuting its columns yields a full rank matrix, we may consider random walks on such random isomorphic copies of the original graph (i.e., copies obtained by relabeling it using a random permutation that leaves T invariant). Hence, we may analyze the matrix that corresponds to a random walk (on Π) that starts at a state that is uniformly distributed in S (rather than starting at the identity permutation). That is, the probability that the matrix B (which represents coordinated random walks on the original graph) is full rank equals the probability that a corresponding random walk on Π misses one of the W_J's (defined as in the proof of Claim 3.1.1), when starting from a uniformly distributed state in S.

Indeed, for every non-empty $J \subset [n]$, we consider the corresponding set W_J, which is the set of all permutations π such that $|\pi(J) \cap T|$ is odd. By the expansion property of the generating set for the symmetric group and the fact that S has density $\Omega(2^{-n})$, a t-step random walk that starts in uniformly distributed state in S passes via W_J with probability at least $1 - \exp(-\Omega(t - O(n)))$, where the first $O(n)$ steps are taken for convergence to the uniform distribution and the remaining steps are used for hitting W_J. Hence, with probability at least $1 - (2^n - 2) \cdot \exp(-\Omega(t - O(n)))$, a random walk that starts at a state that is uniformly distributed in S avoids none of the W_J's. In this case, for every non-empty set $J \subseteq [n]$, the sum of columns (of the corresponding matrix) in positions J is non-zero. □

The CRW Property Does Not Imply that the Set of Relocations is an Expanding Set of Generators for $\mathtt{Sym}_n$. Interpreted in terms of sets of permutations over $[n]$, the CRW property asserts that a random walk on this set passes a specific statistical test (which is specified by the corresponding set T). Theorem 3.1 asserts that if a set of permutations is expanding, then the CRW property is satisfies for *any* set T of odd size $n' \approx n/2$. This holds also if $n' \approx n/4$ (or any odd value in $[0.01n, 0.99n]$).[15] A weaker implication only asserts that if a set of permutations is expanding, then the CRW property is satisfies for *some* set T of odd size $n' \approx n/4$. Here we show that the converse

[15] In this case, for any non-empty set $J \subset [n]$, the density of the corresponding set $W = W_J \subseteq \mathtt{Sym}_n$ may reside in $[0.01, 0.99]$, which suffices for showing that this set is hit with probability $1 - \exp(-\Omega(t) + O(n))$.

of the latter implication does not hold.[16] In other words, we show that *the fact that a relocation graph satisfies the CRW property* (with respect to some set of vertices) *does not imply that the corresponding set of permutations generates the symmetric group* (let alone in an expanding manner).

Theorem 3.2 (on the converse of Theorem 3.1): *There exists a set of permutations, $\{\pi^{(i)} : i \in [3d]\}$, over $[2n]$ that does not generate the symmetric group of $2n$ elements such that the $2n$-vertex graph consisting of the vertex set $[2n]$ and the edge multi-set $\bigcup_{i\in[3d]}\{\{j, \pi^{(i)}(j)\} : j \in [2n]\}$ satisfies the coordinated random walks property* (of Definition 2.1) *with respect to some set of odd size $n' \approx n/2$.*

Proof Sketch: We start with a set of permutations $\Pi = \{\pi^{(i)} : i \in [d]\}$ that generates the symmetric group of n elements and is expanding. We first extend each $\pi^{(i)} \in \Pi$ to the domain $[2n]$ such that $\pi^{(i)}(n + j) = n + \pi^{(i)}(j)$ for every $j \in [n]$ (and $i \in [d]$). Next, we add d copies of the identity permutation and d copies of the permutation that switches $[n]$ and $[2n]\setminus[n]$; that is, for every $i \in [d]$, we have $\pi^{(d+i)}(b\cdot n+j) = b\cdot n+j$ and $\pi^{(2d+i)}(b\cdot n+j) = (1-b)\cdot n+j$ for every $j \in [n]$ and $b \in \{0, 1\}$. Denoting the resulting set of augmented permutations by Π', we consider the $2n$-vertex $6d$-regular relocation graph G' that corresponds to it. This graph consists of two copies of the $2d$-regular n-vertex graph G that corresponds to Π, augmented by d self-loops on each vertex (where each self-loop contributing two units to the vertex's degree) and $2d$ copies of a perfect matching that matches the two copies of each original vertex.

Note that Π' does not generate the symmetric group of $2n$ elements; it rather generates a group of $2\cdot(n!) \ll (2n)!$ permutations; specifically, a permutation $\pi' : [2n] \to [2n]$ is generated by Π' if and only if for some $\pi \in \mathtt{Sym}_n$ either $\pi'(b\cdot n + j) = b\cdot n + \pi(j)$ or $\pi'(b\cdot n + j) = (1 - b)\cdot n + \pi(j)$ for every $(b, j) \in \{0, 1\} \times [n]$. The theorem follows by showing that the ($2n$-vertex) relocation graph G' satisfies the CRW property (with any set $T \subset [n]$ of odd size $n' \approx n/2$).[17] Hence, we focus on proving the following claim.

Claim: *For any set $T \subset [n]$ of odd size $n' \approx n/2$, the $2n$-vertex graph G' satisfies the CRW property with respect to T.*

When analyzing t-step random walks on G', we distinguish steps in which one of the first d permutations is employed from steps in which one of the last $2d$ permutations is employed. We call the latter steps semi-idle, since they either map each vertex to itself or map each vertex to its sibling (i.e., its other copy).

[16] Indeed, we leave open the possibility that the converse of Theorem 3.1 holds. We believe that even if the CRW property is satisfies for any set T of odd size $n' \in [0.01n, 0.99n]$, then it does not necessarily hold that the foregoing set of permutations is expanding.

[17] We stress that T is an arbitrary subset of size n' of $[n]$, whereas the vertex set is $[2n]$. Indeed, picking T of size n' arbitrarily in $[2n]$ will fail; for example, if $T = T' \cup (n + T') \cup \{n\}$, for any $T' \subseteq [n-1]$, then, for every non-empty $J' \subseteq [n]$, the sum of matrix's columns with indices in $J' \cup (n + J')$ is exactly as in the case of $T = \{n\}$, since the contributions of T' and $n + T'$ cancel out (whereas, as shown in Proposition 1.4, the CRW cannot be satisfied with sets of size $o(n)$).

The key observation is that t-step random walks on G' correspond to t-step *lazy* random walks on G in which the walk stays in the current vertex (i.e., is truly idle) with probability 2/3. Indeed, semi-idle steps (on G') correspond to staying in place (on G), whereas in steps that are not semi-idle (on G') the walk moves on the two copies of G in the same manner and identically to the movement of the corresponding walk on G itself. Furthermore, fixing a lazy random walk on G leaves undetermined the type of semi-idle steps taken on the corresponding walk on G': Each of these steps can either be a truly idle step (i.e., stay in place) or a move to the sibling vertex (i.e., the corresponding vertex in the other copy of G).

Turning to the matrices that describe hitting $T \subset [n]$, note that each row in the t-by-n matrix B the describes a walk on G has exactly $|T| = n'$ ones, and the same holds for the matrix B' in walks on G'. Furthermore, each row in the t-by-$2n$ matrix B' (which corresponds to a walk on G') has either the form $0^n r$ or the form $r0^n$, where r is the corresponding row in the matrix B (which represents the hitting pattern in the corresponding walk on G). The choice between $0^n r$ and $r0^n$ is determined by the number of steps (so far) in which the matching permutation (which moves vertices to their sibling in the other copy of G) was selected.

Recall that Claim 3.1.1 asserts that, for every non-empty $J \subseteq [n]$, with probability at least $1 - \exp(-\Omega(t - O(n \log n)))$ over the walks on G, the sum of columns J in the matrix B (which corresponds to a random walk on G) is a non-zero vector. Actually, the same argument applies to a lazy random walk, by focusing on the non-idle steps. Moreover, it can be shown that this vector, denoted $v^{(J)}$, has Hamming weight $t' \stackrel{\text{def}}{=} \Omega(t)$, since expansion implies that, with extremely high probability, sets of constant density are hit with constant frequency (rather than merely hit). Furthermore, with probability at least $1 - \exp(-\Omega(t - O(n \log n)))$, at least $t'/2$ of the 1-entries in $v^{(J)}$ correspond to idle steps. Let us fix a (lazy) random walk on G that has the foregoing properties, and let r_i denote the ith row of the corresponding matrix B.

Turning to the corresponding t-by-$2n$ matrix B' that describes a random walk on G', consider an arbitrary non-empty set of columns $J \subset [2n]$, and let $J' = J \cap [n]$ and $J'' = \{j - n : j \in J \setminus J'\}$. Then, for every $i \in [t]$, if the sum of the entries in r_i and columns J' is 1, then the sum of the entries in row i and columns J of B' is 1 if the ith row of B' equals $r_i 0^n$. Similarly, if the sum of the entries in r_i and columns J'' is 1, then the sum of the entries in row i and columns J of B' is 1 if the ith row of B' equals $0^n r_i$. Recall that the latter event (i.e., where the ith row of B' equals $r_i 0^n$ or $0^n r_i$) depends only on the choices made in the semi-idle steps.

Lastly, we focus on the rows of B that correspond to idle steps and whose sum in columns J' (resp., J'') equals 1. Recalling that if $J'' \neq \emptyset$ (resp., $J'' \neq \emptyset$), then B contains at least $t'/2 = \Omega(t)$ such rows, we note that for each of these rows the sum of the entries in column J of B' is 1 with probability at least 1/2, since with probability 1/2 the ith row of B' equals $r_i 0^n$ (resp., $0^n r_i$). Hence, B'

is full rank with probability at least $1-(2^n-2)\cdot(2^{-\Omega(t-O(n\log n))}+2^{-t'/2})$, and the claim follows. ■

4 A Sufficient and Necessary Condition: The Relaxed CRW Property

We now turn back to the relation between the CRW property (of Definition 2.1) and 1-local expanders. Recall that Theorem 2.3 asserts that graphs satisfying the CRW property yield 1-local expanders. A natural question is whether this sufficient condition is necessary. Leaving this question open, we shall show that a *relaxed* CRW property (of the n-vertex relocation graph) suffices and is necessary for obtaining a 1-local 2^n-vertex expander.

The relaxed CRW property uses a generalization of Definition 2.1 in which several subsets of $[n]$ are considered (rather than one), and at each step of the coordinated random walks hitting is considered with respect to the most beneficial set. That is, given a coordinated random walk, we can select which subset we consider at each step of the walk, and the corresponding row of the matrix is determined accordingly. This freedom of choice is used when proving the "necessity" direction, whereas it can handled in the "sufficiency" direction by using a large number of offsets (i.e., exponential in the number of sets).

Definition 4.1 (a relaxed property of coordinated random walks): *For constants $d, d' \in \mathbb{N}$, consider a d-regular n-vertex graph as in Definition 2.1, and d' sets $T_1, ..., T_{d'} \subseteq [n]$.*

- *As in Definition 2.1, for $t \in \mathbb{N}$, consider a random sequence $\overline{\sigma} = (\sigma_1, ..., \sigma_t) \in [d]^t$ and the n corresponding* coordinate random walks *such that the jth walk starts at vertex j and moves in the ith step to the σ_i^{th} neighbor of the current vertex.*
- *Fixing the random sequence $\overline{\sigma}$, consider an arbitrary sequence $\overline{\tau} = (\tau_1, ..., \tau_t) \in [d']^t$, and let $B^{(\overline{\sigma},\overline{\tau})}$ be the t-by-n Boolean matrix such that its $(i, j)^{\text{th}}$ entry indicates whether the jth walk passed through T_{τ_i} in its ith step.*

The relaxed CRW property *(w.r.t $T_1, ..., T_{d'}$) asserts that, for some $t \in \mathbb{N}$, with probability at least $1-2^{-n-\Omega(t)}$ over the choice of $\overline{\sigma} \in [d]^t$, there exists $\overline{\tau} \in [d']^t$ such that the Boolean matrix $B^{(\overline{\sigma},\overline{\tau})}$ has full rank.*

(Indeed, Definition 2.1 corresponds to the special case of $d' = 1$.)

Theorem 4.2 (constructing 1-local expanders is equivalent to constructing relocation graphs that satisfy the relaxed CRW property (as in Definition 4.1)): *Let $\pi^{(1)}, ..., \pi^{(d)} : [n] \to [n]$ be permutations.*

1. The relaxed CRW property is necessary for 1-local expanders: *If the 1-local $2d$-regular 2^n-vertex graph associated with the permutations $\pi^{(1)}, ..., \pi^{(d)}$ and the offsets $s^{(1)}, ..., s^{(d)} \in \{0,1\}^n$ is an expander, then the corresponding $2d$-regular n-vertex relocation graph satisfies the relaxed CRW property with respect to the sets $T_1, ..., T_{2d}$ such that $T_{2i} = \{j \in [n] : s_j^{(i)} = 1\}$ and $T_{2i-1} = \{\pi^{(i)}(j) : s_j^{(i)} = 1\}$.*

2. The relaxed CRW property is sufficient for 1-local expanders: *Suppose that the 2d-regular n-vertex relocation graph associated with $\pi^{(1)}, ..., \pi^{(d)}$ satisfies the relaxed CRW property with respect to the sets $T_1, ..., T_{d'}$. Then, the $2^{2d'+1} \cdot d$-regular 2^n-vertex 1-local graph having the edge multi-set*

$$\bigcup_{i\in[d],\beta,\gamma\in\{0,1\}^{d'}} \left\{\{x, (x \oplus s^{(\beta)})_{\pi^{(i)}} \oplus s^{(\gamma)}\} : x \in \{0,1\}^n\right\} \tag{3}$$

is an expander, where for every $\alpha \in \{0,1\}^{d'}$ the string $s^{(\alpha)} \in \{0,1\}^n$ denotes the indicator sequence of the set $\bigoplus_{i\in[d']:\alpha_i=1} T_i \subseteq [n]$; that is, the $j^{\rm th}$ bit of $s^{(\alpha)}$ is 1 if and only if $|\{i \in [d'] : \alpha_i = 1 \,\&\, j \in T_i\}|$ is odd.

Theorem 2.3 is a special case of Part 2 (i.e., the case of $d' = 1$). Note that the edge multi-set of Eq. (3) may use $(2^{d'} - 1) \cdot d \cdot 2^{d'} + 2^{d'}$ different offsets (i.e., the offsets $s^{(\gamma)}$ and $s^{(\beta)}_{\pi^{(i)}} \oplus s^{(\gamma)}$ for $i \in [d], \gamma \in \{0,1\}^{d'}$ and $\beta \in \{0,1\}^{d'} \setminus \{0^n\}$).

Proof: We start with the proof of Part 2, which generalizes the proof of Theorem 2.3. Specifically, let $\overline{\sigma} = (\sigma_1, ..., \sigma_t) \in [2d]^t$ be a random walk on the relocation graph such that an even σ_i (resp., an odd σ_i) indicates a forward (resp., reverse) transition using $\pi^{(\lceil\sigma_i/2\rceil)}$. Then, by the hypothesis, with probability at least $1 - \exp(-n - \Omega(t))$ over the choice of $\overline{\sigma}$, there exists $\overline{\tau} = (\tau_1, ..., \tau_t)$ such that $B^{(\overline{\sigma},\overline{\tau})}$ is full rank. Recall that specifying a random walk on the 1-local graph requires specifying also the random choices of $\beta_i, \gamma_i \in \{0,1\}^{d'}$ for each step $i \in [t]$. We do so depending on σ_i mod 2 (i.e., whether σ_i is a forward or reverse transition) and on the value of $\tau_i \in [d']$. Specifically, we consider the following two-stage process of determining the sequence of auxiliary random choices of $\beta_1, ..., \beta_t \in \{0,1\}^{d'}$ and $\gamma_1, ..., \gamma_t \in \{0,1\}^{d'}$.

1. For every $i \in [t]$ such that the $i^{\rm th}$ step is a forward (resp., reverse) transition,
   (a) select β_i (resp., γ_i) uniformly in $\{0,1\}^{d'}$, and
   (b) for every $k \in [d'] \setminus \{\tau_i\}$, select the bit $\gamma_{i,k}$ (resp., $\beta_{i,k}$) uniformly in $\{0,1\}$.
2. For every $i \in [d']$ such that the $i^{\rm th}$ step is a forward (resp., reverse) transition, select γ_{i,τ_i} (resp., β_{i,τ_i}) uniformly in $\{0,1\}$.

Fixing a good $\overline{\sigma}$ and a corresponding good $\overline{\tau}$ (i.e., choices such that $B^{(\overline{\sigma},\overline{\tau})}$ is full rank), consider an arbitrary fixing of the choices in Stage 1. Then, the label of the final vertex in the corresponding random walk on the 1-local graph is a fixed string (determined by $\overline{\sigma}$ and the choices made in Stage 1) that is offset by a random linear combination of the rows of $B^{(\overline{\sigma},\overline{\tau})}$, where the random linear combination is determined in Stage 2. (Specifically, if the $i^{\rm th}$ step is a forward (resp., reverse) transition, then the $i^{\rm th}$ row is included in this offset if and only if $\gamma_{i,\tau_i} = 1$ (resp., $\beta_{i,\tau_i} = 1$).) Thus, when $B^{(\overline{\sigma},\overline{\tau})}$ has full rank, the label of the final vertex is uniformly distributed in $\{0,1\}^n$, and Part 2 follows.

Turning to the proof of Part 1, we start by considering the $4d$-regular 2^n-vertex 1-local expander obtained from the given $2d$-regular 1-local expander by augmenting each transition of the form $x \mapsto x_\pi \oplus s$ with the transition $x \mapsto x_\pi$. (The auxiliary graph is an expander because it contains an expander as a subgraph.) Hence, a step on this auxiliary graph is specified by a pair $(\sigma, b) \in [2d] \times \{0,1\}$, where σ specifies a step on the original 1-local graph and $b = 1$ indicates that the original offset is applied; that is, we shall refer to the edge multi-set $\bigcup_{i\in[d],b\in\{0,1\}}\{\{x, x_{\pi^{(i)}} \oplus (s^{(i)})^b\} : x \in \{0,1\}^n\}$. Consequently, a t-step random walk on the auxiliary $4d$-regular expander corresponds to a sequence $(\sigma_1, b_1), ..., (\sigma_t, b_t) \in ([2d] \times \{0,1\})^t$, and the sequence $\sigma_1, ..., \sigma_t$ corresponds to a walk on the n-vertex relocation graph.

Considering a coordinated random walk $\overline{\sigma} = (\sigma_1, ..., \sigma_t)$ on the relocation graph, we shall use our freedom to determine the τ_i's based on the σ_i's, and doing so we shall obtain a matrix as in Definition 4.1, which we shall show to be of full-rank (with extremely high probability). Specifically, we let $\tau_i = \sigma_i$, while assuming (again, without loss of generality) that an even σ_i (resp., an odd σ_i) indicates a forward (resp., reverse) transition using $\pi^{(\lceil\sigma_i/2\rceil)}$. This assumption is made only in order to match the $2d$ possible transitions with the $2d$ sets defined in the conclusion of Part 1. Indeed, under this assumption, if $\sigma_i = 2k$ (resp., $\sigma = 2k - 1$), then the ith step applied the forward (resp., reverse) transition $x \mapsto x_{\pi^{(k)}} \oplus s^{(k)}$ (resp., $y \mapsto (y \oplus s^{(k)})_{\pi^{(-k)}}$, where $\pi^{(-k)}$ denotes the inverse of $\pi^{(k)}$, whereas $T_{2k} = \{j \in [n] : s_j^{(k)} = 1\}$ and $T_{2k-1} = \{\pi^{(k)}(j) : s_j^{(k)} = 1\} = \{j : s_{\pi^{(-k)}(j)}^{(k)} = 1\}$. Hence, picking $\tau_i = \sigma_i$, the ith row in $B^{(\overline{\sigma},\overline{\tau})} = B^{(\overline{\sigma},\overline{\sigma})}$ indicates hitting the set T_{τ_i}.

As said above, we claim that *if a t-step random walk on the $4d$-regular 1-local graph yields a distribution that is* $\exp(-\Omega(t))$*-close to uniform* (and $t = \Omega(n)$ is large enough), *then the matrix $B^{(\overline{\sigma},\overline{\sigma})}$ must have full rank with probability at least* $1 - \exp(-n - \Omega(t))$. This claim is shown as follows.

Let η denote the probability (over the choice of $\overline{\sigma} \in [2d]^t$) that the matrix $B^{(\overline{\sigma},\overline{\sigma})}$ does not have full rank. Such a choice of $\overline{\sigma}$ determines both the permutation $\pi_{\overline{\sigma}}$ that relates the original locations to the final ones (i.e., $\pi_{\overline{\sigma}} = \pi^{((-1)^{\sigma_t}\cdot\lceil\sigma_t/2\rceil)} \circ \cdots \circ \pi^{((-1)^{\sigma_1}\cdot\lceil\sigma_1/2\rceil)}$) and a non-trivial linear combination $J_{\overline{\sigma}}$ of the columns of the matrix that witnesses the hypothesis that the matrix is not full rank. Hence, there exists a non-empty set $J \subseteq [n]$ such that, with probability $\eta' \geq \eta/(2^n - 1)$ over the choice of $\overline{\sigma}$, the sum of the columns indexed by $\pi_{\overline{\sigma}}^{-1}(J)$ (in the matrix $B^{(\overline{\sigma},\overline{\sigma})}$) equals the all-zero vector (e.g., $\pi_{\overline{\sigma}}^{-1}(J) = J_{\overline{\sigma}}$), whereas in the remaining choices (of $\overline{\sigma}$) this sum does not equals the all-zero vector.[18]

[18] We detail the argument, while highlighting a minor subtle issue: We know that for an η fraction of the $\overline{\sigma}$'s, there exists a $J_{\overline{\sigma}} \neq \emptyset$ such that the sum of the $J_{\overline{\sigma}}$ columns is the all-zero vector (and we may let $J_{\overline{\sigma}} = \emptyset$ otherwise). However, these columns corresponds to locations in the (label of the) initial vertex, whereas we want to analyze locations in the end vertex. Of course, locations $J_{\overline{\sigma}}$ in the initial vertex correspond to locations $\pi_{\overline{\sigma}}(J_{\overline{\sigma}})$ in the final vertex. Hence, there exists a non-empty J (representing locations in final label) such that the sum of the columns in $\pi_{\overline{\sigma}}^{-1}(J)$ (representing locations in initial label) equals the all-zero vector with probability $\eta' \geq \eta/(2^n - 1)$. This lower bound is due to the event $\pi_{\overline{\sigma}}^{-1}(J) = J_{\overline{\sigma}}$, but the sum of these columns may be zero also otherwise. (For this reason, we define η' as the probability that the sum of the columns in $\pi_{\overline{\sigma}}^{-1}(J)$ equals the all-zero vector rather than the probability that $\pi_{\overline{\sigma}}^{-1}(J) = J_{\overline{\sigma}}$.) Needless to say, for the rest of this probability space (of $\overline{\sigma} \in [2d]^t$), this sum is not the all-zero vector.

Looking at the label of the final vertex $v_{\overline{\sigma}}$ in a random walk $\overline{\sigma}$ on the 1-local 2^n-vertex graph that starts at the vertex 0^n, we observe that $v_{\overline{\sigma}}$ equals a random linear combination of the rows of $B^{(\overline{\sigma},\overline{\sigma})}$ permuted by $\pi_{\overline{\sigma}}$; that is, $(v_{\overline{\sigma}})_{\pi_{\overline{\sigma}}^{-1}}$ equals a random linear combination of the rows of $B^{(\overline{\sigma},\overline{\sigma})}$, where this random linear combination is determined by the sequence $(b_1, ..., b_t)$ of choices of whether or not to apply the original offset. This is the case since the ith row permuted by $\pi^{((-1)^{\sigma_i}\cdot\lceil\sigma_i/2\rceil)} \circ \cdots \circ \pi^{((-1)^{\sigma_1}\cdot\lceil\sigma_1/2\rceil)}$ is the offset that is potentially added in the ith step of the walk, whereas this offset is added if and only if $b_i = 1$.

It follows that the sum of $v_{\overline{\sigma}}$'s bits in locations J (equiv., the sum of the bits of $(v_{\overline{\sigma}})_{\pi_{\overline{\sigma}}^{-1}}$ in locations $\pi_{\overline{\sigma}}^{-1}(J)$) is zero with probability exactly $\eta' + (1-\eta') \cdot 0.5 = 0.5 + 0.5\eta'$, since this sum is 0 whenever the sum of the corresponding columns in $B^{(\overline{\sigma},\overline{\sigma})}$ is the all-zero vector (and is uniformly distributed in $\{0,1\}$ otherwise).[19] Hence, the total variation distance between the distribution of the final vertex and the uniform distribution is at least $0.5\eta'$.

Recalling that the hypothesis (i.e., that the 1-local graph is an expander) implies that $\eta' \leq \exp(-\Omega(t))$, it follows that $\eta < 2^n \cdot \eta' = \exp(-n - \Omega(t))$, for sufficiently large $t = \Omega(n)$. This establishes Part 1. ∎

Conclusion. Theorem 4.2 asserts that constructing graphs that satisfy the relaxed CRW property is equivalent to constructing 1-local expanders. One begging question is whether the relaxed CRW property is easier to achieve that the original CRW property. Lacking a positive answer, this raises the following generalization of Problem 2.2.

Open Problem 4.3 (the CRW problem, revised): *For which graphs and which sequences of sets $(T_1, ..., T_{d'})$'s does the relaxed CRW property* (as in Definition 4.1) *hold?*

An appealing conjecture of Benny Applebaum is that every n-vertex expander graph yield a positive instance of Problem 4.3; that is, there exists $d' = O(1)$ sets $T_1, ..., T_{d'} \subset [n]$ such that this n-vertex graph satisfies the relaxed CRW property (of Definition 4.1) with respect to these T_i's. Needless to say, a bolder conjecture is that the foregoing holds with $d' = 1$.

5 Generalization to Non-binary Alphabets

We generalize the main definitions to an arbitrary alphabet of prime size, which is identified with the field GF(p). A function $f : \mathrm{GF}(p)^n \to \mathrm{GF}(p)^n$ is called t-local if each symbol in its output depends on at most t symbol in its input. This yields a generalized notion of a 1-local graph.

Definition 5.1 (1-local graph, generalized): *For a fixed $d \in \mathbb{N}$ and a fixed prime p, let $\{f_1, ..., f_d : \mathrm{GF}(p)^n \to \mathrm{GF}(p)^n\}_{n\in\mathbb{N}}$ be 1-local bijections. Then, the corresponding $2d$-regular p^n-vertex* 1-local graph *consists of the vertex set* $\mathrm{GF}(p)^n$ *and the edge multi-set* $\bigcup_{i\in[d]}\{\{x, f_i(x)\} : x \in \mathrm{GF}(p)^n\}$.

[19] If the sum of these columns is not the all-zero vector, then a random combination of its entries, as determined by $(b_1, ..., b_t)$, is uniformly distributed in $\{0,1\}$.

Note that each f_i is determined by a permutation on the locations $\pi^{(i)} : [n] \to [n]$, called the relocation, and n bijections denoted $h_1^{(i)}, ..., h_n^{(i)} : \mathrm{GF}(p) \to \mathrm{GF}(p)$ such that, for every $j \in [n]$, the jth bit of $f_i(x)$ equals $h_j^{(i)}(x_{\pi^{(i)}(j)})$. Unlike in the binary case (i.e., $p = 2$), where each $h_j^{(i)}$ is affine (i.e., has the form $h_j^{(i)}(z) = z \oplus s_j^{(i)}$), these bijections are not necessarily affine functions. Still, we shall focus on the case that they are affine. Generalizing Theorems 2.3 and 3.1, we obtain.

Theorem 5.2 (a construction of generalized 1-local expanders): *For every constant prime p, there exists a set of $d = O(p^2)$ explicit 1-local bijections, $\{f_1, ..., f_d : \mathrm{GF}(p)^n \to \mathrm{GF}(p)^n\}_{n\in\mathbb{N}}$, such that the $2d$-regular p^n-vertex graph that consists of the vertex set $\mathrm{GF}(p)^n$ and the edge multi-set $\bigcup_{i\in[d]}\{\{x, f_i(x)\} : x \in \mathrm{GF}(p)^n\}$ is an expander. Furthermore, for each $i \in [d]$, there exists a permutation $\pi^{(i)} : [n] \to [n]$ and an offset $s^{(i)} \in \mathrm{GF}(p)^n$ such that $f_i(x) = x_{\pi^{(i)}} + s^{(i)}$.*

The expansion feature holds also for varying $p = p(n)$, but in that case the graph is not of constant degree.

Proof: The overall plan is to use a straightforward generalization of the CRW property for rank defined over $\mathrm{GF}(p)$, and prove adequate generalizations of Theorems 2.3 and 3.1. Specifically, we first show that any n-vertex graph that satisfies the generalized CRW property yields a 1-local p^n-vertex expander, and then show that any generating set for the symmetric group of n elements that is expanding yields an n-vertex graph that satisfies the generalized CRW property (with respect to any set of size $n' \approx n/2$ such that $n' \not\equiv 0 \pmod p$).

Definition 5.2.1 (a property of coordinated random walks, generalized): *For a d-regular n-vertex graph as in Definition 2.1, a set $T \subseteq [n]$ and $t = \Omega(n)$, consider coordinated random walks and Boolean matrices just as in Definition 2.1. The* generalized CRW property *postulates that, with probability at least $1 - p^{-n} \cdot \exp(-\Omega(t))$ over the choice of the t-step random walk, the corresponding matrix has full rank* when the arithmetics is in $\mathrm{GF}(p)$.

We stress that although these random matrices have entries in $\{0, 1\}$, we consider their rank over $\mathrm{GF}(p)$.

Claim 5.2.2 (Theorem 2.3, generalized): *Let $\pi^{(1)}, ..., \pi^{(d)} : [n] \to [n]$ be d permutations and $s = (s_1, ..., s_n) \in \{0, 1\}^n \subseteq \mathrm{GF}(p)^n$. If the $2d$-regular n-vertex graph with the edge multi-set $\bigcup_{i\in[d]}\{\{j, \pi^{(i)}(j)\} : j \in [n]\}$ satisfies the generalized CRW property* (of Definition 5.2.1) *with respect to the set $\{j \in [n] : s_j = 1\}$, then the $2p^2d$-regular p^n-vertex graph with the edge multi-set*

$$\bigcup_{i\in[d], b, c\in\mathrm{GF}(p)} \{\{x, (x - b \cdot s)_{\pi^{(i)}} + c \cdot s\} : x \in \mathrm{GF}(p)^n\}$$

is an expander, where $b \cdot (s_1, ..., s_n) = (bs_1, ..., bs_n)$.

Proof Sketch: We mimic the proof of Theorem 2.3, while noting that in the ith step the vertex's label is randomized by an offset that is a random GF(p)-multiple of the ith row in the corresponding matrix; specifically, in a forward direction the randomization is performed by the value of c (i.e., adding the offset $c \cdot s$), whereas in a reverse direction the randomization is performed by the value of b (i.e., subtracting the offset $-b \cdot s$). Hence, if the matrix has full rank over GF(p), then the label of the final vertex is uniformly distributed in GF$(p)^n$ (since it is randomized by a random linear combination of the rows of the matrix). □

Claim 5.2.3 (Theorem 3.1, generalized): *Let $\Pi = \{\pi^{(i)} : i \in [d]\}$ be a generating set of the symmetric group of n elements and suppose that Π is expanding. Then, the n-vertex graph that consists of the vertex set $[n]$ and the edge multi-set $\bigcup_{i\in[d]}\{\{j, \pi^{(i)}(j)\} : j \in [n]\}$ satisfies the generalized CRW property of Definition 5.2.1 with respect to any set T of size $n' \approx n/2$ such that $n' \not\equiv 0 \pmod p$.*

Proof Sketch: Here we mimic the proof of Theorem 3.1. Specifically, we consider all (non-zero) linear combinations $L : [n] \to \text{GF}(n)$ of the columns of a matrix that corresponds to a random walk, and upper-bound the probability that each such linear combination yields the all-zero vector. That is, fixing any set T of size n', for every such linear combination L, we consider the set W_L of permutations $\pi \in \texttt{Sym}_n$ such that $\sum_{i\in[n]:\pi(i)\in T} L(i) \not\equiv 0 \pmod p$. Once we show that each W_L has constant density, the claim follows as in the binary case by using $t = \Omega(n \log(np))$, where here we use a union bound on all (non-zero) L's. Hence, we focus on proving that for each non-zero $L : [n] \to \text{GF}(p)$, the set W_L has constant density in $\texttt{Sym}_n$.

The case of non-zero constant functions $L : [n] \to \text{GF}(p)$ is handled by the hypothesis that $n' \not\equiv 0 \pmod p$, which implies that $W_L = \texttt{Sym}_n$, and so we focus on non-constant functions L. In this case, one may show that W_L has density at least $0.5 - o(1)$, but we use a simpler argument to show that it has density at least $0.25 - o(1)$. Specifically, considering any $i_1, i_2 \in [n]$ such that $L(i_1) \neq L(i_2)$, we observe that $\Pr_\pi[|\{\pi(i_1), \pi(i_2)\} \cap T| = 1] > 0.5 - o(1)$. On the other hand, conditioned on the values of π on $I \stackrel{\text{def}}{=} [n] \setminus \{i_1, i_2\}$ and on the foregoing event (i.e., $|\{\pi(i_1), \pi(i_2)\} \cap T| = 1$), the value of $\sum_{i\in[n]:\pi(i)\in T} L(i)$ is a random variable that equals $\sum_{i\in I:\pi(i)\in T} L(i) + L(i_1)$ with probability 1/2 (when $\pi(i_1) \in T$) and equals $\sum_{i\in I:\pi(i)\in T} L(i) + L(i_2)$ otherwise. Recalling that $L(i_1) \neq L(i_2)$, we get

$$\Pr_{\pi:[n]\to[n]}\left[\sum_{i\in[n]:\pi(i)\in T} L(i) \not\equiv 0 \pmod p\right] > (0.5 - o(1)) \cdot \frac{1}{2}$$

and the claim follows. □

Combining Claims 5.2.3 and 5.2.2, we get.

Corollary 5.2.4 (obtaining generalized 1-local expanders): *Let $\Pi = \{\pi^{(i)} : i \in [d]\}$ be a generating set of the symmetric group of n elements and suppose that Π is expanding. Then, for any $n' \approx n/2$ such that $n' \not\equiv 0 \pmod p$, the $2p^2d$-regular p^n-vertex graph with the edge multi-set*

$$\bigcup_{i\in[d],b,c\in \mathrm{GF}(p)} \left\{\{x, (x - b^{n'}0^{n-n'})_{\pi^{(i)}} + c^{n'}0^{n-n'}\} : x \in \mathrm{GF}(p)^n\right\}$$

is an expander.

Using Kassabov's result [4] (which asserts that the symmetric group has an explicit generating set that is expanding and of constant size), the theorem follows. ∎

Comment: The foregoing generalizes to any finite field; that is, p may be a prime power. For $p = q^e$, where q is prime, we select $n' \approx n/2$ such that $n' \not\equiv 0 \pmod q$, and proceed as above (while noting that in the proof of Claim 5.2.3 the reductions mod p actually refer to doing arithmetics in $\mathrm{GF}(p)$).

Acknowledgments. I wish to thank Benny Applebaum for helpful discussions and for permission to include his conjecture in this paper. I am also grateful to Roei Tell for numerous commenting on several drafts of this paper, which have significantly improved the presentation.

Appendix: Secondary Observations

This appendix contains proofs of two secondary observations that were mentioned in Sect. 1.1. We also include additional evidence that 1-local expanders may be constructed without using an expanding set of generators for $\mathtt{Sym}_n$ (see Theorem A.2).

Improving over Observation 1.2. A natural way of trying to improve over Observation 1.2 is to use relocation graphs that have shorter cover time. The natural choice is to use n-vertex expander graphs.[20]

Observation A.1 (using an expander for the relocation graph): *Let $\pi^{(1)}, ..., \pi^{(d)} : [n] \to [n]$ be bijections that represent the edges of an 2d-regular expander, and $\mathtt{id} : [n] \to [n]$ denote the identity bijection. Then, the 1-local 2^n-vertex graph associated with the 2d relocation permutation $\pi^{(1)}, ..., \pi^{(d)}, \mathtt{id}, ..., \mathtt{id}$ and the 2d offsets $0^n, ..., 0^n, 0^{n-1}1, ..., 0^{n-1}1$* (i.e,., the ith bijection is $x \mapsto x_{\pi^{(i)}}$ if $i \in [d]$ and $x \mapsto x \oplus 0^{n-1}1$ otherwise) *has second* (normalized) *eigenvalue* $1 - \Theta(1/n \log n)$.

[20] We assume that the edges of this $2d$-regular expander can be represented by d permutations, as in the definition of a relocation graph.

Proof Sketch: In this case, a random walk of length $t = O(t' \cdot n \log n)$ on the n-vertex graph visits all vertices with probability at least $1-2^{-t'}$ (since its cover time is $O(n \log n)$ and we have t' "covering attempts").[21] It follows that taking a random walk of length $O(t' \cdot n \log n)$ on the 1-local graph yields a distribution that is $2^{-t'}$-close to uniform, since (with probability $1-2^{-t'}$) each position in the original n-bit string is mapped to the rightmost position at some time, and at the next step the corresponding value is "randomized" (since the offset is applied with probability 1/2). ■

Proposition 1.4 **(revised):** *Consider a 2d-regular 2^n-vertex graph as in Definition 1.3, and suppose that for every $i \in [d]$ either $|s^{(i)}| = o(n)$ or $|s^{(i)}| = n - o(n)$. Then, this 1-local 2^n-vertex graph is not an expander.*

Proof: For starters, we assume that $|s^{(i)}| = o(n)$ for every $i \in [d]$. We first consider an auxiliary $4d$-regular 2^n-vertex graph in which, for each $i \in [d]$, the ith relocation permutation (i.e., $\pi^{(i)}$) is coupled both with the offset $s^{(i)}$ and with the all-zero offset.

The key observation is that, during a random walk on this 1-local 2^n-vertex graph, bits in the label of the current vertex get randomized by the offsets with too small probability, since at each step only $o(n)$ locations are randomized. Specifically, for a t-step random walk that starts at the vertex 0^n, consider the event this walk does not randomize position $j \in [n]$ (in the initial n-bit string); that is, the corresponding walk on the n-vertex relocation graph that starts at vertex $j \in [n]$ does not go through any vertex in the set $S \stackrel{\text{def}}{=} \bigcup_{i \in [d]} \{k : s_k^{(i)} = 1\}$. This bad event (which refers to a random walk that starts at j) occurs with probability at least $\eta = \exp(-o(t))/n$, because the probability that a walk of length t that starts at a random vertex on any $O(1)$-regular n-vertex graph misses a set of $o(n)$ vertices is at least $(1 - o(1))^t = \exp(-o(t))$.[22]

Note that each randomized bit position is reset to 1 with probability exactly 1/2 (by virtue of the auxiliary construction performed upfront), whereas each non-randomized position maintains the value 0. Considering the expected number of ones in the label of the final vertex of a t-step random walk (on the 2^n-vertex graph), observe that if some bit is not randomized with probability η, then the expected number of ones is at most $(1 - \eta) \cdot 0.5 \cdot n + \eta \cdot 0.5 \cdot (n - 1) =$

[21] The cover time bound was established in [1,2,5].

[22] Note that here we seek a lower bound on the probability of missing the set S (equiv., staying in $\overline{S} = [n] \setminus S$), whereas the usual focus is on good upper bounds (which exists when the graph is an expander). Letting d denote the degree of the n-vertex graph, we observe that there are at most $d \cdot |S|$ edges incident at S, and the worst case is that their other endpoints are distributed evenly among the vertices in $\overline{S}$ (because otherwise, conditioning on not leaving $\overline{S}$ biases the distribution towards vertices that have more neighbors in $\overline{S}$ (equiv., less neighbors in S)). Hence, the probability that the random walk never leaves $\overline{S}$ is at least $(1 - \frac{d \cdot |S|}{d \cdot |\overline{S}|})^t$, whereas in our case $|\overline{S}| = (1 - o(1)) \cdot n$.

$(n - \eta)/2$. It follows that the total variation distance between the distribution of the final vertex and the uniform distribution is at least $\eta' \stackrel{\text{def}}{=} \eta/2n = \exp(-o(t) - \log n)$.[23]

We stress that the foregoing holds for any t, which means that we assume that $n = o(t)$, let alone $\log n = o(t)$. Hence, the convergence rate of the 1-local 2^n-vertex graph is *not* bounded away from 1 (since $\eta' = \exp(-o(t))$ whereas the convergence rate λ must satisfy $2^n \cdot \lambda^t > \eta'$).[24] Lastly, we note that given that the auxiliary graph is not an expander, the original graph (which is a subgraph of it) is also not an expander.

Turning to the case in which also offsets of Hamming weight $n-o(n)$ exist, we note that this is equivalent to using an offset of weight $o(n)$ and complementing all bits in the resulting label. Hence, such offsets can randomize many individual locations but cannot randomize all pairs of locations (i.e., randomize each location independently of its paired location). Hence, we extend the foregoing argument to pairs of locations.

We first observe that there are two positions $j_1 \neq j_2$ such that with probability $\eta' = \exp(-o(t))$ these position are always randomized together (i.e., in each steps either both j_1 and j_2 are in locations that get randomized by some single offset or both are not in such locations).[25] The argument is completed by considering the expected number of pairs of positions that hold the same value. ■

Added in Revision: Additional Evidence Against the Need for an Expanding Set of Generators for $\mathtt{Sym}_n$. Theorem 3.2 asserts that the relocation permutations used in a 1-local 2^n-vertex expander graph need not generate the symmetric group over $[n]$, let alone in an expanding manner. This indicates that sets of expanding generators for $\mathtt{Sym}_n$ may not be essential for the construction of 1-local expanders. Additional evidence in that direction is provided by the following composition of 1-local graphs.

---

[23] We use the fact that if $\mathrm{E}[X] \leq \mathrm{E}[Y] - \epsilon$ and $X, Y \in [0,1]$, then there exists a set of values S such that $\Pr[X \in S] \leq \Pr[Y \in S] - \epsilon$. This can be proved by taking $S = \{v : \Pr[X=v] < \Pr[Y=v]\} \subseteq [0,1]$ and using

$$\begin{aligned}\Pr[Y \in S] - \Pr[X \in S] &= \sum_{v \in S} (\Pr[Y=v] - \Pr[X=v]) \\ &\geq \sum_{v \in [0,1]} (\Pr[Y=v] - \Pr[X=v]) \cdot v\end{aligned}$$

which equals $\mathrm{E}[Y] - \mathrm{E}[X] \geq \epsilon$.

[24] Hence, we have $2^n \cdot \lambda^t > \exp(-o(t))$, which implies $\lambda = \exp(-o(1))$ for $t = \omega(n)$.

[25] To see this, follow the argument in Footnote 22, while defining $S \stackrel{\text{def}}{=} \bigcup_{i \in [d]} \{k : s_k^{(i)} = b^{(i)}\}$, where $b^{(i)}$ is the majority value in the string $s^{(i)}$, while noting that the probability that one of the two coordinated random walks does not stay in $\overline{S}$ is only doubled.

Theorem A.2 (composing 1-local expanders): *Suppose that, for* $j \in \{1,2\}$, *there is a* $2d_j$*-regular 1-local* 2^{n_j}*-vertex expander graph, which uses the 1-local bijections* $f_1^{(j)}, ..., f_{d_j}^{(j)} : \{0,1\}^{n_j} \to \{0,1\}^{n_j}$. *Then, the* $4d_1d_2$*-regular 1-local* $2^{n_1+n_2}$*-vertex graph that uses the set of 1-local bijections*

$$\{f_{i_1,i_2,b} : \{0,1\}^{n_1+n_2} \to \{0,1\}^{n_1+n_2}\}_{i_1 \in [d_1], i_2 \in [d_2], b \in \{\pm 1\}} \tag{4}$$

where $f_{i_1,i_2,b}(yz) = f_{i_1}^{(1)}(y)(f_{i_2}^{(2)})^b(z)$, *is an expander.*

Note that the corresponding set of relocation permutations does not generate the symmetric group of $[n_1+n_2]$, since it generates at most $(n_1!)\cdot(n_2!)$ permutations. We comment that the bijections that correspond to $b = -1$ were added in order to allow coupling moves in one direction (of a bijection $f_{i_1}^{(1)}$) on the n_1-bit long prefix with moves in the opposite direction (of a bijection $f_{i_2}^{(2)}$) on the n_2-bit long suffix.

Proof Sketch: The expanding property of the combined 1-local graph, denoted G, can be seen by considering a sufficiently long random walk on it. The key observation is that a t-step random walk on G corresponds to two independent t-step random walks on the two 1-local graphs of the hypothesis, denoted G_1 and G_2. (In particular, a random step on G specifies a random tuple $(i_1, i_2, b) \in [d_1] \times [d_2] \times \{\pm 1\}$ and a direction $\delta \in \{\pm 1\}$ in which $f_{i_1,i_2,b}$ is applied, and this corresponds to random steps on the two graphs (i.e., a choice of $f_{i_1}^{(1)}$ and direction δ for G_1 and a choice of $f_{i_2}^{(2)}$ and direction $b \cdot \delta$ for G_2).) Hence, the n_1-bit long prefix (resp., the n_2-bit long suffix) of the end-vertex in a t-step random walk on G is $\exp(-\Omega(t))$-close to the uniform distribution on $\{0,1\}^{n_1}$ (resp., on $\{0,1\}^{n_1}$), whereas these two parts are distributed independently of one another. ■

References

1. Broder, A., Karlin, A.: Bounds on the cover time. J. Theoret. Probab. **2**(1), 101–120 (1989)
2. Chandra, A.K., Raghavan, P., Ruzzo, W.L., Smolensky, R., Tiwari, P.: The electrical resistance of a graph, and its applications to random walks. In: 21st STOC (1989)
3. Horry, S., Linial, N., Wigderson, A.: Expander graphs and their applications. Bull. (New Ser.) AMS **43**(4), 439–561 (2006)
4. Kassabov, M.: Symmetric groups and expander graphs. Invent. Math. **170**(2), 327–354 (2007)
5. Rubinfeld, R.: The cover time of a regular expander is $O(n \log n)$. IPL **35**, 49–51 (1990)
6. Viola, E., Wigderson, A.: Local expanders. In: ECCC, TR16-129 (2016)

Worst-Case to Average-Case Reductions for Subclasses of P

Oded Goldreich and Guy N. Rothblum

Abstract. For every polynomial q, we present (almost-linear-time) worst-case to average-case reductions for a class of problems in $\mathcal{P}$ that are widely conjectured not to be solvable in time q. These classes contain, for example, the problems of counting the number of t-cliques in a graph, for any fixed $t \geq 3$.

Specifically, we consider the class of problems that consist of counting the number of local neighborhoods in the input that satisfy some predetermined conditions, where the number of neighborhoods is polynomial, and the neighborhoods as well as the conditions can be specified by small uniform Boolean formulas. We show an almost-linear-time reduction from solving any such problem in the worst-case to solving some other problem (in the same class) on typical inputs. Furthermore, for some of these problems, we show that their average-case complexity almost equals their worst-case complexity.

En route we highlight a few issues and ideas such as sample-aided reductions, average-case analysis of randomized algorithms, and average-case emulation of arithmetic circuits by Boolean circuits. We also observe that adequately uniform versions of $\mathcal{AC}^0[2]$ admit worst-case to average-case reductions that run in almost linear-time.

An early version of this work appeared as TR17-130 of *ECCC*. The current revision offers an improved presentation and includes references to a follow-up work of the authors [17].

1 Introduction

While most research in the theory of computation refers to worst-case complexity (cf., e.g., [13, Chap. 1–10.1] versus [13, Sec. 10.2]), the importance of average-case complexity is widely recognized. Worst-case to average-case reductions, which allow for bridging the gap between the two theories, are of natural appeal (to say the least). Unfortunately, until recently worst-case to average-case reductions were known only either for "very high" complexity classes, such as $\mathcal{E}$ and $\#\mathcal{P}$ (see [4] and [9,15,24][1], resp.), or for problems of "very low" complexity, such as

[1] The basic idea underlying the worst-case to average-case reduction of the "permanent" is due to Lipton [24], but his proof implicitly presumes that the field is somehow fixed as a function of the dimension. This issue was addressed independently by [9] and in the proceeding version of [15]. In the current work, we shall be faced with the very same issue.

O. Goldreich (Ed.): Computational Complexity and Property Testing, LNCS 12050, pp. 249–295, 2020.
https://doi.org/10.1007/978-3-030-43662-9_15

`Parity` (cf. [1,3]). In contrast, presenting a worst-case to average-case reduction for $\mathcal{NP}$ is a well-known open problem, which faces significant obstacles [8,10].

A recent work by Ball, Rosen, Sabin, and Vasudevan [5] initiated the study of worst-case to average-case reductions in the context of fine-grained complexity.[2] The latter context focuses on the exact complexity of problems in $\mathcal{P}$ (see, e.g., survey by Vassilevska Williams [32]), attempting to classify problems into classes of similar polynomial-time complexity (and distinguishing, say, linear-time from quadratic-time and cubic-time). Needless to say, reductions used in the context of fine-grained complexity must preserve the foregoing classification, and the simplest choice – taken in [5] (and followed here) – is to use almost linear-time reductions.[3]

The pioneering paper of Ball *et al.* [5] shows that there exist (almost linear-time) reductions from the worst-case of several natural problems in $\mathcal{P}$, which are widely believed to be "somewhat hard" (i.e., have super-linear time complexity (in the worst case)), to the average-case of some other problems that are in $\mathcal{P}$. In particular, this is shown for the Orthogonal Vector problem, for the 3-`SUM` problem, and for the All Pairs Shortest Path problem. Hence, the worst-case complexity of problems that are widely believed to be "somewhat hard" is reduced to the average-case complexity of some problems in $\mathcal{P}$. Furthermore, the worst-case complexity of the latter problems matches (approximately) the best algorithms *known* for the former problems (but note that the latter problems may actually have higher complexity than the former problems).

1.1 Our Results

In this paper we strengthen the foregoing results in two ways. First, we present worst-case to average-case reductions (of almost linear-time) for a wide class of problems within $\mathcal{P}$ (see Theorem 1.1). Second, we show that the average-case complexity of some of these problems approximately equals their worst-case complexity, which is conjectured to be an arbitrary high polynomial (see Corollary 1.2). Details follow.

Complexity Classes Having Worst-Case to Average-Case Reductions. First, for each polynomial p, we define a worst-case complexity class $\mathcal{C}^{(p)}$ that is a subset

---

[2] Actually, a worst-case to average-case reduction for a problem of this flavor was shown before by Goldreich and Wigderson [18]. They considered the problem of computing the function $f_n(A_1, ..., A_{\ell(n)}) = \sum_{S \subseteq [\ell(n)]} \texttt{DET}(\sum_{i \in S} A_i)$, where the A_i's are n-by-n matrices over a finite field and $\ell(n) = O(\log n)$, conjectured that it cannot be computed in time $2^{\ell(n)/3}$, and showed that it is random self-reducible (by $O(n)$ queries). They also showed that it is downwards self-reducible when the field has the form $GF(2^{m(n)})$ such that $m(n) = 2^{\lceil \log_2 n \rceil}$ (or $m(n) = 2 \cdot 3^{\lceil \log_3 n \rceil}$). We stress that the (subquadratic-time) reduction of [18] does not run in almost-linear time, and that the foregoing problem is not as well-studied as the problems considered in [5].

[3] Furthermore, typically, these reductions make a small number of queries (since their queries are of at least linear length). In any case, composing them with a T-time algorithm for the target problem yields an "almost T-time" algorithm for the reduced problem, provided that $T(n) \in [\Omega(n), \text{poly}(n)]$.

of Dtime($p^{1+o(1)}$), and show, for any problem Π in $\mathcal{C}^{(p)}$, an almost linear-time reduction of Π to the average-case of some problem Π' in $\mathcal{C}^{(p)}$. Loosely speaking, the class $\mathcal{C}^{(p)}$ consists of counting problems that refer to $p(n)$ local conditions regarding the n-bit long input, where each local condition refers to $n^{o(1)}$ bit locations and can be evaluated in $n^{o(1)}$-time. These classes are defined in Sect. 2, and they are related to the class of locally characterizable sets defined by us in [16] (see also Appendix A.1). In particular, for any constant $t > 2$ and $p_t(n) = n^t$, the class $\mathcal{C}^{(p_t)}$ contains problems such as t-CLIQUE and t-SUM (e.g., the number of t-cliques in an n-vertex is expressed as the sum of $\binom{n}{t}$ local conditions, each referring to the subgraph induced by t specific vertices). We show that all theses classes have worst-case to average-case reductions.

Theorem 1.1 (worst-case to average-case reduction, loosely stated):[4] *For every polynomial p, and every* (counting) *problem Π in $\mathcal{C}^{(p)}$, there exists a* (counting) *problem Π' in $\mathcal{C}^{(p)}$ and an almost-linear time randomized reduction of solving Π on the worst-case to solving Π' on the average* (i.e., on at least a 0.76 fraction of the domain).[5]

Hence, worst-case to average-case reductions are shown to be a rather general phenomenon: They exist for each problem in a class that is defined in general terms (i.e., counting the number of local conditions that are satisfied by the input, where the local conditions are specified by formulae of bounded complexity). Moreover, the reductions preserve membership in this class (rather than moving to a problem of a different (arithmeticized) flavor as in [5]). In particular, Theorem 1.1 asserts that the average-case complexity of $\mathcal{C}^{(p)}$ is lower-bounded by the worst-case complexity of $\mathcal{C}^{(p)}$, provided that the latter is at least almost-linear. (Recall that the worst-case complexity of $\mathcal{C}^{(p)}$ is at most $p^{1+o(1)}$, and it is most likely that it is not almost-linear.)

A follow-up work of the authors [17] showed a worst-case to average-case reduction for t-CLIQUE, the problem of counting the number of t-cliques in a graph, for each $t \geq 3$. In contrast, the focus of the current work is on *classes* of problems, which are defined in terms of general complexity bounds. Specifically, the classes of counting problems studied here refer to a polynomially large (uniform) collection of small formulae that are applied to the same input.

Computational Problems with Average-Case Complexity that Almost Equals Their Worst-Case Complexity. In Theorem 1.1 the *average-case* complexity of Π' is sandwiched between the *worst-case* complexities of Π and Π', where the worst-case complexities of Π' is at most $p^{1+o(1)}$. This might leave a rather wide gap between the average-case complexity of Π' and its worst-case complexity (if the worst-case complexity of Π is significantly smaller than the worst-case complexity of Π'). Our second improvement over [5] is in narrowing the potential gap between the complexities of Π and Π' (and, consequently, between the average-case and worst-case complexities of Π').

[4] See Theorem 3.1 for a precise statement.

[5] Here 0.76 stands for any constant greater than 3/4.

We do so by taking advantage of the fact that our reduction preserves membership in the class $\mathcal{C}^{(p)}$. Specifically, for every $\epsilon > 0$, assuming that the worst-case complexity of $\Pi \in \mathcal{C}^{(p)}$ is at least q, and invoking Theorem 1.1 for $O(1/\epsilon)$ times, we obtain a problem in $\mathcal{C}^{(p)} \setminus \text{Dtime}(q)$ whose *average-case* complexity is at most n^ϵ times smaller than its *worst-case* complexity. Hence, the average-case complexity of this problem is approximately equal to its worst-case complexity.

Corollary 1.2 (average-case approximating worst-case): *For polynomials $p > q$, suppose that $\mathcal{C}^{(p)}$ contains a problem of worst-case complexity greater than q. Then, for every constant $\epsilon > 0$, there exists $c' \in [\log_n q(n), \log_n p(n)]$ such that $\mathcal{C}^{(p)}$ contains a problem of average-case complexity at least $n^{c'}$ and worst-case complexity at most $n^{c'+\epsilon}$.*

A result analogous to Corollary 1.2 was proved in our follow-up work [17], while assuming that t-CLIQUE is adequately hard. In contrast, the hypothesis of Corollary 1.2 requires only that *some* problem in the class $\mathcal{C}^{(p)}$ is adequately hard.

Proof: Starting with $\Pi^{(0)} \in \mathcal{C}^{(p)} \setminus \text{Dtime}(q)$, we consider a sequence of $O(1/\epsilon)$ problems in $\mathcal{C}^{(p)}$ such that the problem $\Pi^{(i)}$ is obtained by applying (the worst-case to average-case reduction of) Theorem 1.1 to $\Pi^{(i-1)}$. Recall that the average-case complexity of $\Pi^{(i)}$ is sandwiched between the worst-case complexities of $\Pi^{(i-1)}$ and $\Pi^{(i)}$. Hence, the worst-case complexities of these problems (i.e., the $\Pi^{(i)}$'s) constitute a non-decreasing sequence that is lower-bounded by q and upper-bounded by $p(n)^{1+o(1)}$, which is an upper bound on the complexity of any problem in $\mathcal{C}^{(p)}$. It follows that for some $i \in [O(1/\epsilon)]$, the ratio between the worst-case complexities of $\Pi^{(i)}$ and $\Pi^{(i-1)}$ is at most n^ϵ, and the claim follows. ■

Reductions to Rare-Case. The notion of average-case complexity that underlies the foregoing discussion (see Theorem 1.1) refers to solving the problem on at least 0.76 fraction of the instances. This notion may also be called *typical-case complexity.* A much more relaxed notion, called *rare-case complexity,* refers to solving the problem on a noticeable fraction of the instances (say, on a $n^{-o(1)}$ fraction of the n-bit long instances).[6] Using non-uniform reductions, we show that the rare-case complexity of $\mathcal{C}^{(p)}$ is lower-bounded by its (non-uniform) worst-case complexity (provided that the latter is at least almost-linear).

Theorem 1.3 (worst-case to rare-case reduction, loosely stated):[7] *For every polynomial p, and every problem Π in $\mathcal{C}^{(p)}$, there exists a problem Π' in $\mathcal{C}^{(p)}$*

[6] Here a "noticeable fraction" is the ratio of a linear function over an almost linear function. We stress that this is not the standard definition of this notion (at least not in cryptography).

[7] See Theorem 4.2 for a precise statement. Note that Theorem 4.2 refers to sample-aided reductions, which imply non-uniform reductions. Theorem 1.3 refers to a liberal notion of almost-linearity that includes any function of the form $f(n) = n^{1+o(1)}$. Under a strict notion that postulates that only functions of the form $f(n) = n^{1+o(1)}$ are deemed almost-linear, the rare-case refers to an $1/\text{poly}(\log n)$ fraction of the n-bit long instances.

and an almost-linear time non-uniform *reduction of solving Π on the worst-case to solving Π' on a noticeable fraction of the instances* (e.g., on at least $\exp(-\log^{0.999} n)$ fraction of the n-bit long instances).

We also provide a *uniform* reduction from the worst-case complexity of the problem $\Pi \in \mathcal{C}^{(p)}$ to the rare-case complexity of some problem in Dtime(p'), where $p'(n) = p(n) \cdot n^{1+o(1)}$. For details, see Theorem 4.4.

1.2 Comparison to the Known Average-Case Hierarchy Theorem

We mention the existence of an (unconditional) average-case hierarchy theorem of Goldmann, Grape, and Hastad [12], which is essentially proved by diagonalization. While that result offers no (almost linear-time) worst-case to average-case reduction, it does imply the existence of problems in $\mathcal{P}$ that are hard on the average (at any desired polynomial level), let alone that this result is unconditional. We point out several advantages of the current reductions (as well as those of [5]) over the aforementioned hierarchy theorem.

1. Worst-case to average-case reductions yield hardness results also with respect to probabilistic algorithms, whereas the known hierarchy theorem does not hold for that case. Recall that an honest-to-God hierarchy theorem is not known even for worst-case probabilistic time (cf. [6]).
2. Worst-case to average-case reductions are robust under the presence of auxiliary inputs (or, alternatively, w.r.t non-uniform complexity), whereas the aforementioned hierarchy theorem is not.
3. Our worst-case to average-case reductions (as well as those in [5]) do not depend on huge and unspecified constants.

More importantly, while the standard interpretation of worst-case to average-case reductions is negative (i.e., it establishes hardness of the average-case problem based on the worst-case problem), such reductions have also a positive interpretation (which is not offered by results of the type of [12]): They can be used to actually solve a worst-case problem when given a procedure for the average-case problem. Similarly, they may allow to privately compute the value of a function of a secret instance by making queries that are distributed independently of that instance (see, e.g., [7]).

1.3 Techniques

The proofs of all our results are pivoted at *arithmetic versions* of the (Boolean) counting problems belonging to the class $\mathcal{C}^{(p)}$. These arithmetic versions refer to evaluating a small Arithmetic formula on $p(n)$ different (short) projections of the n-symbol input, which is a sequence over a finite field (of varying size). The counting problem is easily reduced to the arithmetic problem, which is (implicitly) known to have a worst-case to average-case reduction (as well as an average-case to rare-case reduction). The tricky part is presenting a reduction of the arithmetic problem "back to" a Boolean counting problem in $\mathcal{C}^{(p)}$, where the reduction preserves the error rate (i.e., average-case or rare-case complexity).

1.3.1 On the Proof of Theorem 1.1

A first attempt at a reduction from the arithmetic problem to a Boolean (counting) problem in $\mathcal{C}^{(p)}$, captured by Proposition 3.6, increases the error rate of the solver for the Boolean problem by a non-constant factor, yielding average-case hardness only with respect to error rate that tends to zero. Even this reduction is not straightforward; see discussion at the beginning of the proof of Proposition 3.6.

Deriving average-case hardness with respect to constant error rate requires a few refinements of the basic approach. First, rather than reducing the n-bit instance of the original counting problem to a single instance of the Arithmetic problem defined over a finite field of size $\Omega(p(n))$, we reduce it to $O(\log p(n))$ instances defined over different fields of much smaller size.[8] Since all these instances will be reduced to the same average-case complexity problem, some of the values obtained in these reductions may be wrong, and Chinese Remaindering *with errors* (cf. [15]) is used in combining them.

We still face the problem of reducing the Arithmetic problem (over fields of size $n^{o(1)}$ (rather than $n^{\Omega(1)}$)) to a counting problem in the class $\mathcal{C}^{(p)}$, while preserving the average-case complexity (equiv., the error rate of potential solvers). To do so, we use (a non-constant number of) Boolean formulae that compute each of the bits in the representation of the output of the Arithmetic formula (which is a field element). However, we cannot afford to deal with each of the resulting Boolean formulae separately, since this will increase the error rate by a non-constant factor. Instead, we combine these Boolean formulae in a way that allows us to reduce the evaluation of the Arithmetic formula to solving a *single* counting problem in the class $\mathcal{C}^{(p)}$.

Specifically, we construct a single Boolean formula that, when fed with an appropriate auxiliary input, outputs the corresponding bit in the representation of the Arithmetic formula's output (which is a field element). The (weighted) summation of the Boolean outputs on the different auxiliary inputs (one per each bit in the representation of the field element) is performed by the counting problem itself (i.e., by the summation). This is done by using additional variables that emulate the weights (as detailed in Sect. 3.2), and yields a single-query reduction of the Arithmetic problem to the Boolean problem.

1.3.2 On the Proofs of Theorems 1.3 and 4.4

Turning from average-case hardness to rare-case hardness (i.e., from the proof of Theorem 1.1 to the proofs of Theorems 1.3 and 4.4), we employ worst-case to rare-case reduction techniques at the Arithmetic level. Specifically, we use the list decoder of Sudan, Trevisan, and Vadhan [31].

For the proof of Theorem 4.4, we also employ a methodology heralded by Impagliazzo and Wigderson [20], which we distill here. The methodology is

[8] The sizes of these fields are lower-bounded not only by $\Omega(\log p(n))$ but also by the size of the Boolean formula used in the original counting problem. The latter lower bound guarantees that the field is larger than the degree of the polynomial that is computed by the Arithmetic problem used in the reduction.

pivoted at the notion of *sample-aided reductions*, which is a relaxed notion of reduction between problems. Specifically, a sample-aided reduction from problem A to problem B receives random solved instances of problem A, which means that it implies an ordinary non-uniform reduction (of A to B). Furthermore, when coupled with a downwards self-reduction for the problem A, a sample-aided reduction of A to B implies a standard (uniform) reduction of A to B. Before justifying the latter assertion, let us spell out the definition of sample-aided reductions.

In the following definition, a task consists of a computational problem along with a required performance guarantee; for example, A may be "solving problem Π on the worst-case" and B may be "solving Π with success rate ρ". For sake of simplicity, we consider the case that the first task (i.e., A) is a worst-case function evaluation task; that is, in the following definition, Π is a function (and "solving Π" means computing Π).

Definition 1.4 (sample-aided reductions): *Let $\ell, s : \mathbb{N} \to \mathbb{N}$, and suppose that M is an oracle machine that, on input $x \in \{0,1\}^n$, obtains a sequence of $s = s(n)$ pairs of the form $(r, v) \in \{0,1\}^{n+\ell(n)}$ as an auxiliary input. We say that M is an* sample-aided reduction of solving Π in the worst-case to the task $\mathtt{T}$ *if, for every procedure P that performs the task* $\mathtt{T}$, *it holds that*

$$\Pr_{r_1,\ldots,r_s \in \{0,1\}^n}\left[\begin{array}{l}\forall x \in \{0,1\}^n \\ \quad \Pr\left[M^P(x; (r_1, \Pi(r_1)), \ldots, (r_s, \Pi(r_s))) = \Pi(x)\right] \geq \frac{2}{3}\end{array}\right] > \frac{2}{3}$$

where the internal probability is taken over the coin tosses of the machine M and the procedure P.

As stated upfront, a sample-aided reduction implies an ordinary non-uniform reduction. In fact, many known non-uniform reductions (e.g., [31]) are actually sample-aided reductions. Furthermore, coupled with a suitable downwards self-reduction for Π, a sample-aided reduction of solving Π in the worst-case to solving Π' on the average (resp., in the rare-case) implies a corresponding standard reduction (of worst-case to average-case (resp., to rare-case)).

Specifically, letting Π_n denote the restriction of Π to n-bit instances, suppose that solving Π_n on the worst-case reduces to solving Π'_n on the average, when provided a sample of solved instances of Π_n, and that we have an average-case solver for Π'. Then, using the downwards self-reduction of Π, a sample of solved instances of Π_n can be generated by employing the downwards reduction of Π_n to Π_{n-1} and using the worst-case to average-case reduction of Π_{n-1} to Π'_{n-1}, which in turn requires a sample of solved instances of Π_{n-1}. Similarly, generating a sample of solved instances for Π_{n-1} is reduced to generating a sample of solve instances for Π_{n-2}, and ditto for Π_{n-i} via Π_{n-i-1}. We stress that the size of the sample remains unchanged at all levels, and that a sample of solved instances for Π_{n-i} allows for solving all instances of Π_{n-i} (by using the average-case solver of Π'_{n-i}).

The actual process of generating samples of solved instances goes in the other direction; that is, going from $j = 1$ to $j = n - 1$, we use the sample of solved instances of Π_j in order to generate a sample of solved instances of Π_{j+1}. (We do so by using the downwards reduction of Π_{j+1} to Π_j, the sample-aided reduction of solving Π_j on the worst-case to solving Π'_j on the average-case, and the average-case solver of Π'_j.) At the end, we solve the instance of Π_n given to us, by using the sample of solved instances of Π_n (while using the sample-aided reduction of solving Π_n on the worst-case to solving Π'_n on the average-case). Recall that, by the hypothesis, we have an average-case solver for Π', and that we can generate solved instances of Π_1 by ourselves (at the beginning of this process (i.e., for $j = 1$)).

Using the foregoing methodology we establish Theorem 4.4, which provides a uniform reduction from the worst-case complexity of the problem $\Pi \in \mathcal{C}^{(p)}$ to the rare-case complexity of some problem in Dtime(p'), where $p'(n) = p(n) \cdot n^{1+o(1)}$. The proof of Theorem 1.3, which asserts a non-uniform worst-case to rare-case reduction for $\mathcal{C}^{(p)}$, does not use this methodology; it rather uses the straightforward emulation of sample-aided reductions by non-uniform reductions.

We mention that in our follow-up [17] we have extended the notion of sample-aided reductions, allowing the $s(n)$ samples to be drawn from arbitrary distributions and be possibly dependent on one another. We believe that the notion of sample-aided reductions is of independent interest.

1.4 Worst-Case to Average-Case Reduction for Uniform $\mathcal{AC}^0[2]$

One may view our classes of counting problems as classes of (adequately uniform) circuits that consist of a top threshold gate that is fed by very small formulae (on joint variables). Given this perspective, one may ask which other classes of circuits admit worst-case to average-case reductions.

We show that adequately uniform versions of $\mathcal{AC}^0[2]$ (i.e., constant-depth polynomial-size circuits with parity gates) admit worst-case to average-case reductions that run in almost linear-time. Note that, while $\mathcal{AC}^0[2]$ cannot count, these classes contain "seemingly hard problems in $\mathcal{P}$" (e.g., the t-CLIQUE problem for n-vertex graphs can be expressed as a highly uniform DNF with n^t terms (each depending on $\binom{t}{2}$ variables)).

Specifically, letting $\mathcal{AC}^0_d[2]$ denote the class of decision problems regarding n-long inputs and the question of whether such an input satisfies some (adequately uniform) poly(n)-size circuit of depth d, we show an almost linear-time reduction of solving problems in $\mathcal{AC}^0_d[2]$ on the worst-case to solving problems in $\mathcal{AC}^0_{O(d)}[2]$ on 90% of the instances, where the constant in the O-notation is universal and small (but larger than 2).

The foregoing reduction uses some of the ideas that are used in the proof of Theorem 1.1. Again, the pivot is an arithmetic problem that admits a worst-case to average-case reduction, and the point is reducing the Boolean problem to it and reducing it back to a Boolean problem.

For the first reduction we use the approximation method of Razborov [26] and Smolensky [29]. Specifically, we obtain a poly(n)-size depth $O(d)$ Arithmetic circuit, over a sufficiently large extension field of GF(2), so that the standard worst-case to average-case reduction can be used; that is, the size of the extension field should exceed the degree of the polynomial computed by the Arithmetic circuit, which is low (i.e., polylogarithmic).[9] The approximation error is a non-issue, since we are only interested in the value of the original circuit at a given input, and the random choices made in the approximation method can be implemented by pseudorandom (i.e., small-bias) ones.

For the backward reduction we show that (adequately uniform) $\mathcal{AC}^0[2]$-circuits can emulate the computation of the foregoing Arithmetic circuit. In doing so, we rely on the following facts:

1. The multiplication gates (in the resulting Arithmetic circuit) have logarithmic fan-in.
2. The Arithmetic circuit is defined for an extension field of GF(2), and the size of this field is only polylogarithmic.
3. The class of Boolean circuits (i.e., $\mathcal{AC}^0[2]$) allows for parity gates of unbounded arity.

Combining Facts 1 and 2, each multiplication gate in the Arithmetic circuit can be emulated by a constant-depth Boolean circuit of size $n^{o(1)}$. Facts 2 and 3 are used for emulating the addition gates. Finally, we let the Boolean circuit obtain an additional input (representing an index in a codeword), and output the corresponding bit in the Hadamard encoding of the output of the Arithmetic circuit. This allows us to handle a constant error rate (rather than an error rate that is inversely related to the length of the representation of field elements). For more details, see Appendix A.2.

1.5 On Average-Case Randomized Algorithms

Our randomized reduction of solving problem Π on the worst-case to solving problem Π' on the average-case (or rare-case) are naturally interpreted as implying a randomized procedure for solving Π that succeeded (with probability at least 2/3) on each instance of Π, when *given oracle access to any deterministic solver that is correct on a specified fraction of the instances of* Π'. Obviously, the same guaranteed holds when using any randomized procedure that for the same fraction of instances of Π' answers correctly with probability at least 2/3. (Indeed, error reduction may be applied both to the randomized solver of Π' and to the randomized reduction of Π to Π'; in fact, employing error reduction is typically essential before using the randomized procedure for Π' in order to derive a randomized procedure for Π.)[10]

In some of our intermediate results, it will be beneficial to use a seemingly relaxed notion of an average-case randomized solver. Specifically, we say that a randomized algorithm A solving P has success rate $\rho : \mathbb{N} \to [0, 1]$ if *the probability*

[9] Indeed, the original goal of the approximation method is to obtain low degree approximations of $\mathcal{AC}^0[2]$ circuits.

[10] This is the case since typically the randomized reduction of Π to Π' makes several queries to Π'.

that A solves a random n-bit instance of P is at least $\rho(n)$, where the probability is taken both over the uniform choice of an n-bit input and over the internal coin tosses of A. Clearly, a randomized procedure A that solves P on a ρ fraction of the instances can be converted to an (almost as efficient) algorithm for P that has error rate $(1-o(1))\cdot\rho$, by first reducing the error probability of A to $o(1)$. We observe that the converse holds too (essentially). Actually, we prove a stronger result.

Proposition 1.5 (from randomized to deterministic error rate): *Let A be a randomized procedure that solves P with error rate ρ such that $\rho(n) \geq 2^{-n/3}$. Then, there exists a randomized algorithm of about the same complexity as A that, on input 1^n, with probability at least $1-2^{-n/4}$, outputs a Boolean circuit that solves P on an $(1-o(1))\cdot\rho(n)$ fraction of the instances of length n.*

Proof: Letting t denote the running time of A, consider the corresponding residual deterministic algorithm A' that, on input (x,r) such that $|r| = t(|x|)$, invokes A on input x and coins r. Our circuit-generating algorithm selects a pairwise-independent ("hash") function $h : \{0,1\}^n \to \{0,1\}^{t(n)}$ (cf. [13, Apdx. D.2]), and outputs the circuit C_h such that $C_h(x) = A'(x,h(x))$. Using Chebyshev's inequality[11], it follows that, with probability at least $1-\epsilon^{-2}\cdot 2^{-n/3}$ over the choice of h, the success rate of C_h is at least $(1-\epsilon)\cdot\rho(n)$. Using $\epsilon = 2^{-n/24}$, the claim follows. ■

1.6 Terminology and Organization

For a function $f : \mathbb{N} \to \mathbb{N}$, we often use $\exp(f(n))$ to denote $2^{O(f(n))}$. When we neglect to mention the base of a logarithmic function, the reader may assume it is 2. Integrality issues are often ignored. For example, $\log n$ may mean $\lceil \log_2 n \rceil$.

Throughout the paper, n denotes the length of the input (either in terms of bits or in terms of sequences over larger alphabet). For a set $S = \{i_1, ..., i_s\} \subseteq$

[11] For each $x \in \{0,1\}^n$, define a 0–1 random variable $\zeta_x = \zeta_x(h)$ such that $\zeta_x(h) = 1$ if and only if $C_h(x)$ is correct. Then, $\mu_x \stackrel{\text{def}}{=} \mathrm{E}[\zeta_x]$ equals the probability that $A(x)$ is correct, and $\mu \stackrel{\text{def}}{=} \sum_{x\in\{0,1\}^n} \mu_x \geq \rho(n)\cdot 2^n$. Using Chebyshev's inequality, we get

$$\begin{aligned}\Pr\left[\left|\sum_{x\in\{0,1\}^n} \zeta_x - \mu\right| \geq 2^n\cdot\epsilon\cdot\rho(n)\right] &\leq \frac{\mathrm{Var}\left[\sum_{x\in\{0,1\}^n}\zeta_x\right]}{(2^n\cdot\epsilon\cdot\rho(n))^2} \\ &= \frac{\sum_{x\in\{0,1\}^n}\mathrm{Var}[\zeta_x]}{2^{2n}\epsilon^2\rho(n)^2} \\ &< \frac{1}{2^n\epsilon^2\rho(n)^2}\end{aligned}$$

where the equality is due to the pairwise-independence of the ζ_x's. Using $\rho(n) > 2^{-n/3}$, we obtain the desired bound (of $2^{-n/3}/\epsilon^2$). Indeed, the same argument supports $\rho(n) \geq 2^{-\beta n}$, for any constant $\beta \in (0, 0.5)$, and it yields a meaningful result for $\epsilon = 2^{-(1-2\beta)/3}$.

$[n]$ such that $i_1 < i_2 < \cdots < i_s$ and $x = x_1x_2\cdots x_n \in \Sigma^n$, we denote by $x_S = x_{i_1}x_{i_2}\cdots x_{i_s}$ the projection of x at the coordinates S.

Almost Linear Functions. The notion of "almost linear" functions has been given a variety of interpretations in the literature, ranging from saying that a function $f : \mathbb{N} \to \mathbb{N}$ is almost linear if $f(n) = n^{1+o(1)}$ to requiring that $f(n) = \widetilde{O}(n)$. We shall restrict the range of interpretations to the case of $f(n) \le \exp(f'(n)) \cdot n$, where at the very maximum $f'(n) = (\log n)/(\log\log n)^{\omega(1)}$ (which means that $f(n) \le n^{1+(\log\log n)^{-\omega(1)}}$) and at the very minimum $f'(n) = \widetilde{O}(\log\log n)$ (which means that $f(n) \le (\log n)^{(\log\log\log n)^{O(1)}} \cdot n$). In these cases, we shall consider $\exp(f'(n)) = \text{poly}(f(n)/n)$ to be *small*, and $\exp(-f'(n)) = \text{poly}(n/f(n))$ to be *noticeable*.

Organization. In Sect. 2 we define the classes of counting problems that will be studied in this work. In Sect. 3 we present the proof of Theorem 1.1; our presentation proceeds in two steps: First, in Sect. 3.1, we prove a weaker result (which refers to an error rate that is noticeable but tends to zero), and later (in Sect. 3.2) we derive the original claim. The proof of Theorem 1.3 is presented in Sect. 4.1, and in Sect. 4.2 we prove the incomparable result stated in Theorem 4.4.

2 Counting Local Patterns: The Class of Counting Problems

We consider a class of counting problems that are solvable in polynomial time. Such a counting problem specifies a polynomial number of local conditions, where each local condition consists of a short sequence of locations (in the input) and a corresponding predicate, and the problem consists of counting the number of local conditions that are satisfied by the input.

An archetypical example is the problem of counting t-cliques, for some fixed $t \in \mathbb{N}$, where the local conditions correspond to all t-subsets of the vertex set, and each local condition mandates that the corresponding vertices are pairwise adjacent in the graph. Specifically, for an n-vertex input graph, represented by its adjacency matrix $X = (x_{i,j})$, we have local conditions of the form $\bigwedge_{j<k\in[t]} x_{i_j,i_k}$ for every t-subset $\{i_1, ..., i_t\} \in \binom{[n]}{t}$. Hence, counting t-cliques in a graph represented by X corresponds to counting the number of local conditions that are satisfied by X.

In general, for a collection of relatively small and efficiently described subsets $S_1^{(n)}, ..., S_{\text{poly}(n)}^{(n)} \subset [n]$ and formulae $\phi_1^{(n)}, ..., \phi_{\text{poly}(n)}^{(n)}$, we consider the problem of counting, on input $x \in \{0,1\}^n$, the number of subsets $S_i^{(n)}$ such that $\phi_i^{(n)}(x_{S_i^{(n)}}) = 1$. Hence, the $S_i^{(n)}$'s represents (locations of) portions of the input and the $\phi_i^{(n)}$'s represents the (local) conditions imposed on them. (Indeed, this class of counting problems is related to the class of locally characterizable sets defined by us in [16]; see more details in Appendix A.1.)

Key issues that were intentionally left unspecified in the foregoing description are what is the size of the subsets $S_i^{(n)}$'s and formulae $\phi_i^{(n)}$'s and what is meant by efficiently describing them. Starting from the latter question, we postulate that all subsets (i.e., $S_1^{(n)}, ..., S_{\text{poly}(n)}^{(n)} \subset [n]$) are described by an explicit sequence of formulae $\sigma_n : \{0,1\}^{\ell(n)} \to \binom{[n]}{m(n)}$, where $\ell(n) = O(\log n)$, such that $\sigma_n(i) = S_i^{(n)}$ for every $i \in [2^{\ell(n)}] \equiv \{0,1\}^{\ell(n)}$. Indeed, $2^{\ell(n)}$ equals the number of local conditions imposed on inputs of length n, and $m(n)$ is the number of locations considered in each local condition. Likewise, all ("local condition") formulae (i.e., $\phi_1^{(n)}, ..., \phi_{\text{poly}(n)}^{(n)}$) are described by an explicit sequence of formulae $\phi_n : \{0,1\}^{\ell(n)} \times \{0,1\}^{m(n)} \to \{0,1\}$ such that $\phi_n(i,z) = \phi_i^{(n)}(z)$ for every $i \in [2^{\ell(n)}] \equiv \{0,1\}^{\ell(n)}$ and $z \in \{0,1\}^{m(n)}$.

Still, we need to specify the function $m : \mathbb{N} \to \mathbb{N}$, a bound on the size of formulae σ_n and ϕ_n, and be more concrete regarding what we mean by "being explicit" (i.e., the complexity of constructing σ_n and ϕ_n). We actually prefer to leave these issues semi-open and only state minimal conditions regarding these aspects. For sure $m(n)$ and the formulae sizes should be between $n^{o(1)}$ and $\text{poly}(\log n)$, and, by default, *explicit* means that the object can be generated in time that is polynomial in its size.

In addition, since the bounds on the formulae sizes will be used to define complexity classes, we need to define families of bounds that are closed under some operations (so that the complexity classes may be closed under some natural algorithmic compositions). This will be done in Sect. 2.1, and once done we shall spell out the corresponding definition of the counting classes (in Sect. 2.2).

2.1 Admissible Bounding Functions

Our main complexity measure will be formula size (i.e., bounds on the sizes of the formulae σ_n and ϕ_n). Equivalently, we may just discuss formula depth, which is logarithmically related to their size [30] (see, e.g., [22, Sec. 6.1]). Recalling our minimal requirements regarding the sizes of the formulae σ_n and ϕ_n, we focus on depth bounds that are between $o(\log n)$ and $\Omega(\log \log n)$.

Since our reductions may increase the depth of the formulae by a poly-logarithmic factor (i.e., go from $d(n)$ to $\widetilde{O}(d(n))$), we consider classes of (depth) bounding functions that are closed under multiplication by such a factor. Although this choice seems opportunistic, it is quite natural in the context of complexity measures: One typically allows closure under constant factors, and closure under polylogarithmic factors is but the next step. Lastly, we make the standard simplifying assumption that the bounding function is monotonically non-decreasing. Hence, we obtain the following admissible classes of bounding functions.

Definition 2.1 (admissible classes of functions): *A class of functions $\mathcal{D} \subseteq \{f : \mathbb{N} \to \mathbb{N}\}$ is* admissible *if for every $d \in \mathcal{D}$ it holds that*

1. Minimal condition (i.e., d is neither too big nor too small): *The function d is sub-logarithmic and at least double-logarithmic; that is, $d(n) = o(\log n)$ and $d(n) = \Omega(\log\log n)$.*
2. Closure: *The function d' such that $d'(n) = \widetilde{O}(d(n))$ is in $\mathcal{D}$.* (Given that $d(n) = \Omega(\log\log n)$, it follows that $d''(n) = d(n) + \log\log n$ is also in $\mathcal{D}$).
3. Non-decreasing: $d(n+1) \geq d(n)$.

Examples of admissible classes of functions include

$$\begin{aligned}
\mathcal{D}_1 &= \{d \in \mathcal{L} : d(n) \leq \widetilde{O}(\log\log n)\} \\
\mathcal{D}_2 &= \{d \in \mathcal{L} : d(n) \leq \text{poly}(\log\log n)\} \\
\mathcal{D}_3 &= \{d \in \mathcal{L} : d(n) \leq O(\log n)^c\} \text{ for any } c \in (0,1) \\
\mathcal{D}_4 &= \left\{d \in \mathcal{L} : d(n) \leq \frac{\log n}{(\log\log n)^{\omega(1)}}\right\}
\end{aligned}$$

where $\mathcal{L}$ denotes the class of functions that are at least double-logarithmic, and the unspecified constants (in the O-notation) allow for ignoring finitely many n's.

Definition 2.2 (a corresponding notion of almost-linear functions): *For a fixed admissible class $\mathcal{D}$, we say that the function $f : \mathbb{N} \to \mathbb{N}$ is* almost linear *if $f(n) \leq \exp(d(n)) \cdot n$ for some $d \in \mathcal{D}$.*

Indeed, this is shorthand for $\mathcal{D}$-almost-linear. Note that, for any admissible class $\mathcal{D}$, saying that f is $\mathcal{D}$-almost-linear means that $f(n) = n^{1+o(1)}$, although it may be that $f(n) > \widetilde{O}(n)$. (Even saying that f is $\mathcal{D}_1$-almost-linear only means that $f(n) \leq \exp(\widetilde{O}(\log\log n))\cdot n$, which means that $f(n)/n$ is quasi-poly-logarithmic.)

2.2 The Counting Classes

Turning back to the Boolean formulae $\phi_n : \{0,1\}^{\ell(n)+m(n)} \to \{0,1\}$ and $\sigma_n : \{0,1\}^{\ell(n)} \to [n]^{m(n)}$, note that their size is upper-bounded by the time that it takes to construct them. Recalling that we may assume, without loss of generality, that the depth of a formula is logarithmic in its size, we make the running time of the constructing algorithm our main parameter. Hence, for any admissible class $\mathcal{D}$, which provides bounds for the depth of formulae, we shall consider algorithms that on input n run in time that is exponential in some function in $\mathcal{D}$.

For sake of convenience, we replace the Boolean formula $\sigma_n : \{0,1\}^{\ell(n)} \to [n]^{m(n)}$, which specifies $m(n)$-subsets of $[n]$, by the formulae $\sigma_{n,1}, ..., \sigma_{n,m(n)} : \{0,1\}^{\ell(n)} \to [n]$ such that $\sigma_{n,j}(i)$ specifies the jth element in $\sigma_n(i)$, which is the ith subset specified by σ_n. With these preliminaries in place, we are finally ready to state the definition of the class of counting problems that we consider. Actually, we present a family of such classes, each corresponding to a different class of admissible functions and to a different logarithmic function ℓ.

Definition 2.3 (counting local patterns): *Let $\mathcal{D}$ be a set of admissible functions and $\ell : \mathbb{N} \to \mathbb{N}$ be a logarithmic function* (i.e., $\ell(n) = \lceil c \cdot \log n \rceil$ for some constant $c > 0$). *For every $d \in \mathcal{D}$ and $m : \mathbb{N} \to \mathbb{N}$, let A be an algorithm that, on input n, runs for* $\exp(d(n))$*-time and outputs Boolean formulae $\phi_n : \{0,1\}^{\ell(n)} \times \{0,1\}^{m(n)} \to \{0,1\}$ and $\sigma_{n,1}, ..., \sigma_{n,m(n)} : \{0,1\}^{\ell(n)} \to [n]$. The* counting problem associated with *A, denoted $\#_A$, consists of counting, on input $x \in \{0,1\}^*$, the number of $w \in \{0,1\}^{\ell(|x|)}$ such that*

$$\Phi_x(w) \stackrel{\text{def}}{=} \phi_n(w, x_{\sigma_{n,1}(w)}, ..., x_{\sigma_{n,m(n)}(w)}) \tag{1}$$

equals 1. The class of counting problems associated with ℓ and $\mathcal{D}$, *denoted $\mathcal{C}_{\ell,\mathcal{D}}$, consists of all problems associated with A as above.*

Hence, each problem in $\mathcal{C}_{\ell,\mathcal{D}}$ is specified by $2^{\ell(n)}$ local conditions regarding the n-bit input. Each local condition refers to $m(n)$ locations in the input, and the predicate $\Phi_x(w)$ indicates whether or not the local condition associated with the index $w \in \{0,1\}^{\ell(n)} \equiv [2^{\ell(n)}]$ is satisfied by the input x. The corresponding counting problem is to compute the number of local conditions that are satisfies by x (i.e., the number of w's such that $\Phi_x(w) = 1$).

Note that any problem in the class $\mathcal{C}_{\ell,\mathcal{D}}$ can be solved in polynomial-time; specifically, n-bit long instances can be solved in time $2^{\ell(n)} \cdot \exp(d(n)) = 2^{(1+o(1))\cdot\ell(n)}$ on a direct access machine. We mention that the doubly-efficient interactive proof systems presented by us in [16] apply to these counting problems, provided that the verifier is allowed to run in time $\exp(d(n)) \cdot n = n^{1+o(1)}$, for some $d \in \mathcal{D}$.

A Simplified Form. In many natural cases, the local condition may depends only on the values of the bits in the input, and be oblivious of the index of the condition; that is, $\phi_n(w, z)$ may be oblivious of w. In this case, we may write $\phi_n : \{0,1\}^{m(n)} \to \{0,1\}$ (instead of $\phi_n : \{0,1\}^{\ell(n)+m(n)} \to \{0,1\}$), and have $\Phi_x(w) = \phi_n(x_{\sigma_{n,1}(w)}, ..., x_{\sigma_{n,m(n)}(w)})$. This holds for the problem of counting t-cliques, discussed in the beginning of this section, and for the problem of counting t-tuples of integers in the input sequence $x = (x_1, ..., x_n) \in [-b, b]^n$ that sum-up to zero (i.e., the counting version of t-SUM). Specifically, with some abuse of notation, the latter problem is captured by $\Phi_x(i_1, ..., i_t) = 1$ if and only if $\sum_{j\in[t]} x_{i_j} = 0$; that is, we use $\phi_n : [-b, b]^t \to \{0,1\}$ such that $\phi_n(z_1, ..., z_t) = 1$ if and only if $\sum_{j\in[t]} z_j = 0$ (along with $\sigma_{n,j} : \{0,1\}^{t \log n} \to [n]$ such that $\sigma_{n,j}(i_1, ..., i_t) = i_j$ for $j \in [t]$).

3 The Worst-Case to Average-Case Reduction

We are now ready to (re)state our main result.

Theorem 3.1 (worst-case to average-case reduction, Theorem 1.1 formalized): *Let $\mathcal{D}$ be a set of admissible functions and ℓ be a logarithmic function. Then, for every counting problem Π in $\mathcal{C}_{\ell,\mathcal{D}}$, there exists a counting problem Π' in $\mathcal{C}_{\ell,\mathcal{D}}$*

and an almost-linear time randomized reduction of solving Π on the worst case to solving Π' on at least $0.75+\epsilon$ fraction of the domain, where ϵ is any positive constant.

Recall that, for every $d \in \mathcal{D}$, the function $n \mapsto \exp(d(n)) \cdot n = n^{1+o(1)}$ is considered *almost linear*. Actually, we may replace ϵ by $\exp(-d(n)) = o(1)$.

Outline of the Proof. The reduction consists of two main steps.

1. An almost-linear time randomized reduction of solving Π *on the worst case* to evaluating certain Arithmetic expressions over a field of size $\exp(d(n))$, where the evaluation subroutine is correct on at least a $0.5 + o(1)$ fraction of the instances.
2. An almost-linear time reduction of evaluating the foregoing Arithmetic expressions on at least a $0.5 + o(1)$ fraction of the instances to solving Π' on at least $0.75 + o(1)$ fraction of the instances.

We start by presenting the aforementioned arithmetic problem. Recall that an Arithmetic circuit (resp., formula) is a directed acyclic graph (resp., directed tree) with vertices (of bounded in-degree) that are labeled by multiplication-gates and linear-gates (i.e., gates that compute linear combination of their inputs). Actually, since we consider generic Arithmetic circuits, which are well defined for any field, the scalars allowed in the linear gates are only 0, 1 and -1.

Definition 3.2 (evaluating arithmetic expressions of a local type): *For $d \in \mathcal{D}$ and $\ell, m : \mathbb{N} \to \mathbb{N}$ as in Definition 2.3, let A' be an algorithm that, on input n, runs for $\exp(d(n))$-time and outputs an Arithmetic formula $\widehat{\phi}_n : \mathcal{F}^{\ell(n)} \times \mathcal{F}^{m(n)} \to \mathcal{F}$ and Boolean formulae $\sigma_{n,1}, ..., \sigma_{n,m(n)} : \{0,1\}^{\ell(n)} \to [n]$, where $\mathcal{F} \supseteq \{0,1\}$ is a generic finite field. Fixing a finite field $\mathcal{F}$, the* evaluation problem associated with A' and $\mathcal{F}$, *denoted $\mathrm{EV}_{A',\mathcal{F}}$, consists of computing the function $\widehat{\Phi}_{A'} : \mathcal{F}^n \to \mathcal{F}$ such that*

$$\widehat{\Phi}_{A'}(X_1, ..., X_n) \stackrel{\text{def}}{=} \sum_{w \in \{0,1\}^{\ell(n)}} \widehat{\phi}_n(w, X_{i_{n,w}^{(1)}}, ..., X_{i_{n,w}^{(m(n))}}), \tag{2}$$

where $i_{n,w}^{(j)} = \sigma_{n,j}(w)$.

Indeed, as in Eq. (2), we shall often abuse notation and view binary strings as sequences over $\mathcal{F}$ (e.g., in Eq. (2) $w \in \{0,1\}^{\ell(n)}$ is viewed as an $\ell(n)$-long sequence over $\mathcal{F}$). The formally inclined reader should consider a mapping $\xi : \{0,1\} \to \mathcal{F}$ such that $\xi(0) = 0 \in \mathcal{F}$ and $\xi(1) = 1 \in \mathcal{F}$, and write $\widehat{\phi}_n(\xi(w_1), ..., \xi(w_{\ell(n)}), X_{i_{n,w}^{(1)}}, ..., X_{i_{n,w}^{(m(n))}})$ instead of $\widehat{\phi}_n(w, X_{i_{n,w}^{(1)}}, ..., X_{i_{n,w}^{(m(n))}})$.

Organization of the Rest of this Section. We shall first prove a weaker version of Theorem 3.1 in which the tolerated error rate is $1/O(\log n)$ rather than $0.25 - o(1)$. This proof is presented in Sect. 3.1, and it will serves as a basis for the proof of Theorem 3.1 itself, which is presented in Sect. 3.2.

3.1 The Vanilla Version

Slightly detailing the proof outline provided above, we note that it amounts to composing three basic reductions.

1. A (worst-case to worst-case) reduction of the Boolean counting problem to a corresponding Arithmetic evaluation problem.
2. A worst-case to average-case reduction of the Arithmetic evaluation problem to itself.
3. An average-case to average-case reduction of the Arithmetic evaluation problem to a new Boolean counting problem.

All reductions are performed in almost-linear time and preserve the relevant parameters (e.g., ℓ and $d \in \mathcal{D}$). These reduction are presented in Sect. 3.1.1. The reduction deserving most attention is the third one (i.e., reducing the arithmetic problem to a Boolean problem), where the point is that this reduction is performed in the context of average-case analysis. Later, in Sect. 3.1.2, we shall combine these reductions and derive a weak version of Theorem 3.1.

3.1.1 Three Basic Reductions

Our first step is reducing the (Boolean) counting problem to the Arithmetic evaluation problem. This reduction is based on the folklore emulation of Boolean circuits by Arithmetic circuits, which yields a worst-case to worst-case reduction. The validity of the reduction in our context (of counting and summation problems) relies on the choice of the field for the Arithmetic evaluation problem; specifically, we choose a finite field of characteristic that is larger than the number of terms in the counting problem. We note that, for the sake of the worst-case to average-case reduction (for the Arithmetic evaluation problem), which will be performed in the next step, it is important to keep track of the degree of the polynomial that is computed by the Arithmetic circuit that is derived by the Boolean-to-Arithmetic reduction.

Proposition 3.3 (reducing worst-case Boolean counting to worst-case arithmetic evaluation): *Solving the counting problem associated with an algorithm A (and functions ℓ, m and $d \in \mathcal{D}$) is reducible in almost-linear time to the evaluation problem that is associated with a related algorithm A', and any finite field of prime cardinality greater than $2^{\ell(n)}$. Furthermore, the reduction makes a single query, the degree of the polynomial $\widehat{\Phi}_{A'}$ is at most $\exp(d(n))$, and the functions ℓ, m and d equal those in the counting problem.*

Proof: The basic idea is to emulate the Boolean formula ϕ_n by the Arithmetic formula $\widehat{\phi}_n$, where we shall first transform the former formula into an almost-balanced one (i.e., has depth that is logarithmic in its size).[12] The almost-

[12] Recall that, given a formula F of size s, this transformation first locates a subformula F' of size $s' \in [s/3, 2s/3]$. Letting F_b be the formula that results from F when replacing F' by the constant $b \in \{0,1\}$, the transformation is recursively applied to the formulae F_0, F_1 and F', while noting that each of these formulae has size at most $2s/3$. Lastly, we output the formula $(F' \wedge F_1) \vee ((\neg F') \wedge F_0)$.

balanced structure will yield a bound on the degree of the derived Arithmetic formula.

Let $\ell = \ell(n) = O(\log n)$. We may assume, without loss of generality, that the depth the Boolean formula ϕ_n is logarithmic in its size, which is upper-bounded by $s = \exp(d(n))$. Observe that the transformation of arbitrary formula to this (almost-balanced) form can be performed in polynomial (in s) time. Within the same complexity bound, we can construct an Arithmetic formula $\widehat{\phi}_n : \mathcal{F}^{\ell(n)+m(n)} \to \mathcal{F}$ that agrees with ϕ_n on $\{0,1\}^{\ell(n)+m(n)}$; that is, $\widehat{\phi}_n(w,z) = \phi_n(w,z)$ for every $(w,z) \in \{0,1\}^{\ell(n)+m(n)}$. This construction is obtained by replacing each and-gate by a multiplication gate, and replacing each negation-gate by a gate that computes the linear mapping $v \mapsto 1-v$. The crucial point is that $\widehat{\phi}_n$ preserves the depth of ϕ_n, and so the degree of the function computed by $\widehat{\phi}_n$ is upper-bounded by $D = \exp(O(\log s)) = \text{poly}(s) \ll |\mathcal{F}|$, where $\mathcal{F}$ is chosen to be a finite field of prime cardinality that is larger than 2^ℓ (and $s = n^{o(1)}$).

Hence, we reduce the counting problem $\#_A$ (associated with an algorithm A) to the arithmetic evaluation problem $\text{EV}_{A',\mathcal{F}}$ associated with an algorithm A' that computes the foregoing $\widehat{\phi}_n$ and $\sigma_{n,j}$'s, by first invoking algorithm A, and then proceeding as outlined above. Defining $\widehat{\Phi}_{A'}$ as in Eq. (2), for every $x = x_1 \cdots x_n \in \{0,1\}^n \subset \mathcal{F}^n$, it holds that

$$\begin{aligned} \widehat{\Phi}_{A'}(x_1, ..., x_n) &= \sum_{w\in\{0,1\}^\ell} \widehat{\phi}_n(w, x_{i^{(1)}_{n,w}}, ..., x_{i^{(m(n))}_{n,w}}) \\ &= \sum_{w\in\{0,1\}^\ell} \phi_n(w, x_{\sigma_{n,1}(w)}, ..., x_{\sigma_{n,m(n)}(w)}) \\ &= \sum_{w\in\{0,1\}^\ell} \Phi_x(w) \end{aligned}$$

where the equalities hold both over $\mathcal{F}$ and over the integers, because each term is in $\{0,1\}$ and $\mathcal{F}$ is a finite field of prime cardinality that is larger than 2^ℓ. Hence, $\#_A(x) = \text{EV}_{A',\mathcal{F}}(x)$ for every $x \in \{0,1\}^n \subset \mathcal{F}^n$. Recalling that $D = \text{poly}(s) = \exp(d(n))$, the furthermore clause follows. ■

The Second Step: A Worst-Case to Average-Case Reduction. The fact that the arithmetic evaluation problem (i.e., $\text{EV}_{A',\mathcal{F}}$) admits a worst-case to average-case reduction is a special case of the well-known fact that such a reduction holds for any polynomial provided that the field is large enough (i.e., significantly larger than the degree of the polynomial). The latter fact is rooted in Lipton's work [24].

Proposition 3.4 (folklore worst-case to average-case reduction for evaluating polynomials of bounded degree): *Let $P : \mathcal{F}^n \to \mathcal{F}$ be a polynomial of total degree smaller than $D < |\mathcal{F}|/3$. Then, evaluating P on any input can be randomly reduced to evaluating P correctly on at least a $8/9$ fraction of the domain, by invoking the latter evaluation procedure for $3D$ times.*

We stress that the reduction is randomized, and, on each input, it yields the correct answer with probability at least $2/3$.

Proof: On input $x \in \mathcal{F}^n$, we select $r \in \mathcal{F}^n$ uniformly at random, and invoke the evaluation procedure on the points $x + ir$, where $i = 1, ..., 3D$. (This description presumes that $\mathcal{F}$ is a prime field; but otherwise, we may use $3D$ distinct non-zero elements of $\mathcal{F}$ instead of the i's.) Note that the queried points are uniformly distributed in $\mathcal{F}^n$, and so the evaluation procedure is expected to answer correctly on at least an $8/9$ fraction of these queries. It follows that, with probability at least $2/3$, the evaluation procedure answers correctly on at least $2D$ of the queried points. Using the Berlekamp–Welch algorithm, we reconstruct the unique degree $D-1$ polynomial that agrees with at least $2D$ of the answers, and return its free-term (i.e., its value at 0, which corresponds to $P(x)$).[13] Hence, the recovered value is correct with probability at least $2/3$. ■

Towards the Third Step: A Subtle Difficulty. Combining Propositions 3.3 and 3.4, we obtain a reduction of solving the (Boolean) counting problem, on the worst-case, to solving the Arithmetic problem on the average (i.e., for a 8/9 fraction of the instances). In order to prove Theorem 3.1, we have to reduce the latter problem to a (Boolean) counting problem, and this reduction has to be analyzed in the average-case regime. This raises a difficulty, since the Arithmetic problem (denoted $\mathtt{EV}_{A',\mathrm{GF}(p)}$) refers to any fixed prime $p > 2^{\ell(n)}$, and the reduction (provided by Proposition 3.3) does not specify such a prime (i.e., Proposition 3.3 merely states that the reduction is valid for any sufficiently large prime). This becomes a problem when we want to reduce the Arithmetic problem (which refers to an unspecified large field) to a Boolean problem.

Of course, an adequate prime can be selected at random by the reduction (of Proposition 3.3), but in such a case different invocations will yield different primes, and so we shall not have a single Arithmetic problem but rather a distribution over such problems. This difficulty could have been avoided if we had a deterministic $\exp(d(n))$-time algorithm for generating primes that are larger than $2^{\ell(n)}$, but such an algorithm is not known (since $\exp(d(n)) = n^{o(1)}$ whereas $2^{\ell(n)} = n^{\Omega(1)}$).

We resolve this difficulty by considering an extension of Definition 2.3 in which the algorithm that generates the Boolean formulae is given an auxiliary input (in addition to the length parameter n). Indeed, providing the algorithm associated with the Boolean problem with an auxiliary input (that specifies the field used in the Arithmetic problem) allows us to reduce the Arithmetic problem to a Boolean problem. Actually, it suffices to let the algorithm use this auxiliary information for the construction of the local conditions only (whereas the location formulae $\sigma_{n,j}$'s may remain oblivious of it).

Definition 3.5 (Definition 2.3, extended): *For $d \in \mathcal{D}$ and $\ell, m : \mathbb{N} \to \mathbb{N}$ as in Definition 2.3, let B be an algorithm that, on input (n, z) such $|z| \leq \exp(d(n))$, runs for $\exp(d(n))$-time and outputs Boolean formulae $\phi_{n,z} : \{0,1\}^{\ell(n)} \times \{0,1\}^{m(n)} \to \{0,1\}$ and $\sigma_{n,1}, ..., \sigma_{n,m(n)} : \{0,1\}^{\ell(n)} \to [n]$. The* counting problem

[13] This polynomial is unique since we cannot have two different polynomials of degree $D-1$ that each agree with $2D$ of the $3D$ points.

associated with $B(n,z)$ *consists of counting, on input* $x \in \{0,1\}^n$, *the number of* $w \in \{0,1\}^{\ell(n)}$ *such that* $\phi_{n,z}(w, x_{\sigma_{n,1}(w)}, ..., x_{\sigma_{n,m(n)}(w)}) = 1$.

With this definition in place, we can reduce the Arithmetic problem $\mathrm{EV}_{A',\mathrm{GF}(p)}$, which refers to n-long sequences over GF(p), to a counting problem associated with $B(n,p)$, where B is as in Definition 3.5. We warn, however, that the following result reduces solving the arithmetic problem on 8/9 fraction on the instances to solving the Boolean problem on an $1 - \Omega(1/\log n)$ fraction of the instances; that is, the following result refers only to potential solvers (of the Boolean problem) that err on a vanishing yet noticeable fraction of the domain.

Proposition 3.6 (reducing average-case Arithmetic evaluation to average-case Boolean counting): *Let* A' *be as in Definition 3.2, and* $\widehat{\Phi}_{A'}$ *be the corresponding polynomial. Then, for every prime* $p \leq \mathrm{poly}(n)$, *the problem of evaluating* $\widehat{\Phi}_{A'}$ *on at least a* 8/9 *fraction of the domain* $\mathrm{GF}(p)^n$ *is randomly reducible in almost-linear time to solving the counting problem associated with* $B(n'',p)$ *on at least* $1-(1/O(\log p))$ *fraction of the domain* $\{0,1\}^{n''}$, *where* B *is as in Definition 3.5 and* $n'' = \widetilde{O}(n)$. *Furthermore, if* ℓ', d' *and* m' (resp., ℓ'', d'' and m'') *are the functions used in the Arithmetic* (resp., Boolean) *problem, then* $\ell'' = \ell'$, $d''(\widetilde{O}(n)) \leq \mathrm{poly}(\log\log p) \cdot d'(n)$, *and* $m''(\widetilde{O}(n)) = O(\log n) \cdot m'(n)$.

Note that we are only guaranteed that $d''(\widetilde{O}(n)) \leq \mathrm{poly}(\log\log n) \cdot d'(n)$, which means that the Boolean counting problem is in $\mathcal{C}_{\ell,\mathcal{D}}$ only if $\mathcal{D}$ has a stronger closure property than postulated in Definition 2.1 (i.e., we need closure under multiplication by $\mathrm{poly}(\log\log n)$, whereas Definition 2.1 only postulates closure under multiplication by $\mathrm{poly}(\log d'(n))$). This drawback will be removed at a later stage (i.e., in Sect. 3.2), when dealing with the more acute drawback of referring only to vanishing error rates.

Proof: The basic idea is to let the Boolean formula emulate the computation of the Arithmetic formula $\widehat{\phi}_n$ presented in Eq. (2). This will be done by using small Boolean formulae $\phi'_{n,p}$ that emulate the operations of the field GF(p), which are performed in $\widehat{\phi}_n$. This suggestion raises a couple of issues.

1. A straightforward representing of elements of GF(p) by $\lceil \log_2 p \rceil$-bit long strings is good enough for worst-case to worst-case reductions, but in the context of average-case to average-case reductions this may fail badly. The problem is that a constant fraction of the $\lceil \log_2 p \rceil$-bit long strings may have no preimage under the straightforward encoding, which means that the straightforward encoding may map n-long sequences of field elements to an $\exp(-n)$ fraction of the $n \cdot \lceil \log_2 p \rceil$-bit long strings.
   Our solution is to use a randomized encoding of the elements of GF(p); specifically, $e \in \mathrm{GF}(p)$ will be encoded by a randomly selected element of $\{e' \in [2^{O(\log(pn))}] : e' \equiv e \pmod{p}\}$.
2. The Boolean formula outputs a single bit, whereas the Arithmetic formula outputs an element of GF(p). Hence, in order to emulate a computation of the Arithmetic formula, we need to combine $\log_2 p$ Boolean results. Furthermore,

we need to compute the sum (in GF(p)) of many evaluations of an Arithmetic formula, given a procedure that counts the number of times that a Boolean formula evaluates to 1.
Our solution is to compute the sum of the GF(p)-elements over the integers, and do so by computing the sum of each bit in the canonical representation of these elements (rather than in their randomized encodings per Item 1). The Boolean formula that we constructs starts by mapping the random representation to the canonical one, emulates the Arithmetic formula, and then outputs the desired bit, which is specified as part of its input. The corresponding answer to the Boolean counting problem will give us the sum of the values of this bit over the poly(n)-many evaluations of the Arithmetic formula.
The small error rate that we can tolerate is due to the fact that we need to get the correct values to all $\log_2 p$ queries, which correspond to all bits in the binary expansion of the canonical representation of the elements of GF(p). (This can be improved upon using an additional idea, which we postpone to Sect. 3.2.)

Details follow.
We shall represent each element of $\mathcal{F} = \mathrm{GF}(p)$ by an $O(\log n)$-bit long string. Actually, each element of GF(p) will have poly(n)-many possible representations by bit strings of length $\log_2 p + O(\log n) = O(\log n)$. When reducing the Arithmetic evaluation problem to a Boolean counting problem, we shall select at random one of these representations; that is, for each input symbol $e \in \mathrm{GF}(p)$, we shall select at random $e' \in [\mathrm{poly}(n)]$ such that $e' \equiv e \pmod p$. This redundant representation is used in order to guarantee that all field elements have approximately the same number of representations (as $O(\log n)$-long bit strings).

We construct a Boolean formula $\phi'_{n,p}$, which gets as input a sequence of $m'(n)$ redundant representations of elements in GF(p). It first map each such representation to the canonical representation, and then emulate the computation of the Arithmetic circuit (over GF(p)) using the canonical representations all along. Specifically, the canonical representation of $e \in \mathrm{GF}(p)$ is the (zero-padded) binary expansion of e, and the other representations correspond to the binary expansions of $e + i \cdot p$ for each $i \in [\mathrm{poly}(n)]$. Hence, the output will be $t = \lceil \log_2 p \rceil$ bits long, whereas in the input of the Boolean circuit each field element will be represented by a block of $t' = t + O(\log n)$ bits.

Next, the Boolean formula $\phi'_{n,p}$ emulates the computation of the corresponding $\widehat{\phi}_n$ (over GF(p)). Specifically, each arithmetic gate will be replaced by an $\mathcal{NC}$ circuit that emulates the corresponding field operation.[14] Note that these gate-emulation circuits operate on (pairs of) inputs of length $\log_2 |\mathrm{GF}(p)| = \log_2 p$, and so their depth is poly($\log\log p$). Furthermore, the construction of these small circuits depends on the prime p.

In addition, we need to slightly modify the sequence of functions that determines the indices of the field elements fed to $\widehat{\phi}_n$. Suppose that the sequence fed

[14] Recall that integer arithmetics is in $\mathcal{NC}$; see, e.g., [23, Lect. 30].

to $\widehat{\phi}_n(w,\cdot)$ is determined by $\sigma'_{n,1},...,\sigma'_{n,m(n)} : \{0,1\}^{\ell'} \to [n]$. Thus, each $\sigma'_{n,j}$ determines the index of a field element in the input to $\widehat{\phi}_n$, and so it should be replaced by functions $\sigma''_{n,(t'-1)j+1},...,\sigma''_{n,t'j} : \{0,1\}^{\ell'} \to [t'n]$ that determine the corresponding bits in the input to $\phi'_{n,p}$ (i.e., $\sigma''_{n,(t'-1)j+k}(w) = (t'-1)\cdot\sigma'_{n,j}(w)+k$ for $k \in [t']$).

Note that the formula $\phi'_{n,p}$ outputs $t = \log_2 p$ bits, whereas we seek a Boolean formula with a single output bit. Such a Boolean formula is obtained by using an auxiliary input $i \in [t]$, which determines the bit to be output; that is, $\phi''_{n,p}(w,(\cdot,i))$ equals the ith bit of $\phi'_{n,p}(w,\cdot)$. Hence, on input $(y,i) \in \{0,1\}^{n\cdot O(\log n)} \times [t]$, where $y = (y_1,...,y_n)$ is a redundant representation of a sequence of n elements in GF(p), we consider the counting problem that corresponds to $\Phi''_{y,i}$ such that $\Phi''_{y,i}(w) = \phi''_{n,p}(w,(y,i)_{\sigma''_n(w)})$, where σ''_n denote the sequence of $\sigma''_{n,j}$'s (augmented by functions that indicate the bit positions of i in the input (y,i)).

In summary, we reduce the evaluation of $\widehat{\Phi}_{A'}$ on input $x \in \text{GF}(p)^n$ to counting the number of w's that satisfy $\Phi''_{y,i}(w) = 1$, for each $i \in [t]$, where $y \in \{0,1\}^{t'n}$ is a random representation of x. Specifically, the value of $\widehat{\Phi}_{A'}$ on $x = (x_1,...,x_n) \in \text{GF}(p)^n$ is obtained as follows.

1. For each $k \in [n]$, we randomly map $x_k \in \text{GF}(p)$ to a t'-bit string, denoted y_k, that represents it. Recall that $y_k \in \{0,1\}^{t'} \equiv [2^{t'}]$ is a random integer that is congruent to x_k modulo p.
2. For each $i \in [t]$, we compute the number of w's that satisfy $\Phi''_{y,i}(w) = 1$ by invoking the algorithm that supposedly solves the Boolean counting problem (on input $(y,i) = (y_1,...,y_n,i)$), where $\Phi''_{y,i}$ is as above (i.e., $\Phi''_{y,i}(w) = \phi''_{n,p}(w,(y,i)_{\sigma''_n(w)})$, where σ''_n denote the sequence of $\sigma''_{n,j}$'s).
   Recall that, for each $j \in [m(n)\cdot t']$, the formula $\sigma''_{n,j}$ determines integers in $[t'n + \log t]$ such that the value $\sigma''_{n,j}(w)$ determines the $(|w|+j)^{\text{th}}$ bit that will be fed to $\phi''_{n,p}(w,\cdot)$. (Note that the bits in locations $t'n+1,...,t'n+\log t$, which represent $i \in [t]$, will always be fed to $\phi''_{n,p}$.)
3. Denoting by c_i the count obtained by the ith call, we output the value $\sum_{i=1}^{t} c_i \cdot 2^{i-1} \bmod p$.

The key observation is that

$$\begin{aligned}\sum_{w\in\{0,1\}^{\ell'(n)}} \widehat{\Phi}_{A'}(x_1,...,x_n) &= \sum_{w\in\{0,1\}^{\ell'(n)}} \widehat{\phi}_n(w, x_{i_{n,w}^{(1)}},...,x_{i_{n,w}^{(m(n))}}) \\ &\equiv \sum_{w\in\{0,1\}^{\ell'(n)}} \sum_{i\in[t]} \phi''_{n,p}(w,(y,i)_{\sigma_n(w)})\cdot 2^{i-1} \pmod p \\ &\equiv \sum_{i\in[t]} 2^{i-1} \cdot \sum_{w\in\{0,1\}^{\ell'(n)}} \phi''_{n,p}(w,(y,i)_{\sigma_n(w)}) \pmod p.\end{aligned}$$

Hence, $\text{EV}_{A',\text{GF}(p)}(x) = \sum_{i\in[t]} 2^{i-1}\cdot \#_{B(n,p)}(y,i)$, where B is the algorithm that (on input n and p) generates $\phi''_{n,p}$ (and σ_n).

Note that, when given a uniformly distributed $x \in \mathrm{GF}(p)^n$, the ith query made by our reduction is almost uniformly distributed in $S_i \stackrel{\mathrm{def}}{=} \{(y,i) : y \in \{0,1\}^{n \cdot t'}\}$, where the small deviation arises from the fact that some elements in $\mathrm{GF}(p)$ have $\lfloor 2^{t'}/p \rfloor$ representations, whereas others have $\lceil 2^{t'}/p \rceil$ representations.

Now, suppose that we are given an algorithm for the Boolean counting problem associated with $B(n,p)$ such that for every $i \in [t]$ the algorithm errs on an η_i fraction of the inputs in S_i. Then, the probability that our reduction solves $\mathtt{EV}_{A',\mathrm{GF}(p)}$ on a random $x \in \mathrm{GF}(p)^n$ is at least $1 - \sum_{i \in [t]} \eta_i - \frac{n}{p}$ (where the $\frac{n}{p}$ term is due to the aforementioned deviation). Letting $\eta = \mathrm{E}_{i \in [t]}[\eta_i]$ denote the fraction of the instances ($\{0,1\}^{n \cdot t'} \times [t]$) on which the counting algorithm errs, it follows that our reduction has error rate at most $t \cdot \eta + \frac{n}{p}$.

Hence, our reduction errs with probability exceeding $1/3$ on at most an $3 \cdot (t \cdot \eta + \frac{n}{p}) < 4t\eta$ fraction of the instances (in $\mathrm{GF}(p)^n$). Recalling that we seek to solve $\mathtt{EV}_{A',\mathrm{GF}(p)}$ on at least a $8/9$ fraction of the inputs in $\mathrm{GF}(p)^n$, we need a counting algorithm that errs on at most an $\eta = 1/36t = 1/O(\log n)$ fraction of its inputs.

The claim follows, except that our counting problem refers to $2^{\ell'(n)} = 2^{\ell''(n)}$ terms and to input-length of $n'' = t'n + \log t = \widetilde{O}(n)$, whereas the claim asserts a reduction to a counting problem that refers to $2^{\ell''(n'')}$ terms. This formal discrepancy can be fixed by introducing $2^{\ell''(n'')} - 2^{\ell''(n)}$ dummy terms. Specifically, for $\lambda = \ell''(n'') - \ell''(n) \approx c \cdot \log(n''/n) \approx c \cdot \log t'$, we consider counting the number of $(w,v) \in \{0,1\}^{\ell''(n)+\lambda}$ that satisfy $\Phi_{y,i}$, where $\Phi_{y,i}(w,1^\lambda) = \Phi''_{y,i}(w)$ and $\Phi_{y,i}(w,v) = 0$ for every $v \neq 1^\lambda$. ■

3.1.2 An Intermediate Conclusion

Combining Propositions 3.3, 3.4 and 3.6, we obtain a weaker version of Theorem 3.1 in which the tolerated error rate is smaller and the class of admissible functions is slightly more restricted. Specifically:

Corollary 3.7 (weak version of Theorem 3.1): *Let $\mathcal{D}$ be a set of admissible functions that is further closed under multiplication by* $\mathrm{poly}(\log\log n)$ *factors, and ℓ be a logarithmic function. Then, for every counting problem Π in $\mathcal{C}_{\ell,\mathcal{D}}$, there exists a counting problem Π' in $\mathcal{C}_{\ell,\mathcal{D}}$ and an almost-linear time randomized reduction of solving Π on the worst case to solving Π' on at least* $1-(1/O(\log n))$ *fraction of the domain.*

Note that the extra hypothesis regarding $\mathcal{D}$ is satisfied in the case that $\mathcal{D}$ contains a function d such that $d(n) \geq \exp((\log\log n)^{\Omega(1)})$.

Proof: We select a prime number $p \in (2^{\ell(n)}, 2^{\ell(n)+1})$ at random, and observe that (by Proposition 3.3) the original (worst-case) counting problem $\#_A$ is reduced to the Arithmetic evaluation problem $\mathtt{EV}_{A',\mathrm{GF}(p)}$ (i.e., evaluating the Arithmetic formula $\widehat{\Phi}_{A'}$ over $\mathrm{GF}(p)$). We apply the worst-case to average-case reduction of Proposition 3.4 to the evaluation of $\widehat{\Phi}_{A'}$ over $\mathrm{GF}(p)$, and next apply the reduction of Proposition 3.6, which yields a counting problem $\#_{B(p)}$

for some adequate algorithm B (as in Definition 3.5). Hence, we map the parameter $d' \in \mathcal{D}$ of $\widehat{\Phi}_{A'}$ to a parameter d'' of the new counting problem such that $d''(\widetilde{O}(n)) = \text{poly}(\log\log n) \cdot d'(n)$. Actually, since both reduction are randomized, we first apply error reduction so that instances that are originally solved correctly (with probability at least 2/3) are solved correctly with probability at least 0.9.

We stress that the prime p is selected, upfront, by our composed reduction; that is, it is determined prior to applying Propositions 3.3, 3.4 and 3.6. Thus, the algorithm underlying the final counting problem (i.e., B) views p as part of its input (as in Definition 3.5), whereas the algorithm in Definition 2.3 gets no such input. To bridge the gap, we include this part of B's input (i.e., p) in the input to the counting problem to which we reduce; that is, the input to the counting problem is now $(p,(y,i))$ rather than being $y' \stackrel{\text{def}}{=} (y,i)$ as in the proof of Proposition 3.6. Hence, we consider the counting problem $\#_{B'}$ such that $\#_{B'}(p,y') = \#_{B(p)}(y')$.

We stress that B' is a small variation on B (rather than being B itself): This is obvious given that B gets the input (n,p), whereas B' gets the input n only. In addition, $B'(n)$ should construct small circuits that emulate GF(p) operations, when $p \in (2^{\ell(n)}, 2^{\ell(n)+1})$ is part of their input, whereas $B(n,p)$ constructs small circuits that emulate GF(p) operations for a fixed p. Fortunately, the former circuits are still in uniform $\mathcal{NC}$. We also augment the formula $\sigma''_{n''}$ such that p is always fed to the formula $\phi''_{n''}$ (where $\sigma''_{n''}$ and $\phi''_{n''}$ are as in the proof of Proposition 3.6). The minor increase in the length of the input of the counting problem may require additional padding of the index of the summation (but this is needed only if $\ell(n' + \ell(n')) > \ell(n')$, which is quite unlikely since $\ell(n) = O(\log n)$).

The analysis uses the fact that if the final counting problem (i.e., $\#_{B'}$) is solved correctly with probability $1-\eta$, when the probability is take over pairs of the form (p,y'), then, with probability at least 0.9 over the choice of p, the residual counting problem $\#_{B'}(p,\cdot) \equiv \#_{B(p)}$ is solved correctly with probability $1-10\eta$. Hence, the resulting composed reduction reduces solving $\#_A$ on the worst-case to solving $\#_{B'}$ on a $1-\eta$ fraction of the instances, provided that $10\eta \leq 1/36t$, where $t = \lceil \log_2 p \rceil = \ell(n)+1$ is as in the proof of Proposition 3.6. The success probability of the entire (composed) reduction (on each input) is at least $0.9^3 > 2/3$, where one 0.9 factor arises from the choice of p, and the other two factors are due to the two randomized reductions that we use (whose error probability was reduction to 0.1 (see above)). ■

Digest. When tracing the cost of the reduction captured by Corollary 3.7, one should focus on the depth of the various formulae. The reduction in Proposition 3.3 preserves the depth, while the reduction in Proposition 3.6 increases the depth by a $\text{poly}(\log\log p) \leq \text{poly}(\log\log n)$ factor, where p is the size of the field in use. Recall that Proposition 3.4 does not change the formula, and that the depth overhead in Proposition 3.6 is due to the implementation of the field's operations by Boolean formulae. Jumping ahead, we note that using the following Proposition 3.8 allows using $p = \exp(d(n))$ (rather than $p = \text{poly}(n)$), which means that the depth increase is only a $\text{poly}(\log d(n))$ factor.

3.2 Deriving the Original Conclusion (of Theorem 3.1)

The small level of error rate that is allowed by Corollary 3.7 is rooted mainly in the fact that the arithmetic evaluation over a field of size $p > 2^{\ell(n)}$ (which refers to $\widehat{\phi}_n$ (and to the $i_{n,w}^{(j)}$'s)) is emulated by $t = \log_2 p$ invocations of a Boolean counting problem, which refers to the Boolean formula $\phi''_{n,p}$ (and to σ''_n). The key observation, made in the proof of Proposition 3.6, is that

$$\sum_{w\in\{0,1\}^{\ell(n)}} \widehat{\phi}_n(w, x_{i_{n,w}^{(1)}}, ..., x_{i_{n,w}^{(m(n))}})$$
$$\equiv \sum_{w\in\{0,1\}^{\ell(n)}} \sum_{i\in[t]} \phi''_{n,p}(w, (y,i)_{\sigma''_n(w)}) \cdot 2^{i-1} \pmod{p}. \tag{3}$$

In the proof of Proposition 3.6, we computed Eq. (3) by counting, *separately, for each* $i \in [t]$, the number of w's that satisfy $\phi''_n(w, (y,i)_{\sigma''_n(w)})$, and taking a weighted (by 2^{i-1}) sum of these counts (modulo p). But an alternative solution is to replace these weights by auxiliary summations; that is, Eq. (3) can be computed by

$$\sum_{i\in[t]} \sum_{w\in\{0,1\}^{\ell(n)}} \sum_{u\in\{0,1\}^t} \phi'''_{n,p}(w, u, (y,i)_{\sigma''_n(w)}) \tag{4}$$

such that $\phi'''_{n,p}(w, u, (y,i)_{\sigma''_n(w)}) = \phi'''_{n,p}(w, (y,i)_{\sigma''_n(w)})$ if $u \in \{0,1\}^t$ ends with $t-(i-1)$ ones (i.e., $u \in \{u'1^{t-(i-1)} : u' \in \{0,1\}^{i-1}\}$) and $\phi'''_{n,p}(w, u, (y,i)_{\sigma''_n(w)}) = 0$ otherwise.

The problem with the foregoing solution is that it doubles the length of the index of the summation (i.e., $|(i,w,u)| > \ell(n) + t \geq 2\ell(n) = 2|w|$). The source of the trouble is that the reduction of the (Boolean) counting to the Arithmetic evaluation (i.e., Proposition 3.3) uses a setting of $p > 2^{\ell(n)}$ (which implies $t = \log_2 p > \ell(n)$).

As hinted before, we resolve this problem as well as another problem (i.e., the problem of too large depth overhead) by using a reduction of the Boolean counting problem to $O(\ell(n)/d(n))$ Arithmetic problems that refer to fields of size $\exp(d(n)) = n^{o(1)}$ (rather than of size $2^{\ell(n)} = \text{poly}(n)$). (Recall that the field size should be at least $\exp(d(n))$ for the application of Proposition 3.4 (which requires the field to be larger than the degree bound).) The corresponding $O(\ell(n)/d(n))$ values will be combined using *Chinese Remaindering with errors* (cf. [15]) so that we only incur a constant loss in the error rate (rather than a loss factor of $O(\ell(n)/d(n))$). This "CRT with errors" is reminiscent of the Berlekamp–Welch algorithm, which was employed in the proof of Proposition 3.4 (in order to obtain a constant error rate rather than an error rate that is inversely proportional to the degree of the polynomial).

3.2.1 Revisiting the Three Basic Reductions

We start by presenting revised versions of the three reductions presented in Sect. 3.1 (i.e., Propositions 3.3, 3.4 and 3.6). In the revised versions of Propositions 3.3 and 3.6 we use a field of size $\exp(d(n))$ (rather than of size $\text{poly}(n)$). This already yields improvement in some parameters. In the revised version of Proposition 3.4 (i.e., the worst-case to average-case reduction for the arithmetic problem) we use a more refined analysis that allows to reduce the required success rate from 8/9 to $0.5 + \epsilon$, for any $\epsilon > 0$, while requiring a field size that is $O(1/\epsilon^2)$ times larger than the degree of the polynomial. The most important improvement comes from the last reduction (i.e., a revision of Propositions 3.6), which executes the emulation of Eq. (3) by Eq. (4).

Revised Reduction of the Boolean Problem to the Arithmetic One. For sake of simplicity, we first present an alternative to Proposition 3.3 using CRT with no errors. Recall that Proposition 3.3 (as well as the following alternative) is a deterministic reduction from solving the Boolean problem (in the worst-case) to solving the Arithmetic problem (in the worst-case). In the following alternative, we consider the same evaluation problem for several fields, while assuming that the answers obtained for these different fields are all correct. In contrast, we shall use "CRT with errors" when actually proving Theorem 3.1, which will require dealing with a situation in which only the answers obtained for 51% of these fields are correct.

Proposition 3.8 (Proposition 3.3, revised): *Solving the counting problem associated with an algorithm A and functions ℓ, m and $d \in \mathcal{D}$ is reducible in almost-linear time to $O(\ell(n)/d(n)) = o(\log n)$ evaluation problems that are all associated with the same algorithm A', and finite fields of prime cardinality at least $s = \exp(d(n))$. Furthermore, the reduction makes a single query to each problem, the degree of the polynomial $\widehat{\Phi}_{A'}$ is $o(s)$, and the functions ℓ, m and d equal those in the counting problem.*

Note that s is set as small as possible subject to being sufficiently larger than the degree of $\widehat{\Phi}_{A'}$. We wish s to be small because it determines the various overheads that we shall incur when emulating the Arithmetic formula by Boolean formula (in Proposition 3.10). (On the other hand, using Proposition 3.9 requires that s be sufficiently larger than the degree of $\widehat{\Phi}_{A'}$.)

Proof: Following the proof of Proposition 3.3, while using the field GF(p) for an arbitrary prime p, yields a (worst-case) reduction of the "counting (mod p) problem $\#_A$" to the evaluation problem associated with A' and GF(p), where by counting (mod p) problem $\#_A$ we mean computing $\left(\sum_{w\in\{0,1\}^{\ell(n)}} \Phi_x(w)\right) \bmod p$. Invoking this reduction for $\tau = O(\ell(n)/d(n))$ different primes p of size $\exp(d(n))$, we obtain the sum modulo each of these primes. Finally, invoking the Chinese Remainder Theorem, the claim follows provided that $\exp(d(n))^\tau > 2^{\ell(n)}$, which holds for any $\tau > \frac{\ell(n)}{O(d(n))}$. ■

Revised Worst-Case to Average-Case Reduction for the Arithmetic Problem. Foreseeing that it will be more convenient to perform the third reduction when referring to the notion of success rate (see Sect. 1.5), we state the target of the second reduction in these terms. More importantly, we use a more refined analysis, which allows working with success rate $0.5 + o(1)$ rather than 8/9 (or $5/6 + o(1)$).[15]

Proposition 3.9 (Proposition 3.4, revised): *Let $P : \mathcal{F}^n \to \mathcal{F}$ be a polynomial of total degree $D = o(\epsilon^2 \cdot |\mathcal{F}|)$. Then, evaluating P on any input can be randomly reduced to evaluating P with success rate at least $0.5 + \epsilon$, by invoking the latter evaluation procedure for $O(D/\epsilon^2)$ times.*

Recall that the reduction is randomized, and, on each input, it yields the correct answer with probability at least 2/3.

Proof: On input $x \in \mathcal{F}^n$, we select a random curve of degree two that passes through x, and invoke the evaluation procedure on the first $m = O(D/\epsilon^2)$ points of this curve. Note that the queried points are pairwise independent and uniformly distributed in $\mathcal{F}^n$. Hence, with probability at least 2/3, the evaluation procedure answers correctly on at least a $0.5 + 0.5\epsilon$ fraction of the queried points.[16] Using the Berlekamp–Welch algorithm, we reconstruct the unique degree D polynomial that agrees with at least $(0.5 + 0.5\epsilon) \cdot m$ of the answers, and return its value at x. (Uniqueness is guaranteed by $D < \epsilon \cdot m$.) Hence, the returned value is correct with probability at least 2/3. ■

Revised Average-Case Reduction from the Arithmetic Problem to the Boolean One. We finally get to the core of the entire revision, which is the improved reduction that takes us back from the Arithmetic world to the Boolean world. This improved reduction is based on the emulation of Eq. (3) by Eq. (4).

Proposition 3.10 (Proposition 3.6, revised): *Let A' be as in Definition 3.2 and $\widehat{\Phi}_{A'}$ be the corresponding polynomial, where ℓ', d' and m' are the corresponding functions. Then, for every prime $p \leq \text{poly}(n)$, the problem of evaluating $\widehat{\Phi}_{A'} : \text{GF}(p)^n \to \text{GF}(p)$ with success rate at least ρ is randomly reducible in almost-linear time to solving the counting problem associated with $B(n'', p)$ with success rate at least $\rho + (1/\text{poly}(n))$, where B is as in Definition 3.5 and $n'' = \exp(d'(n)) \cdot n$. Furthermore, the reduction makes a single query, and the functions corresponding to the Boolean problem are $\ell'' = \ell'$, $d''(n'') \leq \text{poly}(\log\log p) \cdot d'(n)$, and $m''(n'') = O(\log n) \cdot m'(n)$.*

---

[15] The bound of 8/9 used in the proof of Proposition 3.4 is a consequence of lower-bounding by 2/3 the probability that two third of the points on a random line that passes through a given point are answered correctly. However, a lower-bound of $0.5 + o(1)$ would have been as good (when combined with error reduction).

[16] Indeed, this move from the success rate (of P) to a fraction of correct answers (on a random curve) is analogous to the proof of Proposition 1.5; in the current context we get it "for free".

We shall use the setting $\rho = 0.5 + \epsilon > 0.5$. We stress that Proposition 3.10 improves over Proposition 3.6 in its preservation of the success rate (whereas in Proposition 3.6 solving problem of evaluating $\widehat{\Phi}_{A'}$ with success rate $8/9$ was reduced to solving the counting problem associated with $B(n'', p)$ with success rate $1 - 1/O(\log n)$).

Proof: Following the proof of Proposition 3.6, we derive the same formula $\phi'_{n,p}$, and recall that the formula $\phi'_{n,p}$ outputs $t = \log_2 p$ bits, whereas we seek a Boolean formula with a single output bit. Again, the latter formula is derived by using an auxiliary input $i \in [t]$, which determines the bit to be output; that is, $\phi''_n(w, i, z)$ equals the ith bit of $\phi'_n(w, z)$, where $z = y_{\sigma''_n(w)}$ is a projection of the bits of the input y to the counting problem. Unlike in the proof of Proposition 3.6, we shall not view i as part of the input to the counting problem (which is fed into $\phi''_{n,p}$, just like z), but rather as part of the index of summation (i.e., just as w). The corresponding Boolean formula Φ''_y (per Eq. (1)) is such that $\Phi''_y(w, i) = \phi''_{n,p}(w, i, y_{\sigma''_n(w)})$, where σ''_n denote the sequence of $\sigma''_{n,j}$'s. (That is, the input to this counting problem is y, and the desired output is the number of pairs (w, i) that satisfy Φ''_y.) Recall that, on input $x \in \mathrm{GF}(p)^n$, which is encoded by $y \in \{0,1\}^{t'n}$, we do not wish to obtain $\sum_{w,i} \Phi''_y(w, i)$, but rather wish to obtain the related sum

$$\sum_{w \in \{0,1\}^{\ell'(n)}} \sum_{i \in [t]} 2^{i-1} \cdot \Phi''_y(w, i), \tag{5}$$

where $y \in \{0,1\}^{t'n}$ is a random representation of x. The difficulty is that Eq. (5) does not correspond to a counting problem, but this can be fixed by replacing the scalar multiplication by 2^{i-1} with a sum over 2^{i-1} terms. Specifically, consider $\Phi'''_y : \{0,1\}^{\ell'(n)} \times [t] \times \{0,1\}^t$ such that $\Phi'''_y(w, i, u) = \Phi''_y(w, i)$ if $u \in \{u'1^{t-(i-1)} : u' \in \{0,1\}^{i-1}\}$ and $\Phi'''_y(w, i, u) = 0$ otherwise. Thus, Eq. (5) equals

$$\sum_{w \in \{0,1\}^{\ell'(n)}} \sum_{i \in [t]} \sum_{u \in \{0,1\}^t} \Phi'''_y(w, i, u). \tag{6}$$

Hence, we reduce the evaluation of $\widehat{\Phi}_{A'}$ on $x \in \mathrm{GF}(p)^n$ to counting the number of (w, i, u)'s that satisfy Φ'''_y, where $y \in \{0,1\}^{t'n}$ is a random representation of x. Specifically, the value of $\widehat{\Phi}_{A'}$ on $x = (x_1, ..., x_n) \in \mathrm{GF}(p)^n$ is obtained as follows.

1. As in the proof of Proposition 3.6, for each $k \in [n]$, we randomly map $x_k \in \mathrm{GF}(p)$ to a random t'-bit string, denoted y_k, that represents it. Recall that $y_k \in \{0,1\}^{t'} \equiv [2^{t'}]$ is a random integer that is congruent to x_k modulo p (and that $t' = t + O(\log n)$).
2. Compute the number of (w, i, u)'s that satisfy $\Phi'''_y(w, i, u) = 1$ by invoking the algorithm that supposedly solves the counting problem (on input $y = (y_1, ..., y_n)$), where Φ'''_y is as in Eq. (6).
3. Denoting by c the count obtained by the foregoing invocation, output the value $c \bmod p$.

As outlined in the beginning of this section (see Eq. (3)–(4)), the key observation is that

$$\begin{aligned}\sum_{w\in\{0,1\}^{\ell'(n)}} \widehat{\Phi}_{A'}(x_1,...,x_n) &= \sum_{w\in\{0,1\}^{\ell'(n)}} \widehat{\phi}_n(w, x_{i^{(1)}_{n,w}}, ..., x_{i^{(m(n))}_{n,w}}) \\ &\equiv \sum_{w\in\{0,1\}^{\ell'(n)}} \sum_{i\in[t]} \phi''_{n,p}(w,(y,i)_{\sigma''_n(w)}) \cdot 2^{i-1} \pmod p \\ &\equiv \sum_{i\in[t]} \sum_{(w,u)\in\{0,1\}^{\ell'(n)+t}} \phi'''_{n,p}(w,i,u,y_{\sigma''_n(w)}) \pmod p.\end{aligned}$$

where $\phi'''_{n,p}(w,i,u,z) = \phi''_{n,p}(w,i,z)$ if $u \in \{0,1\}^{i-1} \times \{1\}^{t-i-1}$ and $\phi'''_{n,p}(w,i,u,z) = 0$ otherwise. Hence, $\mathrm{EV}_{A',\mathrm{GF}(p)}(x) = \#_{B(p)}(y,i)$, where B is the algorithm that (on input p) generates $\phi'''_{n,p}$ (and σ''_n).

Note that the single query made by our algorithm is almost uniformly distributed in $\{0,1\}^{n\cdot t'}$, where the small deviation arises from the fact that some elements in GF(p) have $\lfloor 2^{t'}/p \rfloor$ representations, whereas others have $\lceil 2^{t'}/p \rceil$ representations. Hence, if the invoked algorithm has success rate ρ' (on inputs in $\{0,1\}^{nt'}$), then our algorithm will have success rate at least $\rho' - n\cdot(p/2^{t'})$ (on inputs in $\mathcal{F}^n$).

The claim follows, except that our counting problem refers to $2^{\ell'(n)+t+\log t}$ terms and to input-length of $t'n + \log t = \widetilde{O}(n)$, whereas the claim asserts a reduction to a counting problem that refers to $2^{\ell''(n'')}$ terms and to input length $n'' = \exp(d'(n))\cdot n$. This can be fixed by artificially padding the input to length n'' and noting that $\ell''(n'') \geq \ell'(n) + t + \log t$ (where $t = \log p = d'(n)$).[17] ■

3.2.2 Proof of Theorem 3.1

The claim of Theorem 3.1 follows by combining the revised reductions provided by Propositions 3.8–3.10. Specifically, using (a generalization of) Proposition 3.8, we reduce the original counting problem $\#_A$ to $\tau = O(\ell(n)/d(n))$ instances of the Arithmetic evaluation problem, where all instances refer to the same algorithm A' but to different primes $p > s = \exp(d(n))$. Note that we can afford to find and use the first τ such primes (since we can run for poly(s)-time). Hence, in the final counting problem, denoted $\#_B$, the input will be prepended by an index $j \in [\tau]$ of one of these primes, and Propositions 3.9 and 3.10 will be applied to each of these primes.

As hinted above, we shall be using a generalization of Proposition 3.8, rather than Proposition 3.8 itself. This generalization will allow us to solve $\#_A$ on the worst-case even when solving, on the worst-case, only most of the Arithmetic evaluation problems associated with A', rather than all of them. Hence, in this step, we lose a factor of $2+o(1)$ in the error rate, rather than losing a factor of τ. The foregoing generalization is obtained by combining the τ results, which refer to the count modulo different primes, using *Chinese Remaindering with errors*

[17] This uses the hypothesis that $\ell''(n) = c\log n$, which implies that $\ell''(\exp(d'(n))\cdot n) = O(d'(n)) + \ell''(n)$.

(cf. [15]). Indeed, the error-correcting version of the CRT will be used instead of the plain version used in Proposition 3.8, and the number of primes is increased (from $(\ell(n)/d(n)) + O(1)$ to $O(\ell(n)/d(n)) = \tau$) so as to allow for such an error correcting feature.

(We highlight the fact that, since we are using smaller primes (i.e., primes of size $\exp(d(n))$ rather than $\text{poly}(n)$), applying Proposition 3.10 increases the depth of the formula by a factor of $\text{poly}(\log d(n))$, rather than by a factor of $\text{poly}(\log\log n)$, and increases the length of the input by a factor of $\exp(d(n))$.)

Let us spell out the reduction that is obtained by combining all the above. Recall that we are given an input $x \in \{0,1\}^n$ to a counting problem $\#_A$ associated with an algorithm A. For an arbitrary small constant $\epsilon > 0$, we proceed in three steps, which correspond to the three foregoing reductions.

1. Denoting the first $\tau = O(\epsilon^{-1} \cdot \ell(n)/d(n))$ primes that are greater than $s = \exp(d(n))$ by $p_1, ..., p_\tau$, we consider the τ evaluation problems obtained by applying the generalized Proposition 3.8 to algorithm A (i.e., all these problems are associated with algorithm A'). The ith such problem, denoted $\text{EV}_{A',\text{GF}(p_i)}$, consists of evaluating the polynomial $\widehat{\Phi}_{A'}$ at points in $\text{GF}(p_i)^n$, and we shall apply it at the point $x \in \{0,1\}^n$, viewed as an element of $\text{GF}(p_i)^n$.
   The difference between Proposition 3.8 and its generalization will arise when we reconstruct the answer to $\#_A$: We shall be employing CRT with errors and reconstruct $\#_A(x)$ even when obtaining the correct value of $\text{EV}_{A',\text{GF}(p_i)}(x)$ only for $(0.5+\epsilon)\cdot\tau$ of the i's.
2. For each $i \in [\tau]$, we use Proposition 3.9 to reduce the evaluation of $\widehat{\Phi}_{A'}$ at $x \in \text{GF}(p_i)^n$ to its evaluation at $s' = O(s/\epsilon^2)$ points in $\text{GF}(p_i)^n$, denoted $x^{(i,j)}$. Actually, this procedure is repeated for $O(\log\tau)$ times, and the plurality value is used, so that the probability of error is smaller than $0.1/\tau$ (rather than smaller than $1/3$).
   We stress that at this point we only determined the $x^{(i,j)}$'s, where $i \in [\tau]$ and $j \in [s'']$ (where $s'' = O(s' \log\tau)$).
3. For each $i \in [\tau]$, we apply Proposition 3.10 to $\widehat{\Phi}_{A'}$, while providing p_i as an auxiliary input. Denoting the resulting Boolean formulae by $\Phi^{(i)}$'s, we consider the Boolean formula Φ that is given i as auxiliary input and applies the relevant $\Phi^{(i)}$; that is, $\Phi(i,z) = \bigvee_{k\in[\tau]}(i = k \wedge \Phi^{(k)}(z))$. Denoting the corresponding algorithm that produces Φ by B, the foregoing specifies the counting problem $\#_B$.
   At this point, for each $i \in [\tau]$ and $j \in [s'']$, we select uniformly at random a representation $y^{(i,j)} \in \{0,1\}^{n'}$ of $x^{(i,j)} \in \text{GF}(p_i)^n$, where $n' = \exp(d(n)) \cdot n$.

The actual computation (of the reduction) goes in the opposite direction. For every $i \in [\tau]$ and $j \in [s'']$, let $v^{(i,j)}$ denote the answer provided (by a hypothetical solver of the counting problem $\#_B$) for the input $y^{(i,j)}$; that is, $v^{(i,j)}$ is supposed to equal $\#_B(i, y^{(i,j)})$, where by Proposition 3.10 $\#_B(i, y^{(i,j)}) \equiv \text{EV}_{A',\text{GF}(p_i)}(x^{(i,j)}) \pmod{p_i}$ holds with very high probability. Now, for each $i \in [\tau]$, we apply the decoding procedure (of Proposition 3.9, with error reduction) to the values $v^{(i,1)}, ..., v^{(i,s'')}$, and obtain a value $v^{(i)} \in \text{GF}(p_i)$, which is supposed to equal $\#_A(x) \bmod p_i$. Finally, we perform Chinese Remaindering with errors on the $v^{(i)}$'s, and obtain the desired value of $\#_A(x)$, with high probability.

In the analysis, we observe that if some procedure P solves the final counting problem (i.e., $\#_B$) with success rate at least a $0.75 + \epsilon$ (on instances $(i, y) \in [\tau] \times \{0,1\}^{n'}$), then, for at least $0.5 + \epsilon$ of the $i \in [\tau]$, the procedure P solves this counting problem with success rate at least a $0.5+\epsilon$ (on instances of the form $(i, \cdot)$). Hence, for each of these majority i's, Proposition 3.10 yields a procedure that solves $\mathtt{EV}_{A',\mathrm{GF}(p_i)}$ with success rate at least $(0.5+\epsilon-o(1))$ (over instances in $\mathrm{GF}(p_i)^n$). This means that, for each of these i's, the hypothesis of Proposition 3.9 holds, and so (for each of these i's) the decoding procedure (of Proposition 3.9) obtains (w.h.p.) the value of $\#_A(x) \bmod p_i$. In this case, Chinese Remaindering with errors works, since the error rate is below its resiliency rate (i.e., $\frac{\log p_1}{\log p_1 + \log p_\tau} - \frac{\epsilon}{2}$). The theorem follows. ∎

4 The Average-Case to Rare-Case Reduction

Fixing any admissible class $\mathcal{D}$ and a logarithmic function ℓ, and letting $\mathcal{C} = \mathcal{C}_{\ell,\mathcal{D}}$, Theorem 3.1 provides an almost-linear time randomized reduction of solving any problem Π in $\mathcal{C}$ on the worst-case to solving some problem Π' in $\mathcal{C}$ on the average-case, where the notion of average-case requires solving the problem on at least a 0.76 fraction of the instances. In this section we show that, for every $d \in \mathcal{D}$, the latter (average-case) task can be reduced to solving a related problem on at least a $\exp(-d(n))$ fraction of the instances. Combined, these reductions show that solving the latter related problem on a noticeable fraction of its domain is essentially as hard as solving Π on the worst-case.

We actually show two related results of the foregoing flavor. The first result, stated in Theorem 1.3, shows a *non-uniform* reduction from the average-case task Π' to solving some problem Π'' in $\mathcal{C}$ on at least a $\exp(-d(n))$ fraction of the instances. The second result, stated in Theorem 4.4, shows a *uniform* reduction to solving some problem $\widehat{\Pi}$. While $\widehat{\Pi}$ is not in $\mathcal{C}$, it can be solved in $\mathrm{Dtime}(p(n) \cdot n^{1+o(1)})$, where $p(n) = 2^{\ell(n)} \cdot \exp(d(n))$ bounds the worst-case time of solving Π'. On a technical level, both results use *sample-aided reductions*: reductions that are given uniformly-distributed "solved instances" of the problem that they aim to solve. We provide a definition of this relaxed notion of a reduction between problems, which has been implicit in several past works, and which we find to be of independent interest.

Overview. Our starting point is the realization that when referring to such low success rate (i.e., a $\exp(-d(n))$ fraction for some $d \in \mathcal{D}$), we are in the *regime of list decoding*. Hence, for starters, Proposition 3.9 should (and can) be replaced by the list decoding result of Sudan, Trevisan, and Vadhan [31, Thm. 29]. This result essentially asserts the existence of an *explicit list of oracle machines such that, for every low-degree polynomial $P : \mathcal{F}^n \to \mathcal{F}$ and every $F : \mathcal{F}^n \to \mathcal{F}$ that agrees with P on a noticeable fraction of the instances, at least one of these machines computes P correctly on all instances when given oracle access to F.*

Trying to integrate this result in the procedure presented in Sect. 3.2 means that we proceed as follows: First, we reduce solving the Boolean counting problem Π' on the average-case (i.e., on 76% of the instances) to evaluating a polynomial $\widehat{\Phi}$ over $\exp(d(n))$-many different prime fields (on all instances). Next, we apply the foregoing reduction of [31, Thm. 29] in each of these fields, and finally we reduce the oracle calls made by the latter reductions to solving the Boolean counting problem Π'' (on a noticeable fraction of the instances). (Although the statement of [31, Thm. 29] only claims running time that is polynomial in all relevant parameters, it is clear that the running time is linear in the number of variables.)

As in Sect. 3.2, the input to Π'' is a pair of the form (p, y), where p is a prime and y represents an input to the problem of evaluating $\widehat{\Phi}$ in $\mathrm{GF}(p)$. Hence, solving Π'' on a noticeable fraction of its inputs implies that for a noticeable fraction of the primes p, we correctly solve Π'' on a noticeable fraction of the inputs of the form $(p, \cdot)$. This implies that, when given oracle access to such a rare-case solver for $\Pi''(p, \cdot)$, one of the foregoing oracle machines computes $\widehat{\Phi} : \mathrm{GF}(p)^n \to \mathrm{GF}(p)$ correctly on all inputs. If we can identify these primes p and the corresponding oracle machines, then we can solve Π' (on all its instances). (Jumping ahead, we mention that we shall only solve Π' on $1 - o(1) > 76\%$ fraction of its instances, because our identification of the good machines will be approximate in the sense that machines that are correct on almost all their inputs may also pass.)

To see how we can identify these primes and machines, suppose that we have access to uniformly distributed "solved instances" of Π' (i.e., we get a sample of pairs $(r, \Pi'(r))$ for uniformly distributed r's). Using such a sample of solved instances it is easy to estimate the probability that a procedure (e.g., an oracle machine equipt with an adequate oracle) correctly computes Π' mod p: We just compare the value provided for each of the sample points r to the value computed by the procedure. In particular, we can distinguish a procedure that is always correct from a procedure that errs on at least $1/\log n$ of the inputs. Hence, given a list of oracle machines for each of the ($\exp(d(n))$-many) primes, we can pick $\ell(n)/d(n)$ distinct primes (i.e., p's), and an oracle machine for each such prime p such that the chosen machine computes Π' mod p correctly on at least $1 - (1/\log n)$ fraction of the inputs. Using Chinese Remaindering *without errors*, this allows us to compute Π' correctly on at least $1 - \frac{\ell(n)}{d(n)} \cdot \frac{1}{\log n} = 1 - o(1)$ fraction of the inputs. For details see Sect. 4.1.

Three Ways of Obtaining a Sample of Solved Instances. The foregoing discussion leaves us with the problem of obtaining uniformly distributed "solved instances" of Π'. There are two trivial solutions to this problem: The first is that such random solved instances may be available to us in the application, and the second is that such solved instances can be "hard-wired" in the reduction. The second solution yields a non-uniform reduction, which establishes Theorem 1.3. A third solution, detailed in Sect. 4.2, is that such samples can be obtained by a downwards self-reduction, whereas each problem in $\mathcal{C}$ can be reduced to a problem that has a suitable downwards self-reduction.[18] Unfortunately, we were not able to apply this idea to the reduction of Π' to Π''; instead, we apply it to the ("internal") reduction of the Arithmetic problem (i.e., to the reduction between the worst-case and rare-case versions of the Arithmetic problem). This yields the result stated in Theorem 4.4.

4.1 A Sample-Aided Reduction

We start by spelling out the notion of a reduction that obtains uniformly distributed "solved instances" of the problem that it tries to solve. We call such a reduction *sample-aided*, and formulate it in the context of average-case to average-case reductions (while noting that worst-case settings can be derived as special cases).

Definition 4.1 (sample-aided reductions, Definition 1.4 revisited): *Let* $\rho', \rho'' : \mathbb{N} \to [0,1]$ *and* $s : \mathbb{N} \to \mathbb{N}$. *Suppose that* M *is an oracle machine that, on input* $x \in \{0,1\}^n$, *obtains as an auxiliary input a sequence of* $s = s(n)$ *pairs of the form* $(r, v) \in \{0,1\}^{n+\ell(n)}$. *We say that* M *is an* sample-aided reduction of solving Π' on a ρ' fraction of the instances to solving Π'' on a ρ'' fraction of the instances *if, for every procedure* P *that answers correctly on at least a* $\rho''(n'')$ *fraction of the instances of length* n'', *it holds that*

$$\Pr_{r_1,...,r_s \in \{0,1\}^n}\left[\left|\mathtt{corr}_M^{P,\Pi'}(r_1, ..., r_s)\right| \geq \rho'(n) \cdot 2^n\right] > 2/3 \qquad (7)$$

where $\mathtt{corr}_M^{P,\Pi'}(r_1, ..., r_s)$ *denotes the set of inputs on which the reduction returns the correct answer; that is,* $x \in \mathtt{corr}_M^{P,\Pi'}(r_1, ..., r_s)$ *if and only if*

$$\Pr[M^P(x; (r_1, \Pi'(r_1)), ..., (r_s, \Pi'(r_s))) = \Pi'(x)] \geq 2/3, \qquad (8)$$

where the latter probability is taken over the coin tosses of the machine M *and the procedure* P.

The error probability bounds in Eqs. (7) and (8) can be reduced at a moderate cost. This is straightforwards for Eq. (8), but the error reduction for Eq. (7) comes at a (small) cost and requires some care. Specifically, in order to identify good samples (i.e., sequences $(r_1, ..., r_s)$ that satisfy the condition in Eq. (7)), we have to approximate the size of the set $\mathtt{corr}_M^{P,\Pi'}(r_1, ..., r_s)$. This can be done using an auxiliary sample of the solved instances, where $O(k/\epsilon^2)$ samples allow to approximate the density of this set to within a deviation of ϵ, witch probability $1 - 2^{-k}$. Hence, for every $\epsilon > 0$ and $k \in \mathbb{N}$, using $O(k \cdot s) + O(k/\epsilon^2)$ solved samples, we can output a sequence of s solved samples $\overline{r}$ such that, with probability at least $1 - \exp(-k)$, it holds that $\left|\mathtt{corr}_M^{P,\Pi'}(\overline{r})\right| > (\rho'(n) - \epsilon) \cdot 2^n$.

[18] Indeed, this follows a paradigm that can be traced to the work of Impagliazzo and Wigderson [20].

Although Definition 4.1 is stated in terms that fit average-case to rare-case (or average-case) reductions, the definition also applies to reductions from worst-case problems (by setting $\rho' = 1$).

We mention that sample-aided reductions are implicit in many known results (see, for example, [14,21,31]). Furthermore, any average-case to rare-case reduction of the "list-decoding" type (i.e., which outputs a short list of oracle machines that contains the correct one) yields a sample-aided reduction (as in Definition 4.1).[19]

Our Sample-Aided Reduction. Turning back to our specific context, we present a sample-aided reduction of solving any counting problem in $\mathcal{C}_{\ell,\mathcal{D}}$ on at least 99% of the inputs to solving some other problem in $\mathcal{C}_{\ell,\mathcal{D}}$ on 1% of the inputs. Actually, 99% and 1% can be replaced by $1 - \exp(-d(n))$ and $\exp(-d(n))$, respectively, for any $d \in \mathcal{D}$.

Theorem 4.2 (sample-aided reduction of average-case to rare-case for counting problems): *Let $\mathcal{D}$ be a set of admissible functions, ℓ be a logarithmic function, and $d \in \mathcal{D}$. For every counting problem Π' in $\mathcal{C}_{\ell,\mathcal{D}}$, there exists a counting problem Π'' in $\mathcal{C}_{\ell,\mathcal{D}}$ and an almost-linear time sample-aided reduction of solving $\Pi' : \{0,1\}^{n'} \to \mathbb{N}$ on at least $1 - \exp(-d(n'))$ fraction of the domain to solving $\Pi'' : \{0,1\}^{n''} \to \mathbb{N}$ on at least $\exp(-d(n''))$ fraction of the domain.*

In particular, Theorem 4.2 yields an almost-linear time *non-uniform* reduction of solving Π' on at least $1 - \exp(-d(n'))$ fraction of the domain to solving Π'' on at least $\exp(-d(n''))$ fraction of the domain. Combining this non-uniform reduction with Theorem 3.1, we obtain Theorem 1.3.

Proof Sketch: Given an input $x \in \{0,1\}^n$, we seek to output $\Pi'(x)$, and we are required to provide the correct output (with high probability) on at least a $1-\exp(-d(n))$ fraction of the possible x's. The general plan is to mimic the proof of Theorem 3.1, while replacing the worst-case to average-case reduction (of the arithmetic problem) by a worst-case to rare-case reduction, and coping with the difficulties that arise. Hence, we first reduce the counting problem Π' to the task of evaluating a polynomial $\widehat{\Phi}$ in many prime fields, then move from a worst-case version of the Arithmetic problem to a rare-case version of it, and finally reduce the latter to a rare-case version of solving the counting problem Π''.

We start by employing a (Boolean to Arithmetic) reduction as in Proposition 3.8, except that here we use $\text{poly}(D)$ primes of size $\text{poly}(D)$, where $D = \exp(d(n))$ denotes the degree of the polynomial $\widehat{\Phi}$ that is derived in Proposition 3.8; specifically, we refer to the first $\Theta(D \cdot \ell(n))$ primes that are larger

[19] The sample can be used to test the candidate oracle machines (as outlined in the foregoing discussion). Note that this allows to distinguish machines that are correct on all inputs from machines that err on a noticeable fraction of the inputs, but not to rule out machines that err on a negligible fraction of the inputs. Hence, a worst-case to rare-case reduction of the list-decoding type only yields a sample-aided reduction from solving the original problem on a $1 - o(1)$ fraction of the instances.

than poly(D). Note that at this point we only produce the polynomial $\widehat{\Phi}$ and determine the fields in which it will be evaluated. (Recall that $\widehat{\Phi}$ is a generic polynomial that will be evaluated over different finite fields. Jumping ahead, we note that, on input x, we shall only use the evaluation of $\widehat{\Phi}$ at x in $\ell(n)/d(n)$ of these fields, and combine these values using Chinese Remaindering (without errors).)

Next, for each prime p, we invoke the worst-case to average-case reduction of [31, Thm. 29] to our Arithmetic problem (i.e., evaluating $\widehat{\Phi}$ in one of the fields), obtaining $O(1/\epsilon)$ oracle machines that supposedly evaluate $\widehat{\Phi}$ on $\mathrm{GF}(p)^n$, when given oracle access to a procedure that computes $\widehat{\Phi}$ correctly on an ϵ fraction of the inputs, where $\epsilon = \exp(-d(n))$. Now, for each prime p, we invoke the (Arithmetic to Boolean) reduction of Proposition 3.10 so to transform the queries of these oracle machines to queries to the Boolean counting problem Π''. Specifically, the query $q \in \mathrm{GF}(p)^n$ is answered by querying Π'' on the pair $(p, \mathtt{r}(q))$, where $\mathtt{rr}(q)$ denotes a (random) representation of q. We also note that Π'' is actually fed the index of p in the said set of primes (or some other representation that selects such primes almost uniformly).[20]

(Recall that if the counting problem Π'' is solved correctly on an 2ϵ fraction of the inputs, then, for at least an ϵ fraction of the p's, the solution provided for at least ϵ fraction of the fairs $(p, \cdot)$ is correct. For each such p, at least one of the foregoing oracle machines will evaluate $\widehat{\Phi}$ correctly on $\mathrm{GF}(p)^n$ when fed with answers obtained from this "rare-case Π''-solver".)

For each prime p and for each of the corresponding oracle machines M, we approximate the success probability of M in evaluating $\widehat{\Phi}$ over a random element in $\{0,1\}^n$, which is viewed as an element of $\mathrm{GF}(p)^n$. Recall that when restricted to $\{0,1\}^n$, the polynomial $\widehat{\Phi}$ equals the value of the counting problem Π'. Hence, we can approximate the success probability of M in evaluating $\widehat{\Phi}$ over $\{0,1\}^n$ by using the solved sample for Π'. (We stress that while $\widehat{\Phi}$ and M are defined over $\mathrm{GF}(p)^n$, we approximate their agreement rate over $\{0,1\}^n$, which suffices for our application.)[21]

Specifically, for each pair $(r, v) \in \{0,1\}^{n+\ell(n)}$ in the sample (such that $v = \Pi'(r)$), we compare v to the output of M on input r, while answering the queries of M with the values obtained by the reduction to solving the counting problem Π'' such that the query $q \in \mathrm{GF}(p)^n$ is answered by taking the count provided for the input $(p, \mathtt{rr}(q))$ and reducing it modulo p. If any mismatch is found, then M is discarded as a candidate reduction, and if all oracle machines that correspond to p are eliminated then so is p itself.

Note that a (solved) sample of size $\exp(d(n))$ suffices to approximate all poly$(D) = \exp(d(n))$ quantities such that, with high probability, all values that

[20] Note that the formula Φ that underlies the counting problem Π'' can be implemented as a selector function that picks the corresponding formula $\Phi^{(p)}$ that emulates the computation of $\widehat{\Phi}$ in $\mathrm{GF}(p)^n$.

[21] In contrast, it is not clear how to approximate the agreement of $\widehat{\Phi}$ and M over $\mathrm{GF}(p)^n$, since we only have a solved sample of instances that are uniformly distributed in $\{0,1\}^n$.

are below $1 - \exp(-d(n))$ are estimated as smaller than 1. Hence, with high probability, the list of surviving machines will only contain machines that are correct on at least a $1 - \exp(-d(n))$ fraction of the inputs in $\{0,1\}^n$. Needless to say, with high probability, the said list contains all machines that are correct on all inputs, where the small probability of error is due to the error probability of the (randomized) oracle machines. Hence, the said list contains more that $\ell(n)/d(n)$ machines that correspond to different primes, since we can have $\epsilon \cdot \text{poly}(D) > \ell(n)$.

Lastly, we pick any $\ell(n)/d(n)$ surviving primes, and pick any surviving oracle machine for each of them. For each such prime p and machine M, we use M to solve "Π' mod p" on the original input $x \in \{0,1\}^n$, which is viewed as an element of $\text{GF}(p)^n$. As above, this is done while answering the queries of M with the values obtained by the reduction to solving the counting problem Π'' such that the query $q \in \text{GF}(p)^n$ is answered by taking the count-value provided for the input $(p, \texttt{rr}(q))$ and reducing it modulo p. Finally, we combine the values obtained for the count modulo each of these primes, by using Chinese Remaindering (without errors). Assuming that the surviving machines that we picked are correct (w.h.p.) on x, with high probability, the resulting value equals the value of $\Pi'(x)$.

The analysis amounts to showing that, with very high probability, each of the surviving machines is correct on at least a $1 - \exp(-d(n))$ fraction of the inputs in $\{0,1\}^n$, whereas all correct machines survived. Hence, the value computed via the CRT is correct on at least a $1 - \frac{\ell(n)}{d(n)} \cdot \exp(-d(n))$ fraction of $\{0,1\}^n$. The claim follows. ■

Digest. The sample-aided reduction presented in the proof of Theorem 4.2 is the result of composing three reduction, where the second reduction (i.e., the worst-case to average-case reduction of the evaluation of the polynomial $\widehat{\Phi}$) is actually the one that is naturally sample-aided. However, in a natural presentation of the second reduction, it is sample-aided with respect to the set of instances $\text{GF}(p)^n$, where p is a generic prime (which is sufficiently larger than the degree of $\widehat{\Phi}$), and the reduction uses its solved sample in order to estimate the success probability over $\text{GF}(p)^n$ of several candidate solvers. Indeed, this is not the way we used the second reduction in the proof of Theorem 4.2.

If we were to compose this (version of the second) reduction with the (worst-case) reduction of solving the counting problem Π' to evaluating $\widehat{\Phi}$, we would not have obtained a sample-aided reduction of Π' to $\widehat{\Phi}$, since the solved sample would have been of n-bit inputs to Π' whereas the sample-aided reduction for $\widehat{\Phi}$ requires solved instances drawn uniformly from $\text{GF}(p)^n$. Instead, we composed the first two reductions, while using the sample of solved instances for Π' in order to estimate the success rate of candidate solvers for $\widehat{\Phi}$ *in solving random instances in* $\{0,1\}^n$. This yields a sample-aided reduction of Π' to $\widehat{\Phi}$, which reduces solving Π' with success rate $1 - o(1)$ to solving $\widehat{\Phi}$ with success rate $o(1)$. (Combined with the success-rate preserving reduction of $\widehat{\Phi}$ to Π'', we establish Theorem 4.2.)

4.2 Obtaining Solved Samples via Downwards Self-reduction

A general paradigm that can be traced to the work of Impagliazzo and Wigderson [20] asserts that *if Π' is "downward self-reducible" and Π' has a sample-aided reduction to Π'', then Π' has a standard reduction to Π''*. Loosely speaking, on input $x \in \{0,1\}^n$, the standard reduction first generates a (solved) sample for Π', for each length $m \leq n$, where these samples are generated starting at $m = 1$ and going upwards to $m = n$. Specifically, when generating the sample for length m, we produce the answers by using the downwards self-reduction, which generates queries of length $m - 1$, which in turn are answered by the sample-aided reduction of Π' to Π'', while using the (already generated) sample for length $m - 1$. At the end, the answer to the original input x is found using the sample-aided reduction of Π' to Π'', while using the sample for length n.

4.2.1 Difficulties and Resolving Them

The foregoing outline works well in the context of worst-case to rare-case reductions, since in that case the solved sample generated for length $m - 1$ is perfect (i.e., free of errors). But when using an average-case to rare-case reduction (as in Sect. 4.1), we run into a problem: In that case, we can only guarantee that $1 - \rho'$ fraction of the solutions obtained for the sampled $(m - 1)$-bit instances are correct, and in such a case the fraction of errors in the solved sample of m-bit instances generated via downwards self-reduction may increase by a factor that equals the query complexity of the latter reduction, which is typically at least two. The error-doubling effect is disastrous, because the downwards self-reduction is applied many times.

In light of the above, we wish to apply the foregoing process to a worst-case to a rare-case reduction, and recall that the reduction of [31, Thm. 29] is actually of that type. Hence, starting from an arbitrary problem Π in $\mathcal{C}$, we first reduce it to a problem of evaluating low-degree polynomials that is downwards self-reducible, and apply the methodology at that level (i.e., on the worst-case to rare-case of the polynomial evaluation problem). The polynomial evaluation problem will be a generalization of the evaluation problem presented in Definition 3.2, where the generalization is aimed at supporting an adequate notion of downwards self-reducibility.

Refining the Notion of Downwards Self-reducibility. Before presenting this generalization, we introduce a more stringent notion of downwards self-reducibility, which is essential to our application. In particular, we require the reduction to work in almost-linear time (rather than in polynomial-time), and that the iterative process of downward self-reduction terminates after few steps (when reaching input length that is only slightly smaller than the original one). For starters, we say that a problem Π is downward self-reducible if *there exists an almost-linear time oracle machine that, on input $x \in \{0,1\}^n$, makes queries of length $n-1$ only, and outputs $\Pi(x)$ provided that its queries are answered by Π.*

In addition, we require that for a sufficiently dense set of input lengths, the problem can be solved in almost linear time without making any queries to shorter input lengths (i.e., for these input lengths, the oracle machine makes no queries at all). Specifically, *for some* $d \in \mathcal{D}$, *we require that for every* $n \in \mathbb{N}$ *the interval* $[n+1, n+\exp(d(n))]$ *contains such a length.* This additional requirement, hereafter referred to as the Stopping Condition, offers a crucial saving in the foregoing transformation from sample-aided reductions to standard reductions: It implies that the iterative downwards reduction procedure reaches the "base case" rather quickly; that is, on input $x \in \{0,1\}^n$, the iterative downwards reduction stops at length n' such that $n' \geq n - \exp(d(n))$ (i.e., the downwards self-reduction makes no queries on inputs of length n'). Consequently, it suffices for the standard reduction to generate samples for lengths $n', n'+1, ..., n$.

Generalizing the Evaluation Problem. Turning to the generalization of the evaluation problem presented in Definition 3.2, we seize the opportunity to pack together the different problems defined for the various fields. Recalling that the proof of Proposition 3.8 utilizes only $\ell(n)/d(n)$ different fields, which correspond to the first $\ell(n)/d(n)$ primes that are larger than $2^{d(n)}$, we make the index of the field part of the input. Furthermore, we present the index of the field in unary so that inputs that correspond to different fields have different lengths, and rely on the fact that the rare-case solver is supposed to work for all input lengths.[22]

To obtain a downward self-reduction, we reduce the original n-bit long instance, which refers to a summation over the entire range of indices $\{0,1\}^{\ell(n)}$ (i.e., the index w in Eq. (2) of Definition 3.2), to (two instances of) summation over a smaller range $\{0,1\}^{\ell(n)-1}$. This is accomplished in the natural way by fixing the first bit of the summation-index (to 0 or 1), and making this bit part of an augmented input. Similarly, for any $j \in [\ell(n)-1]$, summation over a restricted range $\{0,1\}^{\ell(n)-j}$ is reduced to (two instances of) summation over the smaller range $\{0,1\}^{\ell(n)-j-1}$, which yields inputs that have $j+1$ bits of augmentation. Finally, when the "summation" is over a single element, which happens when the augmented input has length $n + \ell(n)$, the problem can be solved directly in nearly-linear time, which satisfies the additional condition (i.e., the Stopping Condition) of downward self-reducibility.

Note that the foregoing iterative process increases the length of the input in each iteration by one bit, whereas we want the input length to decrease in the iterations. We fix this problem by padding the original input, and removing two bits of this padding in each iteration.

Thus, we divide the input to the problem into three parts, denoted u, v and x. The first part (i.e., u) includes the index of the field in unary, as described

[22] We shall reduce solving the original (worst-case) instance of length n to solving (in the rare-case sense) instances of various lengths, which correspond to different prime fields. Hence, for each input length for the original problem, we rely on being able to rarely solve the reduced problem on several input lengths (rather than on one input length as in the reductions presented so far). We note that this disadvantage is inherent to the downwards self-reduction paradigm of [20], so we may just take advantage of it for this additional purpose.

above, as well as the padding for guaranteeing that the input length shrinks when we fix an additional bit of the summation-index. The second part (i.e., v) represents the prefix of the summation-index (which has been fixed), and x is the "main" input. Lastly, given that the generalized problem is not merely an intermediate methodological locus (but rather appears in the conclusion of Theorem 4.4), we abstract it for the sake of greater appeal.

Definition 4.3 (generalization of Definition 3.2): *For $d \in \mathcal{D}$ and $n \in \mathbb{N}$, let $p_1, ..., p_{\ell'(n)}$ denote the first $\ell'(n) = \ell(n)/d(n)$ primes that are larger than $2^{d(n)}$, and $L_n = \{0, 1, ..., \ell(n)\}$. Let A' be an algorithm that, on input n, runs for $\exp(d(n)) \cdot n$-time and outputs an Arithmetic formula $\widehat{\psi}_n : \mathcal{F}^{\ell(n)+n} \to \mathcal{F}$ that computes a multivariate polynomial of total degree $\exp(d(n))$. The* generalized evaluation problem associated with A' *consists of computing the function*

$$\overline{\Phi}_n : \bigcup_{i \in [\ell'(n)], j \in L_n} \mathrm{GF}(p_i)^{\ell(n)^2 \cdot n + (\ell(n)+1) \cdot i - j} \times \mathrm{GF}(p_i)^n \to \bigcup_{i \in [\ell'(n)]} \mathrm{GF}(p_i)$$

defined as follows: For every $i \in [\ell'(n)]$ and $j \in L_n$, and every $x = (x_1, ..., x_n) \in \mathrm{GF}(p_i)^n$ and $(u, v) \in \mathrm{GF}(p_i)^{\ell(n)^2 \cdot n + (\ell(n)+1) \cdot i - 2j} \times \mathrm{GF}(p_i)^j$, it holds that

$$\overline{\Phi}_n(uv, x) \stackrel{\text{def}}{=} \sum_{w' \in \{0,1\}^{\ell(n)-j}} \widehat{\psi}_n(vw', x) \bmod p_i \tag{9}$$

where vw' is viewed as an $\ell(n)$-long sequence over $\mathrm{GF}(p_i)$.

Note that uv is a sequence of $\ell(n)^2 \cdot n + (\ell(n)+1) \cdot i - j$ elements of $\mathrm{GF}(p_i)$, and that the said length determines both i and j.

4.2.2 Using the Generalized Evaluation Problem

We first observe that Definition 4.3 is suitable for our application, since it is downward self-reducible in an adequate sense and generalizes Definition 3.2. For simplicity, the reader may consider the length of the instance (uv, x) in terms of $\mathrm{GF}(p_i)$-elements (i.e., as equal $(\ell(n)^2 + 1) \cdot n + (\ell(n) + 1) \cdot i - j$), but the following observations remain valid also when defining the length of that instance as $d(n) + 1$ times longer (and observing that all p_i's satisfy $\lceil \log p_i \rceil = d(n) + 1$).

1. The problem in Definition 3.2 (when restricted to any of the fields $\mathrm{GF}(p_i)$) is (worst-case) reducible to the problem in Definition 4.3.
   The reduction consists of mapping the instance $x \in \mathrm{GF}(p_i)^n$ to the instance $(1^{\ell(n)^2 \cdot n + (\ell(n)+1) \cdot i}, x)$, where $1^{\ell(n)^2 \cdot n + (\ell(n)+1) \cdot i}$ is viewed as a sequence of length $\ell(n)^2 \cdot n + (\ell(n) + 1) \cdot i$ over $\mathrm{GF}(p_i)$. (Indeed, this mapping yields an expression as in Eq. (2), but this does yield an expression that fits Eq. (9),

where $\widehat{\psi}_n$ is indeed a multivariate polynomial of degree $\exp(d(n))$.)[23] (Alternatively, one can show that each counting problem in $\mathcal{C}$ is reducible to the generalized evaluation problem, by adapting the proof of Proposition 3.8 in a similarly manner.)

2. The mapping $(n,i,j) \to (\ell(n)^2+1)\cdot n + (\ell(n)+1)\cdot i - j$ is injective when restricted to $i \in [\ell'(n)]$ and $j \in L_n$, since $\ell'(n) < \ell(n)$ (and $L_n = \{0,1,...,\ell(n)\}$).
   This means that instances associated with different (n,i,j)'s have different lengths, as postulated in the motivating discussion (see also Footnote 22).
3. The function $\overline{\Phi}$ satisfies the main condition of downward self-reducibility, since for every $x \in \mathrm{GF}(p_i)^n$ and $(u,v) \in \mathrm{GF}(p_i)^{\ell(n)^2\cdot n+(\ell(n)+1)\cdot i-2j} \times \mathrm{GF}(p_i)^j$, where $i \in [\ell'(n)]$ and $j \in \{0,1,...,\ell(n)-1\}$, it holds that $\overline{\Phi}(uv,x) = \overline{\Phi}(u'0v,x) + \overline{\Phi}(u'1v,x)$, where u' is the $(\ell(n)^2\cdot n + i\cdot\ell(n) - 2(j+1))$-long prefix of u (i.e., $u = u'\tau_1\tau_2$ for some $\tau_1,\tau_2 \in \mathrm{GF}(p_i)$).
   (Indeed, the fact that u' is two field-elements shorter than u allows to extend v by one field-element, while yielding an input that is shorter than the original one.)
4. The function $\overline{\Phi}$ satisfies the additional condition (i.e., the Stopping Condition) of downward self-reducibility, because for every $i \in [\ell'(n)]$, every $x \in \mathrm{GF}(p_i)^n$ and $(u,v) \in \mathrm{GF}(p_i)^{\ell(n)^2\cdot n+(\ell(n)+1)\cdot i-2\ell(n)} \times \mathrm{GF}(p_i)^{\ell(n)}$, it holds that $\overline{\Phi}(uv,x) = \widehat{\psi}_n(v,x) \bmod p_i$, since in this case $j = \ell(n)$.

The foregoing observations will be used when proving the following.

Theorem 4.4 (worst-case to rare-case reduction): *Let $\mathcal{D}$ be a set of admissible functions, ℓ be a logarithmic function, and $d \in \mathcal{D}$. Then, for every counting problem Π in $\mathcal{C} = \mathcal{C}_{\ell,\mathcal{D}}$, there exist a generalized evaluation problem $\overline{\Phi}$* (as in Definition 4.3) *and an almost-linear time randomized reduction of solving Π* (in the worst-case) *to solving $\overline{\Phi}$ on at least* $\exp(-d(n))$ *fraction of the domain.*

Recall that the foregoing evaluation problem can be solved in time $2^{\ell(n)}\cdot n^{1+o(1)}$. Hence, we have reduced the worst-case complexity of $\mathcal{C}$ to the rare-case complex-

[23] Specifically, we use a multi-linear extension $\mathtt{SEL} : \mathrm{GF}(p)^{\log_2 n} \times \mathrm{GF}(p)^n \to \mathrm{GF}(p)$ of the selection function $\mathtt{sel} : \{0,1\}^{\log_2 n} \times \mathrm{GF}(p)^n \to \mathrm{GF}(p)$, which satisfies $\mathtt{sel}(\alpha,x) = x_{\mathtt{int}(\alpha)+1}$, where $\mathtt{int}(\alpha)$ is the integer represented by the binary string α; that is,

$$\mathtt{SEL}(\zeta,x) = \sum_{\beta\in\{0,1\}^{\log_2 n}} \prod_{k\in[\log_2 n]} (\beta_k\zeta_k + (1-\beta_k)(1-\zeta_k)) \cdot x_{\mathtt{int}(\beta)+1}$$

(where a crucial point is that $\mathtt{SEL}(\zeta,x)$ is the sum of n terms such that the ith term is a multilinear function of the ζ_k's and x_i). Then,

$$\widehat{\psi}_n(vw',x) = \widehat{\phi}_n(vw', F_1(vw',x), ..., F_{m(n)}(vw',x))$$

where $F_k(w,x) = \mathtt{SEL}(\widehat{\sigma}_{n,k}(w),x)$ and $\widehat{\sigma}_{n,k} : \mathrm{GF}(p_i)^{\ell(n)} \to \mathrm{GF}(p_i)^{\log n}$ is a low degree polynomial that agrees with $\sigma_{n,k} : \{0,1\}^{\ell(n)} \to [n]$ (analogously to the way $\widehat{\phi}_n$ is derived from ϕ_n (cf. the proof of Proposition 3.3)). Note that, for $w \in \{0,1\}^{\ell(n)}$, it holds that $\mathtt{SEL}(\widehat{\sigma}_{n,k}(w),x) = \mathtt{SEL}(\sigma_{n,k}(w),x) = x_{\sigma_{n,k}(w)}$.

ity of $\overline{\Phi}$, while upper-bounding the worst-case complexity of $\overline{\Phi}$ (alas not reducing $\overline{\Phi}$ to $\mathcal{C}$).

Proof Sketch: By the foregoing discussion, solving Π reduces to solving $\overline{\Phi}$. Specifically, for every $z \in \{0,1\}^n$, the value of $\Pi(z)$ is obtained by applying the Chinese Remaindering (without errors) to the values of $\overline{\Phi}$ on the $\ell'(n)$ instances $y^{(i)} = (1^{\ell(n)^2 \cdot n + (\ell(n)+1) \cdot i}, z)$ for $i = 1, ..., \ell'(n)$. Hence, fixing any ($n \in \mathbb{N}$ and) $i \in [\ell'(n)]$, we focus on reducing the task of computing $\overline{\Phi}$ on *any* instance of length $\widetilde{n}_i = (\ell(n)^2+1) \cdot n + (\ell(n)+1) \cdot i$ to obtaining this value of an $\epsilon = \exp(-d(n))$ fraction of these instances (i.e., members of $\mathrm{GF}(p_i)^{\widetilde{n}_i}$).[24]

(Note that we reduced obtaining the value of Π on an arbitrary (worst-case) $z \in \{0,1\}^n$ to obtaining values of $\overline{\Phi}$ on $\ell'(n)$ related instances, having varying lengths, where the ith instance has length $\ell(n)^2 \cdot n + (\ell(n)+1) \cdot i + n = \widetilde{n}_i$ (i.e., it consists of padding z with $\widetilde{n}_i - n$ ones). Thus, we presented a (worst-case to worst-case) reduction from solving Π for *every* input length on *all inputs* to solving $\overline{\Phi}$ for *every* input length on *all inputs.* We stress that, as stated in Footnote 22, this reduction does not reduce solving Π on all inputs of a specific length n to solving $\overline{\Phi}$ on all inputs of some (possibly other) length n'. The next step, which is based on a downward self-reduction, will have a similar flavour (w.r.t input lengths): We shall present a worst-case to rare-case reduction from solving $\overline{\Phi}$ for *every* input length on *all inputs* to solving $\overline{\Phi}$ for *every* input length on *rare inputs.* Indeed, using many input lengths in the target of the reduction is inherent to the use of downward self-reductions.)

Focusing on the goal of presenting a worst-case to rare-case reduction for the evaluation of $\overline{\Phi}$ over $\mathrm{GF}(p_i)^{\widetilde{n}_i}$, we observe that the list-decoder algorithm of [31, Thm. 29] yields a sample-aided reduction that achieves the desired goal. Details follow.

First, as stated, the list-decoding algorithm of [31, Thm. 29] outputs a list of $O(1/\epsilon)$ oracle machines that contains machines that compute any polynomial that has agreement at least ϵ with the given oracle, and it works in time $\mathrm{poly}(1/\epsilon) \cdot \widetilde{O}(n)$, which is almost-linear time.[25] The foregoing list may contain also other machines. Still, using low-degree tests (e.g., [28]), we can guarantee that all machines on the list are very close (i.e.,ϵ-close) to computing some low degree polynomials, and by employing self-correction procedures (e.g., [11]) we obtain machines that compute low degree polynomials. All this can be done in time $\exp(d(n)) \cdot n$, provided that $\epsilon \geq \exp(-d(n))$.

Next, we convert the list-decoder of [31, Thm. 29] into a sample-aided reduction, by identifying the machines that compute the correct polynomial. Specifically, using a solved sample for the evaluation problem of the polynomial that we seek to compute (i.e., $\overline{\Phi} : \mathrm{GF}(p_i)^{\widetilde{n}_i} \to \mathrm{GF}(p_i)$), we can identify machines that computes the correct polynomial (since the other machines

[24] We stress that we use the foregoing length convention, by which the length refers to the number of field elements in a sequence (rather than to the length of the binary representation of the sequence).

[25] As noted in Sect. 4.1, although the statement of [31, Thm. 29] only claims running time that is polynomial in all relevant parameters, it is clear that the running time is linear in the number of variables.

compute polynomials that disagree with the correct polynomial on most inputs).[26] Although this machine may not be unique, all surviving machines compute the same polynomial, and we may use any of them. Note that a sample of size $O(\log(1/\epsilon)) = O(d(n))$ suffices for identifying all bad machines with high probability.

The solved sample for $\overline{\Phi} : \mathrm{GF}(p_i)^{\widetilde{n}_i} \to \mathrm{GF}(p_i)$ is obtained by the foregoing downwards self-reduction, while using the fact that, for every $j \in L_n$, we can apply the foregoing argument to $\overline{\Phi} : \mathrm{GF}(p_i)^{\widetilde{n}_i - j} \to \mathrm{GF}(p_i)$. Specifically, on input $y \in \mathrm{GF}(p_i)^{\widetilde{n}_i}$, we proceed as follows.

1. Generate a random sample of $O(d(n))$ solved instances for $\overline{\Phi}$: $\mathrm{GF}(p_i)^{\widetilde{n}_i - \ell(n)} \to \mathrm{GF}(p_i)$, while relying on the fact that, for such an input length, the function $\overline{\Phi}$ can be computed in almost-linear time (see the Stopping Condition of downward self-reducibility).
2. For $k = \widetilde{n}_i - \ell(n) + 1, ..., \widetilde{n}_i$ (equiv., using $k = \widetilde{n}_i - j$ for $j = \ell(n) - 1, ..., 0$), generate a random sample of $O(d(n))$ solved instances for $\overline{\Phi} : \mathrm{GF}(p_i)^k \to \mathrm{GF}(p_i)$, where the values of $\overline{\Phi}$ on each (random sample point) $r = (uv, x) \in \mathrm{GF}(p_i)^{k-n} \times \mathrm{GF}(p_i)^n$ is obtained by invoking the downwards self-reducibility on input r and answering its queries by using the sample-aimed reduction for inputs of length $k - 1$ (while using the samples generated in the previous iteration). Recall that on input $r = (uv, x) \in \mathrm{GF}(p_i)^{\widetilde{n}_i - 2(\widetilde{n}_i - k) + (\widetilde{n}_i - k) - n} \times \mathrm{GF}(p_i)^n$ the value of $\overline{\Phi}(r)$ is obtained by querying $\overline{\Phi}$ on $(u'v0, x)$ and on $(u'v1, x)$, where u' is the $(\widetilde{n}_i - 2(\widetilde{n}_i - k) - 2)$-long prefix of u, and that these inputs are of length $\widetilde{n}_i - (\widetilde{n}_i - k) - 1 = k - 1$.
3. Lastly, obtain the value of $\widehat{\Phi}(y)$ by invoking the sample-aimed reduction for inputs of length $\widetilde{n}_i$ (while using the samples generated in the previous step).

We stress that the invocations of the sample-aided reductions are provided with a number of samples that guarantees that the probability of error in the invocation is smaller than $\exp(-d(n)) \ll 1/\ell(n)$. Hence, with very high probability, our reduction provides the correct value of $\overline{\Phi}(y)$. Recalling that $y = y^{(i)}$ was actually derived from the input z such that $\overline{\Phi}(y^{(i)}) \equiv \Pi(z) \pmod{p_i}$, we recover $\Pi(z)$ by applying Chinese Remaindering (without errors) to the values of $\overline{\Phi}$ on the $\overline{\Phi}(y^{(i)})$'s for $i \in [\ell'(n)]$. ■

Digest of the Reduction of Π to $\overline{\Phi}(r)$. The description of the reduction goes top-down (i.e., from Π to the generalized problem, then uses its worst-case to average-case reduction, and then the downwards self-reduction), but the algorithm that computes Π goes bottom-up.

The original input $z \in \{0,1\}^n$ is transformed into $\ell'(n)$ inputs for the generalized problem, denoted $y^{(1)}, ..., y^{(\ell'(n))}$, such that $y^{(i)} = (1^{\widetilde{n}_i - n}, z)$ and $\widetilde{n}_i = (\ell(n)^2 + 1) \cdot n + (\ell(n) + 1) \cdot i$. (The instance $y^{(i)}$ is viewed both as an $(d(n)+1) \cdot \widetilde{n}_i$-bit long string and as an $\widetilde{n}_i$-long sequence over $\mathrm{GF}(p_i)$.) The value

[26] We could not do this in Sect. 4.1, since the solved instances we obtained there were all in $\{0,1\}^{\widetilde{n}_i}$, whereas different low degree polynomials over $\mathrm{GF}(p_i)^{\widetilde{n}_i}$ may agree on $\{0,1\}^{\widetilde{n}_i}$.

$\Pi(z)$ will be computed at the end of the process, based on the values $\overline{\Phi}(y^{(i)})$, which are obtained by the sample-aided worst-case to average-case reduction, where the relevant solved samples are obtained first by using the downwards self-reduction.

Likewise, the process of downwards self-reduction goes from length $\widetilde{n}_i$ to length $\widetilde{n}_i - \ell(n)$, whereas the actual generation of solved samples goes the other way around (i.e., from $\widetilde{n}_i - \ell(n)$ to $\widetilde{n}_i$). Specifically, a sample for length $\widetilde{n}_i - \ell(n)$ is generated by straightforward computation of the value of $\overline{\Phi}$ on random instances length $\widetilde{n}_i - \ell(n)$, while relying on the fact that instances of such length are easy to solve, whereas a sample for length $k = \widetilde{n}_i - j$ is generated by using the downward self-reduction to length $k-1$ and applying the sample-aided worst-case to rare-case reduction for length $k-1$ while using the solved sample generated in the previous iteration. Lastly, the solution to $y^{(i)} = (1^{\widetilde{n}_i - n}, z)$ is found using the sample-aided worst-case to rare-case reduction for length $\widetilde{n}_i$ while using the solved sample generated above. Finally, the solution to z is obtained by combining the solutions to the various $y^{(i)}$'s (using the CRT).

Acknowledgements. We are grateful to Madhu Sudan for many useful discussions regarding list decoding of multivariate polynomials and related issues.

Appendices

Appendix A.1 relates Definition 2.3 to [16, Def. 2], and Appendix A.2 presents a worst-case to average-case reduction for uniform $\mathcal{AC}^0[2]$.

A.1 A Related Class (for Context Only)

In this appendix, we review the definition of locally-characterizable sets [16], and discuss its relation to Definition 2.3.

Definition A.1 (locally-characterizable sets [16, Def. 2]):[27] *A set S is* locally-characterizable *if there exist a constant c, a polynomial p and a polynomial-time algorithm that on input n outputs* poly$(\log n)$*-sized formulae* $\phi_n : \{0,1\}^{c \cdot \log n} \times \{0,1\}^{p(\log n)} \rightarrow \{0,1\}$ *and* $\sigma_{n,1}, ..., \sigma_{n,p(\log n)} : \{0,1\}^{c \cdot \log n} \rightarrow [n]$ *such that, for every $x \in \{0,1\}^n$, it holds that $x \in S$ if and only if for all $w \in \{0,1\}^{c \log n}$*

$$\Phi_x(w) \stackrel{\text{def}}{=} \phi_n(w, x_{\sigma_{n,1}(w)}, ..., x_{\sigma_{n,p(\log n)}(w)}) \tag{10}$$

equals 0.[28]

[27] Actually, this is a slightly revised version, which is essentially equivalent to the original: In [16, Def. 2], instead of w, the formula ϕ_n got as part of its input the sequence $(\sigma_{n,1}(w), ..., \sigma_{n,p(\log n)}(w))$. This is essentially equivalent to the form used here, since, on the one hand, ϕ_n can compute the $\sigma_{n,i}$'s (given w), and on the other hand w can be reconstructed from $\frac{\ell(n)}{\log_2 n} = c$ auxiliary $\sigma_{n,i}$'s.

[28] Indeed, it is required that in case of inputs in S, the predicate ϕ_n evaluates to 0 (rather than to 1). This choice was made in [16] in order to simplify the expansion. We stress that since n is presented in binary, the algorithm runs in poly$(\log n)$-time.

That is, each value of $w \in \{0,1\}^{c\log n}$ corresponds to a local condition that refers to polylogarithmically many locations in the input (i.e., the locations $\sigma_{n,1}(w), ..., \sigma_{n,p(\log n)}(w) \in [n]$). This local condition is captured by ϕ_n, and in its general form it depends both on the selected locations (equiv., on w) and on the value on the input at these locations. A simplified form, which suffices in many case, uses a local condition that only depends on the values of the input at these locations (i.e., $\phi_n : [n]^{ca\cdot\log n} \times \{0,1\}^{p(\log n)} \to \{0,1\}$ only depends on the $p(\log n)$-bit long suffix).

A locally-characterizable set corresponds to the set of inputs for the counting problem (of Definition 2.3) that have value 0 (i.e., the number of satisfied local conditions is 0). This correspondence is not an equality, because the sizes of the formulae in the two definitions are different. Whereas in Definition 2.3 the number of formulae and their sizes are exponential in a function d that is in the admissible class, in Definition A.1 the number and size is poly-logarithmic in n. In both definitions, the formulae are constructed in time that is polynomial in their number and size.

A.2 Worst-Case to Average-Case Reduction for Uniform $\mathcal{AC}^0[2]$

For constants $c, d \in \mathbb{N}$, let $\mathtt{C}^{(d,c)}$ denote the class of decision problems on n-bit inputs that can be solved by uniform families of (unbounded fan-in) Boolean circuits of depth d and size n^c, with `and`, `or`, `parity`, and `not` gates. Specifically, a problem is parameterized by an efficient algorithm that on input n and $i, j \in [n^c]$ returns the type of the ith gate and whether or not the jth gate feeds into it (in the circuit that corresponds to n-bit inputs). The term "efficient" is left unspecified on purpose; possible choices include poly(n)-time and $O(\log n)$-space. Note that any decision problem having sufficiently uniform $\mathcal{AC}^0[2]$ circuits is in $\mathtt{C}^{(d,c)}$ for some constants $c, d \in \mathbb{N}$.

Theorem A.2 (worst-case to average-case reduction for $\mathcal{AC}_0$): *There exists a universal constant γ such that for any $c, d \in \mathbb{N}$, solving any problem in $\mathtt{C}^{(d,c)}$ on the worst-case reduces in almost linear time to solving some problem in $\mathtt{C}^{(\gamma\cdot d, c+o(1))}$ on at least 90% of the instances.*

Proof Sketch: We proceed in three steps: First we reduce the Boolean problem (in $\mathtt{C}^{(d,c)}$) to an Arithmetic problem, next we show that the latter problem supports a worst-case to average-case reduction, and lastly we reduce the Arithmetic problem to a Boolean problem (in $\mathtt{C}^{(\gamma\cdot d, c+o(1))}$). This is very similar to what was done in the main part of this work, except that the first step is fundamentally different.

A straightforward emulation of the Boolean circuit by the Arithmetic circuit would yield multiplication gates of polynomial fan-in, which in turn would mean that the polynomial computed by this circuit is of polynomial degree. In contrast, using the approximation method of Razborov [26] and Smolensky [29], we can get multiplication gates of logarithmic fan-in, and arithmetic circuits that compute polynomials of polylogarithmic degree. The approximation error is of no real concern, since we are actually interested in the value of the circuit at a single point. We need, however, to perform the foregoing randomized reduction using an almost linear amount of randomness, since we need to run in almost linear time, but this is possible using small-bias generators (cf. [25]). Details follow.

The first step is a randomized reduction of solving the Boolean problem in the worst-case to solving a corresponding Arithmetic problem on the worst-case. This reduction uses the ideas underlying the approximation method of Razborov [26] and Smolensky [29], while working with the field GF(2) (as [26], rather than with GF(p) for some prime $p > 2$ (as [29])).[29] When doing so, we replace the random choices made at each gate by pseudorandom choices that are generated by a small bias generator [25]; specifically, we use a "highly uniform" generator $G : \{0,1\}^{O(\log n^c)} \to \{0,1\}^{\widetilde{O}(n^c)}$ such that the individual bits of $G(s)$ are computed by uniform $n^{o(1)}$-size circuits of constant depth (and parity gates) [2,19].[30] We stress that the same pseudorandom sequence can be used for all gates in the circuit, and in each gate we can use $O(\log n^c)$ disjoint portions of the pseudorandom sequence for the $O(\log n^c)$ different linear combinations.[31]

Hence, for a fixed Boolean circuit C_n, on input $x \in \{0,1\}^n$, we uniformly select a seed $s \in \{0,1\}^{O(\log n)}$ for the aforementioned small-bias generator G, and construct the corresponding Arithmetic circuit $A_n^{(s)} : \mathrm{GF}(2)^n \to \mathrm{GF}(2)$, in which or-gates of C_n are replaced by $O(c \log n)$-way multiplications of linear combinations of the original gates that are determined by $G(s)$. For any fixed x, we may have $\Pr_s[A_n^{(s)}(x) \neq C_n(x)] = 1/\mathrm{poly}(n)$. Note that the depth of $A_n^{(s)}$ is only $O(1)$ times larger than the depth of C_n, where the constant is determined by various (local) manipulations (which include replacing and-gates by or-gates, computing inner products of gates' values and generator outputs, and adding $O(\log n)$-wise multiplication gates). Furthermore, $A_n^{(s)}$ uses multiplication gates of $O(c \log n)$ arity, its size is at most $O(\log n)^d \cdot n^c$, and it computes a polynomial of degree $O(c \log n)^d$.

The next step is to embed GF(2) in an extension field of size greater than the foregoing degree so that the standard process of self-correction of polynomials can be performed. Hence, $A_n^{(s)}$ is now viewed as an arithmetic circuit over $\mathcal{F} = \mathrm{GF}(2^\ell)$ (i.e., $A_n^{(s)} : \mathcal{F}^n \to \mathcal{F}$), where $\ell = d \log\log n + O(d)$, since we need $2^\ell \geq O(c \cdot \log n)^d$. Now, a worst-case to average-case reduction is applied to $A_n^{(s)}$

[29] Recall that this construction replaces each or-gate by a conjunction of $O(\log n^c)$ random linear combinations of the values that feed the original or-gate.

[30] Using the third construction in [2], we need to perform exponentiation in a field of size $2^{k/2}$, where $k = O(\log n^c)$ is the length of the seed. By [19, Thm. 4], this operation can be performed by highly uniform constant-depth circuit (with parity gates) of size $\exp(\widetilde{O}(\sqrt{k/2})) = n^{o(1)}$.

[31] Recall that taking t independent linear combinations of the output of an ϵ-bias generator yields a distribution that equal 0^t with probability at most $2^{-t} + \epsilon$. Also recall that the aforementioned generator of [2] produces a n^c-bit long ϵ-biased sequence using a seed of length $O(\log(n^c/\epsilon))$, and so we can set $t = O(c \log n)$ and $\epsilon = 2^{-t}$.

(i.e., evaluating $A_n^{(s)}$ on the worst case reduces to evaluating $A_n^{(s)}$ correctly on at least a 51% fraction of the instances).

Lastly, we wish to get back to a class of Boolean problems. We can do so as follows. First, we replace each (unbounded) GF(2^ℓ)-addition gate by ℓ parity gates (which add-up the ℓ corresponding bits in the sequence of field elements). Next, we replace each $\mathcal{F}$-multiplication gate of arity at most $m = O(\log n)$ by a $\mathcal{F}$-multiplication gate of arity $\sqrt{m}$ that is fed by $\sqrt{m}$ multiplication gates that cover the original m wires. Finally, we implement each of the latter gates by a small Boolean circuit of depth two (via a look-up table of size $|\mathcal{F}|^{\sqrt{m}} = \exp(\ell\cdot\sqrt{m}) = \exp(\widetilde{O}(\sqrt{\log n})) = n^{o(1)}$). Hence, given the Arithmetic circuit $A_n^{(s)} : \mathrm{GF}(2^\ell)^n \to \mathrm{GF}(2^\ell)$, we obtain a Boolean circuit $B_n^{(s)} : \{0,1\}^{n\ell} \to \{0,1\}^\ell$ that emulates $A_n^{(s)}$. Furthermore, using the small circuits that produce the output bits of the small-bias generator G on seed $s \in \{0,1\}^k$, where $k = O(\log n)$, we obtain a Boolean circuit $B_n : \{0,1\}^{n\ell+k} \to \{0,1\}^\ell$ such that $B_n(y,s) = B_n^{(s)}(y)$. Recalling that the aforementioned circuits that compute the bits of G have constant depth and size $n^{o(1)}$, it follows that B_n has depth $O(d) + O(1)$ and size $n^{c+o(1)}$.

We are almost done, except that we need to reduce the evaluation of $B_n : \{0,1\}^{n\ell+k} \to \{0,1\}^\ell$ to the evaluation of a Boolean circuit that has a single output bit, and we need this reduction to work in the average-case setting. We can do so by using the Boolean circuit $B'_n : \{0,1\}^{n\ell+k+\ell} \to \{0,1\}$ that, on input $(z,r) \in \{0,1\}^{n\ell+k} \times \{0,1\}^\ell$, returns the inner product mod 2 of $B_n(z)$ and r. Note that the depth of B'_n exceeds the depth of B_n only by a constant term, and that we can correctly retrieve $B_n(z)$ (with high probability) if we can obtain the correct value of $B'_n(z,r)$ correctly on 0.76% of the r's. Recalling that evaluating C_n on x was reduced to evaluating B_n correctly on 0.51% of the instances, it follows that it suffices to compute B'_n correctly on a ρ fraction of the instances, provided that $\rho \geq 1-0.24\cdot 0.49$. (Of course, 0.24 and 0.49 stand for any constants smaller than 0.25 and 0.5 respectively.)

To summarize, evaluating $C_n : \{0,1\}^n \to \{0,1\}$ on the worst case reduces to evaluating $B'_n : \{0,1\}^{\log\log n + O(\log n)} \to \{0,1\}$ on at least $1 - 0.24 \cdot 0.49 \approx 0.88$ fraction of the instances. Using an adequate indexing of the gates in B'_n, the uniformity of B'_n follows from the uniformity of C_n, since all modifications we have performed are local. ■

References

1. Ajtai, M.: Σ_1^1-formulae on finite structures. Ann. Pure Appl. Logic **24**(1), 1–48 (1983)
2. Alon, N., Goldreich, O., Hastad, J., Peralta, R.: Simple construction of almost k-wise independent random variables. Random Struct. Algorithms **3**(3), 289–304 (1992)
3. Babai, L.: Random oracles separate PSPACE from the polynomial-time hierarchy. IPL **26**, 51–53 (1987)
4. Babai, L., Fortnow, L., Nisan, N., Wigderson, A.: BPP has subexponential time simulations unless EXPTIME has publishable proofs. Complex. Theory **3**, 307–318 (1993)

5. Ball, M., Rosen, A., Sabin, M., Vasudevan, P.N.: Average-case fine-grained hardness. In: The Proceedings of STOC, pp. 483–496 (2017)
6. Barak, B.: A probabilistic-time hierarchy theorem for "slightly non-uniform" algorithms. In: Rolim, J.D.P., Vadhan, S. (eds.) RANDOM 2002. LNCS, vol. 2483, pp. 194–208. Springer, Heidelberg (2002). https://doi.org/10.1007/3-540-45726-7_16
7. Barkol, O., Ishai, Y.: Secure computation of constant-depth circuits with applications to database search problems. In: Shoup, V. (ed.) CRYPTO 2005. LNCS, vol. 3621, pp. 395–411. Springer, Heidelberg (2005). https://doi.org/10.1007/11535218_24
8. Bogdanov, A., Trevisan, L.: On worst-case to average-case reductions for NP problems. SIAM J. Comput. **36**(4), 1119–1159 (2006)
9. Cai, J.-Y., Pavan, A., Sivakumar, D.: On the hardness of permanent. In: Meinel, C., Tison, S. (eds.) STACS 1999. LNCS, vol. 1563, pp. 90–99. Springer, Heidelberg (1999). https://doi.org/10.1007/3-540-49116-3_8
10. Feigenbaum, J., Fortnow, L.: Random-self-reducibility of complete sets. SIAM J. Comput. **22**(5), 994–1005 (1993)
11. Gemmell, P., Lipton, R.J., Rubinfeld, R., Sudan, M., Wigderson, A.: Self-testing/correcting for polynomials and for approximate functions. In: The Proceedings of ACM Symposium on the Theory of Computing, pp. 32–42 (1991)
12. Goldmann, M., Grape, P., Hastad, J.: On average time hierarchies. Inf. Process. Lett. **49**(1), 15–20 (1994)
13. Goldreich, O.: Computational Complexity: A Conceptual Perspective. Cambridge University Press, Cambridge (2008)
14. Goldreich, O., Nisan, N., Wigderson, A.: On Yao's XOR-Lemma. In: ECCC, TR95-050 (1995)
15. Goldreich, O., Ron, D., Sudan, M.: Chinese remaindering with errors. IEEE Trans. Inf. Theory **46**(4), 1330–1338 (2000). Preliminary version in 31st STOC (1999)
16. Goldreich, O., Rothblum, G.N.: Simple doubly-efficient interactive proof systems for locally-characterizable sets. In: The Proceedings of ITCS, pp. 18:1–18:19 (2018)
17. Goldreich, O., Rothblum, G.N.: Counting t-cliques: worst-case to average-case reductions and direct interactive proof systems. In: The Proceedings of FOCS, pp. 77–88 (2018)
18. Goldreich, O., Wigderson, A.: Derandomization that is rarely wrong from short advice that is typically good. In: Rolim, J.D.P., Vadhan, S. (eds.) RANDOM 2002. LNCS, vol. 2483, pp. 209–223. Springer, Heidelberg (2002). https://doi.org/10.1007/3-540-45726-7_17
19. Healy, A., Viola, E.: Constant-depth circuits for arithmetic in finite fields of characteristic two. In: Durand, B., Thomas, W. (eds.) STACS 2006. LNCS, vol. 3884, pp. 672–683. Springer, Heidelberg (2006). https://doi.org/10.1007/11672142_55
20. Impagliazzo, R., Wigderson, A.: Randomness vs time: derandomization under a uniform assumption. J. Comput. Syst. Sci. **63**(4), 672–688 (2001)
21. Impagliazzo, R., Jaiswal, R., Kabanets, V., Wigderson, A.: Uniform direct product theorems: simplified, optimized, and derandomized. SIAM J. Comput. **39**(4), 1637–1665 (2010)
22. Jukna, S.: Boolean Function Complexity: Advances and Frontiers. Algorithms and Combinatorics, vol. 27. Springer, Heidelberg (2012)
23. Kozen, D.: The Design and Analysis of Algorithms. Springer, New York (1991). https://doi.org/10.1007/978-1-4612-4400-4
24. Lipton, R.J.: New directions in testing. In: Feigenbaum, J., Merritt, M. (eds.) Distributed Computing and Cryptography. DIMACS Series in Discrete Mathematics

and Theoretical Computer Science, vol. 2, pp. 191–202. American Mathematics Society, Providence (1991)
25. Naor, J., Naor, M.: Small-bias probability spaces: efficient constructions and applications. SIAM J. Comput. **22**(4), 838–856 (1993). Preliminary version in 22nd STOC (1990)
26. Razborov, A.A.: Lower bounds on the size of bounded-depth networks over a complete basis with logical addition. Matematicheskie Zametki **41**(4), 598–607 (1987). (in Russian). English translation in Math. Notes Acad. Sci. USSR **41**(4), 333–338 (1987)
27. Rubinfeld, R., Sudan, M.: Self-testing polynomial functions efficiently and over rational domains. In: The Proceedings of 3rd SODA, pp. 23–32 (1992)
28. Rubinfeld, R., Sudan, M.: Robust characterization of polynomials with applications to program testing. SIAM J. Comput. **25**(2), 252–271 (1996). Unifies and extends part of the results contained in [11] and [27]
29. Smolensky, R.: Algebraic methods in the theory of lower bounds for Boolean circuit complexity. In: 19th ACM Symposium on the Theory of Computing, pp. 77–82 (1987)
30. Spira, P.M.: On time-hardware complexity tradeoffs for Boolean functions. In: Proceedings of 4th Hawaii Symposium on System Sciences, pp. 525–527 (1971)
31. Sudan, M., Trevisan, L., Vadhan, S.P.: Pseudorandom generators without the XOR Lemma. J. Comput. Syst. Sci. **62**(2), 236–266 (2001)
32. Vassilevska Williams, V.: Hardness of easy problems: basing hardness on popular conjectures such as the strong exponential time hypothesis. In: 10th International Symposium on Parameterized and Exact Computation, pp. 17–29 (2015)

On the Optimal Analysis of the Collision Probability Tester (an Exposition)

Oded Goldreich

Abstract. The collision probability tester, introduced by Goldreich and Ron (*ECCC*, TR00-020, 2000), distinguishes the uniform distribution over $[n]$ from any distribution that is ϵ-far from this distribution using $\text{poly}(1/\epsilon) \cdot \sqrt{n}$ samples. While the original analysis established only an upper bound of $O(1/\epsilon)^4 \cdot \sqrt{n}$ on the sample complexity, a recent analysis of Diakonikolas, Gouleakis, Peebles, and Price (*ECCC*, TR16-178, 2016) established the optimal upper bound of $O(1/\epsilon)^2 \cdot \sqrt{n}$. In this note we survey their analysis, while highlighting the sources of improvement. Specifically:

1. While the original analysis reduces the testing problem to approximating the collision probability of the unknown distribution up to a $1 + \epsilon^2$ factor, the improved analysis capitalizes on the fact that the latter problem needs only be solved "at the extreme" (i.e., it suffices to distinguish the uniform distribution, which has collision probability $1/n$, from any distribution that has collision probability exceeding $(1 + 4\epsilon^2)/n$).
2. While the original analysis provides an almost optimal analysis of the variance of the estimator when $\epsilon = \Omega(1)$, a more careful analysis yields a significantly better bound for the case of $\epsilon = o(1)$, which is the case that is relevant here.

A preliminary version of this exposition was posted in September 2017 as Comment Nr. 1 on TR16-178 of *ECCC*. The current revision is quite minimal, although some typos were fixed and some of the discussions were improved.

1 Introduction

We consider the task of testing whether an unknown distribution X, which ranges over $[n]$, equals the uniform distribution over $[n]$, denoted U_n. On input n, a proximity parameter $\epsilon > 0$, and $s = s(n, \epsilon)$ samples of a distribution $X \in [n]$, the tester should accept (with probability at least 2/3) if $X \equiv U_n$ and reject (with probability at least 2/3) if the statistical distance between X and U_n exceeds ϵ. (This testing task is a central problem in "distribution testing" (see, e.g., [9, Chap. 11]), which in turn is part of property testing [9].)[1]

[1] Although testing properties of distributions was briefly discussed in [10, Sec. 3.4.3], its study was effectively initiated in [4]. The starting point of [4] was a test of uniformity, which was implicit in [11], where it is applied to test the distribution of the endpoint of a relatively short random walk on a bounded-degree graph. Generalizing this tester of uniformity, Batu *et al.* [3,4] presented testers for the property consisting of pairs of identical distributions as well as for all properties consisting of any single distribution (where the property $\{U_n\}$ is a special case).

O. Goldreich (Ed.): Computational Complexity and Property Testing, LNCS 12050, pp. 296–305, 2020.
https://doi.org/10.1007/978-3-030-43662-9_16

The collision probability tester [11] is such a tester. It operates by counting the number of (pairwise) collisions between the s samples that it is given, and accepts if and only if the count exceeds $\frac{1+2\epsilon^2}{n} \cdot \binom{s}{2}$. Specifically, this tester estimates the collision probability of X, and accepts if and only if the estimate exceeds $\frac{1+2\epsilon^2}{n}$. An estimate that is at distance at most $2\epsilon^2/n$ from the correct value (with probability at least 2/3) suffices, since the collision probability of U_n equals $1/n$, whereas the collision probability of any distribution that is ϵ-far from U_n must exceed $\frac{1+4\epsilon^2}{n}$.

The initial analysis of this tester, presented in [11], showed that the collision probability of X can be estimated to within a relative deviation of $\eta > 0$ using $O(\sqrt{n}/\eta^2)$ samples. This yields a tester with sample complexity $O(\sqrt{n}/\epsilon^4)$, where $\epsilon > 0$ is the proximity parameter. Subsequently, it was shown that closely related testers use $O(\sqrt{n}/\epsilon^2)$ samples, and that this upper bound is optimal [13].[2] The fact that $O(\sqrt{n}/\epsilon^2)$ samples actually suffice for the collision probability tester was recently established by Diakonikolas *et al.* [8], and the current note surveys their proof.

The analysis of Diakonikolas *et al.* [8] is based on (1) observing that approximating the collision probability is easier when its value is extremely small, and (2) providing a more tight analysis of the variance of the (empirical) count (i.e., number of collision). The "take home messages" correspond to these two steps: Firstly, one should bear in mind (the well-known fact) that, in many settings, approximating a value is easier when the value is at an extreme (e.g., it is easier to distinguish the cases $\mathbf{Pr}[Y=1] = 1$ and $\mathbf{Pr}[Y=1] = 1-\epsilon$ than to distinguish the cases $\mathbf{Pr}[Y=1] = 0.5$ and $\mathbf{Pr}[Y=1] = 0.5-\epsilon$). Secondly, it often pays to obtain a tighter analysis. Furthermore, a bound that is *optimal in general* may be sub-optimal in extreme cases, which may actually be the cases we care about. (Indeed, this is exactly what happens in the current setting.)

To illustrate and motivate the analysis recall that the s samples of X yield $m = \binom{s}{2}$ votes regarding the collision probability of X, where each vote correspond to a pair of samples. That is, the $(j,k)^{\text{th}}$ vote it 1 if and only if the jth sample yields the same value as the kth sample. Clearly, the expected value of each vote equals the collision probability of X, and having $m = O(n/\eta^2)$ *pairwise independent* votes would have sufficed for approximating the collision probability of X up to a multiplicative factor of $1+\eta$, which would have allowed using $s = O(\sqrt{m}) = O(\sqrt{n}/\eta)$ samples. The problem is that, in general, the votes are not pairwise independent (i.e., the $(j,k)^{\text{th}}$ vote is not independent of the $(k,\ell)^{\text{th}}$ vote), and this fact increases the variance of the count (i.e., number of collision) and leads to the weaker bound of [11]. However, when $X \equiv U_n$, the votes are pairwise independent (e.g., the value of the $(j,k)^{\text{th}}$ vote does not condition the kth sample, and so the value of the $(k,\ell)^{\text{th}}$ vote is statistically independent of the former value). Furthermore, in general, the variance of the count can be upper-bounded by $I+E$, where I represents the value in the ideal case in which the votes are pairwise independent and E is an error term that depends on the

[2] Alternative proofs of these bounds can be found in [5] (see also [7, Apdx.]) and [6, Sec. 3.1.1], respectively.

difference between the collision probability of X and $1/n$. It turns out that the dependence of E on the latter difference is good enough to yield the desired result (see Sect. 3).

2 Preliminaries (Partially Reproduced from [9])

We consider *discrete* probability distributions. Such distributions have a finite *support*, which we assume to be a subset of $[n]$, where the support of a distribution is the set of elements assigned positive probability mass. We represent such distributions either by random variables, like X, that are assigned values in $[n]$ (indicated by writing $X \in [n]$), or by probability mass functions like $p : [n] \to [0,1]$ that satisfy $\sum_{i\in[n]} p(i) = 1$. These two representations are related via $p(i) = \mathbf{Pr}[X=i]$.

The total variation distance (a.k.a. the statistical difference) between distributions X and Y is defined as

$$\frac{1}{2} \cdot \sum_i |\mathbf{Pr}[X=i] - \mathbf{Pr}[Y=i]| = \max_S \{\mathbf{Pr}[X\in S] - \mathbf{Pr}[Y\in S]\}. \quad (1)$$

We say that X and Y are ϵ-close if Eq. (1) is upper-bounded by ϵ, and otherwise we say that they are ϵ-far.

The collision probability of a distribution X is the probability that two samples drawn according to X are equal; that is, the collision probability of X is $\mathbf{Pr}_{i,j\sim X}[i=j]$, which equals $\sum_{i\in[n]} \mathbf{Pr}[X=i]^2$. For example, the collision probability of U_n is $1/n$. Letting $p(i) = \mathbf{Pr}[X=i]$, observe that

$$\sum_{i\in[n]} p(i)^2 = \frac{1}{n} + \sum_{i\in[n]} (p(i) - n^{-1})^2, \quad (2)$$

which means that the collision probability of X equals the sum of the collision probability of U_n and the square of the $\mathcal{L}_2$-norm of $X - U_n$ (viewed as a vector, i.e., $\|X - U_n\|_2^2 = \sum_{i\in[n]} |p(i) - u(i)|^2$, where $u(i) = \mathbf{Pr}[U_n=i] = 1/n$).

Collision Probability vs Distance from Uniformity. The key observation is that, while the collision probability of U_n equals $1/n$, *the collision probability of any distribution that is ϵ-far from U_n is greater than $\frac{1}{n} + \frac{4\epsilon^2}{n}$*. To see the latter claim, let p denote the corresponding probability function, and note that if $\sum_{i\in[n]} |p(i) - n^{-1}| > 2\epsilon$, then

$$\begin{aligned}\sum_{i\in[n]} (p(i) - n^{-1})^2 &\geq \frac{1}{n} \cdot \left(\sum_{i\in[n]} |p(i) - n^{-1}| \right)^2 \\ &> \frac{(2\epsilon)^2}{n}\end{aligned}$$

where the first inequality is due to Cauchy-Schwarz inequality.[3] Indeed, using Eq. (2), we get $\sum_{i\in[n]} p(i)^2 > \frac{1}{n} + \frac{(2\epsilon)^2}{n}$. Hence, *testing whether an unknown distribution $X \in [n]$ equals U_n or is ϵ-far from U_n reduces to distinguishing the case that the collision probability of X equals $1/n$ from the case that the collision probability of X exceeds $\frac{1}{n} + \frac{4\epsilon^2}{n}$.*

In light of the above, we focus on approximating the collision probability of the unknown distribution X. This yields the following test, where the sample size, denoted s, is intentionally left as a free parameter.

Algorithm 1 (the collision probability tester): *On input $(n, \epsilon; i_1, ..., i_s)$, where $i_1, ..., i_s$ are drawn independently from a distribution X, compute*

$$c \leftarrow |\{j < k : i_j = i_k\}|, \tag{3}$$

and accept if and only if $\frac{c}{\binom{s}{2}} \leq \frac{1+2\epsilon^2}{n}$. We call c the empirical collision count.

Algorithm 1 approximates the collision probability of the distribution X from which the sample is drawn, and the issue at hand is the quality of this approximation (as a function of s, or rather how to set s so to obtain good approximation). The key observation is that each pair of sample points provides an unbiased estimator [4] of the collision probability (i.e., for every $j < k$ it holds that $\mathbf{Pr}_{i_j,i_k\sim X}[i_j = i_k] = \sum_{i\in[n]} \mathbf{Pr}[X = i]^2$), and that these $\binom{s}{2}$ pairs are "almost pairwise independent".

Recalling that the collision probability of $X \in [n]$ is at least $1/n$, it follows that a sample of size $O(\sqrt{n})$ (which "spans" $O(n)$ pairs) provides a "good approximation" of the collision probability of X in the sense that, with probability at least 2/3, the value of $c/\binom{s}{2}$ approximates the collision probability up to a multiplicative factor of 1.01. Furthermore, using $s = O(\eta^{-2}\sqrt{n})$ samples suffice for approximating the collision probability up to a factor of $1+\eta$. Recalling that testing requires approximating the collision probability up to a factor of $1+\epsilon^2$. this yield an upper bound of $O(\epsilon^{-4}\sqrt{n})$ on the number of samples.

---

[3] That is, use $\sum_{i\in[n]} |p(i) - n^{-1}| \cdot 1 \leq \left(\sum_{i\in[n]} |p(i) - n^{-1}|^2\right)^{1/2} \cdot \left(\sum_{i\in[n]} 1^2\right)^{1/2}$.

[4] A random variable X (resp., an algorithm) is called an unbiased estimator of a quantity v if $\mathbb{E}[X] = v$ (resp., the expected value of its output equals v). Needless to say, the key question with respect to the usefulness of such an estimator is the magnitude of its variance (and, specifically, the relation between its variance and the square of its expectation). For example, for any NP-witness relation $R \subseteq \bigcup_{n\in\mathbb{N}}(\{0,1\}^n \times \{0,1\}^{p(n)})$, the (trivial) algorithm that on input x selects at random $y \in \{0,1\}^{p(|x|)}$ and outputs $2^{p(|x|)}$ if and only if $(x, y) \in R$, is an unbiased estimator of the number of witnesses for x, whereas counting the number of NP-witnesses is notoriously hard. The catch is, of course, that this estimation has a huge variance; letting $\rho(x) > 0$ denote the fraction of witnesses for x, this estimator has expected value $\rho(x) \cdot 2^{p(|x|)}$ whereas its variance is $(\rho(x) - \rho(x)^2) \cdot 2^{2 \cdot p(|x|)}$, which is typically much larger than the expectation squared (i.e., when $0 < \rho(x) \ll 1/\text{poly}(|x|)$).

The better analysis presented next (in Sect. 3) capitalizes on the fact that we do not need to approximate the collision probability of any distribution up to a factor of $1+\eta$, but rather to distinguish the case that the collision probability equals $1/n$ from the case that the collision probability exceeds $\frac{1}{n}+\frac{2\eta}{n}$. In addition, it uses a more refined analysis of the variance of the empirical collision count c (computed by Algorithm 1), which is presented in Lemma 2.

3 The Actual Analysis

The core of the analysis is captured by the following lemma, which upper-bounds the variance of the empirical collision count in terms of the collision probability of the tested distribution X. Letting μ denote the collision probability of X, and δ its deviation from the collision probability of U_n (i.e., $\delta = \mu - \frac{1}{n}$), the upper bound provided by this lemma improves over the standard upper bound of $O(s^3\mu^{3/2})$, where the improvement is in the dominant term of $O(s^3\delta^{3/2})$. This improvement is significant in case $\delta = o(\mu)$, which is the case that we are interested in (i.e., $\mu = \Theta(1/n)$ and $\delta = 2\epsilon^2/n = o(1/n)$).

Lemma 2 (the variance of the empirical collision count): *Let μ denote the collision probability of the distribution X, and let Z denote the empirical collision count; that is, $Z = |\{1 \le j < k \le s : i_j = i_k\}|$, where $i_1, ..., i_s$ are drawn independently from X. Then, $\mathbb{E}[Z] = \binom{s}{2} \cdot \mu$ and $\mathbb{V}[Z] = O(s^2 \cdot \mu) + O(s^3) \cdot (\delta^{3/2} + \frac{\delta}{n})$, where $\delta = \mu - \frac{1}{n}$.*

The standard upper bound is $\mathbb{V}[Z] = O(s^3\mu^{3/2})$, and it can be obtained by degenerating the refined analysis presented below (as indicated in a couple of notes). Evidently, Lemma 2 implies the standard upper bound: The key point is that $\delta < \mu$ implies that $s^3 \cdot \delta^{3/2} < s^3\mu^{3/2}$, whereas $s^2 \cdot \mu + s^3 \cdot (\delta/n) = O(s^3) \cdot \delta^{3/2}$ holds since $\mu/n < \mu^{3/2}$ (equiv., $\mu^{1/2} > 1/n$)) and $s = \Omega(1/\sqrt{\mu})$.Note that the tighter bound (of Lemma 2) coincides with the standard one when $\delta = \Omega(\mu)$, but we are interested in smaller δ (i.e., $\delta \ll \mu$). For example, when $\delta = 0$ (i.e., $X \equiv U_n$), we get an upper bound asserting $\mathbb{V}[Z] = O(s^2 \cdot \mu)$, which is much better than $\mathbb{V}[Z] = O(s^3 \cdot \mu^{3/2}) = O(s^2 \cdot \mu)^{3/2}$ (assuming $s = \omega(1/\sqrt{\mu})$).

Proof: As noted before, each pair of samples provides an unbiased estimator of μ, and so $\mathbb{E}[Z] = \binom{s}{2} \cdot \mu$. If these pairs of samples were pairwise independent, then $\mathbb{V}[Z] = \binom{s}{2} \cdot (\mu - \mu^2)$ would have followed. But the pairs are not pairwise independent, although they are close to being so in the sense that almost all pairs of samples (i.e., quadruples of samples) are independent (i.e., (i_j, i_k) and $(i_{j'}, i_{k'})$ are independent if $|\{j, k, j', k'\}| = 4$). Hence, the desired bound is obtained by carefully examining the contribution of pairs of samples that are independent and the contribution of pairs of samples that are (potentially) dependent.

Specifically, we consider $m = \binom{s}{2}$ random variables $\zeta_{j,k}$ that represent the possible collision events; that is, for $j, k \in [s]$ such that $j < k$, let $\zeta_{j,k} = 1$ if the

$j^{\rm th}$ sample collides with the $k^{\rm th}$ sample (i.e., $i_j = i_k$) and $\zeta_{j,k} = 0$ otherwise. Then, $\mathbb{E}[\zeta_{j,k}] = \sum_{i\in[n]} \mathbf{Pr}[i_j = i_k = i] = \mu$ and $\mathbb{V}[\zeta_{j,k}] = \mathbb{E}[\zeta_{j,k}^2] - \mu^2 = \mu - \mu^2$. Letting $\overline{\zeta}_{i,j} \stackrel{\rm def}{=} \zeta_{i,j} - \mu$ (and using $\mathbb{V}[Z] = \mathbb{E}[(Z - \mathbb{E}[Z])^2]$), we get:

$$\begin{aligned}\mathbb{V}[Z] &= \mathbb{E}\left[\left(\sum_{j<k} \overline{\zeta}_{j,k}\right)^2\right] \\ &= \sum_{j_1<k_1, j_2<k_2} \mathbb{E}\left[\overline{\zeta}_{j_1,k_1}\overline{\zeta}_{j_2,k_2}\right].\end{aligned}$$

We partition the terms in the last sum according to the number of distinct indices that occur in them such that, for $t \in \{2,3,4\}$, we let $(j_1, k_1, j_2, k_2) \in S_t \subseteq [s]^4$ if and only if $|\{j_1, k_1, j_2, k_2\}| = t$ (and $j_1 < k_1 \wedge j_2 < k_2$). Hence,

$$\mathbb{V}[Z] = \sum_{t\in\{2,3,4\}} \sum_{(j_1,k_1,j_2,k_2)\in S_t} \mathbb{E}\left[\overline{\zeta}_{j_1,k_1}\overline{\zeta}_{j_2,k_2}\right]. \quad (4)$$

The contribution of each quadruple in S_4 to the sum is zero, since the four samples are independent and so $\mathbb{E}[\overline{\zeta}_{j_1,k_1}\overline{\zeta}_{j_2,k_2}] = \mathbb{E}[\overline{\zeta}_{j_1,k_1}] \cdot \mathbb{E}[\overline{\zeta}_{j_2,k_2}] = 0$. Each quadruple in S_2 (which necessarily satisfies $(j_1, k_1) = (j_2, k_2)$) contributes $\mathbb{E}[\overline{\zeta}_{j_1,k_1}^2] = \mathbb{V}[\zeta_{j_1,k_1}] \le \mu$ to the sum, and there are exactly m such quadruples, so their total contribution is at most $m \cdot \mu$. Turning to S_3, we note that each of its $\Theta(s^3)$ quadruples contributes

$$\begin{aligned}\mathbb{E}[\overline{\zeta}_{1,2}\overline{\zeta}_{2,3}] &= \mathbb{E}[\zeta_{1,2}\zeta_{2,3}] - \mathbb{E}[\zeta_{1,2}] \cdot \mathbb{E}[\zeta_{2,3}] \\ &= \sum_{i\in[n]} \mathbf{Pr}[X = i]^3 - \mu^2.\end{aligned}$$

Letting $\tau = \sum_{i\in[n]} \mathbf{Pr}[X = i]^3$ denote the three-way collision probability of X, the total contribution of the quadruples of S_3 is $\Theta(s^3) \cdot (\tau - \mu^2)$. Plugging all of this into Eq. (4), we get

$$\mathbb{V}[Z] = \Theta(s^2) \cdot \mu + \Theta(s^3) \cdot (\tau - \mu^2). \quad (5)$$

(*The standard bound of* $\mathbb{V}[Z] = O(s^3\mu^{3/2})$ *is obtained by giving-up on the* μ^2 *term and using* $\tau \le \mu^{3/2}$, *while assuming* $s = \Omega(1/\sqrt{\mu})$; specifically, note that $\tau = \sum_{i\in[n]} \mathbf{Pr}[X = i]^3$ is upper-bounded by $\max_{i\in[n]}\{\mathbf{Pr}[X = i]\} \cdot \sum_{i\in[n]} \mathbf{Pr}[X = i]^2 \le \sqrt{\mu} \cdot \mu$.)[5]

[5] In fact, one typically derives the standard bound earlier by using $\mathbb{E}[\overline{\zeta}_{1,2}\overline{\zeta}_{2,3}] \le \mathbb{E}[\zeta_{1,2}\zeta_{2,3}] = \tau$ (instead of $\mathbb{E}[\overline{\zeta}_{1,2}\overline{\zeta}_{2,3}] = \tau - \mu^2$), and noting that $\tau \le \mu^{3/2}$.

Letting $p_i \stackrel{\text{def}}{=} \mathbf{Pr}[X=i]$, we upper-bound $\mathbb{V}[Z] = \Theta(s^2)\cdot\mu + \Theta(s^3)\cdot(\tau-\mu^2)$ by upper-bounding τ as follows:

$$\begin{aligned}
\tau &= \sum_{i\in[n]} p_i^3 \\
&= \sum_{i\in[n]} \left(\left(p_i - \frac{1}{n}\right) + \frac{1}{n}\right)^3 \\
&= \sum_{i\in[n]} \left(p_i - \frac{1}{n}\right)^3 + \frac{3}{n}\cdot\sum_{i\in[n]}\left(p_i - \frac{1}{n}\right)^2 + \frac{3}{n^2}\cdot\sum_{i\in[n]}\left(p_i - \frac{1}{n}\right) + \frac{n}{n^3} \\
&\le \left(\sum_{i\in[n]}\left(p_i - \frac{1}{n}\right)^2\right)^{3/2} + \frac{3}{n}\cdot\delta + 0 + \mu^2 \\
&= \delta^{3/2} + 3\cdot(\delta/n) + \mu^2
\end{aligned}$$

where the inequality uses $\sum_i a_i^3 \le \left(\sum_i a_i^2\right)^{3/2}$ as well as $\sum_{i\in[n]}\left(p_i - \frac{1}{n}\right)^2 = \mu - \frac{1}{n} = \delta$ and $\mu \ge 1/n$. Hence, $\mathbb{V}[Z] = \Theta(s^2)\cdot\mu + \Theta(s^3)\cdot(\tau-\mu^2)$ is upper-bounded by $O(s^2\cdot\mu) + O(s^3)\cdot(\delta^{3/2} + (\delta/n))$. ∎

Theorem 3 (distinguishing U_n from distributions of higher collision probability): *Let X and Z be as in Lemma 2. Then, for any $\eta\in(0,1]$ and sufficiently large $s = O(\sqrt{n}/\eta)$, the following holds.*

1. *If $X \equiv U_n$, then $\mathbf{Pr}\left[\frac{Z}{\binom{s}{2}} > \frac{1+\eta}{n}\right] < 1/3$.*
2. *If the collision probability of X exceeds $\frac{1}{n} + \frac{2\eta}{n}$, then $\mathbf{Pr}\left[\frac{Z}{\binom{s}{2}} \le \frac{1+\eta}{n}\right] < 1/3$.*

Hence, for sufficiently large $s = O(\sqrt{n}/\eta)$, with probability at least 2/3, the empirical collision count distinguishes U_n from X having collision probability exceeding $(1+2\eta)/n$. Recall that in the testing application $\eta = 2\epsilon^2$; hence, for $s = O(\sqrt{n}/\epsilon^2)$, with probability at least 2/3, Algorithm 1 distinguishes U_n from any distribution that is ϵ-far from U_n.

Proof: Combining Chebyshev's Inequality with Lemma 2 (while letting $m = \binom{s}{2}$), we get:

$$\begin{aligned}
\mathbf{Pr}\left[\left|\frac{Z}{m} - \mu\right| > \gamma\right] &< \frac{\mathbb{V}[Z]}{m^2\cdot\gamma^2} \\
&= \frac{O(s^2)\cdot\mu + O(s^3)\cdot(\delta^{3/2} + (\delta/n))}{s^4\gamma^2}
\end{aligned}$$

where $\mu = \mathbb{E}[Z]/m$ and $\delta = \mu - (1/n)$. In the case of $X \equiv U_n$ (where $\mu = 1/n$ and $\delta = 0$), we get

$$\begin{aligned}
\mathbf{Pr}\left[\frac{Z}{m} > \frac{1+\eta}{n}\right] &\leq \mathbf{Pr}\left[\left|\frac{Z}{m} - \mu\right| > \frac{\eta}{n}\right] \\
&< \frac{O(s^2)\cdot\mu}{s^4\cdot(\eta/n)^2} \\
&= \frac{O(1/n)}{s^2\cdot(\eta/n)^2} \\
&= \frac{O(1)}{s^2\cdot\eta^2/n}
\end{aligned}$$

which is upper bounded by $1/3$ provided that $s = O(\sqrt{n}/\eta)$ is sufficiently large. Turning to the case that the collision probability of X exceeds $\frac{1}{n} + \frac{2\eta}{n}$ (i.e., $\delta > 2\eta/n$), we get

$$\begin{aligned}
\mathbf{Pr}\left[\frac{Z}{m} \leq \frac{1+\eta}{n}\right] &\leq \mathbf{Pr}\left[\left|\frac{Z}{m} - \left(\frac{1}{n}+\delta\right)\right| > \delta - \frac{\eta}{n}\right] \\
&\leq \mathbf{Pr}\left[\left|\frac{Z}{m} - \mu\right| > \frac{\delta}{2}\right] \\
&< \frac{O(s^2)\cdot\mu + O(s^3)\cdot(\delta^{3/2} + (\delta/n))}{s^4\cdot(\delta/2)^2} \\
&= \left(\frac{O(1/n)}{s^2\cdot\delta^2} + \frac{O(\delta)}{s^2\cdot\delta^2}\right) + \frac{O(1)}{s\cdot\delta^{1/2}} + \frac{O(1)}{s\cdot\delta\cdot n} \\
&= \frac{O(1)}{s^2\cdot\delta^2\cdot n} + \frac{O(1)}{s^2\cdot\delta} + \frac{O(1)}{s\cdot\delta^{1/2}} + \frac{O(1)}{s\cdot\delta\cdot n}
\end{aligned}$$

which is upper bounded by $1/3$ provided that $s = O(\sqrt{n}/\eta)$ is sufficiently large.[6] ■

Comments. The proof of Theorem 3 can be easily adapted to show that *if the collision probability of X is at most $\frac{1}{n} + \frac{\eta}{2n}$, then* $\mathbf{Pr}[Z > (1+\eta)/n] < 1/3$. This implies that, using $s = O(\sqrt{n}/\eta)$, the empirical collision count distinguishes distributions having collision probability at most $(1+0.5\eta)/n$ from distributions having collision probability exceeding $(1+2\eta)/n$. (Note that this does *not* yield an algorithm that, using $O(\sqrt{n}/\epsilon^2)$ samples, distinguishes distributions that are $0.1\cdot\epsilon$-close to U_n from distributions that are ϵ-far from U_n, since a distribution that is 0.1ϵ-close to U_n may have collision probability greater than that of a

[6] Let $s = c\cdot\sqrt{n}/\eta$ for some constant c. Then, when upper-bounding the first and last terms, use $s^2\cdot\delta^2\cdot n > c^2\cdot(n/\eta^2)\cdot(2\eta/n)^2\cdot n = 4c^2$. When upper-bounding the second and third terms, use $s^2\cdot\delta > c^2\cdot(n/\eta^2)\cdot(2\eta/n) \geq 2c^2$, where the last inequality uses $\eta \leq 1$.

distribution that is ϵ-far from U_n.)[7] We note that the proof of Theorem 3 would remain intact if we replaced the bound of Lemma 2 (i.e., $\mathbb{V}[Z] = O(s^2 \cdot \mu) + O(s^3) \cdot (\delta^{3/2} + \frac{\delta}{n})$) by the weaker $\mathbb{V}[Z] = O(s^2 \cdot \mu) + O(s^3) \cdot (\delta^{3/2} + \frac{\delta}{\sqrt{n}})$.

Acknowledgements. I am grateful to Ryan O'Donnell for many helpful discussions regarding the result of [8]. Ryan, in turn, claims to have been benefitting from his collaborators on [2,12], and was also inspired by [1]. Hence, my thanks are extended to all contributors to these works as well as to the contributers to [8].

References

1. Acharya, J., Daskalakis, C., Kamath, G.: Optimal testing for properties of distributions. arXiv:1507.05952 [cs.DS] (2015)
2. Badescu, C., O'Donnell, R., Wright, J.: Quantum state certification. arXiv:1708.06002 [quant-ph] (2017)
3. Batu, T., Fischer, E., Fortnow, L., Kumar, R., Rubinfeld, R., White, P.: Testing random variables for independence and identity. In: 42nd IEEE Symposium on Foundations of Computer Science, pp. 442–451 (2001)
4. Batu, T., Fortnow, L., Rubinfeld, R., Smith, W.D., White, P.: Testing that distributions are close. In: 41st IEEE Symposium on Foundations of Computer Science, pp. 259–269 (2000)
5. Chan, S., Diakonikolas, I., Valiant, P., Valiant, G.: Optimal algorithms for testing closeness of discrete distributions. In: 25th ACM-SIAM Symposium on Discrete Algorithms, pp. 1193–1203 (2014)
6. Diakonikolas, I., Kane, D.: A new approach for testing properties of discrete distributions. arXiv:1601.05557 [cs.DS] (2016)
7. Diakonikolas, I., Kane, D., Nikishkin, V.:Testing identity of structured distributions. In: 26th ACM-SIAM Symposium on Discrete Algorithms, pp. 1841–1854 (2015)
8. Diakonikolas, I., Gouleakis, T., Peebles, J., Price, E.: Collision-based testers are optimal for uniformity and closeness. ECCC, TR16-178 (2016)
9. Goldreich, O.: Introduction to Property Testing. Cambridge University Press, Cambridge (2017)
10. Goldreich, O., Goldwasser, S., Ron, D.: Property testing and its connection to learning and approximation. J. ACM **45**(4), 653–750 (1998). Extended abstract in 37th IEEE Symposium on Foundations of Computer Science, 1996
11. Goldreich, O., Ron, D.: On testing expansion in bounded-degree graphs. ECCC, TR00-020 (2000)

[7] For example, a distribution that assigns mass 0.1ϵ to a single point and is uniform on the other $n-1$ points is 0.1ϵ-close to U_n but has collision probability greater than $(0.1 \cdot \epsilon)^2$, whereas a distribution that with probability $0.5 + 3\epsilon$ is uniform on $[n/2]$ and otherwise is uniform on $\{(n/2)+1, ..., n\}$ is 2ϵ-far from U_n but has collision probability $(1 + 12\epsilon^2)/n$. Actually, the foregoing "tolerant testing" task (i.e., distinguishing distributions that are $0.1 \cdot \epsilon$-close to U_n from distributions that are ϵ-far from U_n) has sample complexity $\Omega(n/\log n)$; see [14].

12. O'Donnell, R., Wright, J.: A primer on the statistics of longest increasing subsequences and quantum states. To appear in SIGACT News
13. Paninski, L.: A coincidence-based test for uniformity given very sparsely-sampled discrete data. IEEE Trans. Inf. Theory **54**, 4750–4755 (2008)
14. Valiant, G., Valiant, P.: A CLT and tight lower bounds for estimating entropy. ECCC, TR10-179 (2010)

On Constant-Depth Canonical Boolean Circuits for Computing Multilinear Functions

Oded Goldreich and Avishay Tal

Abstract. We consider new complexity measures for the model of multilinear circuits with general multilinear gates introduced by Goldreich and Wigderson (*ECCC*, 2013). These complexity measures are related to the size of *canonical constant-depth Boolean circuits*, which extend the definition of canonical *depth-three* Boolean circuits. We obtain matching lower and upper bound on the size of canonical constant-depth Boolean circuits for almost all multilinear functions, and non-trivial lower bounds on the size of such circuits for some explicit multilinear functions.

An early version of this work appeared as TR17-193 of *ECCC*. The presentation was elaborated in the current revision, and a summary of our complexity bounds was added (see Sect. 8). Furthermore, Corollary 6.6 is new.

1 Introduction

Goldreich and Wigderson [4] put forward a model of *depth-three canonical circuits*, with the underlying long-term goal of leading to better lower bounds for general depth-three Boolean circuits computing explicit *multi-linear* functions. Canonical circuits are restricted type of depth-three Boolean circuits, and their study is supposed to be a warm-up and/or a sanity check for the establishing of lower bound on the size of general depth-three Boolean circuits that compute explicit multi-linear functions.

The canonical circuits defined in [4] are *depth-three* Boolean circuits that are obtained by a two-stage process: First, one constructs arithmetic circuits that use arbitrary multilinear gates of parameterized arity (and number of gates), and next one converts these arithmetic circuits to Boolean circuits. As shown in [4], the size of the resulting depth-three Boolean circuits is exponential in the maximum between the arity and the number of gates in the arithmetic circuit.

Hence, a natural complexity measure of such arithmetic circuits arises; specifically, the AN-complexity of a multi-linear function is m if it can be computed by a multilinear circuit (see Sect. 2) having at most m gates of arity at most m, where 'A' stands for 'arity' and 'N' for 'number' (of gates). The immediate challenge posed by [4] is to present explicit t-linear functions on $t \cdot n$ variables that require AN-complexity significantly greater than $(tn)^{1/2}$. Note that a lower bound of $m = \omega(\sqrt{tn})$ on the AN-complexity of such a function f yields a lower bound of $\exp(m)$ on the size of depth-three *canonical* circuits computing f, whereas

O. Goldreich (Ed.): Computational Complexity and Property Testing, LNCS 12050, pp. 306–325, 2020.
https://doi.org/10.1007/978-3-030-43662-9_17

the best bound known on the size of general depth-three Boolean circuits computing an explicit function over $\{0,1\}^n$ is $\exp(\sqrt{n})$. Hence, in the context of the AN-complexity measures of [4], a lower bound of $\omega(\sqrt{tn})$ is considered nontrivial.

In this context, a first nontrivial lower bound on an explicit function was obtained by Goldreich and Tal [5]. They exhibit explicit three-linear and four-linear functions having AN-complexities $\Omega(n^{0.6})$ and $\widetilde{\Omega}(n^{2/3})$, respectfully.

1.1 A Bird's Eye View of the Current Work

Although there is still much to be understood about the foregoing model, which corresponds to depth-three canonical (Boolean) circuits, we dare take another speculative step and put forward a notion of constant-depth canonical (Boolean) circuits along with a corresponding model of arithmetic circuits. In particular:

- We define more permissive "AN-complexity" measures (for multilinear circuits) than those defined in [4], and show a partial correspondence between them and a natural notion of constant-depth canonical circuit. Specifically, the more permissive "AN-complexity" are aimed to accommodate natural constructions of constant-depth (rather than depth-three) Boolean circuits for computing multi-linear functions.
- Extending the results of [4], we obtain matching lower and upper bound on the AN-complexity of almost all multi-linear functions. Correspondingly, for most t-linear functions, the size of canonical circuits of depth d is $\exp(\Theta(tn)^{t/(t+d-2)})$. (Indeed, the results of [4] refer to $d=3$, and assert size $\exp(\Theta(tn)^{t/(t+1)})$.)[1]
- Extending the results of [4] and using the results of [5], we obtain a lower bound on the size of depth-four canonical circuits that compute an explicit trilinear function. The resulting lower bound of $\exp(\widetilde{\Omega}(n^{3/8}))$ should be compared to $\exp(\Omega(n^{1/3}))$, which is the best lower bound known on the size of a general depth-four Boolean circuit computing an explicit function over $\{0,1\}^n$.

The foregoing description is very vague: It does not say what are the "more permissive AN-complexity" measures that are suggested in the current work, let alone justify the specifics of their definition. These obvious gaps will be filled in Sects. 2 and 3.

1.2 Organization

Our conceptual exposition (i.e., Sects. 2 and 3) builds quite heavily on [4], although, technically speaking, the paper is self-contained (with the exception of Sect. 6.2, which merely states an extension of a result of [5]). Familiarity with [4] may be useful also in the other sections. (Note that this volume contains a

[1] In contrast to the notation used here, in the other sections of this paper, the depth of the canonical circuits is denoted $d+1$, whereas d corresponds to the depth of general multilinear circuits.

revised version of [4].) In contrast, the results of [5] are used as a black-box, and so familiarity with that paper is not needed here.

In Sect. 2 we recall the model of multilinear circuits with general multi-linear gates, and present two complexity measures that refer to these circuits. These measures refine and generalize the AN-complexity measures introduced in [4], and are motivated by their relation to the size of canonical Boolean circuits of arbitrary constant-depth (rather than depth three). The latter relation, which actually defines the notion of canonical circuits, is presented in Sect. 3; that is, canonical circuits are constant-depth Boolean circuits that are derived (from multilinear circuits) by the transformation presented in Sect. 3.

In Sect. 4 we present matching lower and upper bounds on the foregoing complexity measures for almost all multilinear functions. These mark the lower bounds we should aim at for explicit functions. While we do not obtain these lower bounds for explicit function, we do obtain non-trivial lower bounds for explicit functions in Sects. 5–7. Specifically, in Sect. 5 we present bounds for an explicit trilinear function, and in Sect. 6 we present higher lower bounds for an explicit 4-linear function. While Sects. 5–6 focus on "canonical depth four", in Sect. 7 we show that non-trivial lower bounds for any "canonical depth" translate to non-trivial lower bounds for larger depths.

In light of the foregoing, a natural place to state our results is right after Sect. 3 (or, alternatively, right after Sect. 2). We chose not to do so, but rather provide a summary of our complexity bounds at the conclusion section (Sect. 8), which may be read right after reading Sect. 3.

2 Definitions

The basic definitions of multilinear circuits are as in [4,5]. Specifically, we focus on multi-linear functions and on multilinear circuits with general gates that compute them. The arity of these gates will serve as a main complexity measure, but we shall also refer to the number of gates in the circuit and to its the depth. The focus on multilinear circuits with such general gates and the complexity measures associated with them are justified by the construction of canonical constant-depth circuits, which are presented in Sect. 3.

Multi-linear Functions. For fixed $t, n \in \mathbb{N}$, we consider t-linear functions of the form $F : (\{0,1\}^n)^t \to \{0,1\}$, where F is linear in each of the t blocks of variables (which contain n variables each). Such a function F is associated with a t-dimensional array, called a tensor, $T \subseteq [n]^t$, such that

$$F(x^{(1)}, x^{(2)}, ..., x^{(t)}) = \sum_{(i_1,i_2,...,i_t)\in T} x^{(1)}_{i_1} x^{(2)}_{i_2} \cdots x^{(t)}_{i_t} \tag{1}$$

where $x^{(j)} = (x^{(j)}_1, ..., x^{(j)}_n) \in \{0,1\}^n$ for every $j \in [t]$. That is, F is linear in the variables of each block.

Multi-linear Circuits with General Gates. We consider multilinear circuits with arbitrary multilinear gates, of bounded arity (where this bound will serve as a complexity measure). The *multilinear requirement* mandates that if two gates have directed paths to them from the same block of inputs, then the results of these two gates are not multiplied together by any other gate. The depth of a circuit is the distance between the input variables and the output gate (e.g., a circuit consisting of a top gate that computes the sum of multilinear gates that are fed by variables only has depth 2).

Complexity Measures. The main complexity measures are the *arity* of the general multilinear gates and the *number* of such gates, where we say that a multilinear circuit C has arity m if m equals the maximum arity of a general gate in C. Specifically, we denote by $\mathtt{AN}(F)$ the minimum m such that there exists a multilinear circuit that computes F with at most m gates that are each of arity at most m. This definition as well as its restriction to depth two multilinear circuits, denoted $\mathtt{AN}_2$, are taken from [4]. Specifically, $\mathtt{AN}_2(F) \leq m$ if there exists a *depth-two* multilinear circuit that computes F with at most m gates that are each of arity at most m.

The definitions of $\mathtt{AN}_2$ and $\mathtt{AN}$ are tailored to fit the emulation of the corresponding multilinear circuits by depth-three Boolean circuits of size that is exponential in these measures. The emulation of depth-two multilinear circuits by Boolean circuits is straightforward: Each gate is emulated by a CNF (or DNF) of size exponential in the gate's arity. This mimics the construction of depth-three Boolean circuits of n-way Parity in which one emulates $\sqrt{n}$-way Parity gates, and the Boolean circuits obtained in this way were called *canonical.*

The same reasoning motivates our generalized complexity measures for multilinear circuits, denoted $\mathtt{A}_d$ and $\mathtt{AN}^{(e)}$. In particular, in Sect. 3 we shall show how to emulate multilinear circuits by Boolean circuits of constant depth, where the size of the derived "canonical" circuits is related to the new complexity measures. The point is that the aim of deriving depth-three Boolean circuits is replaced by deriving constant-depth Boolean circuits, and this relaxation yields a relaxation of the complexity measures for multilinear circuits.

Definition 2.1 (The A-complexity of depth d multilinear circuits): *For a multilinear function F, we denote by $\mathtt{A}_d(F)$ the minimum arity of a multilinear circuit of depth d that computes F.*

We use the notation $\mathtt{A}_d$ (rather $\mathtt{AN}_d$) in order to stress the fact that the definition makes no reference to the number of gates in the circuit.[2] Still, such an upper bound is implied, because the number of gates in a circuit of depth d and arity m is at most $\sum_{i=0}^{d-1} m^i < (m+1)^{d-1}$, since there are at most m^i gates at distance $i \leq d-1$ from the output gate. (Note that gates in a depth d circuit

[2] In contrast, the notation $\mathtt{AN}_d$ used in the revised version of [4] that appears in this volume refers to the maximum between the arity and the number of gates in the circuit; that is, $\mathcal{AN}_d(F) \leq m$ if there exists a multilinear circuit C of depth d that computes F such that C has arity at most m and at most m gates.

are at distance at most $d-1$ from the output gate, whereas only variables may be at distance d from the output gate.) Hence, $\mathtt{A}_2(\cdot)$ matches the notion of AN2-complexity as used in [4,5] (up to a slackness of one unit); that is, $\mathtt{A}_2(F) \leq \mathtt{AN}_2(F) \leq \mathtt{A}_2(F)+1$.

Seeking to generalize the definition AN-complexity, which refers to multilinear circuits of unbounded depth, we use the following definition (in which $e=1$ essentially coincides with AN as reviewed above).

Definition 2.2 (The AN-complexity of multilinear circuits wrt the exponent e): *For a multilinear function F, we denote by $\mathtt{AN}^{(e)}(F)$ the smallest m such that F can be computed by a circuit of arity at most m that has at most $(m+1)^e$ gates.*

Definition 2.2 does look weird at first glance, but as hinted above it is justified by the emulation of such circuits by depth $e+2$ Boolean circuits (described in Sect. 3). At this point we observe that:

- $\mathtt{AN}^{(e)}(F) \leq \mathtt{A}_{e+1}(F)$, since the number of gates in a circuit of depth $e+1$ and arity m is at most $(m+1)^e$.
- $\mathtt{AN}^{(1)}(\cdot)$ matches the notion of AN-complexity as used in [4,5] (again, up to a slackness of one unit); that is, $\mathtt{AN}^{(1)}(F) \leq \mathtt{AN}(F) \leq \mathtt{AN}^{(1)}(F)+1$.

We stress that the definition of $\mathtt{AN}^{(e)}(F)$ makes no reference to the depth of multilinear circuits computing F; it only refers to the arity and number of gates in such circuits, while linking the gate count to the arity in a way that fits their relation in a circuit of depth $e+1$ (i.e., guaranteeing that $\mathtt{AN}^{(e)}(F) \leq \mathtt{A}_{e+1}(F)$ holds).

The definitions of $\mathtt{A}_d$ and $\mathtt{AN}^{(e)}$ are tailored to fit the emulation of the corresponding multilinear circuits by Boolean circuits of size that is exponential in these measures. These emulations are described in Sect. 3, and the circuits obtained by them are called canonical. This fact justifies the definitions of $\mathtt{A}_d$ and $\mathtt{AN}^{(e)}$, which may look weird at first glance.

3 Obtaining Boolean Circuits

A direct emulation of the general multilinear gates in a multilinear circuits of depth d yields a Boolean circuit of depth $d+1$ and size $\exp(O(\mathtt{A}_d(\cdot))$. Specifically, we replace each general gate of arity m by a CNF (resp., a DNF) of size 2^m, where we use CNFs (resp., DNFs) in all even (resp., odd) levels. This allows to combine neighboring levels in the resulting depth $2d$ Boolean circuit, yielding a circuit of depth $d+1$. Hence, we generalize the D-canonical circuits of [4, Cons. 2.6], which constitute the special case of $d=2$.

Proposition 3.1 (D-canonical circuits of depth $d+1$): *For every $d \geq 2$, every multilinear function F can be computed by a Boolean circuit of depth $d+1$ and size* $\exp(O(\mathtt{A}_d(F))$.

For multilinear circuits of unbounded depth, a less direct emulation (i.e., using "Valiant method" [10]) yields depth-three Boolean circuits of size exponential in $\mathtt{AN}(\cdot)$, called ND-canonical circuits [4, Cons. 2.8]. Recalling that $\mathtt{AN}^{(1)}(F) \approx \mathtt{AN}(F)$, we wish to extend this construction to show that *for every* $e \geq 1$, *every multilinear function* F *can be computed by a Boolean circuit of depth* $e+2$ *and size* $\exp(O(\mathtt{AN}^{(e)}(F))$, where the aforementioned result refers to the case of $e = 1$. We were able to obtain such a result only in the special case that the multilinear circuit is "decomposable" in the following sense.

Specifically, we say that a circuit with N gates is (m, τ)-decomposable *if omitting the outgoing edges of at most* $\tau \cdot m$ *of its gates yields* (m, τ)*-decomposable sub-circuits that have each at most* $N/(m+1)$ *gates.*[3] Note that a circuit with $m+1$ gates is trivially $(m, 1)$-decomposable, and this fact underlies [4, Cons. 2.8].

Proposition 3.2 (ND-canonical circuits of depth e+2): *Let* $e \geq 1$, *and suppose that the multilinear function* F *has an* $(m, O(1))$*-decomposable multilinear circuit of arity at most* m *and at most* $(m+1)^e$ *gate, which implies* $\mathtt{AN}^{(e)}(F) \leq m$. *Then,* F *can be computed by a Boolean circuit of depth* $e+2$ *and size* $\exp(O(m))$.

Proof Sketch: The construction proceeds by induction on $e \geq 1$, where the case of $e = 1$ corresponds to [4, Cons. 2.8]. For $e > 1$, given a multilinear circuit C, we proceed as follows.

- Let G_0 denote the output gate of C, and suppose that C can be decomposed by omitting the outgoing edges of the gates $G_1, ..., G_{O(m)}$ such that each G_i (including G_0) is the output gate of a sub-circuit that is $(m, O(1))$-decomposable and contains at most $(m+1)^{e-1}$ gates. Then, by the induction hypothesis, each of the corresponding sub-circuits can be computed by a Boolean circuit of depth $e+1$ and size $\exp(O(m))$.
  We may assume that these Boolean circuits have a top AND-gate, and that they exist both for computing the value of each G_i and each $G_i + 1$; see justification in next item.[4]
- For $m' = O(m)$, consider a DNF that verifies the assertion *there exists* $\alpha \in \{0,1\}^{m'}$ *such that the outputs of* $(G_0, G_1, ..., G_{m'})$ *equal* 1α, where these $m'+1$ outputs correspond to computations that use the values of the original variables and use α_i as the value that replaces the outcome of G_i that is fed to any other gate. Then, combining this $\exp(m')$-sized DNF with the aforementioned Boolean circuits of depth $e+1$, we obtain the desired Boolean circuit (of depth $e+2$).
  By considering the negation of the resulting Boolean circuit (while propagating this negation to the leaves), we get a depth $e+2$ Boolean circuit with a top AND-gate for computing $F+1$. A Boolean circuit with a top AND-gate for computing F itself can be obtained by performing the foregoing for $F+1$.

[3] Indeed, a circuit with a single gate is defined to be $(1, 1)$-decomposable.

[4] This non-homogeneous form (i.e., the added constant 1) may be avoided by using $\prod_{j \in J} x_{n+1}^{(j)}$, for an adequate $J \subseteq [t]$, and setting all $x_{n+1}^{(j)}$'s to 1.

(The induction hypothesis is that *if a multilinear circuit is* (m, τ)*-decomposable and has arity at most* m *and at most* $(m+1)^{e-1}$ *gates, then it can be computed by a Boolean circuit of depth* $e+1$ *and size* $\exp(O((e-1) \cdot \tau \cdot m))$. The induction step starts with a multilinear circuit that is(m, τ)-decomposable and has arity at most m and at most $(m+1)^e$ gates, derives $\tau \cdot m$ multilinear sub-circuits that are each (m, τ)-decomposable with at most $(m+1)^{e-1}$ gates, which yields a Boolean circuit of depth $e+2$ and size $\exp(O((e-1) \cdot \tau m + \tau m))$.) ■

Discussion. Proposition 3.2 leaves open the general case in which we are given a multilinear circuit of arity m that has at most $(m+1)^e$ gates (where this circuit is not necessarily $(m, O(1))$-decomposable). Fortunately, the lower bounds (shown in the next sections) hold also for the general case, which means that we lost nothing by being potentially too permissive in defining $\mathtt{AN}^{(e)}(\cdot)$. Still, we wonder what is the "right" notion of the "AN-complexity of multilinear circuits wrt the exponent e". It is not inconceivable that a measure that requires decomposition is right, since it matches the natural application of the "Valiant method" [10].

4 Guiding Bounds

Analogously to [4], we have tight bounds on the complexities of almost all multilinear functions. These bounds set the ultimate goals we should aim at with respect to explicit functions.

Theorem 4.1 (generic upper bound): *For every* $d, t \in \mathbb{N}$*, every* t*-linear function* F *satisfies* $\mathtt{A}_d(F) = O(tn)^{t/(t+d-1)}$*. In particular,* $\mathtt{AN}^{(d-1)}(F) = O(tn)^{t/(t+d-1)}$.

This generalizes [4, Thm. 3.1], which was stated for $d = 2$.

Proof Sketch: Let $m = t \cdot n^{t/(t+d-1)} \approx (tn)^{t/(t+d-1)}$. Consider a partition of $[n]^t$ into cubes of side-length m/t, note that the tensor corresponding to F is the union of tensors that are each restricted to a different cube, and consider gates that compute the multilinear functions that correspond to these $(n/(m/t))^t = (tn/m)^t$ tensors. Then each of these $(tn/m)^t$ gates has arity $t \cdot (m/t) = m$. By our setting $(tn/m)^t \approx (m^{\frac{t+d-1}{t}-1})^t = m^{d-1}$, whereas the sum of m^{d-1} values can be computed by a linear circuit of depth $d-1$ and arity m. Combining the latter circuit with the aforementioned m^{d-1} gates, we obtain the desired circuit. ■

Theorem 4.2 (non-explicit lower bound): *For every* $d, t \in \mathbb{N}$*, almost all* t*-linear functions* F *satisfy* $\mathtt{AN}^{(d-1)}(F) = \Omega(tn)^{t/(t+d-1)}$*. In particular,* $\mathtt{A}_d(F) = \Omega(tn)^{t/(t+d-1)}$.

This generalizes [4, Thm. 4.1], which was stated for $d = 2$.

Proof Sketch: Letting $e = d - 1$, we upper-bound the number of general multilinear circuits of arity m and size $(m+1)^e$. Ignoring the gates' functionalities, for a moment, we note that the number of relevant DAGs is at most

$$\binom{tn + (m+1)^e}{m}^{(m+1)^e} < (((tn+1)^e)^m)^{(m+1)^e}$$
$$= \exp((m+1)^{e+1} \log(tn+1)^e), \quad (2)$$

where the first expression represents the number of choices of variables and other gates that feed each gate, and the inequality uses $tn + (m+1)^e < (tn+1)^e$. Next, note that (for $t \geq 2$ and $m \gg t \log n$) the quantity in Eq. (2) is dominated by the number of possible gates' functionalities, which is

$$\left(2^{(m/t)^t}\right)^{(m+1)^e} = \exp(m^{t+e}/t^t), \quad (3)$$

since each gate corresponds to a tensor of volume at most $(m/t)^t$. The claim holds since $m^{t+e}/t^t \ll n^t$, provided that $m \ll (tn)^{t/(t+e)}$. ■

5 Lower Bounds on Explicit Functions

The current section as well as the next section focus on the cases of $d = 3$ and $e = 2$ (i.e., $\mathtt{A}_3$ and $\mathtt{AN}^{(2)}$), which are the cases immediately above those studied in [4,5] (which are $d = 2$ and $e = 1$ (equiv., $\mathtt{A}_2 \approx \mathtt{AN}_2$ and $\mathtt{AN}^{(1)} \approx \mathtt{AN}$)).

Using the rigidity results of [5], one can obtain non-trivial lower on the $\mathtt{A}_3$ and $\mathtt{AN}^{(2)}$ complexities of explicit trilinear functions, where by non-trivial we mean *lower bounds significantly higher than* $\Omega(n^{1/3})$.[5] This relies on connections between the $\mathtt{A}_3$ and $\mathtt{AN}^{(2)}$ complexities of bilinear functions and the rigidity of the corresponding matrices, which adapt ideas of [4, Thm. 4.4]. We first recall the relevant definition.

Definition 5.1 (matrix rigidity [9]): *A matrix A (over a field $\mathcal{F}$) has rigidity s for rank r if every matrix of rank at most r (over $\mathcal{F}$) differs from A on more than s entries.*

We shall consider bilinear functions in the variables $x = (x_1, ..., x_n)$ and $y = (y_1, ..., y_n)$, and trilinear functions in the variables x, y and $z = (z_1, ..., z_{2n-1})$.

5.1 The Case of $\mathtt{A}_3$

Recall that $\mathtt{A}_3(F)$ refers to the arity of depth-three multilinear circuits; that is, $\mathtt{A}_3(F) \leq m$ if F can be computed by a depth-three multilinear circuit with gates of arity m. For perspective, recall that [4, Thm. 4.4] implies that *if the corresponding matrix corresponding to a bilinear function F has rigidity m^3 for rank m, then $\mathtt{A}_2(F) > m$.*

[5] Note that it is easy to show that the n-bit parity function, denoted $\mathtt{PAR}_n$, satisfies $\mathtt{A}_d(F) \geq \mathtt{AN}^{(d-1)}(\mathtt{PAR}_n) = \Omega(n^{1/d})$.

Lemma 5.2 (rigidity and $\mathtt{A}_3$): *Let F be a bilinear function and suppose that the corresponding matrix has rigidity m^5 for rank m. Then, $\mathtt{A}_3(F) > m$.*

The proof extends the warm-up of the proof of [4, Thm. 4.4], which referred to the case of $\mathtt{A}_2(F) \leq m$.

Proof: Suppose that $\mathtt{A}_3(F) \leq m$, and consider a depth-three multilinear circuit C of arity m that computes F. Then, without loss of generality, C has the form

$$C(x,y) = G(L_1(x), ..., L_{m_0}, L'_1(y), ..., L'_{m'_0}(y), Q_1(x,y), ...Q_{m''_0}(x,y)),$$

where G is a quadratic gate, $m_0 + m'_0 + m''_0 \leq m$, the $L_i(x)$'s and $L'_j(y)$'s are linear functions computable by depth-two circuits and the $Q_i(x,y)$'s are bilinear functions that are computed by depth-two circuits. Hence, for some $P \subseteq [m_0] \times [m'_0]$ it holds that

$$C(x,y) = \sum_{(i,j)\in P} L_i(x)L'_j(y) + \sum_{i\in[m''_0]} Q_i(x,y), \tag{4}$$

and each Q_i has the form

$$Q_i(x,y) = \sum_{(j,k)\in P_i} L_{i,j}(x)L'_{i,k}(y) + \sum_{j\in[m''_i]} Q_{i,j}(x,y), \tag{5}$$

where $P_i \subseteq [m_i] \times [m'_i]$ and $m''_i \leq m - (m_i + m'_i)$, and the $L_{i,j}(x)$'s and $L'_{i,k}(y)$'s are linear functions computable by depth-one circuits and the $Q_{i,j}(x,y)$'s are bilinear functions that are computed by depth-one circuits. Hence, the $L_{i,j}(x)$'s and $L'_{i,k}(y)$'s are linear gates and the $Q_{i,j}(x,y)$'s are bilinear gates (each taking m variables). Consider the matrix that corresponds to the function computed by Q_i. It is the sum of $|P_i| \leq m_i \cdot m'_i$ matrices of rank one, each being an outer product of two vectors that each has at most m one-entries, and m''_i matrices each having at most m^2 one-entries. Hence, the matrix that corresponds to $\sum_{i\in[m''_0]} Q_i$ has sparsity at most $\sum_{i\in[m''_0]}(m_i m'_i \cdot m^2 + m''_i \cdot m^2) \leq m^5$, since $m''_0 \leq m$ and $m_i + m'_i + m''_i \leq m$. On the other hand, the matrix that corresponds to $\sum_{(i,j)\in P} L_i(x)L'_j(y)$ has rank $\min(m_0, m'_0) < m$. It follows that the matrix that corresponds to F does not have rigidity m^5 for rank m. ■

Corollary 5.3 (an $\mathtt{A}_3$ lower bound for random Toeplitz functions): *Almost all bilinear functions F that correspond to Toeplitz matrices satisfy $\mathtt{A}_3(F) = \widetilde{\Omega}(n^{0.4})$.*

Proof: Using Lemma 5.2 it suffices to show that F has rigidity m^5 for rank $m = \widetilde{\Omega}(n^{0.4})$. This follows from special case of [5, Thm. 1.2], which asserts that a random Toeplitz matrix has rigidity $\Omega(n^2/\log n)$ for rank $\sqrt{n}$. ■

Corollary 5.4 (an $\mathtt{A}_3$ lower bound for an explicit trilinear function): *The trilinear function $F(x,y,z) = \sum_{i,j\in[n]} x_i y_j z_{n+i-j}$ satisfies $\mathtt{A}_3(F) = \widetilde{\Omega}(n^{0.4})$.*

Proof: As in [4,5], this follows from the existence of a bilinear function F' that corresponds to a Toeplitz matrix such that $\mathtt{A}_3(F') = \widetilde{\Omega}(n^{0.4})$, which is asserted in Corollary 5.3. ■

5.2 The Case of $\mathtt{AN}^{(2)}$

Recall that $\mathtt{AN}^{(2)}(F) \leq m$ if F can be computed by a multilinear circuit with at most $(m+1)^2$ gates, each having arity at most m. For perspective, recall that [4, Thm. 4.4] actually asserted that *if the corresponding matrix corresponding to a bilinear function F has rigidity m^3 for rank m, then $\mathtt{AN}^{(1)}(F) \geq m$.*

Lemma 5.5 (rigidity and $\mathtt{AN}^{(2)}$): *Let F be a bilinear function and suppose that the corresponding matrix has rigidity m^4 for rank m^2. Then, $\mathtt{AN}^{(2)}(F) \geq m$.*

Proof: Suppose that $\mathtt{AN}^{(2)}(F) \leq m-1$, and consider a multilinear circuit C of arity $m-1$ that has at most m^2 gates and computes F. We call a bilinear gate hybrid if it fed both by bilinear gates and by either linear gates or variables, and call it a terminal if it is fed by linear gates and/or variables only. We first get rid of hybrid gates by introducing, for each hybrid gate H_i, an auxiliary bilinear gate B_i that "take over" the linear gates and variables that feed H_i, and feeds the modified H_i; that is, suppose that H_i is fed by a sequence of bilinear gates $\overline{Q}$ and a sequence of linear gates and variables $\overline{L}$, then $H_i(\overline{Q}, \overline{L})$ is replaced by $H_i'(\overline{Q}, B_i)$ and $B_i(\overline{L})$. Note that H_i' computes the sum of other bilinear gates, whereas B_i is a terminal.

The resulting number of terminal gates is at most m^2, because each new terminal gate (i.e., the terminal gate B_i introduced by the foregoing process) can be charged to a hybrid bilinear gate in the original circuit (i.e., to H_i). Hence, all the bilinear gates in C are either terminal gates or compute the sum of other bilinear gates (as the top gate and the modified gates M_i'), and so C is the sum of the terminal gates, denoted G_i for $i \in [m^2]$; that is,

$$C(x, y) = \sum_{i \in [m^2]} G_i(x, y),$$

where the each G_i is fed by $m-1$ linear gates and variables.

Considering the sets of linear gates that feed into each of the G_i's, we stress that *these sets are all subsets of a set of at most m^2 linear gates*, since C has at most this number of gates. That is, $G_i(x, y)$ takes the sum of some products of pairs of linear gates and variables; specifically, each product takes one element from $S_i \cup V_i$ and one element from $S_i' \cup V_i'$, where $S_i \subseteq [m^2]$ (resp., $S_i' \subseteq [m^2]$) represents the set of linear gates in x (resp., in y) that feed G_i, and $V_i \subseteq [n]$ (resp., $V_i' \subseteq [n]$) denotes the set of x-variables (resp., y-variables) that feed G_i. Recall that $|S_i| + |S_i'| + |V_i| + |V_i'| \leq m-1$. Hence, G_i has the form

$$G_i(x, y) = \sum_{j \in S_i} L_j(x) M_{i,j}'(y) + \sum_{j \in S_i'} M_{i,j}(x) L_j'(y) + \sum_{(j,k) \in P_i \subseteq V_i \times V_i'} x_j y_k,$$

where the $L_j(x)$'s and $L_j'(y)$'s are linear gates of C, and the $M_{i,j}(x)$'s and $M_{i,j}'(y)$'s are arbitrary linear functions (which may depend on i). Specifically, $M_{i,j}(x)$ (resp., $M_{i,j}'(y)$) is a partial sum of $\sum_{k \in S_i} L_k(x) + \sum_{k \in V_i} x_k$ (resp.,

$\sum_{k\in S'_i} L_k(y)+\sum_{k\in V'_i} y_k)$, where these partial sums are determined by G_i. Denoting $S \stackrel{\rm def}{=} \cup_{i\in[m^2]} S_i$ and $S' \stackrel{\rm def}{=} \cup_{i\in[m^2]} S'_i$, we can express C as

$$C(x,y) = \sum_{i\in[m^2]} \left(\sum_{j\in S_i} L_j(x)M'_{i,j}(y) + \sum_{j\in S'_i} M_{i,j}(x)L'_j(y) + \sum_{(j,k)\in P_i\subseteq V_i\times V'_i} x_j y_k \right)$$
$$= \sum_{j\in S} L_j(x)M'_j(y) + \sum_{j\in S'} M_j(x)L'_j(y) + \sum_{(j,k)\in P} x_j y_k,$$

where $M_j(x) = \sum_{i\in[m^2]} M_{i,j}(x)$ (resp., $M'_j(y) = \sum_{i\in[m^2]} M'_{i,j}(y)$) and P is the multi-set consisting of $\cup_{i\in[m^2]} P_i$. Recalling that $|P_i| \le |V_i| \cdot |V'_i| \le (m-1)^2$, it follows that the matrix corresponding to the function computed by C is the sum of two matrices of ranks $|S|$ and $|S'| \le m^2 - |S|$, respectively, and a matrix of sparsity $m^2 \cdot (m-1)^2$. That is, this matrix does not have rigidity m^4 for rank m^2. ∎

Corollary 5.6 (an $\mathtt{AN}^{(2)}$ lower bound for random Toeplitz functions): *Almost all bilinear functions F that correspond to Toeplitz matrices satisfy* $\mathtt{AN}^{(2)}(F) = \widetilde{\Omega}(n^{3/8})$.

Proof: Using Lemma 5.5 it suffices to show that F has rigidity m^4 for rank m^2, where $m = \widetilde{\Omega}(n^{3/8})$. This follows from [5, Thm. 1.2], which asserts that a random Toeplitz matrix has rigidity $\Omega(n^3/r^2 \log n)$ for rank $r > \sqrt{n}$. Specifically, using $r = m^2 = \widetilde{\Omega}(n^{6/8})$, we get rigidity $\Omega(n^3/r^2 \log n) \ge m^4$, provided that $\Omega(n^3/\log n) \ge m^8$. ∎

Corollary 5.7 (an $\mathtt{AN}^{(2)}$ lower bound for an explicit trilinear function): *The trilinear function $F(x,y,z) = \sum_{i,j\in[n]} x_i y_j z_{n+i-j}$ satisfies* $\mathtt{AN}^{(2)}(F) = \widetilde{\Omega}(n^{3/8})$.

Proof: As in [4,5] (and Corollary 5.4), this follows from the existence of a bilinear function F' that corresponds to Toeplitz matrices such that $\mathtt{A}_3(F') = \widetilde{\Omega}(n^{3/8})$, which is asserted in Corollary 5.6. ∎

6 Better Lower Bounds on Other Explicit Functions

Recall that Corollaries 5.3 and 5.6 establish that *almost all bilinear functions F that correspond to Toeplitz matrices satisfy $\mathtt{A}_3(F) = \widetilde{\Omega}(n^{0.4})$ and $\mathtt{AN}^{(2)}(F) = \widetilde{\Omega}(n^{3/8})$*. In this section we get improved bounds for functions that correspond to matrices that are drawn from any $\exp(-n)$-biased space: Specifically, almost all bilinear functions F whose coefficients are taken from an 2^{-n}-biased space satisfy $\mathtt{A}_3(F) = \widetilde{\Omega}(n^{4/9})$ as well as $\mathtt{AN}^{(2)}(F) = \widetilde{\Omega}(n^{0.4})$. Recall that these results yield similar lower bounds for an explicit 4-linear function [5]. (We shall consider bilinear functions in the variables $x = (x_1, ..., x_n)$ and $y = (y_1, ..., y_n)$, and 4-linear functions in the variables x, y and $(s', s'') \in \{0,1\}^{O(n)+O(n)}$.)

Preliminaries. We recall the definition of an ε-biased distribution (introduced by Naor and Naor [8]).

Definition 6.1 (small-biased distribution): *A distribution Z over $\{0,1\}^N$ is said to be* ε-biased *if for every non-empty set $S \subseteq [N]$, it holds that*

$$\left| \mathrm{E}_{z\sim Z}[(-1)^{\sum_{i\in S} z_i}] \right| \leq \varepsilon \,.$$

We shall use the following property of ε-biased distributions (which is implicit in [8]).

Claim 6.2 (upper-bounding the probability of hitting a linear space [1, Lem. 1]): *Let Z be an ε-biased distribution over $\{0,1\}^N$. Let $\ell_1, \ldots, \ell_t$ be linearly independent linear functions on $z_1, \ldots, z_N$. Then, the probability that all linear functions evaluate to 0 on $z \sim Z$ is at most $\varepsilon + 2^{-t}$; that is,*

$$\mathrm{Pr}_{z\sim Z}[(\forall i \in [t])\ \ell_i(z) = 0] \leq \varepsilon + 2^{-t}.$$

This follows from the observation that $(\ell_1(Z), ..., \ell_t(Z))$ as an ε-biased distribution over $\{0,1\}^t$, and from the representation of distributions as vectors in the Fourier basis (see, e.g., [2, Sec. 1]).

6.1 The Case of $\mathtt{A}_3$

Here we use techniques that that are similar to those used in [5, Sec. 4–5], but the actual argument is different. We call the reader's attention to an argument at the end of Step 2 of the proof, where a union bound on too many values is avoided and the (linear equations satisfied by the) linear span of these values is considered instead.[6]

Theorem 6.3 (a $\mathtt{A}_3$ lower bound for bilinear functions selected from a small-biased sample space): *Almost all bilinear functions F that correspond to matrices drawn from a 2^{-n}-biased distribution on $\mathbb{F}_2^{n\times n}$ satisfy $\mathtt{A}_3(F) \geq \widetilde{\Omega}(n^{4/9})$.*

Proof: Let m and r be non-negative integer parameters smaller than n, which we will set later. Along the way, we shall assume a few inequalities on m and r, which we will eventually satisfy by appropriately choosing m and r.

Our proof will show that the matrices associated with bilinear circuits of arity at most m and depth 3 can be partitioned into at most $O(2^{n/2})$ families such that, for each family of matrices, there exists a system of $r^2/2$ (linearly independent) linear equations in the matrix entries that all matrices in the family satisfy. We will finish the proof by showing that most matrices drawn from a 2^{-n}-biased distribution on $\{0,1\}^{n^2}$ do not belong to any of these families, and hence cannot be computed by a bilinear depth-3 circuits of arity at most m.

---

[6] This technique was used in [5, Sec. 5].

Step 1: Classifying matrices to families. We start by classifying all matrices associated with bilinear functions F that satisfy $\mathtt{A}_3(F) \leq m$ into $O(2^{n/2})$ families of matrices such that in each family all entries in some r-by-r submatrix are linear combinations of $r^2/2$ values. Actually, the current step only identifies the families based on properties that will be useful towards identifying the linear combinations (in Step 2). Consider a depth-three multilinear circuit C of arity m that computes F. As in Lemma 5.2 (see Eqs. (4) and (5)), a generic C has the form

$$C(x,y) = \sum_{(i,j)\in[m]\times[m]} p_{i,j} \cdot \left(\sum_{\ell\in L_i} x_\ell\right) \cdot \left(\sum_{\ell\in L'_j} y_\ell\right) + \sum_{i\in[m]} Q_i(x,y),$$

where $P^{(0)} = (p_{i,j})_{i,j\in[m]} \in \{0,1\}^{m\times m}$, the L_i's and L'_j's are subsets of size at most m^2 of $[n]$, and

$$Q_i(x,y) = \sum_{(j,k)\in[m]\times[m]} p^{(i)}_{j,k} \cdot \left(\sum_{\ell\in L_{i,j}} x_\ell\right) \cdot \left(\sum_{\ell\in L'_{i,k}} y_\ell\right) + \sum_{j\in[m]} Q_{i,j}(x,y),$$

where $P^{(i)} = (p^{(i)}_{j,k})_{j,k\in[m]} \in \{0,1\}^{m\times m}$, the $L_{i,j}$'s and $L'_{i,k}$'s are subsets of size at most m of $[n]$, and the $Q_{i,j}(x,y)$'s are bilinear gates (each taking m variables).

To be even more specific, for each $Q_{i,j}$, we associate two subsets $S^{(i,j)}, T^{(i,j)} \subseteq [n]$ corresponding to the indices of the x and y input variables of $Q_{i,j}$, respectively. We require $|S^{(i,j)}| + |T^{(i,j)}| \leq m$ and write $Q_{i,j}$ as

$$Q_{i,j}(x,y) = \sum_{k\in S^{(i,j)}} \sum_{\ell\in T^{(i,j)}} c_{i,j,k,\ell} \cdot x_k \cdot y_\ell \tag{6}$$

where $c_{i,j,k,\ell}$ are coefficients in $\{0,1\}$ (defined for any $k \in S^{(i,j)}$ and $\ell \in T^{(i,j)}$).

Hence, a concrete depth-three (multilinear) circuit C of arity m is specified in terms of the foregoing generic description by specifying the sets $L_i, L'_i, L_{i,j}, L'_{i,j}$ and $S^{(i,j)}, T^{(i,j)}$, hereafter called the variable wiring (or wiring), as well as the $m+1$ matrices $P^{(i)}$'s (for $i = 0, 1, ..., m$) and the coefficients $c_{i,j,k,\ell}$'s, hereafter called the bilinear forms. Without loss of generality, we may envision C as a *formula* (i.e., a tree), and the elements in the foregoing sequence of sets as its leaves; that is, each leaf corresponds to one of the elements in one of the sets, and each such element is an index of a variable from $x_1, \ldots, x_n, y_1, \ldots, y_n$. This formula has at most $5m^3$ leaves (i.e., $\sum_{i\in[m]}(|L_i|+|L'_i|)+\sum_{i,j\in[m]}(|L_{i,j}|+|L'_{i,j}|+|S^{(i,j)}|+|T^{(i,j)}|) \leq 2m\cdot m^2 + m^2\cdot(2m+m) = 5m^3$), each labeled with a variable from $x_1, \ldots, x_n, y_1, \ldots, y_n$.

Let r be an integer and assume (for simplicity) that r divides n. We partition the x variables into n/r buckets, and similarly we partition the y variables. Specifically, for $a, b \in [n/r]$, let $X_a := \{x_{(a-1)\cdot r+1}, x_{(a-1)\cdot r+2}, \ldots, x_{i\cdot r}\}$ be the ath bucket of the x variables, and let $Y_b := \{y_{(b-1)\cdot r+1}, y_{(b-1)\cdot r+2}, \ldots, y_{b\cdot r}\}$ be the bth bucket of the y variables. For a fixed variable wiring, we call a bucket-pair (X_a, Y_b) typical if the following three conditions (or properties) hold:

1. At most $10 \cdot \frac{5m^3}{n/r} = \frac{50m^3r}{n}$ of the leaves in the formula are labeled with variables from X_a.
   (That is, the number of leaves labeled with a variable in X_a is at most ten times the expectation (for a random pair).)
2. At most $10 \cdot \frac{5m^3}{n/r}$ of the leaves in the formula are labeled with variables from Y_b.
3. There are at most $10 \cdot \frac{m^4}{(n/r)^2}$ quadruples (i,j,k,ℓ) such that $(x_k, y_\ell) \in X_a \times Y_b$ and x_k and y_ℓ are inputs to $Q_{i,j}$ (i.e., $k \in S^{(i,j)}$ and $\ell \in T^{(i,j)}$).

Observing that a random bucket-pair (X_a, Y_b) satisfies each condition (individually) with probability at least 0.9, it follows that most (in fact, at least 70% of the) bucket-pairs satisfy all conditions simultaneously. Hence, for each wiring, most bucket-pairs (X_a, Y_b) are typical.

For each pair $(a, b) \in [n/r] \times [n/r]$, we consider all wirings for which (X_a, Y_b) is typical. Actually, it suffices to consider a partial wiring that specifies only the placing/wiring of variables in $X_a \cup Y_b$. To specify such a partial wiring it suffices to specify which of these variables appears in which leaf of the formula; that is, assign variables of X_a (resp., Y_b) to at most $50m^3r/n$ of the leaves. Hence, we have at most

$$\binom{5m^3}{50m^3r/n} \cdot (|X_a| + 1)^{50m^3r/n} < (n^4)^{50m^3r/n}$$

possibilities for wiring of variables in X_a, where the first factor corresponds to the choice of leaves and the second factor corresponds to the choice of a variable in X_a for each chosen leaf. Ditto for Y_b. Thus, considering wirings for which (X_a, Y_b) is typical, there are at most $(n^4)^{100 \cdot m^3 \cdot r/n}$ possible wirings of variables in $X_a \cup Y_b$ to the gates that read them. We shall assume

$$100 \cdot m^3 \cdot \frac{r}{n} \leq \frac{n}{10 \cdot \log n} \tag{7}$$

giving us at most $(n^4)^{n/10\log n} = 2^{0.4 \cdot n}$ possible wirings.

We partition all bilinear functions F with $\mathtt{A}_3(F) \leq m$ to families according to a choice of a bucket-pair (X_a, Y_b) and a partial wiring of $X_a \cup Y_b$ such that (X_b, Y_b) is typical for this wiring. This gives us an upper bound of $(n/r)^2 \cdot 2^{0.4n} < 2^{n/2}$ on the number of families. (Note that we have used Properties 1–2 of a typical bucket-pair in order to derive an upper bound on the number of families; we shall use Property 3 in the next step.)

Step 2: Associating a system of linear equations with each family of matrices. We consider a fixed family of matrices; that is, we fix a choice of a bucket-pair (X_a, Y_b) and a choice of wirings of $X_a \cup Y_b$ for which the said pair is typical. We focus on the r-by-r submatrices of the matrices in the family whose rows correspond to variables in X_a and columns correspond to variables in Y_b.

For every (k, ℓ) such that $x_k \in X_a$ and $y_\ell \in Y_b$, we consider how the (k, ℓ)-th entry of the matrices in the family looks like. Note that the (k, ℓ)-th entry in the matrix corresponding to the bilinear function equals the value of the bilinear function on the input $e_{k,\ell} \stackrel{\text{def}}{=} (0^{k-1}10^{n-k}, 0^{\ell-1}10^{n-\ell})$ (i.e., the input with all zeros except for x_k and y_ℓ). The key observation is that, for a fixed family, since the wirings of X_a and Y_b are fixed, the (k,ℓ)-th entry is a *fixed* linear combination in the entries that correspond to the $P^{(i)}$'s (with $i \in \{0,1,...,m\}$) and the relevant coefficients $c_{i,j,k,\ell}$ with $i,j \in [m]$, where the relevant coefficients $c_{i,j,k,\ell}$ are those for which $k \in S^{(i,j)}$ and $\ell \in T^{(i,j)}$. To verify this claim, let $\chi_e(A) = 1$ if $e \in A$ and $\chi_e(A) = 0$ otherwise, and consider

$$\begin{aligned} C(e_{k,\ell}) = & \sum_{j,j' \in [m]: \chi_k(L_j) = \chi_\ell(L'_{j'}) = 1} p_{j,j'} \\ & + \sum_{i,j,j' \in [m]: \chi_k(L_{i,j}) = \chi_\ell(L'_{i,j'}) = 1} p^{(i)}_{j,j'} \\ & + \sum_{i,j \in [m]: \chi_k(S^{(i,j)}) = \chi_\ell(T^{(i,j)}) = 1} c_{i,j,k,\ell} \end{aligned}$$

which means that $C(e_{k,\ell})$ is a fixed linear combinations in the entries of $P^{(i)}$'s and the coefficients $c_{i,j,k,\ell}$. for which $k \in S^{(i,j)}$ and $\ell \in T^{(i,j)}$. Hence, as stated above, each entry in the r-by-r submatrix corresponding to $X_a \times Y_b$ is a fixed linear combinations in the entries of $P^{(i)}$'s and the relevant coefficients $c_{i,j,k,\ell}$. The point is that there are at most $(m+1) \cdot m^2$ entries in the $P^{(i)}$'s, and at most $10 \cdot m^4 \cdot r^2/n^2$ relevant coefficients $c_{i,j,k,\ell}$ for each $(k,\ell) \in X_a \times Y_b$ (by Property 3 of a typical bucket-pair). Assuming that

$$(m+1) \cdot m^2 + 10 \cdot m^4 \cdot \frac{r^2}{n^2} \leq \frac{r^2}{2} \tag{8}$$

this means that the r^2 entries of a submatrix in a generic matrix in the family are fixed linear combinations of at most $r^2/2$ values (i.e., the entries of $P^{(i)}$'s and the relevant coefficients $c_{i,j,k,\ell}$). Hence, these r^2 entries must satisfy a fixed system of at least $r^2/2$ independent linear equations.[7]

Step 3: Showing that, w.h.p., small-biased matrices do not belong to any of the families. To finish the proof, we show that a matrix drawn from a 2^{-n}-biased distribution is unlikely to be a member in any of these $2^{n/2}$ families of matrices. Specifically, we first upper-bound the probability that a matrix B drawn from a 2^{-n}-biased distribution belongs to a fixed family, and then take a union bound over all families.

To be included in a fixed family, the matrix B should satisfy at least $r^2/2$ specific independent linear equations. By Claim 6.2, this happens with probability at most $2^{-n} + 2^{-r^2/2} \leq 2 \cdot 2^{-n}$ assuming $r \geq \sqrt{2n}$. Recalling that the number

[7] Formally, we can write each of the r^2 entries as a fixed linear combination of at most $r^2/2$ symbolic variables. Viewing these r^2 entries as an r^2-dimensional vector, we note that this vector must resided in a fixed vector space of dimension at most $r^2/2$ over $\mathbb{F}_2$, which in turn can be characterized by a fixed system of at least $r^2/2$ independent linear equations.

of families is smaller than $2^{n/2}$, it follows that, with very high probability, a random matrix B drawn from a 2^{-n}-biased distribution does not belong to any of these families. Hence, with very high probability, such a random matrix B corresponds to a bilinear function F that satisfies $\mathtt{A}_3(F) \geq m$.

Conclusion. All that is left is picking r and m that satisfy Eqs. (7), (8), and $r \geq \sqrt{2n}$. The choice

$$r \stackrel{\text{def}}{=} \frac{n^{2/3}}{8 \cdot \log^{1/3}(n)} \quad \text{and} \quad m \stackrel{\text{def}}{=} \frac{n^{4/9}}{8 \cdot \log^{2/9}(n)} = \frac{r^{2/3}}{2}$$

satisfies all of the above, assuming n is large enough. ∎

Corollary 6.4 (an $\mathtt{A}_3$ lower bound for an explicit 4-linear function): *There exists an explicit bilinear function $G : \{0,1\}^{O(n)+O(n)} \to \{0,1\}^{n^2}$ such that the 4-linear function $F(x, y, s', s'') = \sum_{i,j \in [n]} G(s', s'')_{i,j} \cdot x_i y_j$ satisfies $\mathtt{A}_3(F) = \widetilde{\Omega}(n^{4/9})$.*

Proof: As in [5], this follows by combining Theorem 6.3 with a construction of a small-biased generator $G : \{0,1\}^{O(n)+O(n)} \to \{0,1\}^{n^2}$ that is a bilinear function (see [7]). By Theorem 6.3, for most settings of $s = (s', s'')$, it holds that the resulting bilinear function $F_s(x, y) = \sum_{i,j \in [n]} G(s)_{i,j} \cdot x_i y_j$ satisfies $\mathtt{A}_3(F_s) = \widetilde{\Omega}(n^{4/9})$, whereas $\mathtt{A}_3(F) \geq \mathtt{A}_3(F_s)$ (for every s). ∎

6.2 The Case of $\mathtt{AN}^{(2)}$

We mention that following the proof of [5, Thm. 5.6], one can get $\mathtt{AN}^{(2)}(F) = \widetilde{\Omega}(n^{0.4})$ for F's as in Theorem 6.3 and Corollary 6.4. We do not present the proof here, since it amounts to reproducing large portions of [5] (i.e., [5, Sec. 4] and [5, Sec. 5.1]), without any new ideas or techniques. The only difference would have been decoupling the number of gates from the arity, and using these two parameters rather than one in the original proof. Specifically, we have

Theorem 6.5 ([5, Thm. 5.6], revised by decoupling size and arity):[8] *Let A be an n-by-n matrix A whose entries are sampled from an ε-biased distribution. Then, the corresponding bilinear function can be computed by a bilinear circuit of arity r having s gates with probability at most*

$$\left(\frac{n}{2s}\right)^2 \cdot \binom{2s^2}{\leq 12s^2 r/n}^4 \cdot \left(\varepsilon + 2^{-s^2 + 24s^3 r^2/n^2}\right).$$

In particular, using $s = r^e > \sqrt{2n}$ (for any constant $e \in \mathbb{N}$) and $\varepsilon = 2^{-n}$, we get a probability bound of

$$\exp(\widetilde{O}(s^2 r/n) - \min(n, s^2 - O(s^3 r^2/n^2))) = \exp(\widetilde{O}(r^{2e+1}/n) - n),$$

[8] Indeed, in [5, Thm. 5.6], $s = m$.

assuming $r = o(n^{2/(e+2)})$. Hence, with high probability, the bilinear function F_A associated with a matrix A whose entries are sampled from an 2^{-n}-biased distribution satisfies $\mathtt{AN}^{(e)}(F_A) = \widetilde{\Omega}(n^{2/(2e+1)})$, since for a sufficiently small $r = \widetilde{\Omega}(n^{2/(2e+1)})$ it holds that $\widetilde{O}(r^{2e+1}/n) < n$.

Corollary 6.6 (an $\mathtt{AN}^{(e)}$ lower bound for an explicit 4-linear function): *There exists an explicit bilinear function $G : \{0,1\}^{O(n)+O(n)} \to \{0,1\}^{n^2}$ such that for every constant $e \in \mathbb{N}$ the 4-linear function $F(x, y, s', s'') = \sum_{i,j\in[n]} G(s', s'')_{i,j} \cdot x_i y_j$ satisfies $\mathtt{AN}^{(e)}(F) = \widetilde{\Omega}(n^{2/(2e+1)})$.*

For $e = 1$ this reproduces the $\widetilde{\Omega}(n^{2/3})$ lower bound of [5], but for $e \geq 2$ we get new bounds; for example, $\mathtt{AN}^{(2)}(F) = \widetilde{\Omega}(n^{0.4})$ and $\mathtt{AN}^{(3)}(F) = \widetilde{\Omega}(n^{2/7})$. In general, Corollary 6.6 provides a non-trivial lower bound for every $e \geq 1$ (i.e., $\widetilde{\Omega}(n^{2/(2e+1)}) = \omega(n^{1/(e+1)})$ for every constant $e \geq 1$).

7 Depth Reductions

In this section, we show connections between $\mathtt{A}_d(\cdot)$ for different depths d. First, we show a simple connection between $\mathtt{A}_{kd}(\cdot)$ and $\mathtt{A}_d(\cdot)$ for any $k \in \mathbb{N}$. As a special case, we get $\mathtt{A}_{2k}(F) \geq \mathtt{A}_2(F)^{1/k}$. Next, we show a less clean connection between $\mathtt{A}_{2k+1}(F)$ and $\mathtt{A}_2(F)$. We note that establishing connections between $\mathtt{AN}^{(e)}(\cdot)$ for different values of e remains open.

Lemma 7.1 (depth reduction – simple case): *For any multilinear function F and $d, k \in \mathbb{N}$, it holds that*

$$\mathtt{A}_d(F) \leq \mathtt{A}_{kd}(F)^k.$$

As a special case, we get $\mathtt{A}_d(F) \geq \mathtt{A}_2(F)^{2/d}$ for every even depth d. Hence, *any non-trivial lower bound for depth 2 implies a non-trivial lower bound for every even depth d*, where a non-trivial lower bound for depth d refers to any lower bound of the form $\mathtt{A}_d(F) = \omega((tn)^{1/d})$ for a t-linear function F. This terminology is justified by the fact that a lower bound of the form $\mathtt{A}_d(F) = \Omega((tn)^{1/d})$ holds trivially for any t-linear function F that depends on all its tn input variables (because otherwise the multilinear circuit cannot even read all the input bits).

Proof Sketch: Starting with any multilinear circuit for F having depth kd and arity $m = \mathtt{A}_{kd}(F)$, collapse every k consecutive layers into one layer, resulting in a t-linear circuit of depth d and arity m^k. Hence, $\mathtt{A}_d(F) \leq \mathtt{A}_{kd}(F)^k$. ■

Since we have non-trivial lower bounds for depth 2, we get from Lemma 7.1 non-trivial lower bounds on $\mathtt{A}_d(\cdot)$ for any even d (see the discussion after Lemma 7.2 for specific details). We would like to get a similar result for odd depths, but the straightforward approach gives $\mathtt{A}_d(F) \geq \mathtt{A}_{d+1}(F) \geq \mathtt{A}_2(F)^{2/(d+1)}$ for every odd d. While this implies non-trivial lower bounds on $\mathtt{A}_d(F)$ for all sufficiently large odd d, at the time of performing this work, it yielded only trivial bounds for $d = 3$:[9] At that time, the best lower bound known for an explicit function F asserted $\mathtt{A}_2(F) = \widetilde{\Omega}(n^{2/3})$, which implies only the trivial bound of $\mathtt{A}_3(F) = \widetilde{\Omega}(n^{1/3})$.

Lemma 7.2 (depth reduction – odd depths to depth 2): *Let $k \in \mathbb{N}$. Then, for any t-linear function F, it holds that*

$$\mathtt{A}_2(F) \leq O(\mathtt{A}_{2k+1}(F)^{k+(t/(t+1))}).$$

Proof Sketch: The main idea is to first split the middle layer into two layers of smaller arity using [4, Thm. 3.1], and then collapse the top $k+1$ (resp., the bottom $k+1$) layers into one layer. Specifically, using [4, Thm. 3.1] (alternatively Theorem 4.1), split each gate in layer $k+1$ to an equivalent sub-circuit with two layers and arity $O(m)^{t/(t+1)}$. After the split, the circuit has $2k+2$ layers, where the first k layers have gates of arity at most m, the next two layers have gates of arity at most $O(m)^{t/(t+1)}$, and the last k layers have of gates with arity at most m. Collapsing the first $k+1$ layers and the last $k+1$ layers, results in a multilinear circuit of depth 2 and arity $O(m^{k+(t/(t+1))})$ computing F. Thus, $\mathtt{A}_2(F) = O(\mathtt{A}_d(F)^{k+(t/(t+1))})$ as required. ■

Corollaries. We use the lower bound from [5, Thm. 1.5], which asserts that the bilinear function associated with a random Toeplitz matrix has $\mathtt{A}_2(F) = \widetilde{\Omega}(n^{2/3})$, with high probability (over the random choices of the $2n-1$ values along the diagonals). Using Lemma 7.1, we get the non-trivial lower bound $\mathtt{A}_d(F) = \widetilde{\Omega}(n^{4/(3d)})$ for even depths d. For odd depths $d = 2k+1$, we use Lemma 7.2 to get the non-trivial lower bound

$$\mathtt{A}_d(F) = \widetilde{\Omega}\left(n^{\frac{2/3}{k+(t/(t+1))}}\right) = \widetilde{\Omega}\left(n^{4/(3d+1)}\right)$$

where the second equality uses the fact that $t = 2$ and $d = 2k+1$. As in [4,5] (and Corollary 5.4), these lower bounds for random Toeplitz matrices imply similar lower bounds for an explicit *tri-linear* function.

Corollary 7.3 (an $\mathtt{A}_d$ lower bound for an explicit trilinear function): *The tri-linear function $F(x,y,z) = \sum_{i,j\in[n]} x_i y_j z_{n+i-j}$ satisfies $\mathtt{A}_d(F) = \widetilde{\Omega}(n^{4/(3d)})$ for even d and $\mathtt{A}_d(F) = \widetilde{\Omega}(n^{4/(3d+1)})$ for odd d.*

In particular, we get $\mathtt{A}_3(F) = \widetilde{\Omega}(n^{0.4})$, just as in Corollary 5.4.

Remark: In light of the above, it may seem that Sect. 5.1 is redundant. However, on top of serving as a warmup for Sects. 5.2 and 6.1, the contents Sect. 5.1 is not exhausted by Corollary 5.4, since it offers a structural result for matrices

[9] Added in revision: This is no longer the case, since [3] presented an explicit poly$(1/\epsilon)$-linear function F_ϵ such that $\mathtt{A}_2(F_\epsilon) \geq n^{1-\epsilon}$ for every constant $\epsilon > 0$. Hence, $\mathtt{A}_d(F_\epsilon) \geq \mathtt{A}_{d+1}(F_\epsilon) \geq n^{(2-2\epsilon)/(d+1)}$ holds for every odd d.

associated with low-complexity depth-3 bilinear circuits (i.e., Lemma 5.2). Furthermore, the proof in Sect. 5.1 relies on a rigidity lower bound of [5, Thm. 1.2], whereas Corollary 7.3 relies on a higher lower bound on "structured rigidity" provided by [5, Thm. 1.5] via a more complex proof.

8 Summary of Bounds on the Generalized AN-Complexity

Our study of the two new complexity measures (i.e., $\mathtt{A}_d$ and $\mathtt{AN}^{(d-1)}$) is guided by the following two facts:

- A generic upper bound that assert that *for every* $d, t \in \mathbb{N}$, *every* t*-linear function* F *satisfies* $\mathtt{AN}^{(d-1)}(F) \leq \mathtt{A}_d(F) = O(tn)^{t/(t+d-1)}$ (see Theorem 4.1).
- A matching lower bound that holds for almost all t-linear functions, which actually asserts that *for every* $d, t \in \mathbb{N}$, *almost every* t*-linear function* F *satisfies* $\mathtt{A}_d(F) \geq \mathtt{AN}^{(d-1)}(F) = \Omega(tn)^{t/(t+d-1)}$ (see Theorem 4.2).

Recall that it is trivial to show that the n-bit parity function, denoted $\mathtt{PAR}_n$, satisfies $\mathtt{A}_d(\mathtt{PAR}_n) \geq \mathtt{AN}^{(d-1)}(\mathtt{PAR}_n) = \Omega(n^{1/d})$. Any lower bound that is greater than that is considered non-trivial.

Our lower bounds for explicit functions are non-trivial, but do not meet the bounds for non-explicit functions. Specifically, we focus on the case of $d = 3$ (i.e., $\mathtt{A}_3$ and $\mathtt{AN}^{(2)}$), whereas the case of $d = 2$ (i.e., $\mathtt{A}_2 \approx \mathtt{AN}_2$ and $\mathtt{AN}^{(1)} \approx \mathtt{AN}$) was studied in [4,5]. Our results include:

- A non-trivial $\mathtt{A}_3$ lower bound for an explicit trilinear function: The trilinear function $F_3(x, y, z) = \sum_{i,j\in[n]} x_i y_j z_{n+i-j}$ satisfies $\mathtt{A}_3(F_3) = \widetilde{\Omega}(n^{0.4})$ (see Corollary 5.4).
- A non-trivial $\mathtt{AN}^{(2)}$ lower bound for an explicit trilinear function: The foregoing trilinear function F_3 satisfies $\mathtt{AN}^{(2)}(F_3) = \widetilde{\Omega}(n^{3/8})$ (see Corollary 5.7)
- A non-trivial $\mathtt{A}_3$ lower bound for an explicit 4-linear function: There exists an explicit 4-linear function F_4 that satisfies $\mathtt{A}_3(F_4) = \widetilde{\Omega}(n^{4/9})$ (see Corollary 6.4).
- A non-trivial $\mathtt{AN}^{(2)}$ lower bound for an explicit 4-linear function: The foregoing 4-linear function F_4 satisfies $\mathtt{AN}^{(2)}(F_4) = \widetilde{\Omega}(n^{0.4})$ (see Corollary 6.6).

Actually, Corollary 6.6 yields a lower bound for every $e \geq 2$ asserting that *the foregoing 4-linear function* F_4 *satisfies* $\mathtt{AN}^{(e)}(F_4) = \widetilde{\Omega}(n^{2/(2e+1)})$ *for every* $e \geq 2$. This implies $\mathtt{A}_d(F_4) = \widetilde{\Omega}(n^{2/(2d-1)})$ for every $d \geq 3$. For $d \geq 4$, a stronger lower bound for $\mathtt{A}_d$ asserts that *the foregoing trilinear function* F_3 *satisfies* $\mathtt{A}_d(F_3) = \widetilde{\Omega}(n^{4/(3d)})$ *for even* d *and* $\mathtt{A}_d(F_3) = \widetilde{\Omega}(n^{4/(3d+1)})$ *for odd* d (see Corollary 7.3). While all these lower bounds are non-trivial for every $d \geq 2$, only the last bound does not approach the trivial bound when d is large enough (but rather approaches a 4/3-power of the trivial bound).

References

1. Alon, N., Goldreich, O., Håstad, J., Peralta, R.: Simple construction of almost k-wise independent random variables. Random Struct. Algorithms **3**(3), 289–304 (1992)
2. Goldreich, O.: Three XOR-lemmas — an exposition. In: Goldreich, O. (ed.) Studies in Complexity and Cryptography. Miscellanea on the Interplay between Randomness and Computation. LNCS, vol. 6650, pp. 248–272. Springer, Heidelberg (2011). https://doi.org/10.1007/978-3-642-22670-0_22
3. Goldreich, O.: Improved bounds on the AN-complexity of multilinear functions. In: ECCC, TR19-171, November 2019
4. Goldreich, O., Wigderson, A.: On the size of depth-three Boolean circuits for computing multilinear functions. In: ECCC, TR13-043, March 2013. See revised version in this volume
5. Goldreich, O., Tal, A.: Matrix rigidity of random Toeplitz matrices. In: ECCC, TR15-079, May 2015
6. Jukna, S.: Boolean Function Complexity: Advances and Frontiers. AC, vol. 27. Springer, Heidelberg (2012). https://doi.org/10.1007/978-3-642-24508-4
7. Mossel, E., Shpilka, A., Trevisan, L.: On ϵ-biased generators in NC^0. In: 44th FOCS, pp. 136–145 (2003)
8. Naor, J., Naor, M.: Small-bias probability spaces: efficient constructions and applications. SIAM J. Comput. **22**(4), 838–856 (1993)
9. Valiant, L.G.: Graph-theoretic arguments in low-level complexity. In: Gruska, J. (ed.) MFCS 1977. LNCS, vol. 53, pp. 162–176. Springer, Heidelberg (1977). https://doi.org/10.1007/3-540-08353-7_135
10. Valiant, L.G.: Exponential lower bounds for restricted monotone circuits. In: 15th STOC, pp. 110–117 (1983)

Constant-Round Interactive Proof Systems for AC0[2] and NC1

Oded Goldreich and Guy N. Rothblum

Abstract. We present constant-round interactive proof systems for sufficiently uniform versions of $\mathcal{AC}^0[2]$ and $\mathcal{NC}^1$. Both proof systems are doubly-efficient, and offer a better trade-off between the round complexity and the total communication than the work of Reingold, Rothblum, and Rothblum (*STOC*, 2016). Our proof system for $\mathcal{AC}^0[2]$ supports a more relaxed notion of uniformity and offers a better trade-off between the number of rounds and the round complexity that our proof system for $\mathcal{NC}^1$. We observe that all three aforementioned systems yield constant-round doubly-efficient proof systems for the All-Pairs Shortest Paths problem.

An early version of this work appeared as TR18-069 of *ECCC*. The current revision follows the high-level strategy employed in the original version, but differs from it in many low-level details (esp., in Sect. 2).

1 Introduction

The notion of interactive proof systems, put forward by Goldwasser, Micali, and Rackoff [9], and the demonstration of their power by Lund, Fortnow, Karloff, and Nisan [12] and Shamir [16] are among the most celebrated achievements of complexity theory. Recall that an interactive proof system for a set S is associated with an interactive verification procedure, V, that can be made to accept any input in S but no input outside of S. That is, there exists an interactive strategy for the prover that makes V accepts any input in S, but no strategy can make V accept an input outside of S, except with negligible probability. (See [3, Chap. 9] for a formal definition as well as a wider perspective.)

The original definition does not restrict the complexity of the strategy of the prescribed prover and the constructions of [12,16] use prover strategies of high complexity. Seeking to make interactive proof systems available for a wider range of applications, Goldwasser, Kalai and Rothblum [8] put forward a notion of *doubly-efficient* interactive proof systems. In these proof systems the prescribed prover strategy can be implemented in polynomial-time and the verifier's strategy can be implemented in almost-linear-time. (We stress that unlike in *argument systems*, the soundness condition holds for all possible cheating strategies, and not only for feasible ones.) Restricting the prescribed prover to run in polynomial-time implies that such systems may exist only for sets in $\mathcal{BPP}$, and thus a polynomial-time verifier can check membership in such sets

O. Goldreich (Ed.): Computational Complexity and Property Testing, LNCS 12050, pp. 326–351, 2020.
https://doi.org/10.1007/978-3-030-43662-9_18

by itself. However, restricting the verifier to run in almost-linear-time implies that something can be gained by interacting with a more powerful prover, even though the latter is restricted to polynomial-time.

The potential applicability of doubly-efficient interactive proof systems was demonstrated by Goldwasser, Kalai and Rothblum [8], who constructed such proof systems for any set that has log-space uniform circuits of bounded depth (e.g., log-space uniform $\mathcal{NC}$). A more recent work of Reingold, Rothblum, and Rothblum [15] provided such (constant-round) proof systems for any set that can be decided in polynomial-time and a bounded amount of space (e.g., for all sets in $\mathcal{SC}$). In our prior works [5,7], we presented simpler and more efficient constructions of doubly-efficient interactive proof systems for some special cases: In particular, in [5] we considered a class of "locally-characterizable sets", and in [7] we considered the problem of counting t-cliques in graphs.

In this work we consider the construction of constant-round doubly-efficient interactive proof systems for (sufficiently uniform) versions of $\mathcal{AC}^0[2]$ and $\mathcal{NC}^1$. We mention that the proof systems for $\mathcal{NC}$ constructed by Goldwasser, Kalai and Rothblum [8] use $O(d(n)\log n)$ rounds, where $d(n)$ is the depth of the nth circuit. Building on their techniques, Kalai and Rothblum have observed the existence of a constant-round proof system for a highly-uniform version of $\mathcal{NC}^1$, but their notion of uniformity was quite imposing and they never published their work [11]. In Sect. 3, we use similar ideas towards presenting a constant-round proof system for a sufficiently uniform version of $\mathcal{NC}^1$, which we believe to be less imposing (see also the overview in Sect. 1.4), but our main contribution is in presenting such a proof system for a sufficiently uniform version of $\mathcal{AC}^0[2]$: The latter proof system is more efficient and refers to a more relaxed notion of uniformity.

1.1 Our Main Result: A Proof System for $\mathcal{AC}^0[2]$

We present constant-round doubly-efficient interactive proof systems for sets acceptable by (sufficiently uniform) constant-depth polynomial-size Boolean circuits of unbounded fan-in with `and`, `or`, and `parity` gates (i.e., the class $\mathcal{AC}^0[2]$). Note that this class contains "seemingly hard problems in $\mathcal{P}$" (e.g., the t-`CLIQUE` problem for n-vertex graphs can be expressed as a highly uniform DNF with n^t terms (each depending on $\binom{t}{2}$ variables)). Postponing, for a moment, a clarification of what is meant by "sufficiently uniform", our result reads.

Theorem 1.1 (constant-round doubly-efficient interactive proofs for $\mathcal{AC}^0[2]$, loosely stated): *For constants $c,d \in \mathbb{N}$, suppose that $\{C_n : \{0,1\}^n \to \{0,1\}\}$ is a sufficiently uniform family of Boolean circuits with unbounded fan-in* `and`, `or`, *and* `parity` *gates such that C_n has size at most n^c and depth d. Then, for every $\delta \in (0,1]$, the set $\{x : C_{|x|}(x)=1\}$ has a $O(cd/\delta)$-round interactive proof system in which the verifier runs in time $O(n^{1+o(1)})$, the prescribed prover can be implemented in time $O(n^{c+o(1)})$, and the total communication is n^{δ}.*

We mention that the work of Reingold, Rothblum, and Rothblum [15] implies that log-space uniform $\mathcal{AC}^0[2]$ (actually, even log-space uniform $\mathcal{NC}^1$)[1] has constant-round doubly-efficient interactive proof systems. One advantage of our construction over [15] is that, being tailored to $\mathcal{AC}^0[2]$, it is much simpler and more transparent. In addition, the round complexity of our proof systems is considerably better than the round-complexity in [15]; specifically, we present a $O(1/\delta)$-round system with total communication n^δ, whereas in [15] obtaining total communication n^δ requires $\exp(\widetilde{O}(1/\delta))$ many rounds.

Corollaries. Using Theorem 1.1, we obtain a constant-round doubly-efficient interactive proof system for the *All Pairs Shortest Path* (APSP) problem (see background in [18]). Such a proof system follows also from the work of [15], but this fact was not observed before. The key observation is that verifying the value of APSP can be reduced to matrix multiplication in the (min, +)-algebra *via a doubly-efficient* NP-proof system.

Recall that matrix multiplication in the (min, +)-algebra refers to the case that multiplication is replace by addition and the sum is replace by the minimum; that is, the product of the matrices $A = (a_{i,j})_{i,j\in[n]}$ and $B = (b_{i,j})_{i,j\in[n]}$, denoted $A * B$, equals $C = (c_{i,j})_{i,j\in[n]}$ such that $c_{i,j} = \min_{k\in[n]}\{a_{i,k} + b_{k,j}\}$ for every $i, j \in [n]$. Given a possibly weighted n-vertex digraph G, we consider the matrix $W = (w_{i,j})_{i,j\in[n]}$ such that $w_{i,j}$ denotes the weight (or length) of the edge from i to j, whereas $w_{i,i} = 0$ and $w_{i,j} = \infty$ if there is no edge from i to j. Then, the shortest paths in G can be read from A^n, and the aforementioned NP-proof consists of the prover sending the matrices $A_1, A_2, ..., A_{\lceil\log_2 n\rceil}$ such that $A_0 = A$ and $A_i = A_{i-1} * A_{i-1}$ for all $i \in [\lceil\log_2 n\rceil]$. Hence, the verification of APSP is reduced to the verification of $\log n$ claims regarding matrix multiplication in the (min, +)-algebra, which can be verified in parallel. Focusing on the latter problem, or rather on the set $\{(A, B, A * B) : A, B \in \bigcup_{n\in\mathbb{N}} \mathbb{R}^{n\times n}\}$, we observe that membership in this set can be recognized in $\mathcal{SC}$ (hence the result of [15] applies) as well as by highly uniform $\mathcal{AC}^0$ circuits.

Corollary 1.2 (a constant-round doubly-efficient interactive proof for APSP): *Let* APSP *consists of pairs (G, L) such that L is a matrix recoding the lengths of the shortest paths between each pair of vertices in the weighted graph G. For every constant $\delta > 0$, the* APSP *has a $O(1/\delta)$-round interactive proof system in which the verifier runs in time $O(n^{2+o(1)})$, and the prescribed prover can be implemented in time $O(n^{4+o(1)})$, where n denotes the number of vertices in the graph and weights are restricted to $[-\exp(n^{o(1)}), \exp(n^{o(1)})]$. Furthermore, with the exception of the first prover message, the total communication is n^δ.*

As with Theorem 1.1, the application of [15] to APSP would have yielded $\exp(\widetilde{O}(1/\delta))$ rounds.

Another problem to which Theorem 1.1 is applicable is the set of graphs having no t-cliques, denoted t-no-CLIQUE. For any constant t, constant-round

[1] Actually, the result of [15] can be applied to $\mathcal{NC}^1$ circuits that can be constructed in polynomial time and $n^{o(1)}$-space.

doubly-efficient interactive proof systems for t-no-CLIQUE are implicit or explicit in several prior works. In particular, such proof systems are implied by the aforementioned result of [15] as well as by [5, Sec. 4.3], and were explicitly presented in [7, Sec. 2]. Noting that the said set can be recognized by highly uniform CNFs of size $O(n^t)$ and using Theorem 1.1, we obtain yet another alternative proof system for t-no-CLIQUE.

Corollary 1.3 (a constant-round doubly-efficient interactive proof for t-no-CLIQUE): *For every constants $t \in \mathbb{N}$ and $\delta > 0$, the set t-no-CLIQUE has a $O(t/\delta)$-round interactive proof system in which the verifier runs in time $O(n^{2+o(1)})$, the prescribed prover can be implemented in time $O(n^{t+o(1)})$, and the total communication is n^δ, where n denotes the number of vertices in the graph.*

In the following Table 1, we compare Corollary 1.3 to the prior proof systems known for the t-no-CLIQUE problem.

Table 1. Comparison of different constant-round interactive proof systems for the t-no-CLIQUE problem, for the constants t and $\delta > 0$, where n (resp., $m > n$) denotes the number of vertices (resp., edges).

Obtained	(in)	# rounds	Total comm.	Verifier time	Prover time
via $\mathcal{SC}$	[15]	$\exp(\widetilde{O}(1/\delta))$	n^δ	$\widetilde{O}(m)$	$\text{poly}(n^t)$
via "local characterization"	[5]	t/δ	n^δ	$\widetilde{O}(m)$	n^{t+1}
directly	[7]	t	$\widetilde{O}(n)$	$\widetilde{O}(m)$	$n^{0.791t}$
via $\mathcal{AC}^0[2]$	(this work)	$O(t/\delta)$	$n^{\delta+o(1)}$	$n^{2+o(1)}$	$n^{t+o(1)}$

Our proof system for t-no-CLIQUE is very similar to the one in [5]. The difference is that we apply the sum-check protocol to an arithmetic circuit defined over an extension field (of size $n^{2\delta}$) of GF(2), whereas in [5] it is implicitly applied to an arithmetic circuit defined over a field of prime characteristic that is larger than $\binom{n}{t}$. Furthermore, here the arithmetic circuit is a pseudorandom linear combination of the $\binom{n}{t}$ tiny circuits that identify specific t-cliques, whereas in [5] the arithmetic circuit counts these t-cliques.

1.2 Notions of Sufficiently Uniform Circuits

Some notion of uniformity (of the circuit family) is essential for a result such as Theorem 1.1, because the set of accepted instances is definitely in $\mathcal{BPP}$. Furthermore, the fact that claims regarding the circuit are verifiable in almost-linear time suggests some local verification features of the circuit. This suggests a notion of uniformity that refers to the complexity of constructing a *succinct representation of the circuit.* Needless to say, we seek the most liberal notion of uniformity that we can support.

We consider three such succinct representations, where in all cases we denote by n the length of the input to the poly(n)-size circuit, and assume that the circuit is *layered* (in the sense detailed below):

Adjacency predicate: Such a predicate indicates, for each pair of gates (u, v), whether or not gate u is fed by gate v. Specifically, dealing with circuits of size $s(n) = \text{poly}(n)$, we consider the adjacency predicate $\texttt{adj} : [s(n)] \times [s(n)] \to \{0,1\}$,

Incidence function: Such a function indicates, for each gate u and index i, the identity of the ith gate that feeds the gate u, where 0 indicates that u is fed by less than i gates. Specifically, for a predetermined fan-in bound $b(n) \leq s(n) - 1$, we consider the incidence function $\texttt{inc} : [s(n)] \times [b(n)] \to [s(n)] \cup \{0\}$.

Input assignment in canonical formulae: Here we consider a fixed structure of the circuit as a formula, and specify only the input bit assigned to each leaf of the formula, where the same bit is typically assigned to many leaves. The assignment is merely a function from leaf names to bit locations in the input. Specifically, we consider the input-assignment function $\texttt{ia} : [s(n)] \to [n + 2]$, where location $i \in \{n + 1, n + 2\}$ is assigned the constant $i \bmod 2$ in the "augmented" n-bit input (which holds $n + 2$ bits).[2]

In all cases, we assume that the (depth d) circuit is layered in the sense that, for each $i \in [d]$, gates at layer $i - 1$ are fed by gates at layer i only, where layer i consists of all gates at distance i from the output gate. Indeed, the output gate is the only gate at layer 0, and the gates at layer d are called leaves, since they are not fed by gates but are rather assigned input bits.[3] (Indeed, for simplicity, we do not allow leaves at other layers.)[4] Furthermore, when using the adjacency predicate and the incidence function representations, we shall assume that (for each $i \in [n]$) the ith leaf is assigned the ith input bit (and for $i \in \{n + 1, n + 2\}$ the constant $i \bmod 2$ is assigned to the ith leaf); but in the canonical formulae representation the assignment of input bits to leaves is the only aspects of the circuit that varies.

In all three cases, we make two additional simplifying assumptions. The first is that the circuit contains no "negation" gates (i.e., `not`-gates). This can be assumed, without loss of generality, because we can replace `not`-gates by `parity`-gates (fed by the desired gate and the constant 1, which is the reason for allowing

[2] The need to feed both constants arises from the following conventions by which the circuit is layered and all gates in the same layer compute the same functionality.

[3] We stress that the term 'leaf' is used here also in the case that the circuit is not a formula (i.e., does not have a tree structure). One may prefer using the terms 'terminal' or 'source' instead.

[4] This can be assumed, without loss of generality, by replacing such a potential leaf at layer i with an auxiliary path of dummy gates that goes from layer d to layer i so that this path indirectly feeds the value of the desired input bit to the corresponding gate at layer i.

to feed leaves with the constant 1). The second assumption is that, *for each* $i \in [d]$, *all gates at layer* $i-1$ *have the same functionality* (gate-type).[5]

Theorem 1.1 holds under each of the three representations, when requiring that the corresponding function, which implicitly describes the poly(n)-size circuit C_n, can be represented by a formula of size $n^{o(1)}$ that can be constructed in time $n^{1+o(1)}$. Recall that the input to the latter formula is of length $O(\log n)$; hence, the foregoing condition is relatively mild (given that C_n is trivially represented by a formula of size poly(n), which is poly(n)-time constructible provided that C_n itself is poly(n)-time constructible).

Our notion of uniformity is incomparable to the notion of log-space uniformity used in [8] (and ditto re [15]). On the one hand, we impose stronger restrictions on the time complexity of constructing a succinct (implicit) representation of the circuit; on the other hand, unlike [8] (and [15]), we do not restrict the space complexity of this task.

1.3 Overview of Our Main Construction

The construction underlying Theorem 1.1 combines a central ingredient of the interactive proof system of Goldwasser, Kalai, and Rothblum [8] with the approximation method of Razborov [14] and Smolensky [17]. Specifically, we first reduce the verification of the claim that the input satisfies the predetermined Boolean circuit to an analogous claim regarding an Arithmetic circuit (over GF(2)) that is derived from the Boolean circuit using the approximation method. The crucial fact is that that *all multiplication gates in the Arithmetic circuit have small fan-in* (whereas the fan-in of addition gates may be large). With high probability, this approximation does not affect the computation on the given input, but it does introduces a "completeness error" in the verification procedure, which we eliminate later (so to obtained perfect completeness).

Next, following [8], we consider a computation of the Arithmetic circuit (on the given input), and encode the values of the gates at each layer by a low degree polynomial over a large (extension) field (of GF(2)). Here we use the fact that, by virtue of the approximation method, the gates in the Arithmetic circuit have bounded fan-in. Using this fact and relying on the foregoing uniformity condition, *we express the relation between the values of the gates at adjacent layers* (of the circuit) *by low degree polynomials*. These polynomials are derived from the small Boolean formulas that compute the adjacency relation.

Lastly, following [8], we reduce the verification of a claim regarding the values at layer $i-1$ in the circuit to a claim regarding the values at layer i, by using the Sum-Check protocol in each reduction step. Specifically, we use the Sum-Check protocol with respect to variables in a relatively large alphabet (of size n^{δ}), so that the number of rounds is a constant (i.e., $O(1/\delta)$). Actually,

[5] This can be assumed, without loss of generality, by replacing each layer by three consecutive layers so that one layer is devoted to `and`-gates, one to `or`-gates, and one to `parity`-gates.

this refers to the way in which addition gates of unbounded fan-in are handled, where each such poly(n)-way addition is written as a sum over a $O(1/\delta)$-long sequence of indices over an alphabet of size n^δ (i.e., $\sum_{i\in[m]} T(i)$ is written $\sum_{i_1,...,i_t\in[m^{1/t}]} T(i_1,..,i_t)$). In contrast, multiplication gates, which are of logarithmically bounded fan-in, are treated in a straightforward manner (i.e., we branch to verify each of the logarithmically many claimed values).[6]

To summarize: Using the approximation method allows us to replace or-gates (and/or and-gates) of unbounded fan-in by multiplication gates of logarithmic fan-in, while introducing parity gates of unbounded fan-in. Each layer of parity gates can be handled by the Sum-Check protocol such that each iteration of this protocol cuts the fan-in of the parity gates by a factor of n^δ. The degree bound on which the Sum-Check protocol relies is due to the uniformity of the original Boolean circuit and to the fact that the multiplication gates have small fan-in. Specifically, a sufficient level of uniformity of the Boolean circuit implies an upper bound on the degree of the polynomials that relate the values of the gates at adjacent layers (of the circuit), whereas the small fan-in of the multiplication gates allows to deal with them in a straightforward manner.

We mention that the idea of using the approximation method towards emulating $\mathcal{AC}^0[2]$ by low degree arithmetic circuits, *in the context of interactive proof systems*, was used before by Kalai and Raz [10]. Both in [10] and here, this causes a (small) error probability (in the completeness condition).[7]

We *regain perfect completeness* by letting the prover point out a gate in which an approximation error occurs (with respect to the input), and prove its claim. That is, we let the verifier accept in case it is convinced of such a claim (of approximation error), which means that we increase the soundness error (rather than introduce a completeness error).

1.4 The Proof System for $\mathcal{NC}^1$

Generalizing and somewhat simplifying the proof systems constructed by Goldwasser, Kalai, and Rothblum [8], we obtain constant-round doubly-efficient interactive proof systems for sufficiently uniform $\mathcal{NC}^1$ (specifically, canonical formulas with a sufficiently uniform input-assignment function as discussed in Sect. 1.2). The simplification is due to relying on a stronger notion of uniformity than the one used in [8], whereas the generalization allows us to reduce the round complexity of [8] by a log-squared factor. Recall that, when handling a (bounded fan-in) circuit $C_n : \{0,1\}^n \to \{0,1\}$ of depth $d(n)$, the proof system of [8] has

[6] Actually, we could reduce the verification of these logarithmically many claims to the verification of a single claim, by using a curve that passes through all the points in these claims, as done in [8]. But since here the number of rounds is a constant, we can afford an overhead that is exponential in the number of rounds (i.e., the overhead is $O(\log n)^{d'}$, where d' is the number of layers having multiplication gates).

[7] In contrast, when using the approximation method in the context of worst-case to average-case reduction for the class $\mathcal{AC}^0[2]$ presented in [6, Apdx A.2], the approximation error is absorbed by the (larger) error rate of the average-case solver.

$O(d(n) \cdot \log n)$ rounds. This is due to invoking the Sum-Check Protocol for each layer in the circuit, and using a version that handles summations over the binary alphabet. Instead, for any constant $\delta > 0$, we invoke the Sum-Check protocol for each block of $\delta \log_2 n$ consecutive layers in the circuit, and use a version that handles summations over an alphabet of size n^δ. Hence, we cut the number of rounds by a factor of $(\delta \log n)^2$.

Theorem 1.4 (constant-round doubly-efficient interactive proofs for $\mathcal{NC}^1$, loosely stated): *Let $\{C_n : \{0,1\}^n \to \{0,1\}\}$ be a sufficiently uniform family of canonical Boolean circuits of fan-in two and logarithmic depth. Then, for every $\delta \in (0,1]$, the set $\{x : C_{|x|}(x) = 1\}$ has a $O(1/\delta^2)$-round interactive proof system in which the verifier runs in time $O(n^{1+o(1)})$, the prescribed prover can be implemented in polynomial-time, and the total communication is $n^{\delta+o(1)}$.*

We stress that Theorem 1.4 does not subsume Theorem 1.1. First, the proof system in Theorem 1.4 uses a larger number of rounds as a function of total communication complexity (i.e., $O(1/\delta^2)$ rather than $O(1/\delta)$ rounds). Second, the uniformity condition in Theorem 1.4 is stronger (cf., Theorem 3.1 and Theorem 2.3).

1.5 Digest and Organization

Our constant-round doubly-efficient interactive proof systems (for $\mathcal{AC}^0[2]$ and $\mathcal{NC}^1$) are based on the proof system of Goldwasser, Kalai and Rothblum [8]. Specifically, these proof systems are designed for proving that an input x satisfies a circuit $C_{|x|}$ that is "efficiently constructable" based on $|x|$ only. The proof systems differ in the specific meaning given to the term "efficiently constructable" (see Sect. 1.2), and they are all pivoted in functions that represent the values of the gates at different layers of the circuit (in its computation on input x).

Typically, the latter functions are too large to be communicated, and their values at specific points cannot be evaluated by the verifier itself (although they are computable in polynomial-time). Still, given x, all these functions are well defined, and they are related by the description of the circuit. The more structured the circuit, the simpler these relations are. In particular, the notions of uniformity defined in Sect. 1.2 yield very simple relations between the functions that describe the values of gates at adjacent layers.

These relations extend also to low-degree extensions of these functions (which constitute error correcting codes of the explicit description of these functions), and they allow for testing the function that corresponds to layer $i - 1$ by using the function that corresponds to layer i. Specifically, the value of the former function at a given point is verified by using the value of the latter function at a few points. Lastly, in contrast to the functions that correspond to higher layers, the function that corresponds to the lowest layer (i.e., the input layer) is known to the verifier (who knows the input x).

In the interactive proof system for $\mathcal{AC}^0[2]$, presented in Sect. 2 (and establishing Theorem 1.1), a strong notion of uniformity is used to directly relate functions that describe the values of gates in adjacent layers of a related arithmetic circuit (in a computation on input x). Here we rely on the hypothesis that the original circuit has constant-depth, and capitalize on the fact that we can obtain a corresponding arithmetic circuit that uses multiplication gates of logarithmic fan-in.

In the interactive proof system for $\mathcal{NC}^1$, presented in Sect. 3 (and establishing Theorem 1.4), an even strong notion of uniformity is used to relate functions that describe the values of gates that are $\delta \log_2 n$ layers apart. Here we rely on the hypothesis that the original circuit has fan-in two and logarithmic depth.

2 The Interactive Proof System for $\mathcal{AC}^0[2]$

Recall that we consider a sufficiently uniform family of layered circuits $\{C_n\}$ of constant depth $d \in \mathbb{N}$ and unbounded fan-in. For simplicity of our presentation, we work with the adjacency predicate representation, while noting that the handling of other representations can be reduced to it (as detailed in Sect. 2.5). We also assume, for simplicity, that the circuit has only gates of the `or` and `parity` type, since `and`-gates can be emulated by these. Letting $s = s(n) = \text{poly}(n)$ be a bound on the number of gates in C_n, for each $i \in [d]$, we consider the $n^{o(1)}$-sized formula $\psi_i : [s] \times [s] \to \{0,1\}$ such that $\psi_i(j,k) = 1$ if and only if gate j resides in layer $i-1$ and is fed by the gate k (which resides in layer i). In doing so, we associate $[s]$ with $\{0,1\}^\ell$, where $\ell = \log_2 s$, and view ψ_i as a function over $\{0,1\}^{2\ell}$.

On input $x \in \{0,1\}^n$, we proceed in three steps: First, we reduce the Boolean problem (of verifying that $C_n(x) = 1$) to an Arithmetic problem (of verifying that a related Arithmetic circuit A_n evaluates to 1 (on the same input x). This reduction uses the approximation method and yields constant-depth arithmetic circuits with multiplication gates of logarithmic fan-in. Next, we express the latter problem as a sequence of $O(d)$ functional equations that relate the value of gates at adjacent layers of the circuit A_n. Here we shall use low degree polynomials that extend the ψ_i's, while deriving (succinct representations of) these polynomials from the corresponding Boolean formulas that compute the ψ_i's. Using the fact that all multiplication gates are of small fan-in, it follows that the resulting equations are all of low degree. Last, we present a constant-round doubly-efficient interactive proof system for the verification of this sequence of functional equations. Hence, we obtain a constant-round doubly-efficient interactive proof system for the set $\{x : C_{|x|}(x)=1\}$.

2.1 Step 1: Approximation by Arithmetic Circuits

The first step is a randomized reduction of solving the Boolean problem to solving a corresponding Arithmetic problem. This reduction follows the ideas underlying the approximation method of Razborov [14] and Smolensky [17], while working with the field GF(2) (as [14], rather than with GF(p) for some prime $p > 2$ as [17]). Recall that this reduction replaces every `or`-gate by a $O(\log n)$-way multiplication of parity gates that each compute a random linear combination of the values of the gates that feed the or-gate in the Boolean circuit.

When following this scheme, we replace the random choices made at each gate by pseudorandom choices that are generated by a small bias generator [13]; specifically, we use a specific small-bias generator that uses a seed of logarithmic (i.e., $O(\log n)$) length such that, for each fixed seed, the bits of the pseudorandom sequence can be succinctly represented by a low-degree function in the bits of the binary extension of the bit's location (e.g., the third construction in [1]).[8] We shall use the same seed to generate all pseudorandom sequences used in the construction, but use different parts of the sequence for each random combination at each gate.

Hence, for a fixed Boolean circuit C_n, on input $x \in \{0,1\}^n$, we randomly reduce the question of whether $C_n(x) = 1$ to the question of whether $A_n^{(\sigma)}(x) = 1$, where $\sigma \in \{0,1\}^{O(\log n)}$ is selected uniformly at random, and $A_n^{(\sigma)}$ is the Arithmetic circuit that results when using σ as the seed for the small-biased generator. Specifically, the choice of σ will be made by the verifier, and we use the fact that $\Pr_\sigma[A_n^{(\sigma)}(x) = C_n(x)] = 1 - s(n) \cdot \exp(-O(\log n)) = 1 - o(1)$. In the rest of the analysis, we assume that the verifier was not extremely unlucky (in its choice of σ), and so that $A'_n(x) \stackrel{\text{def}}{=} A_n^{(\sigma)}(x) = C_n(x)$ holds. Indeed, we shall fix σ for the rest of this description, and will use the shorthand A'_n.

Let us stop for a moment and take a closer look at A'_n. Recall that each or-gate in C_n is essentially replaced by a $O(\log n)$-way multiplication gate that is fed by the inner-product of the values of the original feeding gates and a pseudorandom sequence. Specifically, for $\ell' = O(\log n)$, if the or-gate indexed j (at layer $i-1$ of C_n) was fed by gates indexed $k_1, ..., k_{n'}$ (of layer i), then this or-gate is replaced in A'_n by an arithmetic sub-circuit that computes the function

$$1 + \prod_{j' \in [\ell']} \left(1 + \sum_{t \in [n']} G'((j-1) \cdot s \cdot \ell' + (j'-1) \cdot s + k_t) \cdot y_{k_t} \right) \qquad (1)$$

where y_k represents the output of the gate indexed k (at layer i of C_n), and $G'(k) = G^{(\sigma)}(k)$ is the kth bit in the pseudorandom sequence generated based on the aforementioned fixed seed σ. Note that the ℓ' different linear combinations associated with the same or-gate use different portions of the pseudorandom

[8] In the third construction of [1], the seed is viewed as a pair (ζ, r), where $\zeta \in \mathrm{GF}(2^k)$ and $r \in \{0,1\}^k$, and the ith bit in the output is the inner-product (mod 2) of the binary representation of ζ^i and r. Note that computing ζ^i reduces to computing $\prod_{j \geq 0}(i_j \cdot \zeta^{2^j} + (1 - i_j))$, where $(i_{k'-1}, ..., i_0) \in \{0,1\}^{k'}$ is the binary expansion of $i \in [2^{k'}]$, and that $\zeta^2, ..., \zeta^{2^{k'-1}}$ can be precomputed when ζ is fixed. Note that each element of $\mathrm{GF}(2^k)$ is represented as a k-bit long sequence over $\{0,1\} \equiv \mathrm{GF}(2)$, and so multiplication in $\mathrm{GF}(2^k)$ corresponds to k bilinear forms in these representations (and the product $i_j \cdot \zeta^{2^j}$ corresponds to the products of i_j and the bits in the representation of ζ^{2^j}).

sequence (i.e., for each $j \in [s]$ and $j' \neq j''$, the sets $\{(j-1)\cdot s\cdot \ell' + (j'-1)\cdot s + k : k \in [s]\}$ and $\{(j-1)\cdot s\cdot \ell' + (j''-1)\cdot s + k : k \in [s]\}$ are disjoint). Likewise, different or-gates use different portions of the pseudorandom sequence (but this choice is immaterial).[9]

The analysis of the foregoing reduction, which uses related pseudorandom sequences, is almost identical to the analysis of the original reduction (which uses independent random sequences). On the one hand, if $\bigvee_{t\in[n']} y_{k_t} = 0$, then Eq. (1) always equals $1+(1+0)^{\ell'} = 0$. On the other hand, if $\bigvee_{t\in[n']} y_{k_t} = 1$, then, with probability at most $2^{-\ell'}+\epsilon$ (over the choice of σ), it holds that Eq. (1) equals $1+(1+0)^{\ell} = 0$, where ϵ denote the bias of the small-biased generator. Indeed, we use the fact that ℓ' disjoint and non-zero linear combinations of the bits of an ϵ-bias generator equal $0^{\ell'}$ with probability at most $2^{-\ell'}+\epsilon$, since this holds for any ℓ' linearly independent combinations. Recall that the aforementioned construction of [1] uses a seed of length $O(\log(L/\epsilon))$ in order to generate an ϵ-biased sequence of length L. Hence, we can set $L = \widetilde{O}(s^2)$ and $\epsilon = o(1/s)$, and upper-bound the probability that an approximation-error occurred in any of the gate-replacements by using a union bound; that is, this probability is upper-bounded by $s\cdot(2^{-\ell'}+\epsilon) = o(1)$.

Recall that Eq. (1) represents the function computed by a constant-depth arithmetic sub-circuit that replaces a generic or-gate. It will be convenient to think of this function as being computed by a single gate, which we hereafter call an augmented multiplication gate.

We highlight the following features of the Arithmetic circuit A'_n: Its depth is $O(d)$, its size is $O(\log n)^d \cdot s(n)$, it computes a polynomial of degree $O(\log n)^d$, and it has a succinct representation of size $n^{o(1)}$ that can be constructed in time $n^{1+o(1)}$. (The foregoing assertions use the fact that, given σ, a circuit computing $G^{(\sigma)} : \{0,1\}^{\log_2(s^2\ell')} \to \{0,1\}$ can be constructed in poly($|\sigma|$)-time, and that this circuit corresponds to a multilinear function from $\mathrm{GF}(2)^{\log_2(s^2\ell')}$ to GF(2); see Footnote 8.)

Evaluating A'_n over an Extension Field. Indeed, as defined above, $A'_n : \mathrm{GF}(2)^n \to \mathrm{GF}(2)$ is an arithmetic circuit over GF(2), consisting solely of addition and augmented multiplication gates. But we can view A'_n as an arithmetic circuit over $\mathcal{F} = \mathrm{GF}(2^{2\delta \log_2 n})$, and consider its value at $x \in \{0,1\}^n$, which is viewed as an n-long sequence over $\mathcal{F}$. (It suffices to have $|\mathcal{F}| \geq n^{\delta+\Omega(1)}$; on the other hand, we also use $\log_2|\mathcal{F}| \leq n^{o(1)}$.)

2.2 Step 2: Relating the Values of Layers in the Computation

A key idea of Goldwasser, Kalai and Rothblum [8] consists of representing the values of the gates at various levels of the circuit by functions, and relating

9 This is the case since we bound the approximation error of A'_n by employing a union bound to the errors that may occur in the various gates, and this holds regardless of the dependency between these error events.

these functions by functions that represent the structure of the circuit. (These functions are first viewed as functions from $[s]$ to $\{0,1\}$, and then as ℓ-variate functions from $\{0,1\}^\ell$ to $\{0,1\}$, which serves as basis for considering their low degree extensions over larger fields (i.e., $\mathcal{F}$ as above).)[10]

The Basic Functional Relations. In our context, we use the functions $\alpha_d, ..., \alpha_0 : [s] \to \{0,1\}$ such that $\alpha_{i-1}(j)$ represents the value of gate j (at layer $i-1$) in a computation of $A'_n(x)$. We then relate their values by referring to ψ_i and to the functionality of the gates in layer i. Recall that $\psi_i(j,k) = 1$ if gate j (of layer $i-1$) is fed by gate k (of layer i). In the case of an addition gate (i.e., a layer of addition gates), we have

$$\alpha_{i-1}(j) = \sum_{k\in[s]} \psi_i(j,k) \cdot \alpha_i(k). \tag{2}$$

Hence, Eq. (2) can be viewed as relating functions that range over $[s]$ by using a function that ranges over $[s]^2$. In the case of an augmented multiplication gate (as represented by Eq. (1)), we have

$$\alpha_{i-1}(j) = 1 + \prod_{j'\in[\ell']} \left(1 + \sum_{k\in[s]} G'((j-1)\cdot s\cdot \ell' + (j'-1)\cdot s + k)\cdot \psi_i(j,k)\cdot \alpha_i(k)\right) \tag{3}$$

where (as before) $G'(k)$ represents the kth bit in the output of the generator on the fixed seed. Assuming, without loss of generality, that $\ell' > 2\ell + \log_2 \ell'$ (equiv., $2^{\ell'} > s^2 \cdot \ell'$), we view Eq. (3) as relating functions that range over $[s]$ by using functions that range over $[2^{\ell'}] \supset [s]^2 \cup ([s]^2 \times [\ell'])$.

Revising the Basic Functional Relations by Viewing Gate-Indices as Binary Strings. The next step is viewing all functions as functions over binary strings rather than over natural numbers; that is, we associate $[s]$ with $\{0,1\}^\ell$, and $[2^{\ell'}]$ with $\{0,1\}^{\ell'}$. Furthermore, we consider the arithmetic formula $\widehat{\psi}_i : \mathcal{F}^{\ell+\ell} \to \mathcal{F}$ that is derived from $\psi_i : \{0,1\}^{\ell+\ell} \to \{0,1\}$ in the obvious manner (i.e., replacing and-gates by multiplication gates and negation gates by gates that add the constant 1). Recalling that ψ_i is a formula of size $n^{o(1)}$ (equiv., a bounded fan-in formula of depth $o(\log n)$), it follows that $\widehat{\psi}_i$ computes a polynomial of degree $n^{o(1)}$. Hence, Eq. (2) is replaced by

$$\alpha_{i-1}(j) = \sum_{k\in\{0,1\}^\ell} \widehat{\psi}_i(j,k) \cdot \alpha_i(k), \tag{4}$$

[10] Jumping ahead, we stress that the relation between these functions will be checked by the interactive proof system presented in Sect. 2.3. This will be done by having the verifier ask the prover to provide the values of these functions at few places, while relying on the fact that these functions can be evaluated in polynomial-time. Note that these functions are too large to be communicated to the verifier (see analogous discussion in Sect. 3.1).

where $j \in \{0,1\}^\ell$. Hence, the functions $\alpha_{i-1} : \{0,1\}^\ell \to \{0,1\}$ and $\alpha_i : \{0,1\}^\ell \to \{0,1\}$ are related by an equation that uses a low degree polynomial (i.e., $\widehat{\psi}_i$). The same consideration can be applied to Eq. (3), when recalling that $G' : [2^{\ell'}] \to \{0,1\}$ can be written as an explicit low-degree polynomial by using functions that range over $\{0,1\}^{\ell'} \equiv [2^{\ell'}]$. Specifically, the corresponding low degree polynomial is $\widehat{G'} : \mathcal{F}^{\ell'} \to \mathcal{F}$ such that $\widehat{G'}(z_{\ell'}, ..., z_1)$ is a specific linear combination (i.e., r) of the bits of the field element $\prod_{j\in[\ell']}(z_j \cdot \tau_j + (1 - z_j))$, where r as well as the τ_j's are precomputed based on the fixed seed of the small-biased generator G (see Footnote 8).[11] Hence, we get

$$\alpha_{i-1}(j) = 1 + \prod_{j'\in\{0,1\}^{\log_2 \ell'}} \left(1 + \sum_{k\in\{0,1\}^\ell} \widehat{G'}(j, j', k) \cdot \widehat{\psi}_i(z, k) \cdot \alpha_i(k)\right) \quad (5)$$

where $j \in \{0,1\}^\ell$.

Low Degree Extensions of the Basic Functional Relations. At this point, the standard approach taken in [8] (and followed also in Sect. 3) is to extend Eqs. (4)–(5) to any $j \in \mathcal{F}^\ell$ by using a low-degree extension (see also Eq. (8)). We could have done this, but this is redundant in the case of Eqs. (4)–(5), since the r.h.s. of these equations is well defined also when $j \in \mathcal{F}^\ell$. Hence, we can replace Eq. (4) by

$$\widehat{\alpha}_{i-1}(z) = \sum_{k\in\{0,1\}^\ell} \widehat{\psi}_i(z, k) \cdot \alpha_i(k), \quad (6)$$

where $z \in \mathcal{F}^\ell$. Assuming that α_i is also extended to a low-degree polynomial, denoted $\widehat{\alpha}_i$, we can replace $\alpha_i(k)$ by $\widehat{\alpha}_i(k)$. In this case, the degree of $\widehat{\alpha}_{i-1}$ equals the degree of $\widehat{\psi}_i$, which is $n^{o(1)}$. In the case of Eq. (3), we get, for every $z \in \mathcal{F}^\ell$

$$\widehat{\alpha}_{i-1}(z) = 1 + \prod_{j'\in\{0,1\}^{\log_2 \ell'}} \left(1 + \sum_{k\in\{0,1\}^\ell} \widehat{G'}(z, j', k) \cdot \widehat{\psi}_i(z, k) \cdot \widehat{\alpha}_i(k)\right) \quad (7)$$

In this case the degree of $\widehat{\alpha}_{i-1}$ equals ℓ' times the sum of the degrees of $\widehat{\psi}_i$ and $\widehat{G'}$, where the degree of $\widehat{G'}$ is ℓ' (and the degree of $\widehat{\psi}_i$ is $n^{o(1)}$).

Note, however, that the foregoing can not be applied to α_d, which is supposed to encode the input x to A'_n. This function is well-defined only over $[s] \equiv \{0,1\}^\ell$;

[11] Here we assume that the length of the seed is $2\ell'$, which is justified by the fact that we can afford any $\ell' = O(\log n)$. Recall that, for $(i_{\ell'}, ..., i_1) \in \{0,1\}^{\ell'}$, if $\tau_j = \zeta^{2^{j-1}}$, then $\prod_{j\in[\ell']}(i_j \cdot \tau_j + (1 - i_j)) = \zeta^i$ such that $i = \sum_{j\in[\ell']} i_j \cdot 2^{j-1}$. (Recall the product is over $\mathrm{GF}(2^{\ell'})$, but each multiplication over $\mathrm{GF}(2^{\ell'})$ is emulated by bilinear forms in the bits of the representations of the $\mathrm{GF}(2^{\ell'})$-elements.) In this case, $\widehat{G'}(i_{\ell'}, ..., i_1)$ equals the inner product (mod 2) of r and the binary representation of ζ^i, where $(\zeta, r) = \sigma$ is the seed of the generator.

specifically, recall that $\alpha_d(j) = x_j$ for $j \in [n]$, whereas $\alpha_d(j) = j \bmod 2$ for $j \in \{n+1, n+2\}$, and $\alpha_d(j) = 0$ for $j \in [s] \setminus [n+2]$.[12] Hence, we augment the foregoing definitions by postulating that $\widehat{\alpha}_d$ is a low-degree extension of the values of α_d at $\{0,1\}^\ell$; that is, for $z \in \mathcal{F}^\ell$, we have

$$\widehat{\alpha}_d(z) = \sum_{k \in \{0,1\}^\ell} \mathtt{EQ}(z,k) \cdot \alpha_d(k), \tag{8}$$

where EQ is the bilinear polynomial that extends the function that tests equality over $\{0,1\}^\ell$ (e.g., $\mathtt{EQ}(\sigma_1 \cdots \sigma_\ell, \tau_1 \cdots \tau_\ell) = \prod_{i \in [\ell]}(\sigma_i \tau_i + (1-\sigma_i)(1-\tau_i))$). Note that r.h.s. of Eq. (8) depends only on $n+2$ terms, since $\alpha_d(j) = 0$ for $j \notin [n+2]$.

Revising the Basic Functional Relations by Viewing Gate-Indices as Short Sequences over a Large Alphabet. Lastly, we wish to replace summation over $\{0,1\}^\ell$ by summation over H^m, where $|H| = n^\delta$ and $m = \frac{\ell}{\delta \log_2 n} = O(1/\delta)$. The problem is that the foregoing functions $\widehat{\psi}_i, \widehat{G'}, \widehat{\alpha}_i$ and EQ were also defined as low degree extensions of corresponding functions (i.e., ψ_i, G', α_i and EQ itself) that were defined over binary strings. In particular, the functions ψ_i's are provided by the hypothesis and we must deal with it as is.[13] Hence we need to transform elements of $H \subset \mathcal{F}$ (and sequences over H) to their binary representation, and feed the latter to the foregoing functions. This is quite straightforward, alas cumbersome.

First, we introduce a 1–1 mapping $\mu : H \to \{0,1\}^{\ell''}$, where $\ell'' = \delta \log_2 n = \ell/m$, such that $\mu(h)$ returns the ℓ''-bit long binary representation of $h \in H$. Next, we extended $\mu : H \to \{0,1\}^{\ell''}$ to an ℓ''-long sequence of univariate polynomials of degree $|H|-1$ over $\mathcal{F}$; that is, $\mu : \mathcal{F} \to \mathcal{F}^{\ell''}$ is defined as $\mu(z) = (\mu_{\ell''}(z), ..., \mu_1(z))$ such that $\mu_i(z) = \sum_{h \in H} \prod_{h' \in H \setminus \{h\}} \frac{z-h'}{h-h'} \cdot \mathtt{bin}_i(h)$, where $\mathtt{bin}_i(h)$ is the ith bit in the binary representation of $h \in H \equiv \{0,1\}^{\ell''}$. The next step is extending μ to m-long sequences of such sequences; that is, we use $\widehat{\mu} : \mathcal{F}^m \to (\mathcal{F}^{\ell''})^m$ such that for every $h = (h_1, ..., h_m) \in H^m$ it holds that $\widehat{\mu}(h) = (\mu(h_1), ..., \mu(h_m)) \in \{0,1\}^{m \cdot \ell''}$, whereas $m \cdot \ell'' = \ell$. Finally, we redefine the $\widehat{\alpha}_i$'s so that they range over $\mathcal{F}^m$ (rather than over $\mathcal{F}^\ell$). Starting with Eq. (8), we have

$$\widehat{\alpha}_d(z) = \sum_{k \in H^m} \mathtt{EQ}(\widehat{\mu}(z), \widehat{\mu}(k)) \cdot \alpha_d(\widehat{\mu}(k)), \tag{9}$$

[12] Recall that we need to provide the circuit with the constants 1 and 0; hence, we set $\{\alpha_d(n+1), \alpha_d(n+2)\} = \{0,1\}$. The setting of $\alpha_d(j) = 0$ for $j \in [s] \setminus [n+2]$ is used in order to facilitate the evaluation of r.h.s. of Eq. (8), as discussed below and used in Step 3; that is, this setting ensures that, for every $u \in \mathcal{F}^\ell$, it holds that $\sum_{k \in \{0,1\}^\ell} \mathtt{EQ}(u,k) \cdot \alpha_d(k)$ equals $\sum_{k \in I} \mathtt{EQ}(u,k) \cdot \alpha_d(k)$, where $I \subset \{0,1\}^\ell$ corresponds to $[n+2]$. Alternatively, we could have used the setting $\alpha_d(j) = 1$ for $j \in [s] \setminus [n+2]$, and rely on the fact that $\sum_{k \in \{0,1\}^\ell} \mathtt{EQ}(u,k)$ equals 1.

[13] In contrast, we could have defined G', α_i and EQ over H-sequences from the start, rather than defining them over binary sequences. But doing so would not have allowed us to relate ψ_i to the other functions without using the transformation that we are going to introduce and use now.

where $z \in \mathcal{F}^m$. Turning to Eq. (6), we replace it by

$$\widehat{\alpha}_{i-1}(z) = \sum_{k \in H^m} \widehat{\psi}_i(\widehat{\mu}(z), \widehat{\mu}(k)) \cdot \widehat{\alpha}_i(k) \tag{10}$$

where again $z \in \mathcal{F}^m$. Lastly, Eq. (7) is replaced by

$$\widehat{\alpha}_{i-1}(z) = 1 + \prod_{j' \in \{0,1\}^{\log_2 \ell'}} \left(1 + \sum_{k \in H^m} \widehat{G'}(\widehat{\mu}(z), \widehat{\mu}(j'), \widehat{\mu}(k)) \cdot \widehat{\psi}_i(\widehat{\mu}(z), \widehat{\mu}(k)) \cdot \widehat{\alpha}_i(k)\right) \tag{11}$$

where we assume that $\{0,1\} \subset H$. Denoting the degree of a polynomial p by $\mathtt{deg}(p)$, we have $\mathtt{deg}(\widehat{\alpha}_d) = \mathtt{deg}(\mathtt{EQ}) \cdot \mathtt{deg}(\widehat{\mu}) < \ell \cdot |H|$. For i that is an addition layer (i.e., Eq. (10)), we have $\mathtt{deg}(\widehat{\alpha}_{i-1}) = \mathtt{deg}(\widehat{\psi}_i) \cdot \mathtt{deg}(\widehat{\mu}) < n^{o(1)} \cdot |H|$. Lastly, for i that is an augmented multiplication layer (i.e., Eq. (11)), we have

$$\begin{aligned}\mathtt{deg}(\widehat{\alpha}_{i-1}) &< \ell' \cdot \left(\mathtt{deg}(\widehat{G'}) \cdot |H| + \mathtt{deg}(\widehat{\psi}_i) \cdot |H|\right) \\ &= \ell' \cdot n^{o(1)} \cdot |H|.\end{aligned}$$

Hence, all degrees are upper-bounded by $n^{o(1)} \cdot |H| = n^{\delta + o(1)}$.

2.3 Step 3: Obtaining an Interactive Proof System (with Imperfect Completeness)

On input $x \in \{0,1\}^n$, the verifier selects uniformly $\sigma \in \{0,1\}^{O(\log n)}$, and sends σ to the prover. The prover now attempts to prove that $A'_n(x) \stackrel{\text{def}}{=} A_n^{(\sigma)}(x) = 1$, where a succinct representation of A'_n (which has size $n^{o(1)}$) can be constructed in time $n^{1+o(1)}$. The initial claim is re-interpreted as $\widehat{\alpha}_0(\mu^{-1}(1^\ell)) = 1$, where $\mu^{-1}(1^\ell) = (\mu^{-1}(1^{\ell/m}), ..., \mu^{-1}(1^{\ell/m})) \in H^m \subset \mathcal{F}^m$. (Recall that $\widehat{\alpha}_i : \mathcal{F}^m \to \mathcal{F}$ for every i, and that $\mu : H \to \{0,1\}^{\ell/m}$ is a bijection).

The parties proceed in $O(d)$ steps such that the ith step starts with a claim regarding the value of $\widehat{\alpha}_{i-1}$ at few points, and ends with a claim regarding the value of $\widehat{\alpha}_i$ at few points, where the said number of points may increase by at most a logarithmic factor. We distinguish between the case that the current layer (i.e., $i-1$) is of addition gates and the case that it is of augmented multiplication gates.

Handling a layer of addition gates: (Recall that these gates are supposed to satisfy Eq. (10).) For each claim of the form $\widehat{\alpha}_{i-1}(u) = v$, where $u \in \mathcal{F}^m$ and $v \in \mathcal{F}$ are known, we invoke the Sum-Check protocol on the r.h.s. of Eq. (10). The execution of this (m-round) protocol results in a claim regarding the value $\widehat{\psi}_i(\widehat{\mu}(u), \widehat{\mu}(r)) \cdot \widehat{\alpha}_i(r)$ for a random $r \in \mathcal{F}^m$ selected via the execution. Since the verifier can evaluate $\widehat{\mu}$ and $\widehat{\psi}_i$, it is left with a claim regarding the value of $\widehat{\alpha}_i$ at one point.

Recall that the Sum-Check protocol proceeds in $m = O(1/\delta)$ rounds, where in each round the prover sends the value of the relevant univariate polynomial.

This is a polynomial of degree $|H| \cdot n^{o(1)} = n^{\delta+o(1)}$, where the degree bound is due to the composition of the polynomials $\widehat{\mu}$ and $\widehat{\psi}_i$ (see the end of Sect. 2.2).

Handling a layer of augmented multiplication gates: (Recall that these gates are supposed to satisfy Eq. (11).) For each claim of the form $\widehat{\alpha}_{i-1}(u) = v$, where $u \in \mathcal{F}^m$ and $v \in \mathcal{F}$ are known, we let the prover send the values $(v_1, ..., v_{\ell'})$ such that for every $j' \in [\ell']$

$$v_{j'} \stackrel{\text{def}}{=} \sum_{k \in H^m} \widehat{G'}(\widehat{\mu}(u), \widehat{\mu}(j'), \widehat{\mu}(k)) \cdot \widehat{\psi}_i(\widehat{\mu}(u), \widehat{\mu}(k)) \cdot \widehat{\alpha}_i(k). \quad (12)$$

The verifier checks that $v = 1 + \prod_{j' \in [\ell']}(1 + v_{j'})$ holds, and the parties invoke the ℓ' parallel executions of Sum-Check protocol in order to to verify that each $v_{j'}$ matches the r.h.s. of Eq. (12). The execution indexed by $j' \in [\ell']$ results in a claim regarding the value $\widehat{G'}(\widehat{\mu}(u), \widehat{\mu}(j'), \widehat{\mu}(r)) \cdot \widehat{\psi}_i(\widehat{\mu}(u), \widehat{\mu}(r)) \cdot \widehat{\alpha}_i(r)$ for a random $r \in \mathcal{F}^m$ selected via the execution. Since the verifier can evaluate $\widehat{\mu}, \widehat{\psi}_i$ and $\widehat{G'}$, it is left with $\ell' = O(\log n)$ claims, each regarding the value of $\widehat{\alpha}_i$ at one point.[14]

After $O(d)$ steps, the verifier is left with polylogarithmically (i.e., $O(\log n)^d$) many claims, where each claim refers to the value of $\widehat{\alpha}_d$ at a single point $u \in \mathcal{F}^m$. Such a claim can be checked by the verifier itself using Eq. (9).

Note that a straightforward computation of the r.h.s. of Eq. (9) calls for summing-up $|H^m| = 2^\ell = \text{poly}(n)$ terms, which the verifier cannot afford. However, all but $n+2$ of these terms (of the form $\texttt{EQ}(\widehat{\mu}(u), \widehat{\mu}(k)) \cdot \alpha_d(\widehat{\mu}(k))$), are identically zero, and so the verifier needs to compute only $n+2$ terms. This is the case because $\alpha_d(k) = 0$ for every $k \in [s] \setminus [n+2]$ (whereas $\alpha_d(k) = x_k$ for every $k \in [n]$ and $\{\alpha_d(n+1), \alpha_d(d+2)\} = \{0,1\}$). Hence, per each point $u \in \mathcal{F}^m$, computing $\widehat{\alpha}_d(u)$ reduces to evaluating $\texttt{EQ}(\widehat{\mu}(u), \widehat{\mu}(k)) \cdot \alpha_d(\widehat{\mu}(k))$ at $n+2$ points (only); that is, letting I denote the subset of $H^m \equiv [2^\ell]$ that correspond to $[n+2]$, the verifier just computes $\sum_{k \in I} \texttt{EQ}(\widehat{\mu}(u), \widehat{\mu}(k)) \cdot \alpha_d(\widehat{\mu}(k))$.

Analysis of the Foregoing Interactive Proof System. We first observe that the complexities of the foregoing protocol are as stated in Theorem 1.1. Specifically, the protocol proceeds in $O(d)$ steps and in each step a Sum-Check protocol is invoked on a sum that ranges over H^m, where $m = O(1/\delta)$ and $|H| = n^\delta$. Since the relevant polynomial is of degree $n^{\delta+o(1)}$, the total (m-round) communication is of this order.[15] The total number of rounds is $O(d \cdot m) = O(d \log_{n^\delta} s(n)) = O(d \cdot c/\delta)$, where $s(n) = n^c$. The verification time is dominated by the final check (i.e., evaluating $\widehat{\alpha}_d$ on poly$(\log n)$ points), which runs in time $\widetilde{O}(n) \cdot n^{o(1)} = O(n^{1+o(1)})$. The complexity of the prescribed prover is dominated by its operation in the Sum-Check protocol, which can be implemented in time $|H|^m \cdot n^{o(1)} = 2^\ell \cdot n^{o(1)} =$

[14] Actually, we could afford letting the verifier use the same random choices in all ℓ' parallel executions, which would result in leaving it with a single claim regarding the value of $\widehat{\alpha}_i$ at one point (i.e., r).

[15] This dominates the length of the initial verifier-message $\sigma \in \{0,1\}^{O(\log n)}$.

$s(n)^{1+o(1)}$. Next, we show that this protocol constitutes an interactive proof system (with imperfect completeness) for $\{x : C_{|x|}(x)=1\}$.

Claim 2.1 (imperfect completeness): *If $C_n(x) = 1$ and the prover follows the prescribed strategy, then the verifier accepts with probability $1 - o(1)$.*

Proof: The probability that the value of an or-gate, under a fixed setting of its input wires, is correctly emulated by the multiplication of ℓ' random linear combinations of these wires is at least $1-2^{-\ell'}$, where in case all wires feed 0 the emulation is always correct. The same holds (approximately) when the random linear combinations are replaced by inner products with a small biased sequence; specifically, if the sequence is ϵ-biased, then the emulation is correct with probability at least $1 - 2^{-\ell'} + \epsilon$. Using $\ell' = 2\log_2 s(n) = O(\log n)$ and $\epsilon = 2^{-\ell'}$, and employing a union bound, it follows that with probability $1-o(1)$ over the choice of the random seed $\sigma \in \{0,1\}^{O(\log n)}$, it holds that $A_n^{(\sigma)}(x) = C_n(x)$. Observing that the verifier always accepts when $A_n^{(\sigma)}(x) = 1$ (and the prover follows the prescribed strategy), the claim follows. ■

Claim 2.2 (soundness): *If $C_n(x) = 0$, then, no matter what strategy the prover employs, the verifier accepts with probability at most $o(1)$.*

Proof: As shown in the proof of Claim 2.1, with probability $1 - o(1)$, it holds that $A_n^{(\sigma)}(x) = C_n(x)$. Recalling that the soundness error of the Sum-Check protocol is proportional to the ratio of the degree of the polynomial over the size of the field, it follows that the prover can fool the verifier into accepting a wrong value of $A_n^{(\sigma)}(x)$ with probability $O(dm) \cdot |H| \cdot n^{o(1)}/|\mathcal{F}| = n^{\delta+o(1)-2\delta} = o(1)$, since $|\mathcal{F}| = n^{2\delta}$. The claim follows. ■

2.4 Getting Rid of the Completeness Error

Claims 2.1 and 2.2 assert that the foregoing protocol constitutes a proof system for $\{x : C_{|x|}(x) = 1\}$, but this proof system carries a completeness error (see Claim 2.1). Recalling that this error is only due to the (unlikely) case that $A_n^{(\sigma)}(x) \neq C_n(x)$, the begging fix is to have the prover prove to the verifier that this case has occurred (with respect to the random seed σ chosen by the verifier). Specifically, the (unlikely) case that $A_n^{(\sigma)}(x) \neq C_n(x)$ may occur only when at least one or-gate of C_n is badly emulated by $A_n^{(\sigma)}$; that is, the value of this gate is 1 in the computation of $C_n(x)$ whereas the corresponding (augmented multiplication) gate in $A_n^{(\sigma)}(x)$ evaluates to 0. This means that at least one of the gates (in $A_n^{(\sigma)}$) that correspond to the children of the or-gate in C_n evaluates to 1, whereas the gate (in $A_n^{(\sigma)}$) that corresponds to the or-gate (of C_n) evaluates to 0. So all that the prover needs to do is point out these two gates in $A_n^{(\sigma)}$, and prove that their values are as stated. Hence, we regain perfect completeness, whereas the soundness claim remains valid (since in order to cheat the prover has to prove a false claim regarding the value of a gate in $A_n^{(\sigma)}$). Thus, we obtain:

Theorem 2.3 (Theorem 1.1, restated): *For constants $c, d \in \mathbb{N}$, let $\{C_n : \{0,1\}^n \to \{0,1\}\}$ be a family of layered Boolean circuits with unbounded fan-in* or, and, *and* parity *gates such that C_n has size at most n^c and depth d. Suppose that C_n can be described by an adjacency predicate that is computable by a $n^{o(1)}$-size formula that can be constructed in $n^{1+o(1)}$-time. Then, for every $\delta \in (0,1]$, the set $\{x : C_{|x|}(x) = 1\}$ has a $O(cd/\delta)$-round interactive proof system of perfect completeness in which the verifier runs in time $O(n^{1+o(1)})$, the prescribed prover can be implemented in time $O(n^{c+o(1)})$, and the total communication is $n^{\delta+o(1)}$.*

Note that the foregoing adjacency predicate refers to gates of C_n, which are identified by ℓ-bit long strings, where $\ell = O(\log n)$. Thus, the uniformity condition postulates that this predicate can be computed by a formula of size $\exp(o(\ell))$ (equiv., by a bounded fan-in circuit of depth $o(\ell)$) that can be constructed in time $\exp(\ell/O(1))$.

2.5 Using the Other Two Succinct Representations

The foregoing presentation refers to Boolean circuits C_n that are succinctly represented by their adjacency predicates. Specifically, we referred to the adjacency predicates $\psi_i : \{0,1\}^{2\ell} \to \{0,1\}$, which were extended to $\widehat{\psi}_i : \mathcal{F}^{2\ell} \to \mathcal{F}$. In this section we show that the presentation can be adapted to the other two succinct representations of circuits discussed in Sect. 1.2.

From Incidence Functions to Adjacency Predicate. Suppose that the circuit C_n is represented by incidence functions of the form $\phi_i : [s] \times [s] \to [s] \cup \{0\}$, which we view as $\phi_i : \{0,1\}^{2\ell} \to \{0,1\}^{\ell+1}$, where $[s] \equiv \{0,1\}^\ell$ is identified with $\{1\sigma : \sigma \in \{0,1\}^\ell\} \subset \mathcal{F}^{\ell+1}$ and $0 \equiv 0^{\ell+1} \in \mathcal{F}^{\ell+1}$. Using a multi-linear extension of ϕ_i, denoted $\widehat{\phi}_i : \mathcal{F}^{2\ell} \to \mathcal{F}^{\ell+1}$, for any $j, k \in [s] \equiv \{0,1\}^\ell$, we replace the adjacency value $\widehat{\psi}_i(j,k)$ by the expression $\sum_{p\in\{0,1\}^\ell} \mathtt{EQ}(\widehat{\phi}_i(j,p), 1k)$, since the latter expression equals 1 if and only if $\widehat{\psi}_i(j,p) = 1k$ for a unique $p \in [s]$ (which means that k feeds j). Actually, $\widehat{\psi}_i(j,k)$ is replaced by $\sum_{p\in H^m} \mathtt{EQ}(\widehat{\phi}_i(j, \widehat{\mu}(p)), 1k)$. This means that, in the invocations of the Sum-Check protocol, the relevant summations are over H^{2m} rather than over H^m.

Handling Canonical Circuits. In this case, the s-sized circuit of depth d has the form of a w-ary tree of depth d such that $w = s^{1/d}$, and the input assignment is represented by a function of the form $\pi : \{0,1\}^\ell \to [n+2]$. Hence, we effectively refer to the adjacency predicate $\psi_i : [w]^d \times [w]^d \to \{0,1\}$ such that $\psi_i(j_1 \cdots j_d, k_1 \cdots k_d) = 1$ if and only if $j_1 \cdots j_{i-1} j_{i+1} \cdots j_d = k_1 \cdots k_{i-1} k_{i+1} \cdots k_d$ (or rather $\psi_i : [w]^{i-1} \times [w]^i \to \{0,1\}$ such that $\psi_i(j_1 \cdots j_{i-1}, k_1 \cdots k_i) = 1$ if and only if $j_1 \cdots j_{i-1} = k_1 \cdots k_{i-1}$).[16] In addition, instead of Eq. (8) (or rather Eq. (9)), letting I be a subset of H^m that

[16] Indeed, we can replace Eq. (2) by $\alpha_{i-1}(j_1 \cdots j_d) = \sum_{k_i \in [w]} \alpha_i(j_1 \cdots j_{i-1} k_i j_{i+1} \cdots j_d)$ (or rather by $\alpha_{i-1}(j_1 \cdots j_{i-1}) = \sum_{k_i \in [w]} \alpha_i(j_1 \cdots j_{i-1} k_i)$), and ditto for Eq. (3).

corresponds to $[n+2]$, for $z \in \mathcal{F}^m$, we have

$$\widehat{\alpha}_d(z) = \sum_{k \in H^m} \mathtt{EQ}(\widehat{\mu}(z), \widehat{\mu}(k)) \cdot \sum_{p \in I} \mathtt{EQ}(p, \widehat{\pi}(\widehat{\mu}(k))) \cdot x_p \qquad (13)$$

where $x_p = p \bmod 2$ for $p \in \{n+1, n+2\}$ and $\widehat{\pi}$ is polynomial that is obtained by transforming the Boolean formula that computes π to a corresponding arithmetic formula. The outer sum (in Eq. (13), along with $\mathtt{EQ}(\widehat{\mu}(z), \widehat{\mu}(k)))$ implements a selector of one of the leaves in the canonical circuit (i.e., if $z \in H^m$, then leaf z is selected). In contrast, the inner sum (along with $\mathtt{EQ}(p, \widehat{\pi}(\widehat{\mu}(k))))$ implements a selector of a variable or the constants 0 and 1 for this leaf (i.e., the variable $x_{\pi(k')}$ (or the constant $\pi(k') \bmod 2$) is selected for the leaf with index k'). Hence, for $k \in H^m$, it holds that $\widehat{\alpha}_d(k) = \sum_{p \in I} \mathtt{EQ}(p, \widehat{\pi}(\widehat{\mu}(k))) \cdot x_p$, which equals $x_{\widehat{\pi}(\widehat{\mu}(k)))}$, since $\widehat{\pi}(k') \in [n+2]$ for every $k' \in \{0,1\}^\ell$.

Recall that once the $O(d)$ iterations are completed, the verifier is left with the verification of polylogarithmically many claims, where each claim refers to the value of $\widehat{\alpha}_d$ at a single point $u \in \mathcal{F}^m$. Here we cannot afford having the verifier evaluate $\widehat{\alpha}_d$ at u by itself (since this requires evaluating the $|H|^m = s$ terms of the outer sum). Instead, we instruct the parties to run the Sum-Check protocol on Eq. (13), and the verifier is left with a claim referring to the value of $\mathtt{EQ}(\widehat{\mu}(u), \widehat{\mu}(r)) \cdot \sum_{p \in I} \mathtt{EQ}(p, \widehat{\pi}(\widehat{\mu}(r))) \cdot x_p$ at a random point $r \in \mathcal{F}^m$, which can be verified in time $|I| \cdot n^{o(1)} = n^{1+o(1)}$.

3 The Interactive Proof System for $\mathcal{NC}^1$

In this section we prove the following result.

Theorem 3.1 (Theorem 1.4, restated): *For a logarithmic function $d : \mathbb{N} \to \mathbb{N}$, let $\{C_n : \{0,1\}^n \to \{0,1\}\}$ be a family of canonical Boolean circuits of fan-in two and depth d. Suppose that the input assignment of C_n can be computed by a $n^{o(1)}$-size formula that can be constructed in $n^{1+o(1)}$-time. Then, for every $\delta \in (0,1]$, the set $\{x : C_{|x|}(x) = 1\}$ has a $O(d(n)/\delta \log n)^2$-round interactive proof system of perfect completeness in which the verifier runs in time $O(n^{1+o(1)})$, the prescribed prover can be implemented in polynomial-time, and the total communication is $n^{\delta+o(1)}$.*

We leave open the question of whether the round complexity can be reduced to $O(\delta^{-1} \cdot (d(n)/\log n)^2)$, meeting the bound in Theorem 2.3.

3.1 Overview

The construction generalizes and somewhat simplifies the proof systems constructed by Goldwasser, Kalai, and Rothblum [8]. The simplification is due to working with canonical circuits rather than with general (log-space) uniform circuits as in [8], whereas the generalization allows us to reduce the round complexity of [8] by a log-squared factor. Specifically, the canonical form of the circuit allows us to relate the values of layers in the circuit that are at distance $\delta \log_2 n$ apart, whereas [8] relate values at adjacent layers (only). In addition, we use a version of the sum-check protocol that handles summations over an alphabet of size n^δ (rather than over the alphabet $\{0,1\}$).

Fixing a constant $\delta \in (0,1)$, let $\ell' = \delta \log_2 n$. The core of the proof system asserted in Theorem 3.1 is an iterative process in which a claim about the values of the gates that are at layer $(i-1) \cdot \ell'$ is reduced to a claim about the values of the gates at layer $i \cdot \ell'$. We stress that each of these claims refers to the values of the polynomially many gates at a specific layer of the circuit $C_{|x|}$ during the computation on input x, but these poly$(|x|)$ values are not communicated explicitly but rather only referred to. Nevertheless, in $t = d(|x|)/\ell'$ iterations, the claim regarding the value of the output gate (i.e., the value $C_{|x|}(x)$) is reduced to a claim regarding the values of the bits of the input x, whereas the latter claim (which refers to x itself) can be verified in almost linear time.

Each of the aforementioned claims regarding the values of the gates at layer $i \cdot \ell'$, where $i \in \{0, 1, ..., t\}$, is actually a claim about the value of a specified location in the corresponding encoding of (the string that describes) all the gate-values at layer $i \cdot \ell'$. Specifically, the encoding used is the low degree extension of the said sequence of values (viewed as a function), and the claims are claims about the evaluations of these polynomials at specific points.

The different codewords (or polynomials) are related via the structure of the circuit $C_{|x|}$, which is the case of canonical circuit is straightforward to implement (avoiding a main source of technical difficulty in [8] (see also [4])). Indeed, this reduces a claim regarding one value in the encoding of layer $(i-1) \cdot \ell'$ to $2^{\ell'} = n^\delta$ analogous claims regarding layer $i \cdot \ell'$, but (as in [8]) "batch verification" is possible, reducing these $2^{\ell'}$ claims to a single claim.

3.2 The Actual Construction

For simplicity (and w.l.o.g.), we assume that C_n contains only NAND-gates of (fan-in two), where $\mathtt{NAND}(a, b) = \neg(a \wedge b)$. Viewing this gate as operating in a finite field that contains $\{0,1\}$, we have $\mathtt{NAND}(a, b) = 1 - (a \cdot b)$ for $a, b \in \{0,1\}$. The function computed by a tree of depth i of such gates is given by

$$\mathtt{NAND}_i(b_1, ..., b_{2^i}) = 1 - (\mathtt{NAND}_{i-1}(b_1, ..., b_{2^{i-1}}) \cdot \mathtt{NAND}_{i-1}(b_{2^{i-1}+1}, ..., b_{2^i})), \quad (14)$$

where $b_1, ..., b_{2^i} \in \{0,1\}$ are the values at the tree's leaves and $\mathtt{NAND}_0(b) = b$; indeed, $\mathtt{NAND}_1 = \mathtt{NAND}$.

For sake of simplifying the notation, we fictitiously augment the circuit with gates that are fed by no gate (and feed no gate), where (by convention) gates that are fed nothing always evaluate to 0, so that all layers of the circuits have the same number of gates. Hence, we present the circuit as having $d(n) + 1$ layers of gates such that each layer has exactly $k(n) = 2^{d(n)} = \text{poly}(n)$ gates. As usual, the gates at layer i are only fed by gates at layer $i + 1$, and the leaves (at layer $d(n)$) are input-variables or constants. Recall that the latter assignment is represented by the function $\pi : \{0,1\}^\ell \to [n+2]$, where $\ell = d(n)$ and $[k(n)] \equiv \{0,1\}^\ell$, such that the jth leaf is fed the variable $x_{\pi(j)}$ if $\pi(j) \in [n]$ (and the constant $\pi(j) \bmod n$ otherwise). The output is produced at the first gate of layer zero.

The High Level Protocol. On input $x \in \{0,1\}^n$, the prescribed prover computes the values of all layers. Letting $d = d(n)$ and $k = k(n)$, we denote the values at the ith layer by $\alpha_i \in \{0,1\}^k$; in particular, $\alpha_0 = C_n(x)0^{k-1}$ and α_d is the sequence of values given by $x_{\pi(0^\ell)}, ..., x_{\pi(1^\ell)}$, where $x_j = j \bmod 2$ for $j \in \{n+1, n+2\}$. For a sufficiently large finite field, denoted $\mathcal{F}$, consider an arbitrary fixed set $H \subset \mathcal{F}$ of size $2^{\ell'}$, where $\ell' = \delta \cdot \log_2 n$ (as before),[17] and let $m = \log_{|H|} k = \frac{\log_2 k}{\log_2 |H|} = d/\ell' = O(1/\delta)$. For each $i \in \{0, 1, ..., d-1\}$, viewing α_i as a function from $H^m \equiv [k]$ to $\{0,1\}$, the prover encodes α_i by a low degree polynomial $\widehat{\alpha}_i : \mathcal{F}^m \to \mathcal{F}$ that extends it (i.e., $\widehat{\alpha}_i(\sigma) = \alpha_i(\sigma)$ for every $\sigma \in H^m$); that is,

$$\widehat{\alpha}_i(z_1, ..., z_m) = \sum_{\sigma_1,...,\sigma_m \in H} \mathtt{EQ}(z_1 \cdots z_m, \sigma_1 \cdots \sigma_m) \cdot \alpha_i(\sigma_1, ..., \sigma_m) \quad (15)$$

where $\mathtt{EQ}$ is a low degree polynomial in the z_i's that tests equality over H^m (i.e., $\mathtt{EQ}(z_1 \cdots z_m, \sigma_1 \cdots \sigma_m) = \prod_{i \in [m]} \mathtt{EQ}_{\sigma_i}(z_i)$ and $\mathtt{EQ}_\sigma(z) = \prod_{\beta \in H \setminus \{\sigma\}} \frac{z-\beta}{\sigma-\beta}$). Actually, recalling that all but the first 2^i gates of layer i evaluate to 0, we re-write Eq. (15), for i's that are multiples of ℓ', as

$$\widehat{\alpha}_{i' \cdot \ell'}(z_1, ..., z_m) = \sum_{\sigma_1,...,\sigma_{i'} \in H} \mathtt{EQ}(z_1 \cdots z_m, 1^{m-i'} \sigma_1 \cdots \sigma_{i'}) \cdot \alpha_i(1, ..., 1, \sigma_1, ..., \sigma_{i'}) \quad (16)$$

Either way, $\widehat{\alpha}_i$ is a polynomial of individual degree $|H| - 1$.

In light of the foregoing, proving that $C_n(x) = 1$ is equivalent to proving that $\widehat{\alpha}_0(1^m) = 1$, where $1^m \in H^m$ corresponds to the fixed (e.g., first) location of the output gate in the zero layer. This proof is conducted in $t = d/\ell'$ iterations, where in each iteration a multi-round interactive protocol is employed. Specifically, in i^{th} iteration, the correctness of the claim $\widehat{\alpha}_{(i-1)\cdot\ell'}(\overline{r}_{i-1}) = v_{i-1}$, where $\overline{r}_{i-1} \in \mathcal{F}^m$ and $v_{i-1} \in \mathcal{F}$ are known to both parties, is reduced (via the interactive protocol) to the claim $\widehat{\alpha}_{i\cdot\ell'}(\overline{r}_i) = v_i$, where $\overline{r}_i \in \mathcal{F}^m$ and $v_i \in \mathcal{F}$ are determined (by this protocol) such that both parties get these values. We stress that, with the exception of $i = t$, the $\widehat{\alpha}_{i\cdot\ell'}$'s are not known (or given) to the verifier; still, the claims made at the beginning (and at the end) of each iteration are well defined (i.e., each claim refers to a predetermined low degree polynomial that extends the values assigned to the gates (of a certain layer) of the circuit in a computation of the circuit on input $x \in \{0,1\}^n$).

[17] The fact that the value $\ell' = \delta \cdot \log_2 n$ is used both for $\log_2 |H|$ and for the distance between layers that we relate is a consequence of the fact that both parameters are subject to the same trade-off. Each of these parameters cuts the number of rounds by its value (i.e., ℓ'), while incurring an exponential overhead (i.e., $2^{\ell'}$) in the total volume of communication.

Once the last iteration is completed, the verifier is left with a claim of the form $\widehat{\alpha}_d(\overline{r}_t) = v_t$, where $\widehat{\alpha}_d$ is defined as in Eq. (13). Recall that Eq. (13) has the form $\widehat{\alpha}_d(y) = \sum_{k\in\{0,1\}^\ell} \mathtt{EQ}(y,k)\cdot \mathtt{I}(k)$, where $\mathtt{I}(z) \stackrel{\text{def}}{=} \sum_{v\in[n+2]} \mathtt{EQ}(v,\widehat{\pi}(z))\cdot x_v$ and $\widehat{\pi}$ is polynomial that is obtained by transforming the Boolean formula that computes π to a corresponding arithmetic formula. Hence, the verifier cannot evaluate $\widehat{\alpha}_d$ by itself, but it can verify its value via the Sum-Check protocol, since I is a low degree polynomial that can be evaluated in almost linear (in n) time. So, at this point, the parties run the Sum-Check protocol (see the last paragraph in Sect. 2.5).

A Single Iteration. The core of the iterative proof is the interactive protocol that is performed in each iteration. This protocol is based on the relation between subsequent α_i's, which is based on the canonical structure of the circuit. Specifically, recall that the ith iteration reduces a claim regarding $\widehat{\alpha}_{(i-1)\cdot\ell'}$ to a claim regarding $\widehat{\alpha}_{i\cdot\ell'}$, where these polynomials encode the values of the corresponding layers in the circuit (i.e., layers $(i-1)\cdot\ell'$ and $i\cdot\ell'$). The relation between these layers is given by the following equation that relates the value at a specific (non-dummy) gate of level $(i-1)\cdot\ell'$ to the value of $2^{\ell'} = |H|$ gates of layer $i\cdot\ell'$:

$$\alpha_{(i-1)\cdot\ell'}(1^{m-(i-1)}, u_1, ..., u_{i-1}) = \mathtt{NAND}_{\ell'}((\alpha_{i\cdot\ell'}(1^{m-i}, u_1, ..., u_{i-1}, u))_{u\in H}) \quad (17)$$

where $1, u_1, ..., u_{i-1} \in H \equiv \{0,1\}^{\ell'}$ and $\mathtt{NAND}_{\ell'}$ is as defined in Eq. (14). Combining Eq. (16) with Eq. (17), it holds that $\widehat{\alpha}_{(i-1)\cdot\ell'}(z_1, ..., z_m)$ equals

$$\sum_{u_1,...,u_{i-1}\in H} \mathtt{EQ}(z_1\cdots z_m, 1^{m-i+1}u_1\cdots u_{i-1}) \cdot \mathtt{NAND}_{\ell'}((\widehat{\alpha}_{i\cdot\ell'}(1^{m-i}, u_1, ..., u_{i-1}, u))_{u\in H}). \quad (18)$$

In preparation to applying the Sum-Check protocol to Eq. (18), we observe that In preparation to applying the Sum-Check protocol to Eq. (18), we observe that the corresponding $(i-1)$-variate polynomial is of individual degree $O(2^{\ell'}\cdot|H|) = O(n^{2\delta})$. This is the case because, for any fixed point $(\overline{r}', \overline{r}'') \in \mathcal{F}^{m-i+1}\times\mathcal{F}^{i-1}$, we can write Eq. (18) as

$$\begin{aligned}&\mathtt{EQ}(\overline{r}', 1^{m-i+1}) \cdot \sum_{u_1,...,u_{i-1}\in H} \mathtt{EQ}(\overline{r}'', u_1\cdots u_{i-1}) \cdot \mathtt{NAND}_{\ell'}((\widehat{\alpha}_{i\cdot\ell'}(1^{m-i}, u_1, ..., u_{i-1}, u))_{u\in H})\\ &= \mathtt{EQ}(\overline{r}', 1^{m-i+1}) \cdot \sum_{u_1,...,u_{i-1}\in H} P_{\overline{r}''}(u_1, ..., u_{i-1}),\end{aligned}$$

where $P_{\overline{r}''}(y_1, ..., y_{i-1}) \stackrel{\text{def}}{=} \mathtt{EQ}(\overline{r}'', y_1\cdots y_{i-1}) \cdot \mathtt{NAND}_{\ell'}((\widehat{\alpha}_{i\cdot\ell'}(1^{m-i}, y_1, ..., y_{i-1}, u))_{u\in H})$ is a low degree $(i-1)$-variate polynomial; specifically, its individual degree is dominated by the product of the total degree of $\mathtt{NAND}_{\ell'}$ and the individual degree of $\widehat{\alpha}_{i\cdot\ell'}$, which are $2^{\ell'}$ and $|H|-1$, respectively.

Applying the Sum-Check protocol to Eq. (18) allows to reduce a claim regarding the value of $\widehat{\alpha}_{(i-1)\cdot\ell'}$ at a specific point $\overline{r}_{i-1} = (\overline{r}'_{i-1}, \overline{r}''_{i-1}) \in \mathcal{F}^{m-i+1} \times \mathcal{F}^{i-1}$ to a claim regarding the value of the polynomial $P_{\overline{r}''_{i-1}}$ at a random point $(r''_1, ..., r''_{i-1})$ in $\mathcal{F}^{i-1}$, which in turn depends on the values of $\widehat{\alpha}_{i\cdot\ell'}$ at $2^{\ell'}$ points in $\mathcal{F}^m$ (i.e., the points $((1, ..., 1, r''_1, ..., r''_{i-1}, u))_{u\in H})$).

To reduce this claim to a claim regarding the value of $\widehat{\alpha}_{i\cdot\ell'}$ at a single point, we let the prover send these $2^{\ell'}$ values and perform "batch verification" for them. Specifically, the prover provides a low degree polynomial that describes the value of $\widehat{\alpha}_{i\cdot\ell'}$ on the axis-parallel line that goes through these points, and the claim to be proved in the next iteration is that the value of $\widehat{\alpha}_{i\cdot\ell'}$ at a random point on this line equals the value provided by the polynomial sent by the prover.[18] Hence, the full protocol that is run in iteration i is as follows.

Construction 3.2 (reducing a claim about $\widehat{\alpha}_{(i-1)\cdot\ell'}$ to a claim about $\widehat{\alpha}_{i\cdot\ell'}$): *For known $\overline{r}_{i-1} \in \mathcal{F}^m$ and $v_{i-1} \in \mathcal{F}$, the* entry claim *is $\widehat{\alpha}_{(i-1)\cdot\ell'}(\overline{r}_{i-1}) = v_{i-1}$. The parties proceed as follows.*

1. *Applying the Sum-Check protocol to the entry claim, when expanded according to Eq. (18), determines $\overline{r}' \in \mathcal{F}^{i-1}$ and a value $v \in \mathcal{F}$ such that the residual claim for verification is*

$$\mathtt{EQ}(\overline{r}_{i-1}, 1^{m-(i-1)}\overline{r}') \cdot \mathtt{NAND}_{\ell'}((\widehat{\alpha}_{i\cdot\ell'}(1, ..., 1, \overline{r}', u))_{u\in H}) = v. \tag{19}$$

2. *The prover sends a univariate polynomial p' of degree smaller than $m \cdot |H|$ such that $p'(z) = \widehat{\alpha}_i(1, ..., 1, \overline{r}', z)$.*
3. *Upon receiving the polynomial p', the verifier checks whether v equals*

$$\mathtt{EQ}(\overline{r}_{i-1}, 1^{m-(i-1)}\overline{r}') \cdot \mathtt{NAND}_{\ell'}((p'(u))_{u\in H}), \tag{20}$$

   and continues only if equality holds (otherwise it rejects).
4. *The verifier selects a random $r \in \mathcal{F}$, and sends it to the prover. Both parties set $\overline{r}_i = (1, ..., 1, \overline{r}', r)$ and $v_i = p'(r)$.*

The exit claim *is $\widehat{\alpha}_{i\cdot\ell'}(\overline{r}_i) = v_i$.*

The complexities of Construction 3.2 are dominated by the application of the Sum-Check protocol, which refers to a polynomial of degree $O(2^{\ell'} \cdot |H|) = O(n^{2\delta})$. In particular, this implies that the verifier's strategy can be implemented in time $\widetilde{O}(n^{2\delta})$, provided that $|\mathcal{F}| = \text{poly}(n)$. In this case, the prescribed prover strategy (as defined in Construction 3.2) can be implemented in time $\widetilde{O}(2^{d(n)}) = \text{poly}(n)$.

[18] We mention that the fact that these $2^{\ell'}$ points reside on a line makes the argument simpler, but not in a fundamental way. In general, the prover could have picked a curve of degree $2^{\ell'} - 1$ that goes through any $2^{\ell'}$ points of interest, and provide a low degree polynomial describing the value of $\widehat{\alpha}_{i\cdot\ell'}$ on this curve. In this case, the claim to be proved in the next iteration would have been that the value of $\widehat{\alpha}_{i\cdot\ell'}$ at a random point on this curve equals the value provided by the polynomial sent by the prover.

Recall that after the last iteration of Construction 3.2, the resulting claim is checked by the Sum-Check protocol (applied to Eq. (13)), which leaves the verifier with the task of evaluating $\mathtt{I}$, where $\mathtt{I}(z) \stackrel{\text{def}}{=} \sum_{v\in[n+2]} \mathtt{EQ}(v, \widehat{\pi}(z)) \cdot x_v$. Using the hypothesis regarding π, it follows that the verifier runs in $n^{1+o(1)}$-time. The round complexity of the ith iteration of Construction 3.2 is $i \le m$, and so the total round complexity is $m \cdot m + m = O(d(n)/\delta \log n)^2$.

One can readily verify that if the entry claim is correct, then the exit claim is correct, whereas if the entry claim is false, then with probability at least $1 - O(m \cdot 2^{\ell'} \cdot |H|/|\mathcal{F}|)$ the exit claim is false. Recall that the soundness error of the entire protocol is upper-bounded by the probability that there exists an iteration in which the entry claim is false but the exist claim is true. Hence, the total soundness error is $O(n^{2\delta}/|\mathcal{F}|) = o(1)$, provided $|\mathcal{F}| = \omega(n^{2\delta}$, which we can certainly afford.

Acknowledgements. As noted in the body of the paper, an unpublished work by Kalai and Rothblum [11] proposed a constant-round doubly-efficient proof system for $\mathcal{NC}^1$ under a very strict notion of uniformity. This unpublished work has inspired our own work, and we thank Yael for her contribution to it as well as for many other helpful conversations on these topics.

Appendix: The Sum-Check Protocol

The Sum-Check protocol, designed by Lund, Fortnow, Karloff, and Nisan [12], is a key ingredient in the constructions that we present.

Fixing a finite field $\mathcal{F}$ and a set $H \subset \mathcal{F}$ (e.g., H may be a two-element set), we consider an m-variate polynomial $P : \mathcal{F}^m \to \mathcal{F}$ of individual degree d. Given a value v, the Sum-Check protocol is used to prove that

$$\sum_{\sigma_1,...,\sigma_m \in H} P(\sigma_1, ..., \sigma_m) = v, \tag{21}$$

assuming that the verifier can evaluate P by itself. The Sum-Check protocol proceeds in m iterations, such that in the ith iteration the number of summations (over H) decreases from $m - i + 1$ to $m - i$. Specifically, the ith iteration starts with a claim of the form $\sum_{\sigma_i,...,\sigma_m \in H} P(r_1, ..., r_{i-1}, \sigma_i, ..., \sigma_m) = v_{i-1}$, where $r_1, ..., r_{i-1}$ and v_{i-1} are as determined in prior iterations (with $v_0 = v$), and ends with a claim of the form $\sum_{\sigma_{i+1},...,\sigma_m \in H} P(r_1, ..., r_i, \sigma_{i+1}, ..., \sigma_m) = v_i$, where r_i and v_i are determined in the ith iteration. Initializing the process with $v_0 = v$, in the ith iteration the parties act as follows.

Prover's move: The prover computes a univariate polynomial of degree d over $\mathcal{F}$

$$P_i(z) \stackrel{\text{def}}{=} \sum_{\sigma_{i+1},...,\sigma_m \in H} P(r_1, ..., r_{i-1}, z, \sigma_{i+1}, ..., \sigma_m) \tag{22}$$

where $r_1, ..., r_{i-1}$ are as determined in prior iterations, and sends P_i to the verifier (claiming that $\sum_{\sigma\in H} P_i(\sigma) = v_{i-1}$).

Verifier's move: Upon receiving a degree d polynomial, denoted $\widetilde{P}$, the verifier checks that $\sum_{\sigma \in H} \widetilde{P}(\sigma) = v_{i-1}$ and rejects if inequality holds. Otherwise, it selects r_i uniformly in $\mathcal{F}$, and sends it to the prover, while setting $v_i \leftarrow \widetilde{P}(r_i)$.

If all m iterations are completed successfully (i.e., without the verifier rejecting in any of them), then the verifier conducts a final check. It computes the value of $P(r_1, ..., r_m)$ and accepts if and only if this value equals v_m.

Clearly, if Eq. (21) holds (and the prover acts according to the protocol), then the verifier accepts with probability 1. Otherwise (i.e., Eq. (21) does not hold), no matter what the prover does, the verifier accepts with probability at most $m \cdot d/|\mathcal{F}|$, because in each iteration if the prover provides the correct polynomial, then the verifier rejects (since $\sum_{\sigma \in H} P_i(\sigma) = P_{i-1}(r_{i-1}) \neq v_{i-1}$), and otherwise the (degree d) polynomial sent agrees with P_i on at most d points.[19]

The complexity of verification is dominated by the complexity of evaluating P (on a single point). As for the prescribed prover, it may compute the relevant P_i's by interpolation, which is based on computing the value of P at $(d+1) \cdot |H|^{m-i}$ points, for each $i \in [m]$. (That is, the polynomial P_i is computed by obtaining its values at $d+1$ points, where the value of P_i at each point is obtained by summing the values of P at $|H|^{m-i}$ points.)[20]

References

1. Alon, N., Goldreich, O., Hastad, J., Peralta, R.: Simple construction of almost k-wise independent random variables. Random Struct. Algorithms **3**(3), 289–304 (1992)
2. Barrington, D.A.M., Immerman, N., Straubing, H.: On uniformity within NC1. J. Comput. Syst. Sci. **41**(3), 274–306 (1990)
3. Goldreich, O.: Computational Complexity: A Conceptual Perspective. Cambridge University Press, Cambridge (2008)
4. Goldreich, O.: On the doubly-efficient interactive proof systems of GKR. In: ECCC, TR17-101 (2017)
5. Goldreich, O., Rothblum, G.N.: Simple doubly-efficient interactive proof systems for locally-characterizable sets. In: ECCC, TR17-018 (2017). Extended abstract in 9th ITCS, pp. 18:1–18:19 (2018)
6. Goldreich, O., Rothblum, G.N.: Worst-case to average-case reductions for subclasses of P. In: ECCC TR17-130 (2017). See revised version in this volume

[19] If P_i does not satisfy the current claim (i.e., $\sum_{\sigma \in H} P_i(\sigma) \neq v_{i-1}$), then the prover can avoid upfront rejection only if it sends a degree d polynomial $\widetilde{P} \neq P_i$. But in such a case, $\widetilde{P}$ and P_i may agree on at most d points, since they are both degree d polynomials. Hence, if the chosen $r_i \in \mathcal{F}$ is not one of these points, then it holds that $v_i = \widetilde{P}(r_i) \neq P_i(r_i)$, which means that the next iteration will also start with a false claim. Hence, starting with a false claim (i.e., $\sum_{\sigma \in H} P_1(\sigma) \neq v_0$ since Eq. (21) does not hold), with probability at least $1 - m \cdot d/|\mathcal{F}|$, after m iterations we reach a false claim regarding the value of P at a single point.

[20] Specifically, the value of P_i at p is obtained from the values of P at the points $(r_1, ..., r_{i-1}, p, \sigma)$, where $\sigma \in H^{m-i}$.

7. Goldreich, O., Rothblum, G.N.: Counting t-cliques: worst-case to average-case reductions and direct interactive proof systems. In: ECCC, TR18-046 (2018). Extended abstract in 59th FOCS, pp. 77–88 (2018)
8. Goldwasser, S., Kalai, Y.T., Rothblum, G.N.: Delegating computation: interactive proofs for muggles. J. ACM **62**(4), 271–2764 (2015). Extended abstract in 40th STOC, pp. 113–122 (2008)
9. Goldwasser, S., Micali, S., Rackoff, C.: The knowledge complexity of interactive proof systems. SIAM J. Comput. **18**, 186–208 (1989). Preliminary version in 17th STOC (1985)
10. Kalai, Y.T., Raz, R.: Interactive PCP. In: Aceto, L., Damgård, I., Goldberg, L.A., Halldórsson, M.M., Ingólfsdóttir, A., Walukiewicz, I. (eds.) ICALP 2008. LNCS, vol. 5126, pp. 536–547. Springer, Heidelberg (2008). https://doi.org/10.1007/978-3-540-70583-3_44
11. Kalai, Y.T., Rothblum, G.N.: Constant-round interactive proofs for NC1. Unpublished observation (2009)
12. Lund, C., Fortnow, L., Karloff, H., Nisan, N.: Algebraic methods for interactive proof systems. J. ACM **39**(4), 859–868 (1992). Extended abstract in 31st FOCS (1990)
13. Naor, J., Naor, M.: Small-bias probability spaces: efficient constructions and applications. SIAM J. Comput. **22**(4), 838–856 (1993). Preliminary version in 22nd STOC (1990)
14. Razborov, A.A.: Lower bounds on the size of bounded-depth networks over a complete basis with logical addition. Matematicheskie Zametki **41**(4), 598–607 (1987). (in Russian). English translation in Math. Not. Acad. Sci. USSR **41**(4), 333–338 (1987). https://doi.org/10.1007/BF01137685
15. Reingold, O., Rothblum, G.N., Rothblum, R.D.: Constant-round interactive proofs for delegating computation. In: 48th ACM Symposium on the Theory of Computing, pp. 49–62 (2016)
16. Shamir, A.: IP = PSPACE. J. ACM **39**(4), 869–877 (1992). Preliminary version in 31st FOCS (1990)
17. Smolensky, R.: Algebraic methods in the theory of lower bounds for Boolean circuit complexity. In: 19th ACM Symposium on the Theory of Computing, pp. 77–82 (1987)
18. Williams, R.: Faster all-pairs shortest paths via circuit complexity. In: 46th ACM Symposium on the Theory of Computing, pp. 664–673 (2014)

Flexible Models for Testing Graph Properties

Oded Goldreich

Abstract. The standard models of testing graph properties postulate that the vertex-set consists of $\{1, 2, ..., n\}$, where n is a natural number that is given explicitly to the tester. Here we suggest more flexible models by postulating that the tester is given access to samples the arbitrary vertex-set; that is, the vertex-set is arbitrary, and the tester is given access to a device that provides uniformly and independently distributed vertices. In addition, the tester may be (explicitly) given partial information regarding the vertex-set (e.g., an approximation of its size). The flexible models are more adequate for actual applications, and also facilitates the presentation of some theoretical results (e.g., reductions among property testing problems).

An early version of this work appeared as TR18-104 of *ECCC*. The presentation was elaborated in the current revision.

Introduction

In the last couple of decades, the area of property testing has attracted much attention (see, e.g., a recent textbook [4]). Loosely speaking, property testing typically refers to sub-linear time probabilistic algorithms for deciding whether a given object has a predetermined property or is far from any object having this property. Such algorithms, called testers, obtain local views of the object by making adequate queries; that is, the object is seen as a function and the testers get oracle access to this function (and thus may be expected to work in time that is sub-linear in the size of the object).

A significant portion of the foregoing research has been devoted to testing graph properties in three different models: the dense graph model (introduced in [7] and reviewed in [4, Chap. 8]), the bounded-degree graph model (introduced in [8] and reviewed in [4, Chap. 9]), and the general graph model (introduced in [12,13] and reviewed in [4, Chap. 10]). In all these models, it is postulated that the vertex-set consists of $\{1, 2, ..., n\}$, where n is a natural number that is given explicitly to the tester, and this simplified assumption is made in all studies of these models. The simplifying assumption may be employed, without loss of generality, *provided that (1) the tester can sample the vertex-set, and (2) the tester is explicitly given the size of the vertex-set.*[1]

[1] See Observations 1.2, 2.2 and 3.2.

O. Goldreich (Ed.): Computational Complexity and Property Testing, LNCS 12050, pp. 352–362, 2020.
https://doi.org/10.1007/978-3-030-43662-9_19

Having explicitly stated the two foregoing conditions that allow to extend testers of the simplified model to more general settings, we observe that they are of fundamentally different nature. The first condition (i.e., sampleability of the vertex-set) seems essential to testing any non-trivial property, whereas the second condition (i.e., knowledge of the (exact) size of the vertex-set) may be relaxed and even avoided altogether in many cases. For example, all graph-partition properties (see [7]) and subgraph-free properties are testable in a general version of the dense graph model in which only the first condition holds. This is the case since the original testers (presented in [7] and [1], resp.) use the description of the vertex-set only in order to sample it. Needless to say, it follows that the query complexities of these testers are oblivious of the size of the graph (and depend only on the proximity parameter), but (as observed by [2]) the converse does not hold (i.e., testers of size-oblivious query complexity may depend on the size of the graph for their verdict (see also [11])).

On the other hand, when the query complexity depends on the size of the graph, the tester needs to obtain at least a sufficiently good approximation of the said size. Typically, such an approximation suffices, as in the case of the bipartite tester for the bounded-degree and general graph models [9,12]. Hence, we highlight three cases regarding the (a priori) knowledge of the size of the vertex-set (where in all cases the tester is given access to samples drawn from the vertex-set):

1. The tester is explicitly given the exact size of the vertex-set.
   As shown in Observations 1.2, 2.2 and 3.2, this ("exact size") case is essentially reducible to the simplified case in which the vertex-set equals $\{1, 2, ..., n\}$ and n is explicitly given to the tester.
2. The tester is explicitly given an approximation of the size of the vertex-set, where the quality of the approximation may vary.
3. The tester is not given explicitly any information regarding the size of the vertex-set.

The foregoing three cases are special cases of a general formulation that may be employed in the study of testing graph properties in settings in which the tested graph has an arbitrary vertex-set, which (w.l.o.g.) is a set of strings.

The Flexible Models at a Glance. The general formulation postulates that, when testing a graph with vertex-set $V \subset \{0,1\}^*$, the tester is given access to a device that samples uniformly in V. In addition, the tester is explicitly given some information about V, where this information resides in a set of possibilities, denoted $p(V)$. The point is that the "given information about V" is allowed to be in a predetermined set of possibilities rather than be uniquely determined. For example, the "exact size case" corresponds to $p(V) = \{|V|\}$, the "approximate size case" corresponds to $p(V) \in \{n \in \mathbb{N} : n \approx |V|\}$, and the "no information case" corresponds to $p(V) = \{\lambda\}$. This general formulation is called flexible, since it allows its user to determine the function $p : 2^{\{0,1\}^*} \to 2^{\{0,1\}^*}$ according to the setting at hand.

The benefits of the flexible models are two-fold. First, they narrow the gap between the study of testing graph properties and possible real-life applications. Second, they facilitate the presentation of reductions among property testing problems and models, as will be discussed in the sequel. Examples of such reductions include those reviewed in [4, Thm. 9.22] and [4, Thm. 10.4] and those used in [5].

While flexible models may be applicable also to testing properties of objects that are not naturally viewed as graphs, we focus on testing graph properties in the three aforementioned models (i.e., the dense graph model, the bounded-degree graph model, and the general graph model). In all cases we consider only graph properties, which are sets of unlabeled graphs (equiv., set of label graphs that are closed under the renaming of the vertices).[2]

Subsequent Work. In subsequent work [6], the foregoing flexible models were used as a starting point for more general models in which the tester obtains samples that are arbitrarily distributed in the vertex-set. In this case, the definition of the distance between graphs is modified to reflect this vertex distribution; such a modification is not required in the current paper.

Organization. The following three sections present flexible versions of the three aforementioned models of testing graph properties. In Sect. 1 we consider the dense graph model (a.k.a. the adjacency matrix model), in Sect. 2 we consider the bounded-degree graph model (a.k.a. the bounded incidence lists model), and in Sect. 3 we consider the general graph model. The definitional parts of these sections contain some repetitions in order to enable reading them independently of one another.

1 Testing Graph Properties in the Dense Graph Model

Here we present a more flexible version of the notion of property testing in the dense graph model (a.k.a. the adjacency matrix model, which was introduced in [7] and reviewed in [4, Chap. 8]). The standard version (e.g., as in [4, Def. 8.2]) is obtained as a special case by setting $V = \{1, 2, ..., n\}$ and $p(V) = n$.

In this model, a graph of the form $G = (V, E)$ is represented by its adjacency predicate $g : V \times V \to \{0, 1\}$; that is, $g(u, v) = 1$ if and only if u and v are adjacent in G (i.e., $\{u, v\} \in E$). Distance between graphs (over the same vertex-set) is measured in terms of their foregoing representation; that is, as the fraction of (the number of) entries on which they disagree (over $|V|^2$). The tester is given oracle access to the representation of the input graph (i.e., to the adjacency predicate g) as well as to a device that returns uniformly distributed elements in the graph's vertex-set. As usual, the tester is also given the proximity parameter ϵ, which determined when graphs are considered "far apart" (i.e., see the notion of ϵ-far).

[2] That is, if a graph $G = (V, E)$ has the property, then, for any bijection $\pi : V \to V'$, the graph $G' = (V', \{\{\pi(u), \pi(v)\} : \{u, v\} \in E\}$ has the property.

In addition, the tester gets some partial information about the vertex-set (i.e., V) as auxiliary input, where this partial information is an element of a set of possibilities, denoted $p(V)$. Indeed, two extreme possibilities are $p(V) = \{V\}$, which is closely related to the standard formulation, and $p(V) = \{\lambda\}$, but we can also consider natural cases such as $p(V) \in \{|V|, |V|+1, ..., 2|V|\}$.

For simplicity (and without loss of generality), we assume that the vertex-set is a set of strings (i.e., a finite subset of $\{0,1\}^*$). Hence, p is a function from sets of strings (representing possible vertex-sets) to sets of strings (representing possible partial information about the vertex-set).

Definition 1.1 (property testing in the dense graph model, revised): *Let Π be a property of graphs and $p : 2^{\{0,1\}^*} \to 2^{\{0,1\}^*}$. A* tester for the graph property Π (in the dense graph model) with partial information p *is a probabilistic oracle machine T that is given access to two oracles, an adjacency predicate $g : V \times V \to \{0,1\}$ and a device denoted $\mathsf{Samp}(V)$ that samples uniformly in V, and satisfies the following two conditions:*

1. *The tester accepts each graph $G = (V,E) \in \Pi$ with probability at least $2/3$; that is, for every $g : V \times V \to \{0,1\}$ representing a graph in Π and every $i \in p(V)$ (and $\epsilon > 0$), it holds that $\Pr[T^{g,\mathsf{Samp}(V)}(i,\epsilon)=1] \geq 2/3$.*
2. *Given $\epsilon > 0$ and oracle access to any graph G that is ϵ-far from Π, the tester rejects with probability at least $2/3$; that is, for every $\epsilon > 0$ and $g : V \times V \to \{0,1\}$ that represents a graph that is ϵ-far from Π and $i \in p(V)$, it holds that $\Pr[T^{g,\mathsf{Samp}(V)}(i,\epsilon) = 0] \geq 2/3$, where the graph represented by $g : V \times V \to \{0,1\}$ is* ϵ-far from *Π if for every $g' : V \times V \to \{0,1\}$ that represents a graph in Π it holds that $|\{(u,v) \in V^2 : g(u,v) \neq g'(u,v)\}| > \epsilon \cdot |V|^2$.*

The tester is said to have one-sided error probability *if it always accepts graphs in Π; that is, for every $g : V \times V \to \{0,1\}$ representing a graph in Π (and every $i \in p(V)$ and $\epsilon > 0$), it holds that $\Pr[T^{g,\mathsf{Samp}(V)}(i,\epsilon)=1] = 1$.*

The case of $p(V) = \{V\}$ corresponds to the standard model in which one typically postulates that $V = \{1,2,...,|V|\}$. This is the case because, given V, the tester may use a bijection between V and $\{1,2,...,|V|\}$. The case of $p(V) = \{|V|\}$ is closely related to these cases, except that in this case the bijection can only be constructed on-the-fly. In order to formally state this correspondence, we need to define the query complexity of a tester (as in Definition 1.1). For our purposes, it suffice to define the query complexity of the tester as the total number of queries it makes to both its oracles (i.e., the adjacency predicate and the sampling oracles).[3]

Observation 1.2 (the "exact size case" reduces to the standard case): *Suppose that the graph property Π has a tester of query complexity $q : \mathbb{N}\times(0,1] \to \mathbb{N}$ in the*

[3] A more refined definition, following [3], may consider the number of queries to each of the oracles. In such a case, it makes sense to refer to the number of queries to the adjacency predicate (resp., the sampling device) as the query (resp., sample) complexity of the tester.

dense graph model under its standard formulation (e.g., as in [4, Def. 8.2]). *Then, Π has a tester of query complexity $q' = \widetilde{O}(q)$ in the dense graph model* (as in Definition 1.1) *with partial information p such that $p(V) = \{|V|\}$. Furthermore, one-sided error is preserved, and $q'(n, \epsilon) = O(q(n, \epsilon))$ whenever either $q(n, \epsilon) < n/3$ or $q(n, \epsilon) > 4n \ln n$. The same holds* (simultaneously) *for time complexity.*

Proof: As articulated in [10], graph properties are actually properties of unlabeled graphs, and hence testers of such properties may effectively ignore the labels as long as they can sample the vertex-set. Hence, when testing graph properties, the actual labels of the vertices are immaterial. What matters is whether or not vertices that appear in the current query have appeared in previous queries.

Note that a tester under the standard formulation may easily generate *new* vertices, since the vertex-set equals $[n] \stackrel{\text{def}}{=} \{1, 2, ..., n\}$ and it is explicitly given n. In contrast, the tester that we construct is given $|V|$, but has no other *a priori* information regarding V. Its only way of obtaining *new* vertices is to query $\texttt{Samp}(V)$ for a sample. The overhead of the transformation is due to the fact that obtaining a *new* vertex may require several samples, when the number of queries made so far (by the original tester) is large (i.e., $\Omega(\sqrt{|V|})$). Yet, this is a minor problem that is easily resolved.

Formally, we emulate a tester T of the standard formulation, by invoking this tester, on input (n, ϵ), where $n = |V|$. We answer T's queries by constructing on-the-fly a bijection π between $[n]$ and the (unknown to us) vertex-set V, and querying our own oracle on the corresponding vertex pairs. Specifically, the query $(u, v) \in [n] \times [n]$ is answered by making a corresponding query $(\pi(u), \pi(v))$ such that if π is not defined on $w \in \{u, v\}$, then we assign $\pi(w)$ a new vertex obtained by sampling V (i.e., we repeatedly invoke $\texttt{Samp}(V)$ till we obtain a vertex that is not in the (defined so far) image of π). Hence, our emulation of the tester T proceeds as follows, where π denotes a partial bijection of $[|V|]$ to V.

- On input $(|V|, \epsilon)$, we invoke the standard tester T on this very input, while initializing π to be totally undefined.
  (Recall that T issues queries to an adjacency predicate that is defined over $[|V|] \times [|V|]$.)
- When T issues a query $(u, v) \in [|V|]^2$, we check if $\pi(u)$ and $\pi(v)$ are already defined.
  - If both $\pi(u)$ and $\pi(v)$ are already defined, then we make the query $(\pi(u), \pi(v))$ to our input graph $G = (V, E)$, and answer T accordingly.
  - If for $w \in \{u, v\}$, the value $\pi(w)$ is undefined, then we get a new sample $s \in V$ from the sampling device (i.e., $s \leftarrow \texttt{Samp}(V)$). If $\pi^{-1}(s)$ is undefined, then we define $\pi(w) = s$. Otherwise, we try again, and continue trying till reaching a total number of $q'/3$ (i.e., $q'/3$ invocations of $\texttt{Samp}(V)$), where $q' = \widetilde{O}(q(|V|, \epsilon))$ and q is the query complexity of T.
  Once $\pi(u)$ and $\pi(v)$ are both defined, we proceed as in the previous case.
- If we reached the claimed query complexity (i.e., q') and T has not terminated, then we suspend the execution of T and accept. Otherwise, we output the verdict of T.

Note that if $q(|V|,\epsilon) < |V|/3$, then the probability of obtaining a sample $s \leftarrow \mathtt{Samp}(V)$ on which π^{-1} is undefined is at least 1/3, since the number of vertices in V that were assigned in response to prior queries of T is less than $2 \cdot q(|V|,\epsilon)$. Hence, in this case, with very high probability, we can obtain $2 \cdot q(|V|,\epsilon)$ distinct elements of V by invoking $\mathtt{Samp}(V)$ for $O(q(|V|,\epsilon)$ times, which means that we do not suspend the execution of T. On the other hand, $(4|V| \ln |V|)/3$ samples of $\mathtt{Samp}(V)$ are very likely to cover all of V, so we are fine also in case $q(|V|,\epsilon) \geq |V|/3$. Hence, in both cases, we suspend the execution with very small probability. Our choice to accept in the rare case of suspended executions preserves the one-sided error of T (but slightly increases the error on graphs that are far from Π).

Note that the foregoing tester can be efficiently implemented (w.r.t time complexity) by maintaining dynamic sets (of the values) on which π and π^{-1} are defined. ■

The No-Information Case. We mention that the testers for the various graph-partition problems presented in [7] satisfy the requirements of Definition 1.1 with $p(V) = \{\lambda\}$ (i.e., the "no partial information" case). Indeed, these (low complexity) testers use the description of the vertex-set only in order to sample it, and so this auxiliary input can be replaced (in them) by a vertex sampling device. The same holds for many other testers (in the dense graph model), including the subgraph-freeness testers presented in [1].

We also observe that applying the transformation of [10, Thm. 2] to a tester that satisfies Definition 1.1 with $p(V) = \{\lambda\}$, yields a canonical tester of the same type; that is, the (auxiliary) property that the induced subgraph should satisfy is oblivious of the size of the input graph (cf., [10,11]). That is:

Observation 1.3 (canonical testers for the "no partial information case"): *Suppose that Π has a tester of query complexity $q : (0,1] \to \mathbb{N}$ in the dense graph model* (of Definition 1.1) *with no partial information* (i.e., $p(V) = \{\lambda\}$). *Then, there exists a graph property Π' and a tester that, on input ϵ, accepts if and only if the subgraph induces by $O(q(\epsilon))$ random vertices is in Π'.*

We mention that this special case of Definition 1.1 (i.e., $p(V) = \{\lambda\}$) is pivotal to the reduction used in the proof of [5, Thm. 4.5]. In fact, this special case of Definition 1.1 appears as [5, Def. 4.3], and triggered us to write the current paper.

2 Testing Graph Properties in the Bounded-Degree Graph Model

Here we present a more flexible version of the notion of property testing in the bounded-degree graph model (a.k.a. the bounded incidence lists model, which was introduced in [8] and reviewed in [4, Chap. 9]). The standard version (e.g., as in [4, Def. 9.1]) is obtained as a special case by setting $V = \{1, 2, ..., n\}$ and $p(V) = n$.

The bounded-degree graph model refers to a fixed (constant) degree bound, denoted $d \geq 2$. In this model, a graph $G = (V, E)$ of maximum degree d is represented by the incidence function $g : V \times [d] \rightarrow V \cup \{\perp\}$ such that $g(v, j) = u \in V$ if u is the jth neighbor of v and $g(v, j) = \perp \notin V$ if v has less than j neighbors.[4] Distance between graphs is measured in terms of their foregoing representation; that is, as the fraction of (the number of) different array entries (over $d|V|$).

As in the dense graph model, the tester is given oracle access to the representation of the input graph (i.e., to the incidence function g) as well as to a device that returns uniformly distributed elements in the graph's vertex-set. As usual, the tester is also given the proximity parameter ϵ. In addition, the tester gets some partial information about the vertex-set (i.e., V) as auxiliary input, where this partial information is an element of a set of possibilities, denoted $p(V)$. (Again, two extreme possibilities are $p(V) = \{V\}$, which is closely related to the standard formulation, and $p(V) = \{\lambda\}$, but we can also consider natural cases such as $p(V) \in \{|V|, |V|+1, ..., 2|V|\}$). Again, we assume that the vertex-set is a set of strings (i.e., a finite subset of $\{0,1\}^*$).

Definition 2.1 (property testing in the bounded-degree graph model, revised): *For a fixed $d \in \mathbb{N}$, let Π be a property of graphs of degree at most d, and $p : 2^{\{0,1\}^*} \rightarrow 2^{\{0,1\}^*}$. A* tester for the graph property Π (in the bounded-degree graph model) with partial information p *is a probabilistic oracle machine T that is given access to two oracles, an incidence function $g : V \times [d] \rightarrow V \cup \{\perp\}$ and a device denoted $\mathtt{Samp}(V)$ that samples uniformly in V, and satisfies the following two conditions:*

1. *The tester accepts each graph $G = (V, E) \in \Pi$ with probability at least $2/3$; that is, for every $g : V \times [d] \rightarrow V \cup \{\perp\}$ representing a graph in Π and every $i \in p(V)$ (and $\epsilon > 0$), it holds that $\Pr[T^{g,\mathtt{Samp}(V)}(i, \epsilon) = 1] \geq 2/3$.*
2. *Given $\epsilon > 0$ and oracle access to any graph G that is ϵ-far from Π, the tester rejects with probability at least $2/3$; that is, for every $\epsilon > 0$ and $g : V \times [d] \rightarrow V \cup \{\perp\}$ that represents a graph that is ϵ-far from Π and $i \in p(V)$, it holds that $\Pr[T^{g,\mathtt{Samp}(V)}(i, \epsilon) = 0] \geq 2/3$, where the graph represented by $g : V \times [d] \rightarrow V \cup \{\perp\}$ is* ϵ-far from Π *if for every $g' : V \times [d] \rightarrow V \cup \{\perp\}$ that represents a graph in Π it holds that $|\{(v, j) \in V \times [d] : g(v, j) \neq g'(v, j)\}| > \epsilon \cdot d|V|$.*

The tester is said to have one-sided error probability *if it always accepts graphs in Π; that is, for every $g : V \times [d] \rightarrow V \cup \{\perp\}$ representing a graph in Π* (and every $i \in p(V)$ and $\epsilon > 0$), *it holds that $\Pr[T^{g,\mathtt{Samp}(V)}(i, \epsilon) = 1] = 1$.*

Defining the query complexity as in Sect. 1, we make analogous observations regarding the cases of $p(V) = \{V\}$ and $p(V) = \{|V|\}$. In particular,

[4] For simplicity, we adopt the standard convention by which the neighbors of v appear in arbitrary order in the sequence $(g(v, 1), ..., g(v, \deg(v)))$, where $\deg(v) \stackrel{\text{def}}{=} |\{j \in [d] : g(v, j) \neq \perp\}|$.

Observation 2.2 (the "exact size case" reduces to the standard case): *Suppose that the graph property Π has a tester of query complexity $q : \mathbb{N} \times (0,1] \to \mathbb{N}$ in the bounded-degree graph model under its standard formulation* (e.g., as in [4, Def. 9.1]). *Then, Π has a tester of query complexity $q' = \widetilde{O}(q)$ in the bounded-degree graph model* (as in Definition 2.1) *with partial information p such that $p(V) = \{|V|\}$. Furthermore, one-sided error is preserved, and $q'(n, \epsilon) = O(q(n, \epsilon))$ whenever $q(n, \epsilon) < n/3$. The same holds* (simultaneously) *for time complexity.*

Proof Sketch: We follow the proof of Observation 1.2, except that here "new vertices" are such that have not appeared in previous queries or in previous answers (to such queries). Furthermore, when we answer a query $(v, j) \in [|V|] \times [d]$ of the standard tester T by making the query $(\pi(v), j)$ to our own input graph, we may obtain as an answer either an old or a new vertex, denoted $\alpha \in V$. In the former case, the value of $\pi^{-1}(\alpha) \in [|V|]$ is already defined, and we provide this value as answer. Otherwise, we answer with a random $w \in [|V|]$ such that $\pi(w)$ is yet undefined, and set $\pi(w) = \alpha$. Indeed, this new vertex w is obtained from the sampling device, and this may require repeated sampling as in the proof of Observation 1.2. ■

The No-Information Case. As in the dense graph model, natural testers that have query complexity that depends only on the proximity parameter are easily adapted to the bounded-degree graph model (as in Definition 2.1) with no partial information (i.e., p such that $p(V) = \{\lambda\}$). The list includes testers that operate by local searchers (reviewed in [4, Sec. 9.2])[5] and testers that operate by constructing and utilizing partition oracles (reviewed in [4, Sec. 9.5]).

The Approximate-Size Case. Obviously, testers of query complexity that depend on the size of the graph must obtain some information regarding this size, and a constant-factor approximation will typically do (see, e.g., the bipartite tester of [9]). We mention that testers of query complexity that is at least the square root of the size of the graph can obtain such an approximation by sampling the vertex-set, but this method does not preserve one-sided error probability (and only yield probabilistic bounds on the complexity).[6]

The approximate-size version of Definition 2.1 is implicit in the reduction that underlies the presentation of the proof of [4, Thm. 9.22]. The original presentation lacks this notion of a reduction, and so it proceeds by emulating a specific tester for bipartiteness (i.e., the one of [9]) on an auxiliary graph that is derived from the input graph. Using Definition 2.1, we can now say that

[5] In some cases (e.g. [4, Sec. 9.2.3]), these testers use an estimate of $|V|$ in order to avoided pathological problems that arise when $|V| < B \stackrel{\text{def}}{=} O(1/\epsilon)$. But determining whether not $|V| < B$ holds can be done by using $\widetilde{O}(1/\epsilon)$ samples of V, let alone that it actually suffices to distinguish between $|V| < B$ and $|V| \geq 2B$.

[6] The obvious procedure is to keep sampling till seeing, say, 100 pairwise collisions, and then outputting the square of the number of trials (divided by 200).

(one-sided error) testing of cycle-freeness in the bounded-degree graph model is randomly reducible to (one-sided error) testing of bipartiteness in the model of Definition 2.1 with $p(V) = \{\Omega(|V|), ..., O(|V|)\}$. The same holds also w.r.t the proof of [4, Thm. 10.4], which can be presented as a reduction of testing bipartitness in the general graph model to testing bipartiteness in the model of Definition 2.1 with $p(V) = \{\Omega(|V|), ..., O(|V|)\}$.

3 Testing Graph Properties in the General Graph Model

Here we present a more flexible version of the notion of property testing in the general graph model (which was introduced in [12,13] and reviewed in [4, Chap. 10]). The standard version (e.g., as in [4, Def. 10.2]) is obtained as a special case by setting $V = \{1, 2, ..., n\}$ and $p(V) = n$.

Unlike in the previous two models, here the representation of the graph is decoupled from the definition of the (*relative*) distance between graphs. Following the discussion in [4, Sec. 10.1.2], we define the relative distance between $G = (V, E)$ and $G' = (V, E')$ as the ratio of the symmetric difference of E and E' over $\max(|E|, |E'|) + |V|$.

In this model, a graph $G = (V, E)$ is redundantly represented by both its incidence function $g_1 : V \times \mathbb{N} \to V \cup \{\bot\}$ (alternatively, we may consider $g_1 : V \times [b(|V|)] \to V \cup \{\bot\}$, where $b : \mathbb{N} \to \mathbb{N}$ is some degree-bounding function)[7] and its adjacency predicate $g_2 : V \times V \to \{0, 1\}$; indeed, as before, $g_1(v, j) = u \in V$ if u is the jth neighbor of v (and $g_1(v, j) = \bot$ if v has less than j neighbors), and $g_2(u, v) = 1$ if and only if $\{u, v\} \in E$. The tester is given oracle access to the two representations of the input graph (i.e., to the functions g_1 and g_2) as well as to a device that returns uniformly distributed elements in the graph's vertex-set. As usual, the tester is also given the proximity parameter ϵ.

In addition, the tester gets some partial information about the vertex-set (i.e., V) as auxiliary input, where this partial information is an element of a set of possibilities, denoted $p(V)$. (Again, two extreme possibilities are $p(V) = \{V\}$, which is closely related to the standard formulation, and $p(V) = \{\lambda\}$, but we can also consider natural cases such as $p(V) \in \{|V|, |V| + 1, ..., 2|V|\}$). Again, we assume that the vertex-set is a set of strings (i.e., a finite subset of $\{0, 1\}^*$).

Definition 3.1 (property testing in the general graph model, revised):[8] *Let Π be a property of graphs and $p : 2^{\{0,1\}^*} \to 2^{\{0,1\}^*}$. A* tester for the graph property

[7] In a previous version of this paper, we considered $g_1 : V \times [|V| - 1] \to V \cup \{\bot\}$, where $|V| - 1$ served as a trivial degree bound. In retrospect, we feel that using such an upper bound is problematic, because it may allow the tester to determine the number of vertices in the graph (assuming that querying g_1 on an input that is not in its domain results in a suitable indication). On the other hand, allowing an infinite representation of finite graphs is not problematic, because the representation is not used as a basis for the definition of the relative distance between graphs.

[8] Here we follow [4, Def. 10.2], rather than [4, Def. 10.1]. See discussion in [4, Sec. 10.1.2].

Π (in the general graph model) with partial information p *is a probabilistic oracle machine* T *that is given access to three oracles, an incidence function* $g_1 : V \times \mathbb{N} \to V \cup \{\perp\}$, *an adjacency predicate* $g_2 : V \times V \to \{0,1\}$, *and a device denoted* $\mathtt{Samp}(V)$ *that samples uniformly in* V, *and satisfies the following two conditions:*

1. *The tester accepts each graph* $G = (V, E) \in \Pi$ *with probability at least* $2/3$*; that is, for every* $g_1 : V \times \mathbb{N} \to V \cup \{\perp\}$ *and* $g_2 : V \times V \to V \cup \{\perp\}$ *representing a graph in* Π *and every* $i \in p(V)$ (and $\epsilon > 0$), *it holds that* $\Pr[T^{g_1,g_2,\mathtt{Samp}(V)}(i,\epsilon)=1] \geq 2/3$.
2. *Given* $\epsilon > 0$ *and oracle access to any graph* G *that is* ϵ*-far from* Π, *the tester rejects with probability at least* $2/3$*; that is, for every* $\epsilon > 0$ *and* (g_1, g_2) *such that* $g_1 : V \times \mathbb{N} \to V \cup \{\perp\}$ *and* $g_2 : V \times V \to V \cup \{\perp\}$ *represent a graph that is* ϵ*-far from* Π, *and every* $i \in p(V)$, *it holds that* $\Pr[T^{g_1,g_2,\mathtt{Samp}(V)}(i,\epsilon)=0] \geq 2/3$, *where the graph* $G = (V, E)$ *is* ϵ-far from Π *if for every* $G' = (V, E')$ *that represents a graph in* Π *it holds that the symmetric difference of* E *and* E' *is greater than* $\epsilon \cdot (\max(|E|, |E'|) + |V|)$.

The tester is said to have one-sided error probability *if it always accepts graphs in* Π.

Defining the query complexity as in the previous sections, we make analogous observations regarding the cases of $p(V) = \{V\}$ and $p(V) = \{|V|\}$. In particular,

Observation 3.2 (the "exact size case" reduces to the standard case): *Suppose that the graph property* Π *has a tester of query complexity* $q : \mathbb{N} \times (0,1] \to \mathbb{N}$ *in the general graph model under its standard formulation* (e.g., as in [4, Def. 10.2]). *Then,* Π *has a tester of query complexity* $q' = \widetilde{O}(q)$ *in the general graph model* (as in Definition 3.1) *with partial information* p *such that* $p(V) = \{|V|\}$. *Furthermore, one-sided error is preserved, and* $q'(n,\epsilon) = O(q(n,\epsilon))$ *whenever* $q(n,\epsilon) < n/3$. *The same holds* (simultaneously) *for time complexity.*

References

1. Alon, N., Fischer, E., Krivelevich, M., Szegedy, M.: Efficient testing of large graphs. Combinatorica **20**, 451–476 (2000)
2. Alon, N., Shapira, A.: A separation theorem in property testing. Combinatorica **28**(3), 261–281 (2008)
3. Balcan, M., Blais, E., Blum, A., Yang, L.: Active property testing. In: 53rd FOCS, pp. 21–30 (2012)
4. Goldreich, O.: Introduction to Property Testing. Cambridge University Press, Cambridge (2017)
5. Goldreich, O.: Hierarchy theorems for testing properties in size-oblivious query complexity. Comput. Complex. **28**(4), 709–747 (2019). https://doi.org/10.1007/s00037-019-00187-2
6. Goldreich, O.: Testing graphs in vertex-distribution-free models. In: 51st STOC, pp. 527–534 (2019)
7. Goldreich, O., Goldwasser, S., Ron, D.: Property testing and its connection to learning and approximation. J. ACM **45**, 653–750 (1998)

8. Goldreich, O., Ron, D.: Property testing in bounded degree graphs. Algorithmica **32**(2), 302–343 (2002)
9. Goldreich, O., Ron, D.: A sublinear bipartitness tester for bounded degree graphs. Combinatorica **19**(3), 335–373 (1999)
10. Goldreich, O., Trevisan, L.: Three theorems regarding testing graph properties. Random Struct. Algorithms **23**(1), 23–57 (2003)
11. Goldreich, O., Trevisan, L.: Errata to [10]. Manuscript, August 2005. http://www.wisdom.weizmann.ac.il/~oded/p_ttt.html
12. Kaufman, T., Krivelevich, M., Ron, D.: Tight bounds for testing bipartiteness in general graphs. SIAM J. Comput. **33**(6), 1441–1483 (2004)
13. Parnas, M., Ron, D.: Testing the diameter of graphs. Random Struct. Algorithms **20**(2), 165–183 (2002)
14. Rubinfeld, R., Sudan, M.: Robust characterization of polynomials with applications to program testing. SIAM J. Comput. **25**(2), 252–271 (1996)

Pseudo-mixing Time of Random Walks

Itai Benjamini and Oded Goldreich

Abstract. We introduce the notion of pseudo-mixing time of a graph, defined as the number of steps in a random walk that suffices for generating a vertex that looks random to any polynomial-time observer. Here, in addition to the tested vertex, the observer is also provided with oracle access to the incidence function of the graph.

Assuming the existence of one-way functions, we show that the pseudo-mixing time of a graph can be much smaller than its mixing time. Specifically, we present bounded-degree N-vertex Cayley graphs that have pseudo-mixing time t for any $t(N) = \omega(\log\log N)$. Furthermore, the vertices of these graphs can be represented by string of length $2\log_2 N$, and the incidence function of these graphs can be computed by Boolean circuits of size $\text{poly}(\log N)$.

An early version of this work appeared as TR19-078 of *ECCC*. The presentation was somewhat elaborated in the current revision.

1 Introduction

A popular way to sample a huge set that is endowed with a group structure is to start at a fixed element of the set, which is typically easy to find, and take a random walk on the corresponding Cayley graph. If the length of the random walk exceeds the graph's mixing time, then the end-point of the walk is almost uniformly distributed in the corresponding set.

A couple of comments are in place. First, the aforementioned sampling procedure is beneficial only when the original set S is non-trivial in the sense that its elements can be represented by bit-strings of a certain length, denoted ℓ, but S may occupy only a negligible fraction of $\{0,1\}^\ell$ (i.e., $|S| < \mu(\ell) \cdot 2^\ell$, where $\mu : \mathbb{N} \to [0,1]$ is a negligible function (e.g., tends to zero faster than the reciprocal of any polynomial)). In such cases, it is infeasible to sample S (almost) uniformly by selecting uniformly few ℓ-bit long strings and taking the first string that falls in S, and so one needs an alternative.

Second, the set S should be endowed with a feasible group operation, and the group should be generated be a small number of generators that are easy to find (along with their inverses). If this is that case, then we can sample S almost uniformly by taking a sufficiently long random walk on the corresponding Cayley graph, starting at the vertex that corresponds to the identity element (which we assume, without loss of generalization, to be represented by 0^ℓ).[1] Of course, the

[1] Given an arbitrary group $(S,*)$, where $S \subseteq \{0,1\}^\ell$, with i representing the identity element, consider the group $(S\oplus i, \diamond)$ such that $x \diamond y = ((x\oplus i) * (y\oplus i))\oplus i$, where $\oplus$ denotes the bit-by-bit exclusive-or of bit strings.

O. Goldreich (Ed.): Computational Complexity and Property Testing, LNCS 12050, pp. 363–373, 2020.
https://doi.org/10.1007/978-3-030-43662-9_20

length of the random walk should slightly exceed the graph's mixing time (i.e., the minimum t such that the total variation distance between the end-vertex of a t-step random walk and a uniformly distributed vertex is negligible).

The requirement that the random walk be longer than the mixing time of the graph is aimed at obtaining a distribution that is *statistically close to the uniform distribution over the set of vertices.* However, for actual applications, which may be modeled as efficient procedures that run in time that is polynomial in the length of the description of a vertex in the graph, it suffices to require that *it is infeasible to distinguish the distribution of the end-point of the walk from a uniformly distributed vertex.* Indeed, here we adopt the notion of computational indistinguishability, which underlies much of modern cryptography and the computational notion of pseudorandomness (cf. [3, Apdx. C] and [3, Chap. 8], resp.), and consider the following notion of "pseudo-mixing time" (which is formulated, as usual in the theory of computation, using asymptotic terms).

Definition 1 (pseudo-mixing time, a naive definition): *For a constant d, let $\{G_\ell = (V_\ell, E_\ell)\}_{\ell \in \mathbb{N}}$ be a sequence of d-regular graphs such that $0^\ell \in V_\ell \subseteq \{0,1\}^\ell$. For a function $t : \mathbb{N} \to \mathbb{N}$, the graphs $\{G_\ell\}_{\ell \in \mathbb{N}}$ have* pseudo-mixing time *at most t if for every probabilistic polynomial-time algorithm A it holds that*

$$|\Pr[A(W_t(G_\ell))=1] - \Pr[A(U(G_\ell))=1]| = \mathtt{negl}(\ell),$$

where $W_t(G_\ell)$ denotes the distribution of the end-point of a $t(|V_\ell|)$-long random walk on G_ℓ starting at 0^ℓ, and $U(G_\ell)$ denotes the uniform distribution over V_ℓ. Indeed, negl *denotes a negligible function* (i.e., one that vanishes faster than $1/p$, for any positive polynomial p).

Clearly, if the total variation distance between $W_t(G_\ell)$ and $U(G_\ell)$ is negligible (in terms of ℓ), then G_ℓ has pseudo-mixing time at most t. The point is that this sufficient condition is not necessary; as we shall see, G_ℓ may have pseudo-mixing time at most t also in case that the total variation distance between $W_t(G_ell)$ and $U(G_\ell)$ is large (say, larger than 0.99).

Definition 1 is titled "naive" because algorithm A, which represents an observer that examines (or uses) the sampled vertex, does not get (or use) any auxiliary information about the graph G_ℓ. In contrast, the motivating discussion referred to the case that one can take a random walk on the graph. Hence, it is natural to augment the (efficient) observer by providing it with a device that lists the neighbors of any vertex of its choice. In other words, we provide the observer with oracle access to the incidence function of the graph; this oracle answers the query (v, i) with the ith neighbor of vertex v in the graph. Indeed, such a device constitutes a representation of the graph, and we are most interested in the case that this representation is succinct; that is, the incidence function can be computed by an efficient algorithm (given some short auxiliary information) or equivalently by a small Boolean circuit. This leads to the following definition.

Definition 2 (succinct representation of graphs and pseudo-mixing time): *For a constant d, let $\{G_\ell = (V_\ell, E_\ell)\}_{\ell \in \mathbb{N}}$ be a sequence of d-regular graphs such* *that $0^\ell \in V_\ell \subseteq \{0,1\}^\ell$. The graph G_ℓ is represented by its incidence function $g_\ell : V_\ell \times [d] \to V_\ell$, where $g_\ell(v,i)$ is the ith neighbor of v in G_ℓ.*

Succinct representation: *The graphs $\{G_\ell\}_{\ell\in\mathbb{N}}$ have a* succinct representation *if there exists a family of* poly(ℓ)*-size Boolean circuits that compute their incidence functions.*

Pseudo-mixing time (revised)*: For a function $t : \mathbb{N} \to \mathbb{N}$, the graphs $\{G_\ell\}_{\ell\in\mathbb{N}}$ have* pseudo-mixing time *at most t if for every probabilistic polynomial-time oracle machine M it holds that*

$$|\Pr[M^{g_\ell}(W_t(G_\ell))=1] - \Pr[M^{g_\ell}(U(G_\ell))=1]| = \mathtt{negl}(\ell),$$

where $M^g(x)$ denotes the output of M on input x when making queries to the oracle g, and the other notations (e.g., $W_t(G_\ell)$ and $U(G_\ell)$) *are as in Definition 1.*

We stress that pseudo-mixing refers to observers that are efficient in terms of the length of the description of vertices in the graph (i.e., they run for poly(ℓ)-time). Recall that our focus is on the case that the size of V_ℓ is exponential in ℓ. Hence, a polynomial-time machine cannot possibly explore the entire graph G_ℓ. On the other hand, if G_ℓ is rapidly mixing (i.e., its mixing time is $O(\log|V_\ell|)$), then a polynomial-time machine may obtain many samples of $W_t(G_\ell)$ and $U(G_\ell)$ (or rather many samples that are distributed almost as $U(G_\ell)$).[2] At this point, we can state our main result.

Theorem 3 (main result): *Assuming the existence of one-way functions, there exists a sequence of bounded-degree Cayley graphs $\{G_\ell = (V_\ell, E_\ell)\}_{\ell\in\mathbb{N}}$ such that the following properties hold:*

1. *The graphs $\{G_\ell\}_{\ell\in\mathbb{N}}$ have a* succinct representation.
2. *For every $t : \mathbb{N} \to \mathbb{N}$ such that $t(N) = \omega(\log\log N)$, the graphs $\{G_\ell\}_{\ell\in\mathbb{N}}$ have* pseudo-mixing time at most t.
3. The vertex set is hard to sample "without walking on the graph": *Any probabilistic polynomial-time algorithm that is given input 1^ℓ, fails to output a vertex of the graph other than 0^ℓ, except with negligible probability.*[3]

Furthermore, the graph G_ℓ has $2^{(0.5+o(1))\cdot\ell}$ vertices, and its mixing time is $\Theta(\ell) = \Theta(\log|V_\ell|)$.

---

[2] Here we assume that t does not exceed the mixing time; otherwise, the entire discussion is moot.

[3] We stress that this algorithm (unlike the observer in the pseudo-mixing condition) is not given oracle access to the succinct representation of the graph. Evidently, when given such a representation one can find other vertices in the graph. The corresponding condition in Theorem 6 asserts that such an oracle machine cannot output vertices in the graph other than 0^ℓ and those obtained as answers to incidence queries.

We stress that these bounded-degree graphs have size $N = \exp(\Theta(\ell))$, and so their pseudo-mixing time (i.e., $t(N) = \omega(\log \log N) = \omega(\log \ell)$) is only slightly larger than logarithmic in their (standard) mixing time (i.e., $\Theta(\ell) = \Theta(\log N)$). Indeed, the pseudo-mixing time of these graphs is only slightly larger than double-logarithmic in their size (i.e., $|V_\ell|$). In contrast, *the pseudo-mixing time of bounded-degree graphs cannot be* (strictly) *double-logarithmic* in $|V_\ell|$, since an observer can explore the $O(\log \log |V_\ell|)$-neighborhood of 0^ℓ in poly(ℓ)-time (and so distinguish vertices in this poly(ℓ)-sized set from all other $\exp(\Theta(\ell))$ vertices of the graph).

We comment that, as shown in Theorem 7, the existence of one-way functions is essential for the conclusion of Theorem 3. Essentially, the existence of one-way function, which implies the existence of pseudorandom generators, allows to construct Cayley graphs G_ℓ with succinct representation in which all vertices (except for 0^ℓ) look random. Of course, once we reach a vertex in G_ℓ by following a walk from the vertex 0^ℓ, we learn the label of this vertex, but the labels of vertices that we did not visit in such walks look random to us. Hence, for $t(N) = \omega(\log \log |V_\ell|)$, both the random variable $W_t(G_\ell)$ and $U(G_\ell)$ are unlikely to hit vertices that were visited by such poly(ℓ)-many explorations; that is, $W_t(G_\ell)$ and $U(G_\ell)$ hit explored vertices with negligible probability, and otherwise they look random.

A Stronger Notion of the Pseudo-mixing Time. Given that our focus is on Cayley graphs, one may wonder whether our upper bound on the pseudo-mixing time holds also when the observer is given access to the group operation (as well as to the corresponding inverse operation).[4] A necessary condition for such a stronger notion of pseudo-mixing time is that, for every (poly(ℓ)-time computable)[5] word $w = w[X]$ over the set of generators and a variable X, if w equals the identity when X is in the support of $W_t(G_\ell)$, then it is trivial (i.e., it equals the identity over the entire group). Hence, for starters, we suggest the following problem (where $r(\ell)$ replaces $t(\exp(\ell))$, which was used above).

Open Problem 4 (vanishing over a large ball implies vanishing in the group): *For some* $r : \mathbb{N} \to \mathbb{N}$ *such that* $r(\ell) \in [\omega(\log \ell), o(\ell)]$, *is it possible to construct an* $\exp(\ell)$*-sized group having succinct representation that satisfies the following condition: For every word over the group's generators and a variable* X *that vanishes over a ball of radius* $r(\ell)$ *centered at the group's identity, it holds that the word vanishes over the entire group.*

[4] Note that giving oracle access to the incidence function of the graph is equivalent to giving access to an oracle that performs the group operation only when the second operand is in a fixed set of generators (of the group).

[5] By $T(\ell)$-time computable word, we mean a word that is computable by a (uniform) arithmetic circuit of size $T(\ell)$, where the leaves in this circuit are fed by the fixed constants and variables and the internal gates perform the group operations (of the Cayley graph G_ℓ). Note that such a circuit can compute a word of length exponential in $T(\ell)$, but only a negligible fraction of words of $\exp(T(\ell))$ length are computable by such (poly($T(\ell)$)-size) circuits..

As hinted above (see also footnote 5), we are actually interested in words that are computable by arithmetic circuits of size poly(ℓ).

2 The Main Result and Its Proof

Theorem 3 is proved by combining the following facts:

1. Arbitrarily relabelling the elements of any finite group yields a group that is isomorphic to the original group (Proposition 5).
2. Assuming the existence of one-way functions, the foregoing relabeling can be pseudorandom and succinct (in the sense of Property 1 of Theorem 3).
3. There are explicit Cayley graphs with $2^{(1+o(1))\cdot\ell/2}$ vertices such that the $o(\ell)$-neighborhood of each vertex is a tree.[6]

Indeed, the graphs asserted in Theorem 3 are obtained by starting with a Cayley graph as in Fact 3, and relabeling its vertices as suggested by Facts 1–2. Hence, the pivot of our proof is performing such relabeling, and so we first prove Fact 1.

Proposition 5 (relabeling a group): *Let $(S,*)$ be a group such that $S \subset \{0,1\}^\ell$ and 0^ℓ is the identity element, and let π be a permutation over $\{0,1\}^\ell$ such that $\pi(0^\pi) = 0^\pi$. Then, the set $S_\pi = \{\pi(e) : e \in S\}$ combined with the operation $\diamond_\pi : S_\pi \times S_\pi \to S_\pi$ such that $\diamond_\pi(\alpha,\beta) = \pi(\pi^{-1}(\alpha) * \pi^{-1}(\beta))$ forms a group that is isomorphic to $(S,*)$.*

Proof: One can readily verify that the group axioms hold.

- For every $\alpha,\beta,\gamma \in S_\pi$ it holds that

$$\begin{aligned}
\diamond_\pi(\diamond_\pi(\alpha,\beta),\gamma) &= \pi(\pi^{-1}(\diamond_\pi(\alpha,\beta)) * \pi^{-1}(\gamma)) \\
&= \pi(\pi^{-1}(\pi(\pi^{-1}(\alpha) * \pi^{-1}(\beta))) * \pi^{-1}(\gamma)) \\
&= \pi((\pi^{-1}(\alpha) * \pi^{-1}(\beta)) * \pi^{-1}(\gamma)) \\
&= \pi(\pi^{-1}(\alpha) * (\pi^{-1}(\beta) * \pi^{-1}(\gamma))) \\
&= \diamond_\pi(\alpha, \diamond_\pi(\beta,\gamma))
\end{aligned}$$

  where the last equality is established analogously to the first three equalities.
- For every $\alpha \in S_\pi$ it holds that $\diamond_\pi(\alpha, 0^\ell) = \pi(\pi^{-1}(\alpha) * 0^\ell) = \alpha$ and $\diamond_\pi(0^\ell,\alpha) = \pi(0^\ell * \pi^{-1}(\alpha)) = \alpha$. Indeed, 0^n is the identity element of the new group.
- Denoting by $\mathtt{inv}(a)$ the inverse of $a \in S$ in the original group, we can verify that $\pi(\mathtt{inv}(\pi^{-1}(\alpha)))$ is the inverse of $\alpha \in S_\pi$ in the new group. Specifically,

$$\begin{aligned}
\diamond_\pi(\alpha, \pi(\mathtt{inv}(\pi^{-1}(\alpha)))) &= \pi(\pi^{-1}(\alpha) * \pi^{-1}(\pi(\mathtt{inv}(\pi^{-1}(\alpha))))) \\
&= \pi(\pi^{-1}(\alpha) * \mathtt{inv}(\pi^{-1}(\alpha))) \\
&= 0^\ell
\end{aligned}$$

  and similarly $\diamond_\pi(\pi(\mathtt{inv}(\pi^{-1}(\alpha))), \alpha) = 0^\ell$.

Lastly, note that π is an isomoprphism of $(S,*)$ to $(S_\pi, \diamond_\pi)$, since $\pi(a * b) = \diamond_\pi(\pi(a), \pi(b))$. ■

[6] Actually, a much weaker condition suffices.

Restating Theorem 3. For sake of clarity, we restated Theorem 3, while strengthening a few of its aspects. First, the succinct representation of an $\exp(\Theta(\ell))$-vertex graph G_ℓ is provided by an ℓ-bit string (called a seed) coupled with a uniform (polynomial-time) algorithm. Second, we will consider a distribution $\mathcal{D}_\ell$ over such graphs rather than a single graph per each value of ℓ; actually, the distribution $\mathcal{D}_\ell$ will correspond to a uniformly selected ℓ-bit long seed. Lastly, the difficulty-of-sampling condition is stated with respect to oracle machines that have access to the incidence function of the graph, and it discards vertices obtained as answers to incidence queries. As before, we use $\texttt{negl}$ to denote a generic negligible function; that is, a function that tends to zero faster than the reciprocal than any positive polynomial.

Theorem 6 (main result, restated): *Assuming the existence of one-way functions, there exists a distribution ensemble of bounded-degree Caley Graphs, denoted* $\{\mathcal{D}_\ell\}_{\ell\in\mathbb{N}}$*, such that the following properties hold:*

1. Succinct representation: *For some* $d \in \mathbb{N}$ *and every sufficiently large* $\ell \in \mathbb{N}$*, each graph in the support of* $\mathcal{D}_\ell$ *is a* d*-regular graph with* $2^{(0.5+o(1))\cdot\ell}$ *vertices. Furthermore, the vertex set is a subset of* $\{0,1\}^\ell$ *and contains* 0^ℓ*. These graphs are strongly explicit in the sense that each of these graphs is represented by an* ℓ*-bit long string, called its* seed*, and there exists a polynomial-time algorithm that on input a seed* s*, a vertex* v *in the graph represented by* s*, and an index* $i \in [d]$*, returns the* i^{th} *neighbor of* v *in the graph.*
2. Pseudo-mixing time: *For every* $t : \mathbb{N} \to \mathbb{N}$ *such that* $t(N) = \omega(\log\log N)$*, any probabilistic polynomial-time oracle machine* M *that is given oracle access to the incidence function of a graph* $G = (V, E)$ *selected according to* $\mathcal{D}_\ell$ *cannot distinguish the uniform distribution over* G*'s vertices from the distribution of the end-vertex in a* $t(|V|)$*-step random walk on* G *that starts in* 0^ℓ*. Furthermore, with probability* $1 - \texttt{negl}(\ell)$ *over the choice of* G *in* $\mathcal{D}_\ell$*, it holds that*
$$|\Pr[M^G(W_t(G))=1] - \Pr[M^G(U(G))=1]| = \texttt{negl}(\ell),$$
where $U(G)$ *denotes the uniform distribution on* G*'s vertices, and* $W_t(G)$ *denotes the distribution of the end-vertex of a* $t(|V|)$*-step random walk on* G *that starts at* 0^ℓ*.*
3. Difficulty of sampling "without walking on the graph": *Any probabilistic polynomial-time oracle machine* M *that is given oracle access to the incidence function of a graph* G *selected according to* $\mathcal{D}_\ell$ *produces, with probability* $1 - \texttt{negl}(\ell)$*, an output that is either not a vertex or is a vertex obtained as answers to one of its queries or* 0^ℓ*. Furthermore, with probability* $1 - \texttt{negl}(\ell)$ *over the choice of* $G = (V, E)$ *in* $\mathcal{D}_n$*, it holds that*
$$\Pr[M^G(1^\ell) \in V \setminus (0^\ell \cup A_M^G)] = \texttt{negl}(\ell),$$
where A_M^G *is the set of answers provided to* M*'s queries during the execution of* $M^G(1^\ell)$*.*

Moreover, these graphs are expanders.

Theorem 3 follows by fixing an arbitrary typical graph in each distribution $\mathcal{D}_\ell$. Recall that Property 2, which asserts the non-triviality of the notion of pseudo-mixing time (i.e., $t(|V|) = o(\log|V|)$), is the focus of this work, whereas Property 1 asserts that this fact may hold also for succinctly represented graphs. Property 3 asserts that we are in a case of interest in the sense that taking walks on G is the only feasible way to obtain vertices of G (other than 0^ℓ).

Proof: Our starting point is a family of explicit d-regular Cayley Graphs $\{G_\ell = (V_\ell, E_\ell)\}_{\ell\in\mathbb{N}}$ such that $0^\ell \in V_\ell \subset \{0,1\}^\ell$ and $|V_\ell| = 2^{(0.5+o(1))\cdot\ell}$. Furthermore, we assume that $W_t(G_\ell)$ has min-entropy $\omega(\log \ell)$, where the min-entropy of a random variable X equals $\min_x\{\log_2(1/\Pr[X=x])\}$. This certainly holds if G_ℓ is an expander graph, which is the choice made to satisfy the "moreover clause".

Next, we consider the distribution $\mathcal{D}_\ell$ obtained by selecting a permutation π over $\{0,1\}^\ell$ that is pseudorandom subject to $\pi(0^\ell) = 0^\ell$. It is instructive to view π as selected as follows: First, we select a pseudorandom permutation ϕ of $\{0,1\}^\ell$, and then we let $\pi(x) = \phi(x) \oplus \phi(0^\ell)$, where $\oplus$ denotes the bit-by-bit exclusive-or of bit strings. Recall that pseudorandom permutations can be constructed based on any one-way function (cf. [4–6]). These permutations of $\{0,1\}^\ell$ are represented by ℓ-bit long strings, and are coupled with efficient algorithms for evaluating the permutation and its inverse.

Applying Proposition 5 to the group $(V_\ell, *)$ that underlies the Cayley graph G_ℓ, using the foregoing permutation π, we obtain a group $(S_\pi, \diamond_\pi)$ that is isomorphic to $(V_\ell, *)$, and it follows that $\pi(G_\ell) = (S_\pi, E_\pi)$, where $E_\pi = \{\{\pi(u), \pi(v)\} : \{u,v\} \in E_\ell\}$, is a Cayley Graph of the group $(S_\pi, \diamond_\pi)$. Specifically, letting $g : V_\ell \times [d] \to V_\ell$ denote the incidence function of G_ℓ, the incidence function of $\pi(G_\ell)$, denoted g_π, satisfies $g_\pi(\alpha, i) = \diamond_\pi(\alpha, \pi(\gamma_i))$ (which equals $\pi(\pi^{-1}(\alpha) * \gamma_i)$), where γ_i is the ith generator of the set underlying the definition of G_ℓ. (Here and below, we include both the generator and its inverse in the set of generators, so to avoid inverse notations.) This establishes Property 1.

To prove Properties 2 and 3, we analyze an ideal construction in which $\pi : \{0,1\}^\ell \to \{0,1\}^\ell$ is a uniformly distributed permutation that satisfies $\pi(0^\ell) = 0^\ell$; equivalently, ϕ is a totally random permutation. By the pseudorandomness of the permutations used in the actual construction, if Property 2 (resp., Property 3) holds for the ideal construction, then it holds also for the actual construction, since otherwise we obtain a poly(ℓ)-time oracle machine that distinguishes the truly random permutations from the pseudorandom ones. Hence, we focus on the analysis of the ideal construction.

Starting with Property 3, observe that, by making oracle calls to $G' = \pi(G_\ell)$, a machine that runs in poly(ℓ)-time may encounter poly(ℓ) many vertices of G' by taking walks (of various lengths) from the vertex 0^ℓ. As far as such a machine is concerned, the name of each other vertex is uniformly distributed among the remaining $2^\ell - \text{poly}(\ell)$ strings of length ℓ. Hence, if such a machine outputs a string that is neither 0^ℓ nor any of the vertices encountered by it, then this string is a vertex of G' with probability at most $\frac{|V_\ell|}{2^\ell} = \exp(-\Omega(\ell)) = \texttt{negl}(\ell)$.

Turning to Property 2, note that the relevant machine may encounter vertices of $G' = \pi(G_\ell)$ by taking walks (of various lengths) either from 0^ℓ or from the test vertex (which is distributed either according to $W_t(G')$ or according to $U(G')$). Each newly encountered vertex looks as being uniformly distributed among the unencountered vertices, and so the difference between the two cases (regarding the tested vertex) amount to the difference between pattern of collisions that the machine sees, where all walks are oblivious of the names of the encountered vertices (since these are random). Collisions between walks that start at the same vertex do not matter, since these are determined by the corresponding sequence of steps obliviously of the start vertex.[7] Hence, the only collisions that may contribute to distinguishing the two cases are collisions between a walk from the tested vertex and a walk from the vertex 0^ℓ (equiv., a collision between the tested vertex and a walk from vertex 0^ℓ).[8]

The key observation is that both $W_t(G')$ and $U(G')$ have min-entropy $\omega(\log \ell)$; specifically, $\max_v\{\Pr[W_t(G') = v]\} = \exp(-\omega(\log \ell))$ and $\max_v\{\Pr[U(G') = v]\} = \exp(-\Omega(\ell))$. Hence, such a collision (between the tested vertex and a walk from vertex 0^ℓ) occurs with probability at most $\exp(-\omega(\log \ell)) = \mathtt{negl}(\ell)$. It follows that a poly(ℓ)-query machine cannot distinguish between $W_t(G')$ or $U(G')$; that is, its "distinguishing gap" is at most $\text{poly}(\ell) \cdot \mathtt{negl}(\ell) = \mathtt{negl}(\ell)$.

The foregoing analysis refers to what happens *in expectation* over the choice of $G' \sim \mathcal{D}_\ell$, whereas the two furthermore claims refer to what happens on typical graphs drawn from $\mathcal{D}_\ell$. In the case of Property 3, applying Markov inequality will do, since we actually upper-bounded (by $\frac{|V_\ell|}{2^\ell} + \mathtt{negl}(\ell)$) the probability that the machine outputs a vertex in $\pi(V_\ell) \setminus \{0^\ell\}$ that was not encountered in its queries. The case of Property 2 requires a slightly more refined argument, since the gap $\Pr[M^G(W_t(G)) = 1] - \Pr[M^G(U(G)) = 1]$, for each G, may be either positive or negative.[9] Using the fact that (for any G in the support of $\mathcal{D}_\ell$)[10] we can approximate both $\Pr[M^G(W_t(G)) = 1]$ and $\Pr[M^G(U(G)) = 1] \approx \Pr[M^G(W_{O(\ell)}(G)) = 1]$, we can translate a gap on a non-negligible measure of $\mathcal{D}_\ell$ to a gap in the expectation, which means that the main claim of Property 3 implies its furthermore claim. ∎

Digest: Why is Theorem 6 *true?* Essentially, *a priori* (and with the exception of 0^ℓ), the vertices of the (distribution of) graphs that we construct look like random

[7] Specifically, suppose that two walks that start at vertex v collide, and that these walks correspond to the sequences of steps $i_1, ..., i_m$ and $j_1, ..., j_n$. Hence, $v \diamond \gamma_{i_1} \diamond \cdots \diamond \gamma_{i_m} = v \diamond \gamma_{j_1} \diamond \cdots \diamond \gamma_{j_n}$, which implies $\gamma_{i_1} \diamond \cdots \diamond \gamma_{i_m} = \gamma_{j_1} \diamond \cdots \diamond \gamma_{j_n}$.

[8] Specifically, suppose that a walk from the tested vertex v collides with a walk from vertex $w = 0^\ell$, and that these walks correspond to the sequences of steps $i_1, ..., i_m$ and $j_1, ..., j_n$. Then, $v \diamond \gamma_{i_1} \diamond \cdots \diamond \gamma_{i_m} = w \diamond \gamma_{j_1} \diamond \cdots \diamond \gamma_{j_n}$, which implies $v = w \diamond \gamma_{j_1} \diamond \cdots \diamond \gamma_{j_n} \diamond \gamma_{i_m}^{-1} \diamond \cdots \diamond \gamma_{i_1}^{-1}$.

[9] Hence, even if the absolute value of the gap, on each specific G, is large, it may be the case that the average gap is negligible due to cancellations. See [1] for a detailed discussion of this issue.

[10] Recall that each graph G in the support of $\mathcal{D}_\ell$ is an expander, and so the statistical difference between $U(G)$ and $W_{O(\ell)}(G)$ is negligible (e.g., $\exp(-\ell)$). Hence, we can obtain poly(ℓ) samples of both $W_t(G)$ and $U(G) \approx W_{O(\ell)}(G)$ in poly(ℓ)-time.

ℓ-bit strings. By taking walks from the vertex 0^ℓ, the observer may discover poly(ℓ) vertices of the graph, but (in both cases) these vertices look random and are unlikely to include the tested vertex. The latter assertion relies on the fact that $\max_v\{\Pr[W_t(G)=v]\}$ is negligible, which requires $t(|V_\ell|) = \omega(\log \ell)$. (Recall that the support of $W_t(G)$ can be found in time $\exp(O(t(|V_\ell|)))$).

3 Additional Comments

We first note that our choice to use graphs with $2^{(0.5+o(1))\cdot\ell}$ vertices is quite immaterial. It is merely a natural intermediate point that balanced between having too many vertices and too little vertices. Likewise, we could have used seeds of length $\ell^{\Omega(1)}$ rather than length ℓ.

3.1 The Pseudorandomness of Multiple Walks

Recall that each graph G in the support of $\mathcal{D}_\ell$ is an expander, and so the statistical difference between $U(G)$ and $W_{O(\ell)}(G)$ is negligible (e.g., $\exp(-\ell)$). Hence, essentially, we can generate samples of both $W_{O(\ell)}(G))$ and $G_t(G)$ in poly(ℓ)-time, provided that $t(N) = \text{poly}(\log N)$. It follows that, given oracle access to $G \sim \mathcal{D}_\ell$, a poly(ℓ)-time observer cannot distinguish between multiple samples of the t-step random walk (i.e., samples of $W_t(G)$) and multiples samples of the graph's vertices (i.e., samples of $U(G)$).

The foregoing result is a special case of a general result that asserts that indistinguishability of a single sample (drawn from one of two distributions) implies indistinguishability of multiple samples (that are all drawn from one of the two distributions). This generic result presupposes that the (single sample) distinguisher can obtain additional samples of each of the two distributions, which holds in our case by the foregoing discussion. The general result can be proved using a "hybrid argument" (see, e.g., [3, Sec. 8.2.3.3]).

3.2 On the Necessity of One-Way Functions

A natural question is whether using one-way function is necessary towards establishing Theorem 6. We show that the answer is affirmative, essentially because *the existence of efficiently sampleable distribution ensembles that are far apart but are computationally indistinguishable implies the existence of one-way functions* [2]. Next, we present a more direct argument (tailored to the current application).

Theorem 7 (one-way functions are necessary for Theorem 6): *The existence of a distribution ensemble of expander graphs that satisfies Conditions 1 and 2 of Theorem 6 implies the existence of one-way functions.*

The hypothesis that the graphs in the distribution are expanders is used only in order to infer that computational indistinguishability holds also with respect to multiple samples of the tested distribution (as discussed in Sect. 3.1). This is required in order to assert that a sequence of samples of $U(G)$ has higher entropy than the length of the succinct representation of $G \sim \mathcal{D}_\ell$ (i.e., an m-long sequence of samples of $U(G)$ has entropy at least $m \cdot \ell/2$, whereas each graph in $\mathcal{D}_\ell$ is described using a seed of length ℓ). Alternatively, if we assume that this succinct representation is shorter than the min-entropy of a single sample of $U(G)$, then the hypothesis that G is an expander is not needed.

Proof: Using the hypothesis, we present a "false entropy generator" (as defined in [5]), and use the fact that such a generator implies a pseudorandom generator (see [5]), which in turn implies a one-way function. Details follow.

Fixing a function $t : \mathbb{N} \to \mathbb{N}$ such $t(N) = o(\log N)$, let $t'(\ell) = t(2^{(0.5+o(1))\cdot\ell}) = o(\ell)$. For $m = 8$ and every ℓ, we consider a function $F : \{0,1\}^\ell \times ([d]^{t'(\ell)})^m \to \{0,1\}^{m\cdot\ell}$ that takes as input a seed s for a graph in the support of $\mathcal{D}_\ell$ along with the descriptions of m random $t'(\ell)$-step walks, where each description is an $t'(\ell)$-long sequence over $[d]$, and outputs a sequence of m corresponding end-vertices; that is, $F(s, w_1, ..., w_m) = (v_1, v_2, ..., v_m)$ if v_i is the end-vertex of the ith random walk, described by $w_i \in [d]^{t'(\ell)}$, on the graph represented by s.

By Condition 1 of Theorem 6, the function F is computable in polynomial-time. On the other hand, by Condition 2 (and the foregoing comment about the pseudorandomness of multiple walks), it follows that the output of the function is indistinguishable (in polynomial-time) from m uniformly and independently distributed vertices of the graph. The point is that the length of the input to F is $\ell' \stackrel{\text{def}}{=} \ell + m \cdot O(t'(\ell)) = \ell + O(1) \cdot o(\ell) < 2\ell$, whereas the output is computationally indistinguishable from an m-long sequence of samples of $U(G)$, which has min-entropy at least $k \stackrel{\text{def}}{=} m \cdot \ell/2 > 2\ell'$ (i.e., each possible outcome occurs with probability at most 2^{-k}). Hence, $F : \{0,1\}^{\ell'} \to \{0,1\}^{m\cdot\ell}$ is a "false entropy generator" (as defined in [5]).

At this point, applying [5] yields the claimed result. For sake of self-containment we outline the rest of the proof, while assuming basic familiarity with the notions of pseudorandom generators and randomness extractors (see, e.g., [3, Sec. 8.2] and [3, Apdx. D.4], resp.). The direct proof capitalizes on the fact that the output of F has high min-entropy (rather than high entropy only). Specifically, we obtain a pseudorandom generator by using either a randomness extractor of seed length $d = o(\ell)$ or a *strong* randomness extractor of seed length $\text{poly}(\ell)$ (see, e.g., [3, Def. D.8] for both versions).[11] Either way, we extract $0.75 \cdot k$ bits (from an input of min-entropy k), while incurring an error of $\exp(-\Omega(k)) = \exp(-\Omega(\ell))$. Applying the extractor on the output of F, we obtain a pseudorandom generator, since the output of the extractor is computationally indistinguishable from a distribution that is $\exp(-\Omega(\ell))$-close to being uniform on $\{0,1\}^{0.75k}$, whereas its input is shorter than its output.

For example, suppose that $E : \{0,1\}^d \times \{0,1\}^{m\cdot\ell} \to \{0,1\}^{0.75k}$ is such an extractor, and consider $G(s', s'') = E(s'', F(s'))$, then $|G(s', s'')| =$

[11] The required extraction parameters are very weak, and explicit constructions of extractors of both types were known for decades.

$0.75 \cdot |F(s')| \geq 1.5 \cdot |s'|$, whereas $|s''| = o(|s'|)$. Alternatively, consider $G(s', s'') = (s'', h_{s''}(F(s')))$, where $h_{s''} : \{0,1\}^{O(|s''|)} \to \{0,1\}^{1.5 \cdot |s''|}$ is a pairwise-independent hash function (see, e.g., [3, Apdx. D.2]), which constitutes a strong extractor.[12] ■

Acknowledgments. We are grateful to Tsachik Gelander, Nir Avni, and Chen Meiri for sharing with us their knowledge and conjectures regarding Problem 4.

References

1. Brakerski, Z., Goldreich, O.: From absolute distinguishability to positive distinguishability. In: Goldreich, O. (ed.) Studies in Complexity and Cryptography. Miscellanea on the Interplay between Randomness and Computation. LNCS, vol. 6650, pp. 141–155. Springer, Heidelberg (2011). https://doi.org/10.1007/978-3-642-22670-0_17
2. Goldreich, O.: A note on computational indistinguishability. Inf. Process. Lett. **34**(6), 277–281 (1990)
3. Goldreich, O.: Computational Complexity: A Conceptual Perspective. Cambridge University Press, Cambridge (2008)
4. Goldreich, O., Goldwasser, S., Micali, S.: How to construct random functions. J. ACM **33**(4), 792–807 (1986)
5. Hastad, J., Impagliazzo, R., Levin, L.A., Luby, M.: A pseudorandom generator from any one-way function. SIAM J. Comput. **28**(4), 1364–1396 (1999)
6. Luby, M., Rackoff, C.: How to construct pseudorandom permutations from pseudorandom functions. SIAM J. Comput. **17**, 373–386 (1988)

[12] Note that $|G(s', s'')| = 0.75 \cdot |F(s')| + |s''| \geq 1.5 \cdot |s'| + |s''|$, whereas $|s''| = \text{poly}(|s'|)$.

On Constructing Expanders for Any Number of Vertices

Oded Goldreich

Abstract. While typical constructions of explicit expanders work for certain sizes (i.e., number of vertices), one can obtain constructions of about the same complexity by manipulating the original expanders. One way of doing so is detailed and analyzed below.

For any $m \in [0.5n, n]$ (equiv., $n \in [m, 2m]$), given an m-vertex expander, G_m, we construct an n-vertex expander by connecting each of the first $n - m$ vertices of G_m to an (otherwise isolated) new vertex, and adding edges arbitrarily to regain regularity. Our analysis of this construction uses the combinatorial definition of expansion.

A preliminary version of this memo was posted on the author's web-site in October 2019.[1] The current revision corrects various typos.

1 The Story, Which Can Be Skipped

Expander graph have numerous applications in the theory of computation (see, e.g., [3]), which explains the extensive interest in constructing these objects. Actually, when talking about expander graphs, one typically refers to families of regular graphs of fixed degree, for a varying number of vertices, that are $\Omega(1)$-expanding, where the expansion factor is fixed for the entire family. That is, there exists a constant $c > 0$ such that for every graph $G = (V, E)$ in the family, and every $S \subseteq V$ of size at most $|V|/2$, it holds that

$$|\{v \in V \setminus S : \exists u \in S \text{ s.t. } \{u, v\} \in E\}| \geq c \cdot |S|.$$

While designers of expanders focus on optimizing various parameters, their users tend to care most of having explicit expanders for any number of vertices (i.e., for any size). The most popular notions of being explicit are a *minimal* notion that requires that the graph be constructed in time that is polynomial in its size, and a *stronger* notion that requires that the neighbors of each vertex in each graph can be identified in time that is poly-logarithmic in the size of the graph (equiv., polynomial in the length of the description of the vertices, assuming a non-redundant representation).[2]

Unfortunately, typical constructions of explicit expanders work only for certain sizes (i.e., number of vertices). Yet, fortunately, one can obtain constructions

[1] See http://www.wisdom.weizmann.ac.il/~oded/p_ex4all.html.
[2] See [3, Def. 2.3] or [2, Apdx. E.2.1.2].

O. Goldreich (Ed.): Computational Complexity and Property Testing, LNCS 12050, pp. 374–379, 2020.
https://doi.org/10.1007/978-3-030-43662-9_21

of the same level of explicitness (or complexity) by manipulating the original expanders. One such construction was presented recently by Murtagh *et al.* [4]. It reminded me of a different construction, which I heard from Noga Alon many years ago. (In fact, checking something else in [2, Apdx. E.2], I noticed that I used Noga's construction there (see last paragraph in [2, Apdx. E.2.1.2]), but forgot about this.)

The starting point in both cases is a construction for a "dense" set of sizes M; that is, we have explicit m-vertex expanders for every $m \in M$ and for every $n \in \mathbb{N}$ it holds that $M \cap [n, 2n] \neq \emptyset$ (alternatively, $M \cap [0.5n, n] \neq \emptyset$). The aim is to obtain an explicit n-vertex expander, for any given $n \in \mathbb{N}$.

The construction of Murtagh *et al.* [4] takes an m-vertex graph, where $m \in M \cap [n, 2n]$, designates $m - n$ pairs of vertices in it, merges each such pair to a single vertex (doubling the degree), and adds self-loops on the other $m - 2 \cdot (m - n) = n - (m - n)$ vertices to regain regularity. The analysis of this construction is conducted in terms of the algebraic definition of expansion (i.e., eigenvalues), and is presented in [4, Apdx. B]. Assuming that the m-vertex graph has a second eigenvalue smaller (in absolute value) than $\beta < 1/3$, the resulting n-vertex graph has a second eigenvalue smaller than $(1 + 3\beta)/2$.

Noga Alon's construction starts by picking $m_1 \in M \cap [0.5n, n]$. Discarding the fortunate case of $m_1 = n$, note that if $m_1 = n/2$ we are done by connecting two copies of the m_1-vertex graph by a matching. The resulting n-vertex graph is shown to be an expander using the combinatorial definition of expansion (i.e., the expansion of vertex-sets). In general, we set $r_1 = n - m_1 \in (0, 0.5n]$, and proceed by picking $m_2 \in M \cap [0.5r_1, r_1]$, setting $r_2 = r_1 - m_2$, and so on; that is, in iteration i we pick $m_i \in M \cap [0.5r_{i-1}, r_{i-1}]$ and set $r_i = r_{i-1} - m_i$, till we get to $r_t = O(1)$. At this point we connect the vertices of the $t - 1$ smaller graphs to $\sum_{i=2}^{t} m_i$ vertices of the m_1-vertex graph by using a matching (and add self-loops to maintain regularity).

The analysis of Noga's construction is less trivial than it seems. The source of trouble is that, when analyzing the expansion of sets, one needs to consider sets of size at most $n/2$ and such sets may have more than $m_1/2$ vertices in the large (m_1-vertex) expander. This difficulty can be resolved by using a definition that guarantees expansion also for larger sets (actually, it suffices to guarantee expansion for sets that have density at most $3/4$). Furthermore, the standard definition of expansion does imply expansion also for larger sets (as needed above).

Thinking a little more about Noga's suggestion, I realized that, if one does not care about the specific expansion parameters, then the smaller expanders play no real role. Hence, the added small expanders can be replaced by isolated vertices; that is, wishing to have an n-vertex expander and given an m-vertex expander such that $m \in [0.5n, n]$, we connect $n - m$ auxiliary vertices (which are otherwise isolated) to $n - m$ vertices of the original expander (and then add edges arbitrarily to recover regularity). The analysis uses the combinatorial definition of expansion, with the aforementioned caveat.

2 The Actual Construction and Its Analysis

While typical constructions of explicit expanders work for certain sizes (i.e., number of vertices), one can obtain construction of about the same complexity by manipulating the original expanders. One way of doing so is detailed and analyzed below.

The construction. For $m \in [0.5n, n]$ (equiv., $n \in [m, 2m]$), given an m-vertex expander, G_m, we construct an n-vertex expander by connecting each of the first $n - m$ vertices of G_m to an (otherwise isolated) new vertex, and add edges arbitrarily to regain regularity. Hence, we obtain a construction of expanders for all sizes, provided we are given a construction of expanders for a sufficiently dense set of sizes (which is effectively accessible as assumed below).

Construction 1 (padding and matching with isolated vertices): *For $d \in \mathbb{N}$, suppose that $M \subseteq \mathbb{N}$ and $\{G_m\}_{m \in M}$ is a set of d-regular graphs such that the following two conditions hold.*

1. *Given any $m \in M$, we can construct the m-vertex graph $G_m = ([m], E_m)$.*
2. *Given any $n \in \mathbb{N}$, we can determine an $m \in M$ such that $m \in [0.5n, n]$ (equiv., $n \in [m, 2m]$).*

Then, given any $n \in \mathbb{N}$, we construct a d'-regular n-vertex graph $G_n = ([n], E_n)$ by picking $m \in M \cap [0.5n, n]$, constructing $G_m = ([m], E_m)$, and letting

$$E_n = E_m \cup \{\{i, m+i\} : i \in [n-m]\} \cup E_{m,n},$$

where $d' \in \{d+1, d+2\}$ and $E_{m,n}$ is an arbitrary set of $\frac{(d' \cdot n - d \cdot m}{2} - (n-m)$ edges that is added so to make G_n be d'-regular. Specifically, $d' = d+1$ may be used if either n is even or d is odd, and $d' = d+2$ is used otherwise.

We say that a graph $G = (V, E)$ is (ρ, c)-expanding if for every $S \subset V$ such that $|S| \leq \rho \cdot |V|$ it holds that $|\partial(S)| \geq c \cdot |S|$, where $\partial(S) = \{u \in V \setminus S : \exists v \in S \text{ s.t. } \{v, u\} \in E\}$ is the boundary of S. The standard definition of expansion corresponds to $(0.5, \Omega(1))$-expansion, but it implies $(\rho, \Omega(1))$-expansion for any constant $\rho < 1$.[3] Hence, when showing that G_n is an expander, we may assume that G_m is $(0.75, \Omega(1))$-expanding, rather than $(0.5, \Omega(1))$-expanding.

Theorem 2 (analysis of Construction 1): *If G_m is $(0.75, c)$-expanding, then G_n is $(0.5, c/2)$-expanding.*

---

[3] Assume that the graph is $(0.5, c)$-expanding, and let $S \subset V$ be an arbitrary set such that $0.5 \cdot |V| < |S| \leq \rho \cdot |V|$. Then, $R \stackrel{\text{def}}{=} V \setminus (S \cup \partial(S))$ has cardinality smaller than $0.5 \cdot |V|$, and it follows that $|\partial(R)| \geq c \cdot |R|$. On the other hand, $\partial(R) \subseteq \partial(S)$, and so $|\partial(S)| \geq c \cdot |R| = c \cdot (|V| - |S| - |\partial(S)|)$. Hence, $|\partial(S)| \geq \frac{c}{1+c} \cdot (|V| - |S|) \geq \frac{c}{1+c} \cdot \frac{1-\rho}{\rho} \cdot |S|$, and it follows that the graph is (ρ, c')-expanding for $c' = \frac{c \cdot (1-\rho)}{(1+c) \cdot \rho}$. .

The proof does not use the edges in $E_{m,n}$, which makes sense given their arbitrary choice. Yet, it is quite likely that a more careful analysis of other aspects will yield a stronger result. In particular, assuming that G_m is $(0.5, c)$-expanding, we only conclude that G_n is $(0.5, c/12)$-expanding[4] (so the real challenged is to establish a higher expansion bound for G_n, when assuming that G_m is $(0.5, c)$-expanding).

Proof: Recall that $0 \leq n - m \leq m$. For an arbitrary set $S \subset [n]$ of size at most $0.5n$, we consider the following four disjoint subsets of S:

$$S' \stackrel{\text{def}}{=} \{i \in [n-m] : i \in S \;\&\; m+i \in S\}$$
$$S'' \stackrel{\text{def}}{=} \{i \in ([m] \setminus [n-m]) : i \in S\}$$
$$S''' \stackrel{\text{def}}{=} \{i \in [n-m] : i \in S \;\&\; m+i \notin S\}$$
$$R \stackrel{\text{def}}{=} \{m+i \in S : i \notin S\}$$

Note that (S', S'', S''') is a partition of $S \cap [m]$ whereas $(m + S', R)$ is a partition of $S \setminus [m]$. We may assume, without loss of generality, that $S''' = \emptyset$, because moving $i \in S'''$ to $m + i$ (i.e., replacing S by $(S \setminus \{i\}) \cup \{m + i\}$) can only decrease the $\partial(\cdot)$-value.[5]

Next, we show that $|S| \leq n/2$ implies $|S' \cup S''| \leq 0.75 \cdot m$. This holds because $|S''| \leq m - (n - m) = 2m - n$, which implies

$$\begin{aligned}
|S'| + |S''| &= \frac{|S| - |S''| - |R|}{2} + |S''| \\
&\leq \frac{|S| - |S''|}{2} + |S''| \\
&\leq \max_{s \leq 2m-n} \left\{ \frac{|S| - s}{2} + s \right\} \\
&= \frac{|S| + 2m - n}{2} \\
&\leq \frac{2m - 0.5n}{2} \\
&\leq 0.75 \cdot m,
\end{aligned}$$

where the third (resp., last) inequality is due to $|S| \leq n/2$ (resp., $m \leq n$).

Having established $|S' \cup S''| \leq 0.75 \cdot m$ and using the $(0.75, c)$-expansion of G_m, we get $|\partial(S' \cup S'')| \geq c \cdot (|S'| + |S''|)$. Turning to R, and using the matching

[4] We first infer that G_m is $(0.75, c')$-expanding for $c' = \frac{0.25 \cdot c}{0.75 \cdot (1+c)}$ (see Footnote 3). Hence, G_n is $(0.5, c'')$-expanding for $c'' = \frac{c}{6(1+c)}$. Using $c \leq 1$, we get $c'' \geq c/12$.

[5] Suppose that $i \in [m] \cap S$ and $m + i \in [n] \setminus S$, and let $T = (S \setminus \{i\}) \cup \{m + i\}$. Then, $i \notin \partial(S)$ and $m + i \in \partial(S)$, whereas $i \in \partial(T)$ and $m + i \notin \partial(T)$, which means that $|\partial(T) \cap \{i, m+i\}| = 1 = |\partial(S) \cap \{i, m+i\}|$. However, $\partial(T) \setminus \{i, m+i\} \subseteq \partial(S) \setminus \{i, m+i\}$, since the move may only eliminate a contribution of i to $\partial(S) \setminus \{i, m + i\}$ (whereas $m + i$ does not contribute to $\partial(T) \setminus \{i, m + i\}$).

edges (i.e., the set $\{\{i, m+i\} : i \in [n-m]\}$), we have $|\partial(R)| = |R| \geq c \cdot |R|$, since $c \leq 1/3$. Note that $\partial(S' \cup S'') \cap R = \emptyset$ and $\partial(R) \cap (S' \cup S'') = \emptyset$, since the vertices in R are matched to vertices in $[m] \setminus (S' \cup S'')$. Hence, $|\partial(S' \cup S'' \cup R)| \geq c \cdot (|S'| + |S''| + |R|)/2$, since each vertex may contribute at most twice to the sum $|\partial(S' \cup S'')| + |\partial(R)|$. Noting that $|S'| + |S''| + |R| \geq |S|/2$, we infer that $|\partial(S)| \geq c \cdot |S|/4$.

Using a more careful analysis, we note that $|\partial(S' \cup S'' \cup R)| \geq |\partial(S' \cup S'')| + |\partial(R)| \geq c \cdot (|S'| + |S''|) + 0.5 \cdot |R|$, since the double contribution may occur only on elements of $\partial(R)$ and $|\partial(R)| = |R|$. Using $|S| = 2 \cdot |S'| + |S''| + |R|$, we get

$$\begin{aligned} |\partial(S' \cup S'' \cup R)| &\geq c \cdot (|S'| + |S''|) + 0.5 \cdot |R| \\ &= c \cdot \left(\frac{|S| - (|S''| + |R|)}{2} + |S''| \right) + 0.5 \cdot |R| \\ &= c \cdot \frac{|S| + |S''|}{2} + \frac{1-c}{2} \cdot |R|, \end{aligned}$$

and the claim follows (since $c \leq 1/3$). ■

3 Postscript

As mentioned in Sect. 1, it turns out that I did mention Noga Alon's construction in [2, Apdx. E.2.1.2] (but forgot of this). Also, it seems that Noga has mentioned the construction (and/or variants of it) in some old papers of his. Asking him about this in October 2019, he suggested a few alternative constructions, which are aimed at better expansion parameters. My favorite one, starts with an m-vertex d-regular graph, G_m, for $m \in [n - o(n), n]$, and obtains an n-regular d'-regular graph by connecting each of the $n-m$ new vertices to d' different old vertices.

A combinatorial analysis of the resulting graph, $G_n = ([n], E_n)$, maintains much of the expansion features of $G_m = ([m], E_m)$. Specifically, assume that, for some monotone non-decreasing function $\mathtt{X} : [m] \to [m]$ (e.g., $\mathtt{X}(s) = \Omega(d \cdot s)$ for $s < m/2d$), every s-subset of vertices of G_m has at least $\mathtt{X}(s)$ neighbors (in G_m) that are outside it. Consider an arbitrary set $S \subset [n]$ of vertices in G_n, and let $S' \stackrel{\text{def}}{=} S \cap [m]$ and $S'' = S \setminus S'$. If $|S''| > \mathtt{X}(|S|)/2d'$, then $|\partial(S)| \geq |\partial(S'') \setminus S'| \geq d' \cdot |S''| - |S'| > 0.5 \cdot \mathtt{X}(|S|) - |S|$, where the second inequality is due to the fact that each vertex in $S'' \subseteq [n] \setminus [m]$ has d' neighbors in $[m]$. Otherwise (i.e., $|S''| \leq \mathtt{X}(|S|)/2d'$), we use $|\partial(S)| \geq |\partial(S') \cap [m]| \geq \mathtt{X}(|S'|) \geq \mathtt{X}(|S| - (\mathtt{X}(|S|)/2d')) \geq \mathtt{X}(|S|/2)$, since $\mathtt{X}(|S|) \leq d \cdot |S|$. Hence, the expansion of G_n is given by the function $\mathtt{X}'(s) = \min(0.5 \cdot \mathtt{X}(s) - s, \mathtt{X}(s/2))$.

Noga is currently writing a paper with a spectral analysis of some of these alternative construction [1]. The tentative abstract reads as follows.

An (n, d, λ)-graph is a d-regular graph on n vertices in which the absolute value of any nontrivial eigenvalue is at most λ.

- For any constant $d \geq 3$ and $\epsilon > 0$, and all sufficiently large n we show that there is a deterministic poly(n)-time algorithm that outputs an (n, d, λ)-graph (on exactly n vertices) with $\lambda \leq 2\sqrt{d-1} + \epsilon$.
- For any $d = p + 2$ with $p \equiv 1 \bmod 4$ prime and all sufficiently large n, we describe a strongly explicit construction of an (n, d, λ)-graph (on exactly n vertices) with $\lambda \leq \sqrt{2(d-1)} + \sqrt{d-2} + o(1)$ $(< (1 + \sqrt{2})\sqrt{d-1} + o(1))$, with the $o(1)$ term tending to 0 as n tends to infinity.
- For every $\epsilon > 0$, $d > d_0(\epsilon)$ and $n > n_0(d, \epsilon)$ we show a strongly explicit construction of an (m, d, λ)-graph with $\lambda < (2+\epsilon)\sqrt{d}$ and $m = n+o(n)$.

All constructions are obtained by starting with known ones of Ramanujan or nearly Ramanujan graphs, modifying or packing them in an appropriate way. The spectral analysis relies on the delocalization of eigenvectors of regular graphs in cycle-free neighborhoods.

References

1. Alon, N.: Explicit expanders of every degree and size. In: Preparation (2019). (Added in revision: See https://m.tau.ac.il/~nogaa/PDFS/expexp.pdf)
2. Goldreich, O.: Computational Complexity: A Conceptual Perspective. Cambridge University Press, Cambridge (2008)
3. Hoory, S., Linial, N., Wigderson, A.: Expander graphs and their applications. Bull. (New Series) Am. Math. Soc. **43**(4), 439–561 (2006)
4. Murtagh, J., Reingold, O., Sidford, A., Vadhan, S.: Deterministic approximation of random walks in small space. In: 23rd RANDOM, LIPIcs 145, Schloss Dagstuhl - Leibniz-Zentrum für Informatik (2019)

About the Authors

Oded Goldreich (oded@wisdom.weizmann.ac.il) is a Meyer W. Weisgal Professor at the Faculty of Mathematics and Computer Science of the Weizmann Institute of Science, Israel. Oded was born in 1957, and completed his graduate studies in 1983 under the supervision of Shimon Even. He was a post-doctoral fellow at MIT (1983–86), a faculty member at the Technion (1986–94), a visiting scientist at MIT (1995–98), and a Radcliffe fellow at Harvard (2003/04). Since 1994, he is a member of the Computer Science and Applied Mathematics Department of the Weizmann Institute. He is the author of "Modern Cryptography, Probabilistic Proofs and Pseudorandomness" (Springer, 1998), the two-volume work "Foundations of Cryptography" (Cambridge University Press, 2001 and 2004), "Computational Complexity: A Conceptual Perspective" (Cambridge University Press, 2008), and "Introduction to Property Testing" (Cambridge University Press, 2017).

Itai Benjamini (itai.benjamini@gmail.com) is a professor of mathematics at the Weizmann institute. He graduated in 1992 from the Hebrew university.

Scott Decatur (scott_decatur@alum.mit.edu) is the Director of Quantitative International Strategies at investment manager Segall Bryant & Hamill LLC. Scott received his B.S. and M.S. in Electrical Engineering and Computer Science from MIT in 1989, and his Ph.D. in Computer Science from Harvard University in 1995 under the supervision of Les Valiant. He has also held postdoctoral research positions at MIT and DIMACS.

Maya Leshkowitz (mleshkowitz@gmail.com) completed her master thesis at the Weizmann Institute of Science in 2017. She is currently working on her PhD in Psychology at the Hebrew University.

Or Meir (ormeir2@gmail.com) is a member of the computer science department at the university of Haifa. He completed his Ph.D. in 2011 under the supervision of Oded Goldreich, and was a postdoctoral fellow at Stanford university, the institute for Advanced Study at Princeton, and the Weizmann Institute of Science.

Dana Ron (danar@eng.tau.ac.il) is the Lazarus Brothers Chair of Computer Engineering at the School of Electrical Engineering of Tel Aviv University, Israel. Dana was born in 1964, and completed her graduate studies in 1995 under the supervision of Naftali Tishby. She was an NSF post-doctoral fellow at MIT (1995–97) and a science scholar at the Bunting Institute, Radcliffe (1997–98). She was also a Radcliffe fellow at Harvard (2003/04). Since 1998 she is a faculty member in Tel Aviv University.

Guy Rothblum (guy.rothblum@gmail.com) is a faculty member in the Faculty of Mathematics and Computer Science at the Weizmann Institute of Science. He has wide interests in theoretical computer science, with a focus on cryptography, complexity theory, and societal concerns in algorithms and data analysis. Guy completed his Ph.D. at MIT, where his advisor was Shafi Goldwasser, and his M.Sc. at The Weizmann Institute of Science, where his advisor was Moni Naor. Before joining the faculty at Weizmann, he completed a postdoctoral fellowship at Princeton University and was a researcher at Microsoft Research's Silicon Valley Campus and at Samsung Research America.

Avishay Tal (atal@berkeley.edu) is an Assistant Professor in the Department of Electrical Engineering and Computer Sciences at the University of California, Berkeley. Avishay received his M.Sc. degree in 2012 under the supervision of Amir Shpilka and his Ph.D. in 2015 under the supervision of Ran Raz. He was a post-doc fellow at the Institute for Advanced Study (2015–2017), Stanford University (2017–2019), and Simons Institute for the Theory of Computing (Fall 2018). Avishay received the STOC'19 best paper award and the ITCS'13 best student paper award. His research interests lie within theoretical computer science and include complexity theory at large, analysis of Boolean functions, circuit complexity, pseudorandomness, computational learning theory, and quantum computing

Liav Teichner (liav.teichner@gmail.com) is a software developer working for Amazon on the Amazon Go project. Liav was born in 1987 in Haifa, Israel. He completed his MSc in Mathematics and Computer Science at the Weizmann Institute of Science in 2016 advised by Irit Dinur. After finishing his studies Liav worked as a software developer at the cyber security company Illusive Networks from 2016 to 2018. Since 2018 Liav has been working for Amazon.

Roei Tell (roeitell@gmail.com) is currently a PhD student in the Mathematics and Computer Science Department in the Weizmann Institute of Science, under the supervision of Oded Goldreich. His research focuses on computational complexity.

Avi Wigderson (avi@ias.edu) is the Herbert Maass Professor at the School of Mathematics of the Institute for Advanced Study in Princeton. Avi was born in 1956, and completed his graduate studies in 1983 at Princeton University under the supervision of Dick Lipton. After postdoctoral positions at UC Berkeley, IBM labs at San Jose and at MSRI he became a faculty member at the Hebrew University at 1986. In 1999 he moved to the Institute for Advanced Study, where he is leading a program on theoretical computer science and discrete mathematics.

Printed in the United States
By Bookmasters